Lecture Notes in Computer Science 15981

Founding Editors

Gerhard Goos
Juris Hartmanis

AF173791

The series Lecture Notes in Computer Science (LNCS), including its subseries Lecture Notes in Artificial Intelligence (LNAI) and Lecture Notes in Bioinformatics (LNBI), has established itself as a medium for the publication of new developments in computer science and information technology research, teaching, and education.

LNCS enjoys close cooperation with the computer science R & D community, the series counts many renowned academics among its volume editors and paper authors, and collaborates with prestigious societies. Its mission is to serve this international community by providing an invaluable service, mainly focused on the publication of conference and workshop proceedings and postproceedings. LNCS commenced publication in 1973.

Giuseppa Castiglione · Sabrina Mantaci
Editors

Implementation and Application of Automata

29th International Conference, CIAA 2025
Palermo, Italy, September 22–25, 2025
Proceedings

 Springer

Editors
Giuseppa Castiglione ⓘ
University of Palermo
Palermo, Italy

Sabrina Mantaci ⓘ
University of Palermo
Palermo, Italy

ISSN 0302-9743   ISSN 1611-3349 (electronic)
Lecture Notes in Computer Science
ISBN 978-3-032-02601-9   ISBN 978-3-032-02602-6 (eBook)
https://doi.org/10.1007/978-3-032-02602-6

Preface

The 29th International Conference on Implementation and Application of Automata (CIAA 2025) was organized by members of the University of Palermo. The conference took place from September 22–25, 2025, in Palermo, Italy.

This event was part of the CIAA conference series, a major international venue that brings together researchers in the field of automata theory and implementation. This year, for the first time, the CIAA conference took place in Italy. The previous editions of the conference were held in various locations over five continents: Akita (2024), Famagusta (2023), Rouen (2022), Bremen (2021), Košice (2019), Charlottetown (2018), Marne-la-Vallée (2017), Seoul (2016), Umeå (2015), Giessen (2014), Halifax (2013), Porto (2012), Blois (2011), Winnipeg (2010), Sydney (2009), San Francisco (2008), Prague (2007), Taipei (2006), Nice (2005), Kingston (2004), Santa Barbara (2003), Tours (2002), Pretoria (2001), London, Ontario (2000), Potsdam (WIA 1999), Rouen (WIA 1998), and London, Ontario (WIA 1997 and WIA 1996). Due to the CoVid-19 pandemic, CIAA 2020, planned to be held in Loughborough, was canceled.

This volume of Lecture Notes in Computer Science contains the papers presented at CIAA 2025 and the abstracts of the invited speakers.

The 22 regular papers were selected from 32 submissions covering various fields in the application, implementation, and theory of automata and related structures. Each paper was single-blindly reviewed by three members of the Program Committee, with support from external reviewers, and was thoroughly discussed. Papers were submitted by authors from various countries: Belgium, Canada, China, France, Germany, Italy, Japan, Netherlands, Pakistan, Poland, Russia, Slovakia, Sweden, South Africa, South Korea, UK, USA.

CIAA 2025 featured three invited talks:

- Marie-Pierre Béal, Université Gustave Eiffel, France: *Symbolic dynamics and automata*
- Dirk Nowotka, University of Kiel, Germany: *On Decision Problems of Pattern Languages*
- Alberto Policriti, University of Udine, Italy: *Wheeler languages and automata*

We would like to express our sincere gratitude to everyone who contributed to the success of this conference: the authors for submitting their carefully prepared manuscripts; the invited speakers for their excellent presentations of topics related to the theme of the conference; the Program Committee members and external referees for evaluating the submitted manuscripts; the session chairs; and the participants who made CIAA 2025 possible.

The administration of the reviews and discussions was handled using the EasyChair system, which made the process flawless.

We are grateful to the members of the Organizing Committee: Chiara Epifanio, Gabriele Fici, Estéban Gabory, Dora Giammarresi, Giuseppe Romana, and Marinella

Sciortino, who took care of all local aspects of the planning and coordination, ensuring the smooth running of the conference.

We wish to thank the University of Palermo, which partially funded the conference and also provided a beautiful room inside the Botanical Garden where the sessions took place.

We finally thank the editorial team of Lecture Notes in Computer Science at Springer for the opportunity to publish the proceedings in the series, and for the guidance and support during the publication process.

July 2025

Giuseppa Castiglione\
Sabrina Mantaci

Organization

Steering Committee

Markus Holzer (Chair)	Justus Liebig University Giessen, Germany
Oscar Ibarra	University of California, Santa Barbara, USA
Sylvain Lombardy	Université de Bordeaux, France
Nelma Moreira	University of Porto, Portugal
Kai T. Salomaa (Co-chair)	Queen's University, Canada
Hsu-Chun Yen	National Taiwan University, Taiwan

Program Committee Chairs

Giuseppa Castiglione	University of Palermo, Italy
Sabrina Mantaci	University of Palermo, Italy

Program Committee

Marcella Anselmo	University of Salerno, Italy
Frédérique Bassino	Université Paris 13 - CNRS, France
Martin Berger	University of Sussex, UK
Luca Breveglieri	Politecnico di Milano, Italy
Pascal Caron	University of Rouen, France
Giuseppa Castiglione	University of Palermo, Italy
Erzsébet Csuhaj-Varjú	Eötvös Loránd University, Hungary
Frank Drewes	Umeå University, Sweden
Szilard Zsolt Fazekas	Akita University, Japan
Dora Giammarresi	University of Roma "Tor Vergata" Italy
Yo-Sub Han	Yonsei University, South Korea
Markus Holzer	Justus Liebig University Giessen, Germany
Galina Jirásková	Slovak Academy of Sciences, Slovakia
Natasha Jonoska	University of South Florida, USA
Jarkko Kari	University of Turku, Finland
Ondrej Klima	Masaryk University, Czech Republic
Martin Kutrib	Justus Liebig University Giessen, Germany
Julien Leroy	University of Liège, Belgium
Andreas Malcher	Justus Liebig University Giessen, Germany

Andreas Maletti	Universität Leipzig, Germany
Florin Manea	University of Göttingen, Germany
Sebastian Maneth	Universität Bremen, Germany
Sabrina Mantaci	University of Palermo, Italy
Victor Mitrana	Universidad Politécnica de Madrid, Spain
Nelma Moreira	University of Porto, Portugal
Cyril Nicaud	Université Gustave Eiffel, France
Giovanni Pighizzini	University of Milan, Italy
Luca Prigioniero	Loughborough University, UK
Rogério Reis	University of Porto, Portugal
Michel Rigo	University of Liège, Belgium
Giuseppe Romana	Università di Palermo, Italy
Kai Salomaa	Queen's University, Canada
Shinnosuke Seki	University of Electro-Communications, Japan
Mikhail Volkov	Ural Federal University, Russia
Bruce Watson	National Security Centre of Excellence, Canada
Hsu-Chun Yen	National Taiwan University, Taiwan

Organizing Committee

Giuseppa Castiglione	University of Palermo, Italy
Chiara Epifanio	University of Palermo, Italy
Gabriele Fici	University of Palermo, Italy
Esteban Gabory	University of Palermo, Italy
Dora Giammarresi	University of Roma "Tor Vergata", Italy
Sabrina Mantaci	University of Palermo, Italy
Giuseppe Romana	University of Palermo, Italy
Marinella Sciortino	University of Palermo, Italy

Additional Reviewers

Amrane, Amazigh
Berglund, Martin
Björklund, Henrik
Bourotte, Coda
Bozga, Marius
Broda, Sabine
Bruyère, Véronique
Carayol, Arnaud
David, Julien

Duarte, Guilherme
Fijalkow, Nathanaël
Fujiyoshi, Akio
Gruber, Hermann
Heam, Pierre-Cyrille
Ho, Hsi-Ming
Ise, Daihei
Kapoutsis, Christos
Kim, Sungmin

Koechlin, Florent
Kunc, Michal
Lefebvre, Arnaud
Lombardy, Sylvain
Madonia, Maria
Marsault, Victor
Martens, Willeke
Nakano, Keisuke
Ng, Timothy
Olejár, Viktor
Ollinger, Nicolas
Otop, Jan
Paul, Erik
Perez, Guillermo

Policriti, Alberto
Rampersad, Narad
Rauch, Christian
Rosenfeld, Matthieu
Rotondo, Pablo
Ryzhikov, Andrew
Santini, Massimo
Sauer, Fabian
Schmid, Markus
Schnoebelen, Philippe
Specht, Timo
Sung, Sicheol
Yen, Di-De

Invited Talks Abstracts

Symbolic Dynamics and Automata

Marie-Pierre Béal

Univ. Gustave Eiffel, CNRS, LIGM, F-77454 Marne-la-Vallée, France

Symbolic dynamics is a well-studied field at the intersection of discrete mathematics and theoretical computer science. Shift spaces are sets of bi-infinite sequences of symbols over a finite alphabet, characterized by their sets of forbidden blocks. Two shifts are conjugate, or isomorphic, if there exists a bijection between them that is a block map, i.e., a map defined by a sliding window of bounded length. A conjugacy can be seen as a recoding, and conjugate shifts are essentially the same. One-sided shift spaces are sets of right-infinite sequences characterized by their sets of forbidden blocks.

Shift spaces with a regular set of blocks are called sofic. When they are irreducible, they have a unique minimal presentation. Shifts of finite type are sofic shifts that can be characterized by a finite set of forbidden blocks. Two irreducible shifts of finite type are conjugate if there exists a sequence of local operations—specifically in-splittings, out-splittings, in-mergings, and out-mergings—that transforms a minimal presentation of the first into a minimal presentation of the second. Although it is not known whether it is decidable if two irreducible shifts of finite type are conjugate, many dynamical properties can be decided using this minimal deterministic automaton.

In the context of one-sided shift spaces of finite type, Williams' theory enables the decidability of conjugacy between one-sided shifts of finite type by using another minimal presentation of the shift space, known as its total amalgamation. Specifically, two irreducible shift spaces of finite type are conjugate if there exists a sequence of out-splittings and out-mergings that transforms the minimal presentation of the first shift space into the minimal presentation of the second shift space. It turns out that two out-mergings commute, which allows us to compute a common total amalgamation when the shifts are conjugate.

In this talk, I will present how automata are used in symbolic dynamics for shifts of sequences and shifts of trees. I will focus on finite-type shifts of sequences and trees, as well as on the computation of minimal presentations and amalgamations for one-sided shifts. I will also present some of the ongoing work with Alexi Block Gorman on Hom shifts.

On Decision Problems of Pattern Languages

Dirk Nowotka

Kiel University, Germany

Patterns are words with terminals and variables. The language of a pattern is the set of words obtained by uniformly substituting all variables with words that contain only terminals. A pattern language is called erasing or non-erasing if substitutions include or do not include the empty word, respectively. Length constraints restrict valid substitutions of variables by associating the variables of a pattern with a system (or finite disjunction of systems) of linear diophantine inequalities. Pattern languages with length constraints contain only words in which all variables are substituted by words with lengths that fulfill such a given set of length constraints. Pattern languages with regular constraints contain only words in which all variables are substituted by words from a regular language given for the respective variable.

We give an overview on decision problems of pattern languages. We elaborate on what is known and what are the open problems. We consider membership, inclusion, and equivalence problems for erasing and non-erasing pattern languages with and without length and/or regular constraints.

Wheeler Languages and Automata

Alberto Policriti

University of Udine, Italy

Wheeler languages form a robust and expressive subclass of regular languages characterized by a strong structural alignment between automata and (co-)lexico-graphic ordering of strings. More specifically, automata accepting Wheeler languages are equipped with an ordering of states constrained in such a way that a state q precedes a state q' if and only if the *set* of strings reaching q precedes—in a suitable and very simple sense—the *set* of strings reaching q'.

In this talk we will explore the theoretical foundations and touch practical implications of the definition of Wheeler languages and automata, building from the core notion of Wheeler graphs—a class of labeled and directed graphs admitting a total node-order propagating coherently through edge labels. We will enrich the presentation by an alternative (equivalent) view on Wheeler automata, obtained by embedding strings into aperiodic rational numbers and embedding states into open/closed intervals bounded by rationals.

The Wheeler framework sets the stage for a new, order-aware perspective on automata theory, as well as for the introduction of a new complexity measure on general automata. We will illustrate the formal characterization of languages recognized by Wheeler automata, presenting the basic results in this area. In particular, we will briefly go through an adaptation to the Wheeler (ordered) case of classic results such as the Myhill-Nerode theorem and determinization and will then report on our investigation relative to closure properties and expressiveness. Turning to the issue of efficient representations, we will briefly present some results on the relative size of Wheeler and non-Wheeler representations of a given Wheeler language.

Wheeler languages turn out to constitute the simplest stratum of a hierarchy of automata encompassing the full class of regular languages. Such a hierarchy is obtained when the requirement of *totally* ordering the collection Q of states is replaced by the weaker requirement of *partially* ordering Q. The weaker requirement allows the introduction of a parameter—the *width* of an automaton, i.e. the smallest p such that Q can be partitioned into p linear and "Wheeler-compliant" parts—that turns out to be central in bounding many of the complexities of classical automata-manipulation procedures.

We will conclude examining recent results and open problems that challenge the understanding of bridging concepts from formal language theory and order theory.

Contents

Complementable Normal Form of Parametrized Automata 1
  Franziska Alber and Philipp Rümmer

Toward the Glushkovization of Automata: The Strong Stabilization 15
  Samira Attou, Ludovic Mignot, Clément Miklarz, and Florent Nicart

Constructing Compact BPE Token DFAs 27
  Martin Berglund, Anna Jonsson, Willeke Martens,
  and Brink van der Merwe

Epsilon Automata on Linear Orderings 41
  Bernard Boigelot, Thomas Braipson, and Tom Clara

Multi-entry DFA with Reduced Initial States to Speedup Parallel
Recognition 55
  Angelo Borsotti, Luca Breveglieri, Stefano Crespi Reghizzi,
  and Angelo Morzenti

Two-Way Automata and Bounded Languages 73
  Alessandro Clerici Lorenzini, Giovanni Pighizzini, and Luca Prigioniero

An Algebraic Approach to the Equivalence Checking of Deterministic
Top-down Tree Transducers 86
  Deng Zhibo, Tang Tianxiang, and Vladimir A. Zakharov

An Active Learning Algorithm for Bidirectional Deterministic Finite
Automata 99
  Simon Dieck and Sicco Verwer

Dynamically Weighted Tree Transducers 115
  Frank Drewes, Marco Kuhlmann, and Olle Torstensson

Engineering an LTL$_f$ Synthesis Tool 129
  Alexandre Duret-Lutz, Shufang Zhu, Nir Piterman,
  Giuseppe De Giacomo, and Moshe Y. Vardi

Subsequence Matching and Analysis Problems for Automata
with Translucent Letters 148
  Szilárd Zsolt Fazekas, Béla Klein, Tore Koß, Florin Manea,
  Robert Mercaş, and Timo Specht

Shape Preserving Tree Transducers 165
   Paul Gallot and Sebastian Maneth

Simulating Two-Way Nondeterministic Finite Automata Over Small
Alphabets by One-Way Nondeterministic Automata 180
   Viliam Geffert and Alexander Okhotin

A New Approach for Showing Termination of Parameterized Transition
Systems 193
   Roland Herrmann and Philipp Rümmer

An Earley-Based Universal Error-Correcting Parser 208
   Maurice Herwig, Norbert Hundeshagen, and Martin Lange

More on Language Families with a Decidable Pumping-Problem
(Extended Abstract) 223
   Markus Holzer and Christian Rauch

Self-verifying Predicates in Büchi Arithmetic 237
   Mazen Khodier, Luke Schaeffer, and Jeffrey Shallit

State-Freezing Pushdown Automata 252
   Martin Kutrib, Andreas Malcher, and Priscilla Raucci

From Regular Expressions to Deterministic Finite Automata:
$2^{\frac{n}{2}+\sqrt{n}(\log n)^{\Theta(1)}}$ States Are Necessary and Sufficient 267
   Olga Martynova and Alexander Okhotin

A First Taste of MeSCaL, a Tool for Solving Membership Problems
for Regular Languages 281
   Thomas Place and Marc Zeitoun

In Orbit with MeSCaL: Higher in Concatenation and Navigational
Hierarchies of Regular Languages 299
   Thomas Place and Marc Zeitoun

A Hierarchy of Reversible Finite Automata 316
   Maria Radionova and Alexander Okhotin

Author Index 331

Complementable Normal Form of Parametrized Automata

Franziska Alber[1]([envelope])[ID] and Philipp Rümmer[1,2][ID]

[1] University of Regensburg, Regensburg, Germany
franziska.alber@informatik.uni-regensburg.de
[2] Uppsala University, Uppsala, Sweden

Abstract. Parametrized automata (PA) are an extension of two existing kinds of automata, symbolic automata and variable automata. In PA, transitions are labeled with formulas that may contain variables. PA are a powerful tool for modeling systems over infinite alphabets, but complementation of PA has been proven to be challenging: not every PA has a complement, and complementation in general may be non-computable. This paper presents a new notion of normal form for PA, called complementable normal form (CFPA). CFPA is sufficiently expressive to completely characterize the class of PA that can be complemented. We show that all Boolean operations (including complementation) can be computed efficiently on CFPA, and key problems such as the universality and non-emptiness problem are decidable for CFPA. Based on CFPA, we propose a new method for complementing PA.

Keywords: Parametrized Automata · Infinite Alphabets · Complementation

1 Introduction

Parametrized Automata (PA, [10]) are an extension of finite-state automata to infinite alphabets: the transitions are labeled with logical formulas that may contain (non-reassignable) parameters. As such, PA subsume both symbolic automata [4] and variable automata [8] and are more expressive than either. As an example, we may choose the real numbers as an alphabet and allow relations and operations on the real numbers to occur in transition formulas. Then, PA can compare two different input letters, or measure the distance between letters. PA over the real numbers can identify unsorted words, words where the last letter is largest, or words where all letters fall within a fixed range. PA find application in verification tasks, for example verifying invariant properties of Dijkstra's self-stabilizing protocol or Quicksort [5].

PA are not closed under complementation: there are languages that can be represented by PA while their respective complement languages cannot [1]. In the context of software verification, this is a serious limitation. Proving that a system satisfies a property often requires showing that no execution exists that

G. Castiglione and S. Mantaci (Eds.): CIAA 2025, LNCS 15981, pp. 1–14, 2026.
https://doi.org/10.1007/978-3-032-02602-6_1

violates the property, therefore involving complement operations. If the complement language of a PA exists and can be represented by another PA, we call that PA complementable. Because the universality problem for PA is undecidable, even the simple task of deciding whether a given PA is complementable might only be possible for limited subclasses of PA.

Contributions. As our first contribution, we present Complementable Normal Form, a new kind of normal form for PA. The subclass of PA in complementable normal form (CFPA) is both easy to complement and sufficiently expressive to represent all complementable PA. In a CFPA, the set of states is partitioned into three subsets: accepting states, complement-accepting states, and weak states. If the run of a word terminates in an accepting state, it is part of the PA's language. If the run terminates in a complement-accepting state, the word is part of the complement language. If the run terminates in a weak state, we do not gain any information. In Sect. 3, we formally define CFPA and show that every complementable PA is equivalent to a CFPA.

As the second contribution, we survey the properties of CFPA, in particular closure under Boolean operations, efficient computation thereof and decidability of key decision problems such as universality. We also compare CFPA to the related concept of strong determinism in Subsect. 3.2. We propose CFPA as the foundation for a new method for complementing PA: every complementable PA is equivalent to some CFPA, and complementation of CFPA is then straightforward. The legwork, therefore, lies in finding an equivalent CFPA.

Our third and most important contribution is found in Sect. 4, where we describe a method for transforming arbitrary complementable PA into CFPA. The key idea behind the method is to exploit the relationship between words and accepting parameter assignments that is implicitly defined by a PA. We show that the method is applicable for all complementable PA.

Related Work. More information on PA can be found in [1,10]. There are two classes of automata that have a particularly close relationship to PA: symbolic automata [4], where transitions are labeled using logical formulas, and variable automata [8], which can compare input letters to non-reassignable variables for equality and inequality. PA are strictly more expressive than either of these classes, because symbolic automata cannot compare different input letters, while variable automata are blind to the logical "structure" of the input alphabet. Variable automata and PA are incomparable to register automata [11], as the latter can overwrite stored values while variable automata and PA use fixed parameter assignments.

PA can be considered a generalization of parametric semilinear data automata, which are described in [6]. The latter use a specific extension of linear temporal logic while PA allow a wider range of first-order theories. PA should also be distinguished from symbolic register automata [3], which combine symbolic automata [4] with register automata [11] instead of variable finite automata [8].

We further point out differences to the known notion of ambiguity in automata [7]. In an unambiguous finite-state automaton, every word has at most one accepting run (although multiple non-accepting runs are permitted). In CFPA, by contrast, there is no limit on the number of distinct acceptable runs. Instead, CFPA have limitations on which states are "reachable" by words that lie in the CFPA's language.

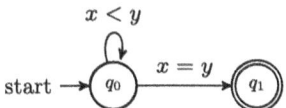

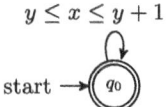

(a) A_2, identifying words whose last letter is largest.

(b) A_3, identifying words whose letters fall within an interval of length 1.

Fig. 1. Examples of parametrized automata using the theory of real numbers. In both examples, y denotes the single parameter.

2   Parametrized Automata

2.1   Definition and Notation

A finite-state automaton (FSA, [9]) is a tuple $A = (\Sigma, Q, q_0, \delta, F)$, where Σ is a finite alphabet, Q a set of states, $q_0 \in Q$ an initial state, $\delta \subseteq Q \times \Sigma \times Q$ a transition relation, and $F \subseteq Q$ a set of accepting states. A word w is accepted by A (i.e., is part of the regular language corresponding to A) if there is a sequence of transitions $(q_0, w_1, q_1), \ldots, (q_{n-1}, w_n, q_n) \subseteq Q$, called a run of w, such that $w_1 w_2 \ldots w_n = w$ and $q_n \in F$.

Parametrized automata (PA) extend FSA in the following manner: instead of letters, the transitions in PA are labeled with logical formulas which also contain a finite number of variables. PA are therefore equipped to handle infinite alphabets. A PA is always defined in relation to a fixed underlying first-order theory T (represented by a structure M), and we require T-satisfiability to be decidable. We refer to [2] for a brief revision of the terminology.

Examples in this paper use the theory of real numbers as defined in [2], i.e., input letters are real numbers and transition formulas may contain the operations $+$, $-$ and $\cdot$ and the predicates $=$ and $\leq$.

Definition 1 (parametrized automata). *Let D be an infinite alphabet and $M = (D, I)$ be a structure. Let $Y = \{y_1, y_2, \ldots\}$ be an infinite set of variables, or parameters, and x be a distinguished variable representing the current input letter. Let Φ be the set of first-order formulas over M using variables from $\{x\} \cup Y$. A parametrized automaton is a tuple $A = (M, Q, q_0, \delta, F)$, where*

- *Q is a finite set of states,*

- $q_0 \in Q$ is the initial state,
- $F \subseteq Q$ is the set of accepting states, and
- $\delta \subseteq_{fin} Q \times \Phi \times Q$ denotes the finite transition relation.

Since δ is finite, each PA uses only a finite subset of parameters $Y_A \subset Y$. For a PA A, we call a function $\mu : Y_A \to D$ a parameter assignment. The set of all possible parameter assignments for A is denoted Θ_A.

A word $w = (w_1, \ldots, w_k) \in D^*$ is accepted by A if there exists a parameter assignment $\mu \in \Theta_A$ and a sequence of transitions $(q_0, \varphi_1, q_1), \ldots, (q_{k-1}, \varphi_k, q_k)$, called a complete run, such that $q_k \in F$ and for each i, φ_i evaluates to true when x is assigned w_i and parameters are assigned according to μ. The assignment μ is fixed throughout the run, and different words may be accepted using different parameter assignments.

We write $L(A)$ for the language accepted by A. Two PA A and A' are equivalent if $L(A) = L(A')$.

A symbolic automaton is therefore a PA in which no variables occur and predicates have to be drawn from an effective Boolean algebra. It can be shown (see [1]) that every variable automaton is equivalent to a PA which, aside from variables and constant symbols, only uses the predicates $=$, $\wedge$, $\vee$ and $\neg$ (and vice versa: every such PA is equivalent to a variable automaton). PA over finite alphabets identify exactly the regular languages, because all possible parameter assignments can be enumerated.

For a fixed parameter assignment μ, A_μ denotes the symbolic automaton obtained by replacing each parameter y in A's transition formulas with its value $\mu(y)$. The language $L(A)$ can thus be alternatively described as the union $\bigcup_{\mu \in \Theta} L(A_\mu)$.

We also point out that there is no straightforward way of defining determinism in PA, and introduce two different notions of determinism: we say that A is deterministic per assignment if each A_μ is a deterministic symbolic automaton. Because the corresponding algorithms (see [12]) can be lifted straight from symbolic automata, every PA can effectively be transformed to a PA that is deterministic per assignment. Determinism per assignment is a useful property that we will need later.

Observe that, in a PA that is deterministic per assignment, two different parameter assignments may still cause a word to complete two different runs. A different way to define determinism in PA would therefore be to demand that every word completes exactly one run over all possible parameter assignments. We call this property "strong determinism," and will explore the notion in more detail in Subsect. 3.2.

2.2   The Complementation Problem

Throughout this paper, we use $\cdot^c$ for the complement operation. We say that a PA A is complementable if there exists another PA A^c such that $L(A)^c = L(A^c)$. This leads to a first theorem.

Theorem 1. *Not every PA is complementable.*

Proof. Consider example A_1 seen in Fig. 2, a PA using the theory of real numbers which randomly assigns a letter w_i of a word $w = (w_1, \dots, w_k)$ to its parameter y and accepts the word if the succeeding letter is smaller than $w_i = y$. As such, A_1 identifies unsorted words: $L(A_1) = \{w = (w_1, \dots, w_k) \in D^* \mid \exists i.1 \leq i < k \wedge w_i > w_{i+1}\}$. Note that, by this definition, a word is sorted if its letters occur in *non-decreasing* order. An example of a sorted word is $(1, 2, 3)$, while $(1, 3, 2)$ is not a sorted word. The complement of $L(A_1)$ is the set of all sorted words $w = (w_1, \dots, w_k)$ with the property $i < j \Rightarrow w_i \leq w_j$.

We will prove that a complement automaton identifying $L(A_1)^c$ cannot exist using a proof by contradiction that has a similar flavor as the pumping lemma (see [9, section 4.1]). Assume for contradiction that such an automaton A_1^c exists and has n states. Let $w = (1, 2, 3, \dots, 2n + 1)$. Since $w \in L(A_1)^c$, there is a parameter assignment μ such that w completes an accepting run in $(A_1)_\mu$. Then by a counting argument, we can argue that the run of w in A_1^c traverses some state q thrice, creating a loop of length greater than 1. This loop is traversed by a subword $(i, i + 1, \dots, i + k)$ of w where $k \geq 1$.

Therefore, the word $w' = (1, 2, \dots, i+k-1, i+k, i, i+1, \dots, 2n+1)$ obtained by repeating the found loop completes an accepting run in A_1^c. Now we see why it was necessary for w to traverse some state thrice: w' is not a sorted word because $i + k > i$. Therefore, A_1^c does not correctly identify the complement language of A_1. A PA identifying all sorted words cannot exist. □

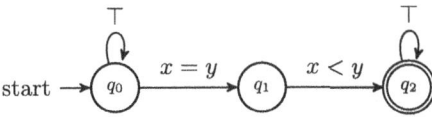

Fig. 2. A_1, a PA that cannot be complemented.

Other Operations and Decision Problems. It can be shown that PA are closed under the Boolean operations of union and intersection, using a product construction similar to symbolic automata, see [12]. Here, for both automata to operate independently we have to ensure that their sets of parameters do not overlap; this can be achieved by renaming the parameters in one PA. We call this type of product construction the *direct product*.

We are interested in two particular decision problems, checking whether a PA accepts at least one word (the non-emptiness problem) and checking whether a PA accepts all words (the universality problem). The non-emptiness problem is decidable for PA because the finite set of paths without loops terminating in accepting states can be checked for T-satisfiability. The universality problem is not decidable, a trait that is inherited from variable automata (see [8]).

This, in turn, makes the complementation problem more challenging because of the following conclusion:

Theorem 2. *At least one of the following problems is non-computable:*

- *Deciding whether an arbitrary PA is complementable.*
- *For an arbitrary complementable PA, find a complement automaton.*

Proof. Assume that both problems are computable. Then we can solve the universality problem for PA: the universal language and its complement can both be represented by PA. If an arbitrary PA is not complementable, it is not universal. If an arbitrary PA is complementable and its complement can be computed, then the PA is universal if and only if its complement is empty.   □

3   Complementable Normal Form

3.1   Definition and Examples

In PA, not all non-accepting states are created equal: some may be reached both by words in the language and words in the complement language, depending on the parameter assignment. Others can only be reached by words in the complement language, and this distinction is the idea behind complementable normal form.

Definition 2 (complementable normal form). *A PA* $A = (M, Q, q_0, \delta, F)$ *is in complementable normal form (called a CFPA) if there is a subset* $F_c \subseteq Q \backslash F$ *such that the automaton* $C = (M, Q, q_0, \delta, F_c)$ *identifies the complement of A, i.e.,* $L(C) = L(A)^c$.

In a CFPA, the set of states Q is therefore partitioned into three pairwise disjoint subsets:

- F is the set of accepting states,
- F_c is the set of complement-accepting states,
- Any state not in F or in F_c is called a weak state. If the run of a word terminates in a weak state, we cannot deduce whether the word is part of $L(A)$ or $L(A)^c$.

We illustrate CFPA using an example, and afterwards show that every complementable PA is equivalent to a CFPA.

Example 1. Consider the PA A_3, which can be seen in Fig. 1b. The automaton corresponds to the language L_3 of all words whose letters fall within an interval of length 1. The automaton A_3 has an accepting state but no weak or complement-accepting states.

A_3 can be made deterministic per assignment by adding a sink state q_1 and redirecting a run to q_1 as soon as the condition $y \leq x \leq y + 1$ is broken by a letter (Fig. 3a). The new state q_1 is a weak state.

We can obtain an equivalent CFPA by reasoning about the relationship between accepted words and the corresponding parameter assignment. For instance, every word in $L(A_3)$ is accepted if y is assigned the minimum letter of the word. If on the other hand, y is assigned the minimum letter of a word and then the condition $y \leq x \leq y + 1$ is broken, we know that word could not have been part of $L(A_3)$.

Based on this observation, we construct the equivalent CFPA C_3 seen in Fig. 3c. There are no transitions permitting $x < y$: this forces the value assigned to y to be less than or equal to all letters of the word. If a letter satisfying $x < y$ is encountered, there is no viable exiting transition and the run is aborted before proper termination. The state p_3 can only be entered if y corresponds to the minimum letter and another letter satisfying $x > y+1$ has been encountered, thus p_3 identifies the complement language correctly. If a letter satisfying $x > y + 1$ has been encountered before confirmation that y corresponds to the minimum letter, the weak state p_2 is entered and can be exited as soon as there is a confirmation that y corresponds to the minimum letter. Otherwise, as long as the condition $y \leq x \leq y + 1$ is not violated, we remain in the accepting states p_0 or p_1 depending on whether the minimum letter has already been encountered.

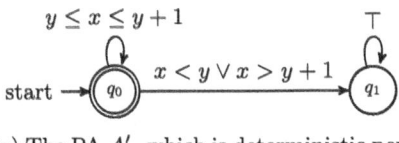

(a) The PA A_3', which is deterministic per assignment.

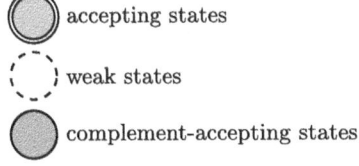

accepting states

weak states

complement-accepting states

(b) Symbols used in CFPA.

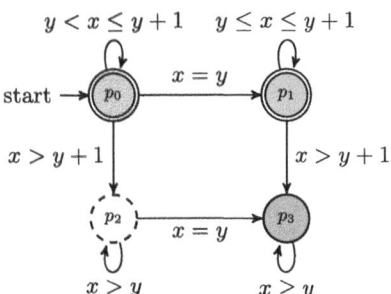

(c) A PA C_3 for L_3 in complementable normal form. State p_3 identifies the complement language.

Fig. 3. The language L_3 of words with letters within an interval of length 1.

Theorem 3. *For every complementable PA, there is an equivalent CFPA.*

Proof. Let $A = (M, Q, q_0, \delta, F)$ be a PA and $A^c = (M, P, p_0, \delta_c, F_c)$ be a complement automaton of A. Assume without loss of generality that $Q \cap P = \varnothing$. We introduce a new initial state $r_0 \notin Q \cup P$ and copy each of the transitions exiting either q_0 or p_0 so they also exit r_0: let $\delta' = \{(r_0, \varphi, r) \mid (q_0, \varphi, r) \in \delta \vee (p_0, \varphi, r) \in \delta_c\}$. Exactly one of the automata A and A^c accepts the empty word ε, so either $q_0 \in F$ or $p_0 \in F_c$ holds. We need to assign r_0 to either the language-accepting or to the complement-accepting set accordingly. Without loss of generality, let $q_0 \in F$. Then the PA $A' = (M, Q \cup P \cup \{r_0\}, r_0, \delta \cup \delta_c \cup \delta', F \cup \{r_0\})$ is equivalent to A and the PA $(M, Q \cup P \cup \{r_0\}, r_0, \delta \cup \delta_c \cup \delta', F_c)$ is equivalent to A^c. Therefore, A' is a CFPA. □

Unfortunately, this expressivity comes at a cost that has already been discussed in Theorem 2: there is no algorithm that can both identify and then complement all complementable PA. The same restriction has to apply to CFPA.

Corollary 1. *There is no algorithm that can either transform a PA into an equivalent CFPA, or return that no such CFPA exists.*

CFPA are closed under union, intersection and complementation, all of which can be computed efficiently because the direct product of two CFPA is again a CFPA. In order to show that the universality problem is decidable for CFPA, we need to mind a little detail: the definition of CFPA requires the existence of a set of states F_c which accepts exactly the complement language, but knowing exactly which states belong to F_c is not required.

We do not need to communicate information about F_c along with a CFPA $A = (M, Q, q_0, \delta, F)$ because this information can be recovered algorithmically: we assign a state q to F_c if the intersection of A and $(M, Q, q_0, \delta, \{q\})$ is empty. Note that F_c may not be unique, and this algorithm identifies the largest possible set of complement-accepting states.

At this point, the undecidability of the universality problem throws another wrench into our gears. Every universal PA is naturally a CFPA whose set of complement-accepting states is empty. If there was an algorithm for identifying CFPA, we could take any universal PA, verify that it is a CFPA, identify the set F_c of complement-accepting states and then confirm that no words have runs terminating in F_c. This is a recipe for identifying universal PA, which would contradict the undecidability of the universality problem.

Corollary 2. *There is no algorithm that can decide whether a given PA is a CFPA.*

Example 2 (Finite-State Automata). Complementable normal form can also be applied to finite-state automata (FSA), and can lead to automata that are much smaller than their deterministic equivalent while still being easy to complement. For a fixed $n \in \mathbb{N}$, consider the automaton A for the regular language $(a + b)^*b(a + b)^n$.

The automaton will be familiar to the reader, because it is a standard example to illustrate the exponential blowup when transforming a nondeterministic finite-state automaton into a deterministic one. Both A and its smallest nondeterministic complement automaton have $O(n)$ states, but a deterministic FSA that is equivalent to A will have $O(2^n)$ states. In the FSA in Fig. 4b, we have added two additional "tracks" for all words that cannot be accepted by A: one for words where the nth letter from the last is an a, and one for all words that have fewer than n letters. The resulting FSA is in complementable form. Just like deterministic FSA, it can accept either $L(A)$ or $L(A)^c$ depending on how the set of accepting states is chosen. The size of this automaton is still $O(n)$, a significant improvement compared to deterministic FSA.

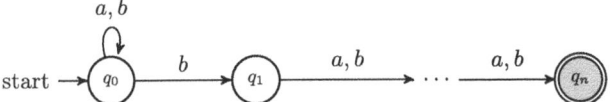

(a) A finite-state automaton A accepting all words in which the nth letter from the end is a b.

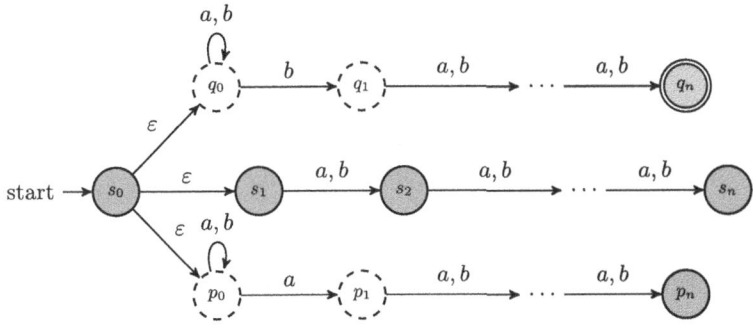

(b) This FSA is equivalent to A and in complementable normal form. The states that are not weak states accept the complement language of A.

Fig. 4. Complementable Normal Form in finite-state automata.

3.2 Strong Determinism

Complementation of FSA is easy if the FSA in question is deterministic. In a deterministic FSA, every word completes exactly one run, so the word is part of the complement if and only if its unique run terminates in a non-accepting state. We will briefly explore what happens if the same notion of determinism is applied to PA, and why this approach might be inferior to CFPA.

Definition 3. *A PA is called* strongly deterministic *(or an SDPA) if every word completes exactly one run.*

In contrast to determinism per assignment, in an SDPA every word has to complete a unique run considering *all possible parameter assignments*. The two notions are strictly orthogonal: SDPA are in general not deterministic per assignment, and vice versa. An example of an SDPA can be seen in Fig. 5.

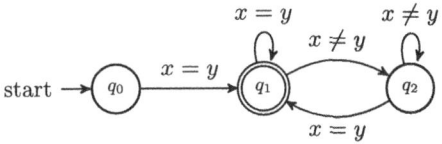

Fig. 5. D, an SDPA identifying all words whose first and last letter coincide.

SDPA are a subset of CFPA in which no weak states occur. As a consequence, SDPA can be complemented in the same manner as deterministic FSA.

Obviously, not every PA is equivalent to some SDPA, because SDPA are always complementable. In fact, this limitation goes even further, because there are even some complementable PA that are not equivalent to any SDPA.

Theorem 4. *Strongly deterministic PA form a strict subset of complementable PA. There are complementable PA that do not have a strongly deterministic equivalent.*

Proof. We claim that the language $K = \{w = (w_1, \ldots, w_k) \in D^* \mid \exists 1 \leq i < j \leq k : |w_i - w_j| > 1\}$, which is the complement of L_3 and therefore recognized by a PA, cannot be recognized by any SDPA.

Consider the sequence of words $((1, \frac{1}{2}, \ldots, \frac{1}{n}, 1 + \frac{1}{n-1}))_{1 < n \in \mathbb{N}}$, whose first few elements are $(1, \frac{1}{2}, 2)$, $(1, \frac{1}{2}, \frac{1}{3}, 1 + \frac{1}{2})$, $(1, \frac{1}{2}, \frac{1}{3}, \frac{1}{4}, 1 + \frac{1}{3})$, ...

Assume that there is an SDPA A identifying K, and let A consist of k states. Each word of the sequence has to be accepted by A upon reaching the last letter, and all runs of prefixes have to terminate in non-accepting states. We make use of a peculiar property of SDPA: If two words w and v complete runs in A using the parameter assignments μ_w and μ_v, respectively, and if w is a prefix of v, then the runs of w in A_{μ_w} and A_{μ_v} have to be identical. It can be shown by induction (which we skip here; the full version of the proof is available in [1]) that all runs of proper prefixes of the sequence, which have the form $(1, \frac{1}{2}, \ldots, \frac{1}{n})$ for some n, have to traverse n distinct states in A.

The final state is accepting and therefore cannot coincide with any state that has been previously traversed. Therefore, the run of the word $(1, \frac{1}{2}, \ldots, \frac{1}{n}, 1 + \frac{1}{n-1})$ traverses $n + 1$ distinct states.

A run of the word $(1, \frac{1}{2}, \ldots, \frac{1}{k}, 1 + \frac{1}{k-1})$ needs to traverse $k + 1$ distinct states, contradicting the assumption of A only having k states.   □

In conclusion, SDPA only represent a fraction of all complementable PA. Therefore, CFPA are the superior choice, as they are both strictly more expressive and share the same pleasant properties regarding closure under Boolean operations and decidability of important decision problems.

4   Construction of CFPA

Idea. We will now present a method for transforming arbitrary complementable PA into CFPA. We generalize the approach of Example 1, where the relationship between words and accepting parameter assignments was exploited.

For the algorithm, we introduce two new concepts:

Definition 4 (Skolem automaton). *Let A be a PA. Let B be a PA such that:*

- *B is universal,*
- *the parameters of A are a subset of the parameters of B: $Y_A \subseteq Y_B$, where Y_A and Y_B are the finite sets of parameters used by A resp. B, and*

  – *for all $w \in L(A)$, if $w \in L(B_\mu)$ for some $\mu \in \Theta$ then $w \in L(A_\mu)$.*

Then B is called a Skolem automaton of A.

We can use Skolem automata for complementation thanks to the contra-positive of the third statement: if a word is accepted by B using a parameter assignment μ, but is not accepted by A_μ, this implies that the word is part of the complement language.

Definition 5 (synchronized product). *Let $A = (M, Q, q_0, \delta_A, F_A)$ and $B = (M, P, p_0, \delta_B, F_B)$ be two PA whose parameter sets may intersect, i.e., $Y_A \cap Y_B \neq \varnothing$. Let $F \subseteq F_A \times F_B$ be arbitrary, and let $\delta = \{((q, p), \varphi_1 \wedge \varphi_2, (q', p')) \mid (q, \varphi_1, q') \in \delta_A, (p, \varphi_2, p') \in \delta_B\}$. Then the PA $(A \otimes B, F) := (M, Q \times P, (q_0, p_0), \delta, F)$ is called a synchronized product of A and B. The parameter set of $(A \otimes B, F)$ is $Y_A \cup Y_B$.*

The synchronized product allows a PA and its Skolem automaton to exchange information. Constraints placed on the parameter assignment by B also have to hold in A.

Theorem 5. *Let $A = (M, Q, q_0, \delta, F_A)$ be a PA that is deterministic per assignment. Let $B = (M, P, p_0, \delta', F_B)$ be a Skolem automaton of A. Then the synchronized product $(A \otimes B, F_A \times F_B)$ is equivalent to A and is in complementable normal form. The complement of $L(A)$ is accepted by the set of states $(Q \setminus F_A) \times F_B$.*

Proof. Let $w \in L(A)$. Since B is universal, there is a parameter assignment $\mu \in \Theta$ such that $w \in L(B_\mu)$. By the definition of B, this means $w \in L(A_\mu)$, and therefore, $L(A) \subseteq L((A \otimes B, F_A \times F_B))$.

Let now $w \in L((A \otimes B, F_A \times F_B))$. Then there exists a parameter assignment μ such that $w \in L(B_\mu)$ and $w \in L(A_\mu)$. Therefore, $L((A \otimes B, F_A \times F_B)) \subseteq L(A)$. This concludes the proof that $L(A) = L((A \otimes B, F_A \times F_B))$.

In order to prove that $(A \otimes B, F_A \times F_B)$ is in complementable normal form, it is sufficient to prove that $(A \otimes B, (Q \setminus F_A) \times F_B)$ identifies the complement of $L(A)$. Let $w \in L(A)^c$. As B is universal, there is a parameter assignment $\mu \in \Theta$ such that $w \in L(B_\mu)$. Since A is deterministic per assignment, w completes a run in A_μ which terminates in a non-accepting state since $w \notin L(A)$. Therefore, w completes a run in $(A \otimes B)_\mu$ which terminates in $(Q \setminus F_A) \times F_B$.

Vice versa, let a word w terminate in $(p, b) \in (Q \setminus F_A) \times F_B$ for some parameter assignment μ. Thus, $w \in L(B_\mu)$. If w were in $L(A)$, then the third condition would force the run of w in A_μ to terminate in an accepting state. However, because w terminates its run in A_μ in a non-accepting state, w cannot lie in $L(A)$. $\qquad\square$

Example 3. Figure 6 illustrates the method with an example based on A_2 seen in Fig. 1, the automaton which accepts all words in which the last letter is largest.

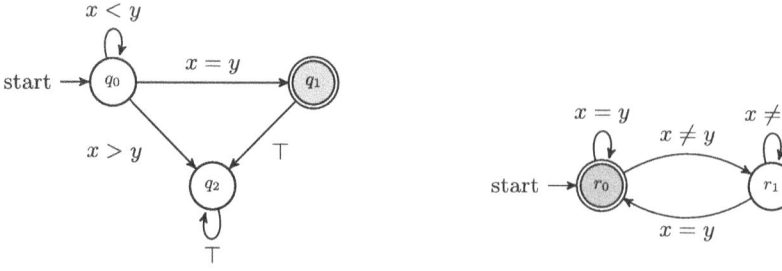

(a) A_2', which is equivalent to A_2 and deterministic per assignment.

(b) B, a suitable Skolem automaton for A_2'.

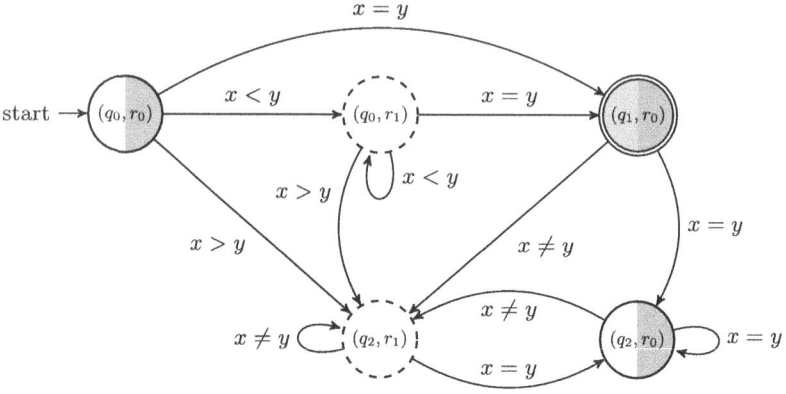

(c) The synchronized product of A_2' and B, a CFPA.

Fig. 6. An example of our method. Weak states are indicated by dashed lines, complement-accepting states are half blue and accepting states are half red and half blue and marked as accepting. (Color figure online)

Applicability. According to Theorem 3, for every complementable PA, there is an equivalent CFPA. A similarly structured proof can show the existence of a Skolem automaton for every complementable PA:

Proposition 1. *For every complementable PA A, a suitable Skolem automaton B exists.*

Proof. The product automaton in the proof of Theorem 3 is a Skolem automaton when picking the set of accepting states $F \cup F_c \cup \{r_0\}$. □

Skolem automata are not unique: for example, a CFPA equivalent to A_2 can also be obtained by demanding that y is the maximum letter of the word.

A Skolem automaton has to fulfill three properties, which each can be confirmed with widely differing degrees of difficulty: the universality problem is undecidable, while $Y_A \subseteq Y_B$ is easily checked. The third property can be verified as well, hinging on the complexity of the non-emptiness problem.

Proposition 2. *Given a PA A that is deterministic per assignment and a universal PA B, it can be decided whether B is a Skolem automaton of A.*

Proof. We need to prove that for every $w \in L(A)$ and every parameter assignment μ, $w \in L(B_\mu)$ implies $w \in L(A_\mu)$. The condition is breached if there is a $w \in D^*$ and a μ such that $w \in L(A) \cap L(B_\mu) \cap L(A_\mu)^c$. Such a w exists if and only if the intersection of $L(A)$ and $\bigcup_{\mu \in \Theta}(L(B_\mu) \cap L(A_\mu)^c)$ is non-empty, and $\bigcup_{\mu \in \Theta}(L(B_\mu) \cap L(A_\mu)^c)$ corresponds to the synchronized product of B and A. □

5 Conclusion

In our search for new approaches to the complementation problem for PA, we have identified CFPA as a formalism that sits in a particularly sweet spot. CFPA are powerful enough to completely characterize the class of complementable PA, yet at the same time complementation of CFPA is easy. Their computational properties are pleasant and comparable in complexity to those of SDPA, a model that is far more restrictive.

Because of the undecidability of the universality problem for PA in general, per Theorem 2, there can be no algorithm that either complements a given PA or returns that a complement PA does not exist. Because of this restriction, we have closely monitored the effects of Theorem 2. This has allowed us to identify the parts of the problem that are decidable to a reasonable degree.

We have shown that every complementable PA has a Skolem automaton, and studied how to confirm whether a given PA is a Skolem automaton (barring the universality property, which is of course undecidable). Given a Skolem automaton, the synchronized product with the PA as well as the complement of the PA can be constructed. The big, remaining, unmendable hole in the method is therefore the construction of suitable Skolem automata.

This is interesting, because the relationship between complementable PA and Skolem automata goes both ways: a PA is complementable iff it has a Skolem automaton as in Definition 4. The complementation problem can thus be rephrased "compute a Skolem automaton, or return that no such Skolem automaton exists." We point out that one part of the problem might be computable, as long as the other part is not. Future research might identify the computable part, or establish that both parts of the problem are non-computable.

The gap can also be closed with an application-oriented approach. We are delighted to announce that a library for PA is currently being implemented. The library will support all operations and algorithms described in this paper, make PA more accessible for applications and open up new avenues of research.

Acknowledgement. This work was supported by the Swedish Research Council through grant 2021-06327, and by the Knut and Alice Wallenberg Foundation through project UPDATE. We are grateful for the useful feedback by the anonymous referees!

Disclosure of Interests. The authors have no competing interests to declare that are relevant to the content of this article.

References

1. Alber, F.: Parametrized automata over infinite alphabets: properties and complementation. Master thesis (2024)
2. Bradley, A.R., Manna, Z.: The Calculus of Computation - Decision Procedures with Applications to Verification. Springer, Heidelberg (2007). https://doi.org/10.1007/978-3-540-74113-8
3. D'Antoni, L., Ferreira, T., Sammartino, M., Silva, A.: Symbolic register automata. In: Dillig, I., Tasiran, S. (eds.) CAV 2019, Part I. LNCS, vol. 11561, pp. 3–21. Springer, Cham (2019). https://doi.org/10.1007/978-3-030-25540-4_1
4. D'Antoni, L., Veanes, M.: The power of symbolic automata and transducers. In: Majumdar, R., Kunčak, V. (eds.) CAV 2017, Part I. LNCS, vol. 10426, pp. 47–67. Springer, Cham (2017). https://doi.org/10.1007/978-3-319-63387-9_3
5. Dijkstra, E.W.: Self-stabilizing systems in spite of distributed control. Commun. ACM **17**(11), 643–644 (1974)
6. Figueira, D., Lin, A.W.: Reasoning on data words over numeric domains. In: Baier, C., Fisman, D. (eds.) LICS 2022: 37th Annual ACM/IEEE Symposium on Logic in Computer Science, Haifa, Israel, 2–5 August 2022, pp. 37:1–37:13. ACM (2022)
7. Goldstine, J., Kappes, M., Kintala, C.M.R., Leung, H., Malcher, A., Wotschke, D.: Descriptional complexity of machines with limited resources. J. Univers. Comput. Sci. **8**(2), 193–234 (2002)
8. Grumberg, O., Kupferman, O., Sheinvald, S.: Variable automata over infinite alphabets. In: Dediu, A.-H., Fernau, H., Martín-Vide, C. (eds.) LATA 2010. LNCS, vol. 6031, pp. 561–572. Springer, Heidelberg (2010). https://doi.org/10.1007/978-3-642-13089-2_47
9. Hopcroft, J.E., Motwani, R., Ullman, J.D.: Introduction to Automata Theory, Languages, and Computation, 3rd edn. Pearson International Edition. Addison-Wesley (2007)
10. Jez, A., Lin, A.W., Markgraf, O., Rümmer, P.: Decision procedures for sequence theories. In: Enea, C., Lal, A. (eds.) CAV 2023, Part II. LNCS, vol. 13965, pp. 18–40. Springer, Cham (2023). https://doi.org/10.1007/978-3-031-37703-7_2
11. Kaminski, M., Francez, N.: Finite-memory automata. Theor. Comput. Sci. **134**(2), 329–363 (1994)
12. Veanes, M., de Halleux, P., Tillmann, N.: Rex: symbolic regular expression explorer. In: Third International Conference on Software Testing, Verification and Validation, ICST 2010, Paris, France, 7–9 April 2010, pp. 498–507. IEEE Computer Society (2010)

Toward the Glushkovization of Automata: The Strong Stabilization

Samira Attou, Ludovic Mignot$^{(\boxtimes)}$, Clément Miklarz, and Florent Nicart

GR²IF, Université de Rouen Normandie, Avenue de l'Université, 76801
Saint-Étienne-du-Rouvray, France
{samira.attou2,ludovic.mignot,clement.miklarz1,
florent.nicart}@univ-rouen.fr

Abstract. Several algorithms exist to transform an automaton into a regular expression that denotes its recognized language. Some of these algorithms, such as those by Brzozowski and McCluskey, McNaughton and Yamada, and Arden, can produce expressions whose width grows exponentially with the number of states in the automaton. Caron, Champarnaud, and Mignot proposed an algorithm that transforms any acyclic automaton into an extended regular expression with linear width with respect to the number of states. Few years before, Caron and Ziadi introduced another algorithm that also produces regular expressions with linear width. However, this algorithm is restricted to so-called Glushkov automata, which are isomorphic to those obtained through the Glushkov construction. These automata are characterized by specific structural properties, including the strong stability of their maximal cycles. In this paper, we present a novel algorithm for strongly stabilizing an automaton, which serves as a foundational step toward the *Glushkovization* of any automaton. The key advantage of this approach is that strong stabilization addresses the non-acyclic parts of the automaton, allowing other algorithms to handle the acyclic parts with optimal efficiency.

Keywords: Automata · Regular Expressions · Glushkov
Construction · Stabilization

1 Introduction

There exist several algorithms to transform an automaton into a regular expression that denotes the recognized language. Some of them (Brzozowski and McCluskey [2], McNaugton and Yamada [7], Arden [1]) produce an expression whose width can be exponentially larger than the number of states of the automaton. Notice that this bound is tight, in the sense that there exists automata for which the smallest regular expression denoting their language has an exponential width with respect to the number of states [5]. Caron, Champarnaud and Mignot [3] proposed an algorithm to transform any acyclic automaton into an extended regular expression whose width is linear with respect to the number of states, using extended operators (multi-tildes-bars). Caron and Ziadi [4]

G. Castiglione and S. Mantaci (Eds.): CIAA 2025, LNCS 15981, pp. 15–26, 2026.
https://doi.org/10.1007/978-3-032-02602-6_2

submitted an algorithm producing a (classical) regular expression whose width is linear. However, this algorithm can only be applied to so-called Glushkov automata, which are isomorphic to the ones obtained by the Glushkov construction [6]. These automata have been characterized and must satisfy some very specific structural properties. Among them the strong stability[1] of their maximal cycles.

In this paper, we present a new algorithm that allows to strongly stabilize an automaton, first step toward the *Glushkovization* of any automaton. The main advantage of this technique is that the strong stabilization deals with the non-acyclic part of the automaton, leaving the acyclic parts to other algorithms that can be chosen to be the most efficient ones.

Some preliminaries definitions are given in Sect. 2. Section 3 presents the gate isolation operation that allows us to stabilize any automaton, which is the first step of the strong stabilization. Then, Sect. 4 is devoted to the strong stabilization algorithm, that can be applied after stabilization. A Haskell implementation is presented in Sect. 5.

2   Preliminaries

We denote by $\subset$ the (non-strict) subset relation. For two sets S_1 and S_2, we denote by $S_1 + S_2$ the *disjoint union* of S_1 and S_2, which is isomorphic to the set $(S_1 \times \{0\}) \cup (S_2 \times \{1\})$. An *automaton* is a 5-tuple $(\Sigma, Q, I, F, \delta)$ where Σ is the *alphabet*, Q is the (finite) set of *states*, $I \subset Q$ is the set of *initial* states, $F \subset Q$ is the set of *final* states, and $\delta : Q \times \Sigma \to 2^Q$ is the *transition function*. The function δ can be extended to the function δ' in $2^Q \times \Sigma \to 2^Q$, computed for any $P \subset Q$ as $\delta'(P, a) = \bigcup_{p \in P} \delta(p, a)$. Finally, the function δ' can be extended to the function δ'' in $2^Q \times \Sigma^* \to 2^Q$: $\delta''(P, \varepsilon) = P$ and $\delta''(P, aw) = \delta''(\delta'(P, a), w)$, for any subset P of Q, for any symbol a in Σ and for any word w in Σ^*. By notation abuse, we will use δ indistinctly to also denote δ' and δ''. The *language* defined by an automaton $A = (\Sigma, Q, I, F, \delta)$ is the set $L(A) \subset \Sigma^*$ defined by $L(A) = \{w \in \Sigma^* \mid \delta(I, w) \cap F \neq \emptyset\}$. The *direct successors* of a state s is the set $\bigcup_{a \in \Sigma} \delta(s, a)$. The *successors* of a state s is the set $\bigcup_{w \in \Sigma^*} \delta(s, w)$. The *direct predecessors* of a state s is the set $\{q \in Q \mid \exists a \in \Sigma, s \in \delta(q, a)\}$. The *predecessors* of a state s is the set $\{q \in Q \mid \exists w \in \Sigma^*, s \in \delta(q, w)\}$.

An automaton $(\Sigma, Q, I, F, \delta)$ is said to be

- *accessible* if for any state q in Q, there exists a word w in Σ^* sending an initial state into q, *i.e.*, $p \in \delta(I, w)$; in other words, if an initial state is a predecessor of q;
- *coaccessible* if for any state q in Q, there exists a word w in Σ^* sending q into a final state, *i.e.*, $\delta(p, w) \cap F \neq \emptyset$; in other words, if a final state is a successor of q;
- *trim* if it is accessible and coaccessible;

---

[1] In this paper, we merge the two notions defined in [4], stability and transversality, into the notion of (internal and external) stability.

- *standard* if for any state q in Q, for any symbol a in Σ, there is one and only one initial state i and it satisfies $i \notin \delta(q, a)$;
- *homogeneous* if for any three states p, q_1 and q_2 in Q, for any two symbols a and b in Σ, it holds: $p \in \delta(q_1, a) \cap \delta(q_2, b) \Rightarrow a = b$.

An automaton computed by the Glushkov construction [6] is, among other characteristics, a trim, standard and homogeneous automaton [4]. Any automaton can be turned into an equivalent trim, standard and homogeneous automaton.

Example 1. Let us consider the automaton A in Fig. 1. The standardized automaton $A' = \text{std}(A)$ is represented in Fig. 2. The homogenized automaton $A'' = \text{hom}(A')$ is represented in Fig. 3.

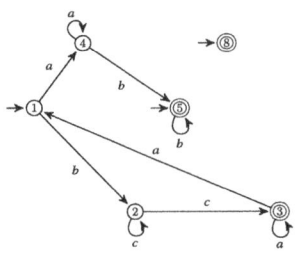

Fig. 1. The automaton A.

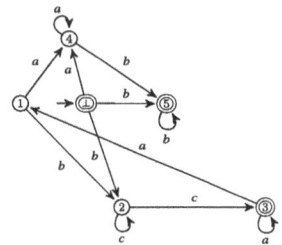

Fig. 2. The automaton std(A).

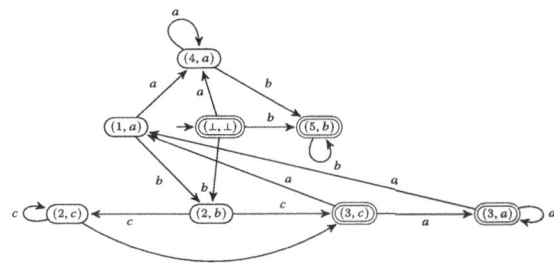

Fig. 3. The automaton hom(A').

3   Stabilization *via* Gates Isolation

Let $A = (\Sigma, Q, I, F, \delta)$ be an automaton. An *orbit* O is a subset of Q such that

1. (Cycle) for any two states p and q in O, there exists a word w in Σ^* such that $q \in \delta(p, w)$,
2. (Maximality) for any two states p in O and q not in O, p is not a successor of q or q is not a successor of p.

Let O be an orbit of A. An *ingate* of O is a state of O that has a direct predecessor not in O. An *outgate* of O is a state of O that is final or has a direct successor not in O. The set of ingates (resp. outgates) of O is denoted by $\text{In}(O)$ (resp. $\text{Out}(O)$). The orbit O is *trivial* if it is a singleton $\{o\}$ where o is not one of its direct successors. The orbit O is said to be *externally stable* if

1. any direct predecessor of an ingate of O that is not in O is also a direct predecessor of every other ingate of O,
2. any direct successor of an outgate of O that is not in O is also a direct successor of every other outgate of O,
3. either all outgates are final or none of them are.

The orbit O is said to be *internally stable* if all of its ingates are direct successors of all of its outgates. The orbit O is said to be *stable* if it is internally and externally stable. An automaton is *stable* if all its orbits are stable. Notice that these definitions are straightforward adaptations to our needs of the definitions given by Caron and Ziadi [4]. Although our definitions differ in that they do not constrain transition labels, they coincide with theirs when invoked on a homogeneous automaton.

We say that an orbit satisfies the **P** property if it is an orbit with only one ingate and only one outgate, and this ingate is a direct successor of the outgate. Obviously, an orbit satisfying the **P** property is stable. Let us show that any automaton can be transformed into a stable one where the orbits satisfy the **P** property *via* gates isolations.

3.1   Outgate Isolation

The *outgate isolation* is an operation consisting in cloning an orbit O with at least two outgates into a second orbit O', to then make the clone g' of an outgate g of O the only outgate of O' while removing g as an outgate of O. This can be performed by removing the finality of g and its transitions to states out of O, and then by doing this same operation in O' for the clones of the outgates of O distinct from g.

Definition 1. *Let $A = (\Sigma, Q, I, F, \delta)$ be an automaton, O be an orbit of A and g be an outgate of O. The outgate isolation of g is the operation producing the trim part of the automaton $(\Sigma, Q', I', F', \delta')$ where*

- $Q' = Q + \{o' \mid o \in O\}$,
- $I' = I + \{i' \mid i \in O\}$,
- $F' = \begin{cases} F & \text{if } g \notin F, \\ (F \setminus \{g\}) + \{g'\} & \text{otherwise,} \end{cases}$
- $\delta'(p, a) = \begin{cases} (\delta(g, a) \setminus O) + \{o' \mid o \in O \cap \delta(g, a)\} & \text{if } p = g', \\ \{o' \mid o \in O \cap \delta(r, a)\} & \text{if } p = r' \wedge r \in O \setminus \{g\}, \\ \delta(p, a) & \text{if } p \in O \setminus \{g\}, \\ \delta(g, a) \cap O & \text{if } p = g, \\ \delta(p, a) + \{o' \mid o \in O \cap \delta(p, a)\} & \text{otherwise.} \end{cases}$

Example 2. Let us consider the automaton $A'' = \text{hom}(A')$ in Fig. 3. Let us consider the orbit $O = \{(1,a), (2,b), (2,c), (3,c), (3,a)\}$ whose outgates are $(1,a)$, $(3,a)$ and $(3,c)$. The isolation of the outgate $(1,a)$ is represented in Fig. 4, creating a new orbit O' the only outgate of which is $(1,a)'$. The isolation of the outgate $(3,a)$ is represented in Fig. 5. To avoid name clashing, the disjoint union allows us to rename states by adding ticks.

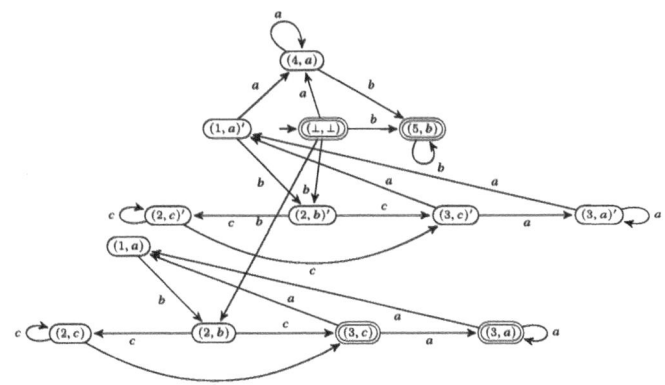

Fig. 4. The isolation of the outgate $(1,a)$.

Proposition 1. *Let A be an automaton, O be an orbit of A with at least two outgates and g be an outgate of O. Then the outgate isolation of g produces an automaton recognizing $L(A)$, where g is not anymore an outgate of O and where g' is the only outgate of the orbit $\{o' \mid o \in O\}$. Moreover, this transformation preserves homogeneity, standardness and trimness.*

3.2 Ingate Isolation

To isolate an ingate, we could apply an equivalent transformation to the one used for the outgate isolation by cloning the whole orbit. However, we want to compute stable orbits, and therefore we must be able to select as an ingate a successor of the remaining outgate after the outgate isolation. Therefore, we perform another operation that allows us to select any state in the orbit as the unique ingate.

We first clone the states o of the orbit O into new states o', except for the selected new ingate g. Then, the direct successors of a clone o'_1 include the same direct successors as its original o_1 if those are not in $O \setminus \{g\}$, plus the clones of every direct successors of o_1 that are in $O \setminus \{g\}$. Finally, the incoming transitions of the previous ingates of O (distinct from g) from outside O are redirected to their clones o'.

Notice that this operation may produce new orbits. However, these orbits are necessarily smaller than O, since the selected gate g is not cloned.

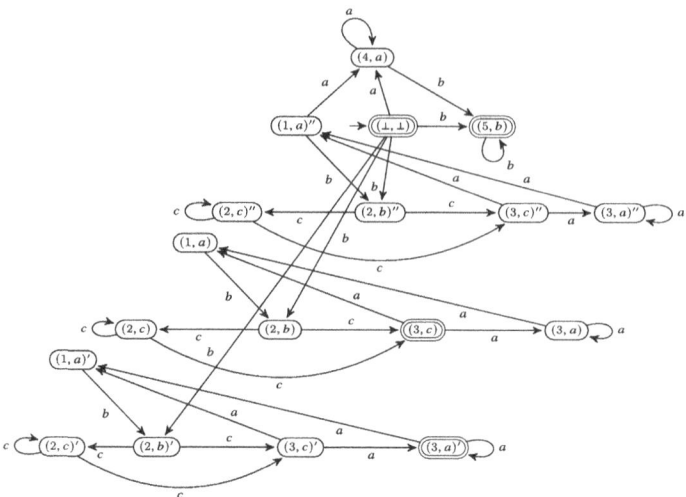

Fig. 5. The isolation of the outgate $(3, a)$.

Definition 2. *Let $A = (\Sigma, Q, I, F, \delta)$ be an automaton, O be an orbit of A and g be a state of O. The ingate isolation of g is the operation producing the trim part of the automaton $(\Sigma, Q', I', F', \delta')$ where*

- $Q' = Q + \{o' \mid o \in O_{-g}\}$,
- $I' = I + \{i' \mid i \in O_{-g}\}$,
- $F' = F + \{o' \mid o \in O_{-g} \cap F\}$,

$$
\delta'(p, a) = \begin{cases}
\delta(p, a) & \textit{if } p \in O, \\
\{o' \mid o \in O_{-g} \cap \delta(r, a)\} \cup \{g \mid g \in \delta(r, a)\} & \\
\quad \cup\, \delta(r, a) \cap Q \setminus O & \textit{if } p = r' \wedge r \in O, \\
\{o' \mid o \in O_{-g} \cap \delta(p, a)\} \cup \{g \mid g \in \delta(p, a)\} & \\
\quad \cup\, \delta(p, a) \cap Q \setminus O & \textit{otherwise,}
\end{cases}
$$

where $O_{-g} = O \setminus \{g\}$.

Example 3. Let us consider the automaton B in Fig. 6, subautomaton of the automaton in Fig. 5, with the orbit $\{(1, a)', (2, b)', (2, c)', (3, c)', (3, a)'\}$ that admits one ingate $((2, b)')$ and one outgate $((3, a)')$. However, the current ingate is not a successor of the outgate. The ingate isolation of $(1, a)'$ is represented in Fig. 7. Alternatively, the ingate isolation of $(3, a)'$ is represented in Fig. 8.

Proposition 2. *Let A be an automaton, O be an orbit of A and g be a state of O. Then the ingate isolation of g produces an automaton recognizing $L(A)$ where g is the only ingate of O. Moreover, this transformation preserves outgates of O, homogeneity, standardness and trimness, and may produce new orbits strictly smaller than O.*

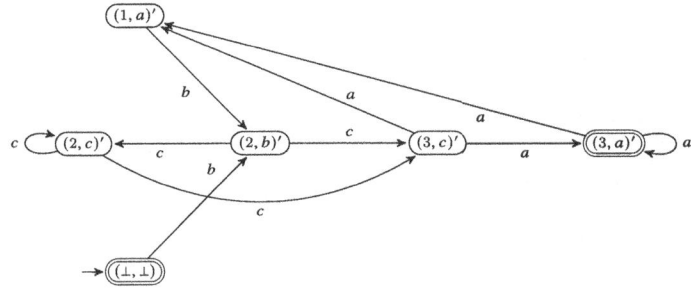

Fig. 6. The subautomaton B.

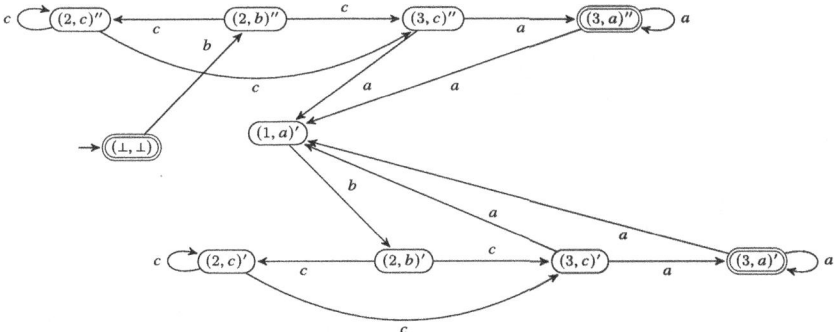

Fig. 7. The ingate isolation of $(1, a)'$.

3.3   Stabilization

Let us show how to obtain from an automaton an equivalent automaton whose orbits satisfy the **P** property.

Definition 3. *The* orbital isolation *of an automaton is the composition of the following operations:*

1. *repeat outgate isolation as long as there are orbits with at least two gates,*
2. *for each orbit O (with only one outgate g), choose a successor s of g and compute the ingate isolation of g,*
3. *recursively compute the orbital isolation of the resulting automaton if the ingate isolations have produced new orbits.*

Proposition 3. *The orbital isolation halts and preserves the standardness, the homogeneity, the trimness and the recognized language of an automaton. Moreover, the resulting automaton is stable, each orbit satisfying the **P** property.*

4   Strong Stabilization

As we have seen in the previous section, it is possible to stabilize any automaton. Let us show how to recursively apply the stabilization process in order to obtain a stronger property, satisfied by the Glushkov automata, and defined as follows.

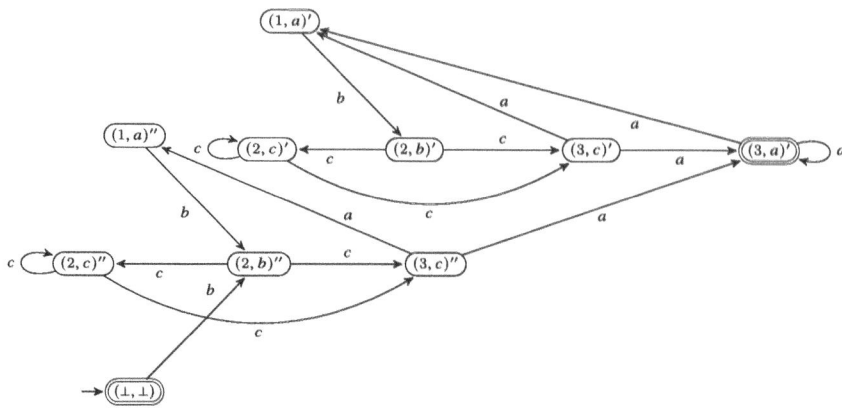

Fig. 8. The ingate isolation of $(3, a)'$.

An orbit O of an automaton $(\Sigma, Q, I, F, \delta)$ is *strongly stable* if it is stable and if each orbit $O' \subset O$ of the automaton $(\Sigma, Q, I, F, \delta')$ are strongly stable, where

$$\delta'(p, a) = \begin{cases} \delta(p, a) & \text{if } p \notin \text{Out}(O), \\ \delta(p, a) \setminus \text{In}(O) & \text{otherwise.} \end{cases}$$

Let O be a subset of a set Q. A *standard orbital automaton over* O is a standard automaton $A = (\Sigma, O + \{\bot\}, \{\bot\}, F, \delta)$ where O is an orbit.

For s a state of a homogeneous automaton, we denote by $\text{Symb}(s)$ the symbol labelling the transitions entering in s, if this symbol exists.

For O a non-trivial stable orbit of a standard and homogeneous automaton $A = (\Sigma, Q, I, F, \delta)$, the *standard orbital automaton* associated with O in A is the automaton $(\Sigma, O + \{\bot\}, \{\bot\}, \text{Out}(O), \delta')$ where

$$\delta'(p, a) = \begin{cases} \delta(p, a) \cap O & \text{if } p \neq \bot, \\ \{s \in \text{In}(O) \mid \text{Symb}(s) = a\} & \text{otherwise.} \end{cases}$$

Example 4. The automaton C in Fig. 9 is the standard orbital automaton associated with the orbit $O = \{(1, a)', (2, b)', (2, c)', (3, c)', (3, a)'\}$ in the automaton of Fig. 8. Notice that the orbit O is stable, but not strongly stable, since after removing the loop on the state $(3, a)'$ (see Fig. 10), the resulting automaton is not internally stable, the outgate $(3, a)'$ not having any ingate as direct successor.

In order to strongly stabilize a stable automaton, we substitute any orbit O with an equivalent strongly stable orbital automaton obtained by recursively applying the stabilization operation on the standard orbital automaton associated with O in A, after removal of the transitions from the outgates to the ingates. The substitution removes the old orbit, to replace it with its equivalent strongly stable orbital automaton by adding edges as follows. If a state not in O was linked to an ingate of O, it is now linked to the direct successors of the

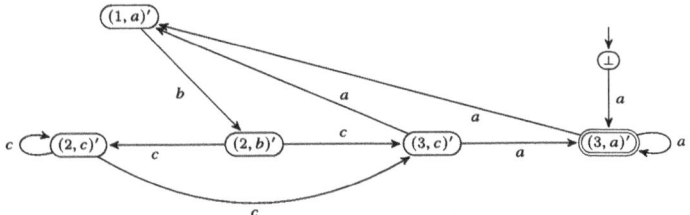

Fig. 9. A standard orbital automaton C.

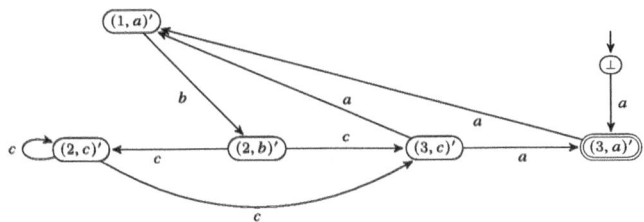

Fig. 10. The automaton computed from C by removing a loop.

initial state of the new orbital automaton. The final states of the orbit of the orbital automaton are now linked to the successors of the outgates of O. Let us formalize the substitution operation.

Definition 4. *Let* $A = (\Sigma, Q, \{i\}, F, \delta)$ *be a standard and homogeneous automaton,* O *be a stable orbit of* A *and* $A' = (\Sigma, Q' + \{i'\}, \{i'\}, F', \delta')$ *be a standard orbital automaton. The substitution of* O *by* A' *is the trim part of the automaton* $(\Sigma, Q'', \{i\}, F'', \delta'')$ *where*

- $Q'' = (Q \setminus O) + Q'$,
- $F'' = \begin{cases} F & \text{if } O \cap F = \emptyset, \\ (F \setminus O) + F' & \text{otherwise.} \end{cases}$
- $\delta''(p, a) = \begin{cases} \delta(p, a) & \text{if } p \notin O \land \delta(p,a) \cap \mathrm{In}(O) = \emptyset, \\ \delta(p, a) + \delta'(i', a) & \text{if } p \notin O \land \delta(p,a) \cap \mathrm{In}(O) \neq \emptyset, \\ \delta'(p, a) & \text{if } p \in Q' \setminus F', \\ (g, a) \setminus O) + \delta'(p, a) & \text{otherwise if } p \in F' \text{ with } g \in \mathrm{Out}(O). \end{cases}$

Proposition 4. *Let* A *be a standard and homogeneous automaton,* O *be a stable orbit of* A, A' *be a standard orbital automaton over an orbit* O' *and* A'' *be the substitution of* O *by* A' *in* A. *If* A' *is equivalent to the standard orbital automaton associated with* O *in* A, *then* A'' *is equivalent to* A. *Moreover, if* O' *is strongly stable in* A', *so it is in* A''.

Then, using this operation, we can define the strong stabilization as follows.

Definition 5. *Let A be a standard and homogeneous automaton. The strong stabilization of a stable automaton A is the operation defined from the two following mutually recursive operations:*

- *if A is not a standard orbital automaton, then the strong stabilization of A is performed by applying, for any non-trivial orbit O of A, the substitution of O by the strong stabilization of the standard orbital automaton of O;*
- *if A is a standard automaton over an orbit O, then the strong stabilization of A is the automaton A'''' obtained by the composition of the following operations:*
  1. *computation of A' obtained from A by removing the edges from $\text{Out}(O)$ to $\text{In}(O)$,*
  2. *computation of A'' the stabilization of A' using orbital isolation,*
  3. *computation of A''' the strong stabilization of A'',*
  4. *computation of A'''' by addition of the edges from the final states to the successors of the initial state of A'''.*

Example 5. As an example, let us compute the strong stabilization of the automaton in Fig. 9. It is a standard orbital automaton, so we first remove the loop, leading to Fig. 10. We then need to select a successor of the outgate $(3, a)'$ in order to compute the ingate isolation, the only choice being the state $(1, a)'$. The resulting automaton D is represented in Fig. 11. This automaton is strongly stable, and the recursive application of the strong stabilization is not needed. Finally, the strong stabilization of the automaton in Fig. 9 is represented in Fig. 12 by adding in D the edges from the final states to the successor of the initial state.

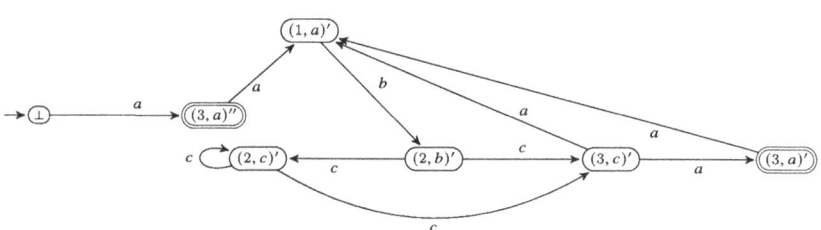

Fig. 11. The ingate isolation of $(1, a)'$.

Theorem 1. *The strong stabilization of a standard, homogeneous, and trim automaton halts and produces an equivalent, standard, homogeneous, trim and strongly stable automaton.*

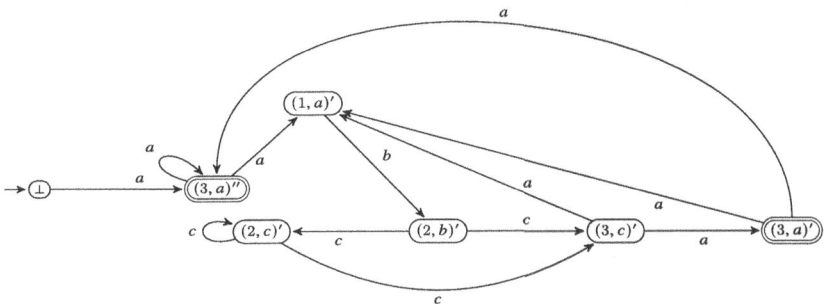

Fig. 12. The strongly stabilized automaton.

5   Haskell Implementation

The strong stabilization algorithm has been implemented in Haskell, using advanced functional programming techniques, such as free monads to compute the sets of states of repeated operations or GADT to hide the concrete types of the states.

A web interface has been developed to allow the reader to visualize the automata and their transformations, using SVG conversion *via* the Dot language. This interface is made using functional reactive programming *via* Reflex, and has been converted into JavaScript using the GHC WASM backend.

The sources are available on GitHub [9]. The web application is available on the web [8].

6   Conclusion and Perspectives

In this paper, we have shown how to strongly stabilize a standard and homogeneous automaton. This process can directly be transformed into a conversion algorithm to produce a regular expression. During the strong stabilization process, once an acyclic automaton obtained, it can be turned into a regular expression, using Caron and Ziadi technique [4] if the automaton is a Glushkov one, or Caron *et al.* one [3] in the general case. Once this regular expression computed, the unary operator $^{+}$ can be applied in order to obtain an expression for a standard orbital automaton. Replacing the orbits by expression-labelled states and transitions allows us to obtain a new acyclic automaton on which another acyclic conversion method can be applied, until the end of the process. We plan to explicit this algorithm in an extended version of this paper.

References

1. Arden, D.N.: Delayed-logic and finite-state machines. In: SWCT, pp. 133–151. IEEE Computer Society (1961)
2. Brzozowski, J.A., McCluskey, E.J.: Signal flow graph techniques for sequential circuit state diagrams. IEEE Trans. Electron. Comput. **12**(2), 67–76 (1963)

3. Caron, P., Champarnaud, J., Mignot, L.: Multi-tilde-bar expressions and their automata. Acta Informatica **49**(6), 413–436 (2012)
4. Caron, P., Ziadi, D.: Characterization of Glushkov automata. Theor. Comput. Sci. **233**(1–2), 75–90 (2000)
5. Ehrenfeucht, A., Zeiger, H.P.: Complexity measures for regular expressions. J. Comput. Syst. Sci. **12**(2), 134–146 (1976)
6. Glushkov, V.M.: The abstract theory of automata. Russian Math. Surveys **16**, 1–53 (1961)
7. McNaughton, R., Yamada, H.: Regular expressions and state graphs for automata. IRE Trans. Electron. Comput. **9**(1), 39–47 (1960)
8. Mignot, L.: Application: strong stabilization. http://ludovicmignot.free.fr/programmes/glushkovization. Accessed 16 Apr 2025
9. Mignot, L.: An implementation of the strong stabilization: Glushkovization–Haskell (2023). https://github.com/LudovicMignot/glushkovization--Haskell

Constructing Compact BPE Token DFAs

Martin Berglund[1]([✉]), Anna Jonsson[1]([✉]), Willeke Martens[1]([✉]),
and Brink van der Merwe[2,3]([✉])

[1] Department of Computing Science, Umeå University, Umeå, Sweden
{mbe,aj,wms}@cs.umu.se
[2] Department of Computer Science, Stellenbosch University,
Stellenbosch, South Africa
abvdm@cs.sun.ac.za
[3] National Institute for Theoretical and Computational Sciences,
Stellenbosch, South Africa

Abstract. Byte pair encoding (BPE) tokenization is a popular technique for subdividing text into relevant subwords and is frequently used in large language model systems. Since the tokenization procedure is deterministic and produces regular languages, it is beneficial to characterize it in terms of finite automata to allow for further automata-aided processing, such as pattern matching. In this paper, we demonstrate how to build such automata efficiently in practice, applying a series of optimization techniques to represent them compactly.

1 Introduction

Byte pair encoding (BPE) tokenization [13] is a key preprocessing step in prominent large language model systems like OpenAI's ChatGPT. BPE tokenization iteratively merges adjacent tokens according to a dictionary of merge rules, presented in order of priority, to produce a subword tokenization of a text. For example, when tokenizing the word "Palermo" using the OpenAI GPT-3 dictionary [11], we begin with each symbol as a separate token: $\boxed{P}\boxed{a}\boxed{l}\boxed{e}\boxed{r}\boxed{m}\boxed{o}$. The algorithm then repeatedly picks the highest-priority applicable rule from the dictionary and merges the left-most matching token pair. In our example, the first rule applied is $e \wr r$, resulting in $\boxed{P}\boxed{a}\boxed{l}\boxed{er}\boxed{m}\boxed{o}$. Subsequent applications of rules $a \wr l$, $P \wr al$ and $m \wr o$ yield the final tokenization $\boxed{Pal}\boxed{er}\boxed{mo}$. As no further rules apply, this is the *correct* GPT-3 tokenization of "Palermo".

To support formal reasoning over BPE tokenizations, it is helpful to construct finite automata operating over token sequences instead of raw text. Algorithm 1, adapted from [2], builds a deterministic finite automaton (DFA) that accepts exactly the valid tokenizations for a given BPE dictionary. However, both state and transition blowup present practical challenges. The algorithm creates one state per useful dictionary rule, which is optimal in certain theoretical cases, yet problematic given the size of real-world dictionaries. Moreover, since the token alphabet grows linearly with the dictionary size, the total number of transitions

G. Castiglione and S. Mantaci (Eds.): CIAA 2025, LNCS 15981, pp. 27–40, 2026.
https://doi.org/10.1007/978-3-032-02602-6_3

grows quadratically in both theory *and* practice. For instance, the Phi-4 dictionary [10], used in Microsoft Phi 4 models, contains 100 000 rules, resulting in a DFA with 100 000 states and nearly 10 billion transitions.

This paper explores four optimizations to reduce state and transition explosion in practice: two prevent the creation of redundant states, and two compress transition representation. The most significant optimization is the use of *default transitions*, which specify a fallback state when no explicit transition exists for the next input symbol. This automata extension was introduced as *delayed-input automata* in Kumar et al. [8], but the technique occurs in many forms, notably as the failure function in the Aho-Corasick algorithm [1]. This in turn inspired an equivalent formalism under the name *failure automata* [5,7]. Bille et al. [4] offers a good overview of the recent literature in this area. This technique is especially effective here because Algorithm 1 copies a vast number of transitions across states, resulting in substantial behavioural overlap. Figure 1 previews the automata produced by the optimized procedure. Observe the repetitive pattern of transitions to the sink state, which motivates the second optimization that focuses on compact transition representation. Together, these two optimizations enable an efficient and scalable automaton representation.

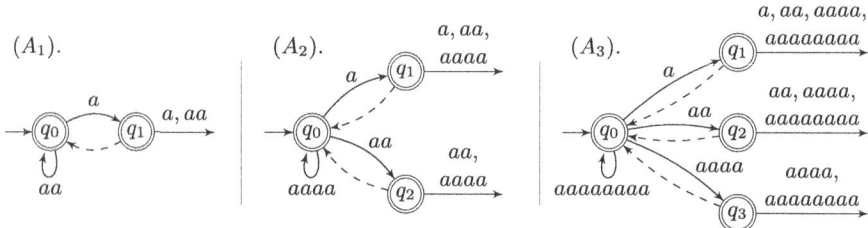

Fig. 1. Algorithm 2 is applied iteratively, starting from the single-state DFA accepting any number of as. At each iteration, the next rule from the dictionary $[a \wr a, aa \wr aa, aaaa \wr aaaa]$ is incorporated into the DFA. Transitions with no target go to a sink state (not drawn). The dashed transitions are *default transitions:* if the current state has no transition on the next input symbol, we may take the default transition. For example, reading the token a in state q_2 in A_2 brings us to q_1 by first defaulting to q_0.

It is theoretically possible to first construct the full BPE token DFA using Algorithm 1, then minimize and insert default transitions using heuristics. However, both the initial construction and subsequent minimization and default-transition insertion steps are prohibitively expensive [5,7]. Even given a fast heuristic for adding default transitions the size of the intermediary automaton creates problems. Instead, our approach performs all optimizations on-the-fly, integrating one dictionary rule at a time while maintaining a compact DFA.

2   Background

Let $\mathbb{N} = \{0, 1, 2, \ldots\}$. For a partial function $f : A \to B$, we write $f(a) = \bot$ when f is undefined for $a \in A$. The symbol $\bot$ is also used to refer to the partial empty function, i.e., the partial function that is undefined for all its arguments.

An *alphabet* Σ is a finite set of symbols. To distinguish between alphabets over indivisible symbols and tokens, we call Σ a *base alphabet*. The set Σ^* consists of all strings over Σ, including the empty string ε. A *token alphabet* Γ over Σ is a finite set $\Gamma \subset \Sigma^*$ such that $\varepsilon \notin \Gamma$, $\Sigma \subseteq \Gamma$, and for every token $u \in \Gamma$ with $|u| > 1$, u can be written as vw for some $v, w \in \Gamma$. A tokenization over Γ is a sequence $u_1 \wr \cdots \wr u_k$ with each $u_i \in \Gamma$. This sequence is distinct from the string $u_1 \cdots u_k \in \Sigma^*$ obtained by concatenation. To keep notation clear, we use Σ for base alphabets and Γ for token alphabets. Furthermore, the symbols a, b, c denote elements of Σ and u, v, w elements of Γ. This includes also all super- and subscripted variants (so e.g. $a_3 \in \Sigma'$ is a symbol in a base alphabet).

Definition 1. *A byte pair dictionary D over Σ is a sequence of token pairs or rules $u_i \wr v_i$ with $u_i, v_i \in \Sigma^*$, written as $D = [u_1 \wr v_1, \ldots, u_n \wr v_n]$. A rule $u_i \wr v_i$ has* higher priority *than $u_j \wr v_j$ if $i < j$. A dictionary is* proper *if for each j with $|u_j| > 1$ there exists some $i < j$ such that $u_j = u_i v_i$, and, symmetrically, for each j with $|v_j| > 1$ there exists some $i < j$ such that $v_j = u_i v_i$. A proper dictionary D induces the token alphabet $\Gamma = \Sigma \cup \{u_1 v_1, \ldots, u_n v_n\}$. The rule $u_i \wr v_i$ is* useful *in D if it can be applied to some input string.*

We strictly consider proper dictionaries in which all rules are useful and use D and all its variants to refer to such dictionaries.

Definition 2. *A* delayed input deterministic finite automaton *(D^2FA) is a tuple $A = (Q, \Gamma, q_0, \delta, \nu, F)$ where:*

- *Q is a finite set of* states,
- *Γ is a finite token alphabet,*
- *$q_0 \in Q$ is the* initial state,
- *$\delta : Q \times \Gamma \to Q$ is a (possibly partial)* transition function,
- *$\nu : Q \to Q$ is a (possibly partial)* default transition function, *which is acyclic (i.e. $\nu(\cdots \nu(q) \cdots) \neq q$ for all q and non-zero number of applications of ν),*
- *$F \subseteq Q$ is the set of* accept states.

The extended transition function δ^ν of A is defined recursively as:

$$\delta^\nu(q, u) = \begin{cases} \delta(q, u) & \text{when } \delta(q, u) \text{ is defined,} \\ \delta^\nu(\nu(q), u) & \text{when } \delta(q, u) = \bot \text{ and } \nu(q) \neq \bot, \\ \bot & \text{otherwise.} \end{cases}$$

A path *is a sequence $q_1, u_1, \ldots, q_{n-1}, u_{n-1}, q_n$, where $n \in \mathbb{N}$, $q_i, q_n \in Q$ and $u_i \in \Gamma$, so that $\delta^\nu(q_i, u_i) = q_{i+1}$ for all $i \in \{1, \ldots, n-1\}$. We write $q_1 \xrightarrow{u_1} q_2 \xrightarrow{u_2} \cdots \xrightarrow{u_{n-1}} q_n$. If the path begins at the initial state and ends in an accept state, A accepts the tokenization $u_1 \wr \cdots \wr u_{n-1}$. The language accepted by A, denoted $\mathcal{L}(A)$, consists of all such accepted tokenizations. Two D^2FAs A and A' are* equivalent *if $\mathcal{L}(A) = \mathcal{L}(A')$. If $\nu(q) = \bot$ for all $q \in Q$, then A is a DFA.*

Observation 1. Let $A = (Q, \Gamma, q_0, \delta, \nu, F)$ be a D^{2}FA. Define the DFA $A' = (Q, \Gamma, q_0, \delta^\nu, \perp, F)$. Then $\mathcal{L}(A') = \mathcal{L}(A)$, since the extended function δ^ν simulates the effect of the default transitions in A.

In some of the constructions below, it is convenient to assume the existence of a *unique sink state* $q_s \notin F$ so that $\delta(q_s, u) = q_s$ for all $u \in \Gamma$ and $\nu(q_s) = \perp$. By writing the state set as $Q \cup \{q_s\}$, we make this assumption explicit.

We recall the original procedure from [2] for constructing a BPE token DFA from a DFA A over base alphabet Σ and dictionary D over Γ. The construction proceeds by incorporating one dictionary rule at a time, in order of priority, using Algorithm 1. Incorporating the rule $u \wr v$ consists of the following main steps: First, for each state triple (s_1, s_2, s_3) such that $s_1 \xrightarrow{u} s_2 \xrightarrow{v} s_3$, a new transition $s_1 \xrightarrow{uv} s_3$ is added. Second, to prevent A from reading u followed by v without affecting other uses of s_2, a new state $\mathit{fresh}(s_2)$ is created. This new state inherits all outgoing transitions of s_2, except those on v (and on uv if $u = v$). Finally, all current transitions on u to s_2 are redirected to $\mathit{fresh}(s_2)$.

Algorithm 1. Original algorithm from [2]

1: **Input:** a DFA $A = (Q, \Gamma, q_0, \delta, \perp, F)$, and a rule $u \wr v$
2: let $S = \{(s_1, s_2, s_3) \in Q^3 \mid \delta(s_1, u) = s_2, \delta(s_2, v) = s_3\}$
3: let $S_2 = \{s_2 \mid (s_1, s_2, s_3) \in S\}$
4: **for** $(s_1, s_2, s_3) \in S$ **do**
5:    add new transition by defining $\delta(s_1, uv) = s_3$
6: add uv to Γ
7: **for** $s_2 \in S_2$ **do**
8:    create a fresh state from s_2, denote it $\mathit{fresh}(s_2)$
9:    add $\mathit{fresh}(s_2)$ to Q, and if $s_2 \in F$ add $\mathit{fresh}(s_2)$ to F as well
10:   **if** $u \neq v$ **then**
11:       add new transition by defining $\delta(\mathit{fresh}(s_2), \alpha) = \delta(s_2, \alpha)$ for all $\alpha \in \Gamma \setminus \{v\}$
12:   **else**
13:       add transition by def. $\delta(\mathit{fresh}(s_2), \alpha) = \delta(s_2, \alpha)$ for all $\alpha \in \Gamma \setminus \{v, uv\}$
14: **for** $q \in Q$ with $\delta(q, u) \in S_2$ **do**
15:    replace the transition by defining $\delta(q, u) = \mathit{fresh}(\delta(q, u))$
16: output the resulting DFA

---

3   Constructing Compact Automata

We propose four optimizations to improve both the size and efficiency of the BPE token DFA construction. This section covers the three optimizations that directly modify Algorithm 1. A fourth optimization, targeting the compact encoding of sink transitions, is described in Sect. 5.

Our optimizations operate on D^{2}FAs, enabling significant reductions of the transition count. Optimization *O1* replaces explicit copying of transitions from

s_2 to $fresh(s_2)$ with a default transition from $fresh(s_2)$ to s_2, only adding explicit sink transitions on v (and possibly uv). While yielding a sparser transition table, the shift also requires other minor adjustments to Algorithm 1, such as in state triplet identification.

The shift to D^2FAs reduces the number of transitions, but not the state count. This is addressed by optimizations $O2$ and $O3$: the former avoids creating $fresh(s_2)$ when s_2 is not *multiple-use* while the latter re-uses an existing *clean-copy* state with respect to s_2 and the relevant rule. Definition 3 and 4 below characterize the states affected by $O2$ and $O3$, respectively. The reader may want to look ahead to Sect. 4 for further motivation of these definitions.

Definition 3. *Let $A = (Q, \Gamma, q_0, \delta, \nu, F)$ be a D^2FA. For $q \in Q$ and $u \in \Gamma$, let $P_{q,u} \subseteq Q$ be the smallest set such that $q \in P_{q,u}$, and for all $p \in Q$, if $\nu(p) \in P_{q,u}$ and $\delta(p, u) = \bot$, we have $p \in P_{q,u}$. Let $S_{q,u} = \{v \in \Gamma \mid q' \in Q, \delta(q', v) \in P_{q,u}\}$. Then q is* multiple-use *with respect to u if $|S_{q,u}| > 1$.*

Definition 4. *Let $A = (Q \cup \{q_s\}, \Gamma, q_0, \delta, \nu, F)$ be a D^2FA. For a rule $u \wr v$, define $\Lambda = \{u, uv\}$ if $u = v$ and $\Lambda = \{u\}$ otherwise. For $q, p \in Q$ we say that q is a* clean-copy *state with respect to p and $u \wr v$ iff $\nu(q) = p$, $q \in F \Leftrightarrow p \in F$, $\delta(q, w) = \bot$ for all $w \in \Gamma \setminus \Lambda$, and $\delta(q, w) = q_s$ for $w \in \Lambda$.*

We now present the modified algorithm in Algorithm 2. Note that removing all blue-highlighted lines disables $O2$ and all red-highlighted lines disables $O3$.

Algorithm 2 will only ever be applied to a string DFA with a distinguished but unused sink state (i.e. initially there are no transitions to the sink), or to a D^2FA which was itself produced by Algorithm 2.

Example 1. Consider the dictionary $D = [a \wr b, ab \wr b, abb \wr b]$ and the universal DFA over the token alphabet $\{a, b\}$. Figure 2 shows the token D^2FA after each merge by Algorithm 2. The set S always only contains (q_0, q_0, q_0), so $S_2 = \{q_0\}$. Besides being the start state, q_0 is multiple-use with respect to v, disabling $O2$. The first merge creates the new state q_1, but then $O3$ reuses it for all subsequent merges.

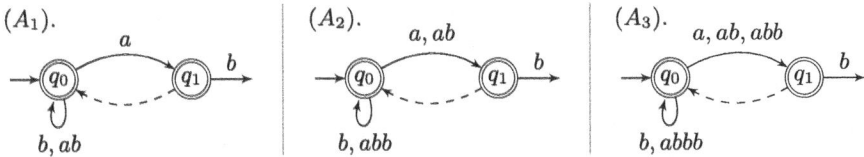

Fig. 2. Let A_1 be the D^2FA after merging $a \wr b$ into the universal DFA. A_2 and A_3 result from merging $ab \wr b$ and $abb \wr b$ subsequently. In each case, $O3$ avoids creating a new state, while $O2$ has no effect.

Algorithm 2. Optimized BPE Rule Merge

1: **Input:** a D^2FA $D = (Q \cup \{q_s\}, \Gamma, q_0, \delta, \nu, F)$ and rule $u \wr v$
2: **for** $q \in Q$ such that $\delta(q, u) = q_s$ **do**
3:      add new transition by defining $\delta(q, uv) = q_s$ ⊢
4: $S = \{(s_1, s_2, s_3) \in Q^3 \mid \delta(s_1, u) = s_2, \ \delta^\nu(s_2, v) = s_3\}$ —— $\boxed{\textit{O4, see Section 5}}$
5: **for** $(s_1, s_2, s_3) \in S$ **do**
6:      add new transition by defining $\delta(s_1, uv) = s_3$
7: add uv to Γ
8: $S_2 = S_2' = \{s_2 \in Q \mid (s_1, s_2, s_3) \in S\}$
9: redefine $S_2' = \{s_2 \in S_2 \mid s_2$ is multiple-use with respect to v or $q = q_0\}$ ⊢ $\boxed{\textit{O2.}}$
10: **for** $s_2 \in S_2'$ **do**
11:      $S' = \{q \in Q \mid q$ is a clean-copy with respect to s_2 and $u \wr v\}$ ⊢ $\boxed{\textit{O3.}}$
12:      **if** $S' = \varnothing$ **then**
13:          create a fresh state from s_2, denote it $fresh(s_2)$
14:          add $fresh(s_2)$ to Q, if $s_2 \in F$ add $fresh(s_2)$ to F
15:          add new default transition $\nu(fresh(s_2)) = s_2$ ⊢—— $\boxed{\textit{O1.}}$
16:          add new transition by defining $\delta(fresh(s_2), v) = q_s$
17:          **if** $u = v$ **then**
18:              add new transition by defining $\delta(fresh(s_2), uv) = q_s$
19:      **else**
20:          denote some $s \in S'$ as $fresh(s_2)$
21: **for** $s_2 \in S_2 \setminus S_2'$ **do**
22:      redirect transition by defining $\delta(s_2, v) = q_s$
23:      **if** $u = v$ **then**
24:          define $\delta(s_2, uv) = q_s$
25: **for** $(s_1, s_2, s_3) \in S$ with $s_2 \in S_2'$ **do**
26:      add new transition by defining $\delta(s_1, u) = fresh(s_2)$

Example 2. Similarly, with the dictionary $D = [a \wr b, a \wr ab, a \wr aab, \ldots]$ and the universal DFA over the token alphabet $\{a, b\}$, the first merge creates state q_1, but all subsequent merges modify q_1 directly via *O2* since it is only reachable on a.

Example 3. Consider the dictionary $\{a \wr a, aa \wr aa, aaaa \wr aaaa, \ldots\}$ and the universal DFA over $\{a\}$. Figure 1 shows the result of the first three merges. Neither *O2* nor *O3* saves any states here: q_0 repeatedly takes the role of s_2, and no states share the identical local behaviour for reuse. Thus, the worst case state complexity remains the same as in Algorithm 1, producing $\mathcal{O}(|Q||D|)$ states.

4   Proof of Correctness

We first demonstrate the correctness of Algorithm 2 with *O2* and *O3* disabled (i.e. all coloured lines are removed), referring to this variant as Algorithm 2^X. We then show that enabling *O2* and *O3* preserves correctness. Observe that *O4*, which deals with transition compression, is treated separately in Sect. 5.

Lemma 1. *Take D^2FA A and A', and DFAs B, B', and C fulfilling the following. A' and B' are obtained from A and B using Algorithm 2^X and Algorithm 1 respectively (merging the rule $u \wr v$). B and C are the DFA corresponding to A and A', respectively, applying Observation 1 and removing the sink and any transitions to the sink. Then B' is isomorphic to C.*

$$
\begin{array}{ccc}
A & \xrightarrow{\quad\quad Observation\ 1 \quad\quad} & B \\
{\scriptstyle Algorithm\ 2}\Big\downarrow & & \Big\downarrow {\scriptstyle Algorithm\ 1} \\
A' & \xrightarrow[\quad]{Observation\ 1} \quad C \xeq{shown\ to\ be\ isomorphic} & B'
\end{array}
$$

Proof. Let e.g. δ_A denote the transition function of A, $\delta_{A'}^{\nu}$ the extended transition function for A' and so on. Let Q_A denote the states of A *without the sink*. Observe that $Q_C = Q_{A'}$, $Q_B = Q_A$, and Q_A is a subset of all the other state sets. By definition, B' and C are isomorphic iff there exists a bijection $f : Q_{A'} \to Q_{B'}$ such that $\delta_{B'}(f(q), w) = f(\delta_C(q, w))$ for all $q \in Q_{A'}$ and $w \in \Gamma \cup \{uv\}$.

Let $h : Q_B \to (Q_{B'} \setminus Q_B)$ be the "fresh" mapping in the run of Algorithm 1 and $h' : Q_A \to (Q_{A'} \setminus Q_A)$ the one from the run of Algorithm 2^X (i.e. $h(q)$ is defined if q occurs as s_2 in the algorithm, and in that case $h(q) = fresh(q)$). Define $f(q) := h(q')$ if $q = h'(q')$ for some $q' \in Q_A$, and $f(q) := q$ otherwise.

We first show that f is bijective. Suppose $h'(q') \in Q_{A'} \setminus Q_A$. Then there exist $s_1, s_3 \in Q_A$ such that $\delta_A(s_1, u) = q'$ and $\delta_A^{\nu}(q', v) = s_3$, implying $\delta_B(s_1, u) = q'$ and $\delta_B(q', v) = s_3$, so Algorithm 1 adds $h(q')$ to $Q_{B'}$. Since h and h' are injective, f is injective on these new states.

Conversely, if $h(q') \in Q_{B'} \setminus Q_B$, then for some $s_1, s_3 \in Q_A$, $\delta_B(s_1, u) = q'$ and $\delta_B(q', v) = s_3$. Because $\delta_B = \delta_A^{\nu}$, there exists $p \in Q_A$ such that $s_1 \in P_{p,u}$ (recall Definition 3) and $\delta_A(p, u) = q'$. Therefore, Algorithm 2^X adds $h'(q')$ to $Q_{A'}$, so $f(h'(q')) = h(q')$. Thus, f is also surjective.

To complete the proof, we show that f is a homomorphism. That is, for all $w \in \Gamma \cup \{uv\}$ and $q \in Q_{A'}$ we have $\delta_{B'}(f(q), w) = f(\delta_C(q, w))$.

First, consider $q \in Q_A$. Algorithm 2^X defines the transition function $\delta_{A'}$ in terms of δ_A as follows:

$$
\delta_{A'}(q, w) = \begin{cases}
h'(\delta_A(q, u)) & \text{if } w = u \text{ s.t. } \delta_A^{\nu}(\delta_A(q, u), v) \in Q_A, \\
\delta_A^{\nu}(\delta_A(q, u), v) & \text{if } w = uv \text{ s.t. } \delta_A^{\nu}(\delta_A(q, u), v) \in Q_A, \\
q_s & \text{if } w = uv \text{ and } \delta_A(q, u) = q_s, \\
\delta_A(q, w) & \text{otherwise.}
\end{cases}
$$

The actions preserve the default paths on Γ for states in Q_A, thus if $q \in P_{p,w}$ for some $p \in Q_A$ before applying Algorithm 2^X, then still $q \in P_{p,w}$ afterwards. Thus, the transition function of C satisfies

$$
\delta_C(q, w) = \begin{cases}
h'(\delta_A^{\nu}(q, u)) & \text{if } w = u \text{ such that } \delta_A^{\nu}(\delta_A^{\nu}(q, u), v) \in Q_A, \\
\delta_A^{\nu}(\delta_A^{\nu}(q, u), v) & \text{if } w = uv \text{ such that } \delta_A^{\nu}(\delta_A^{\nu}(q, u), v) \in Q_A, \\
\bot & \text{if } \delta_A^{\nu}(q, u) = q_s, \\
\delta_A^{\nu}(q, w) & \text{otherwise.}
\end{cases}
$$

Algorithm 1 applies similar changes to B to obtain B', retargeting transitions on u and adding transitions on uv only when $\delta_B(\delta_B(q, u), v) \in Q_A$. While no sink transitions are introduced, if B cannot read u from q, then q also has no uv-transition in B'. Thus, for all $q \in Q_A$, we have $\delta_{B'}(f(q), w) = f(\delta_C(q, w))$.

Now, consider a new state $h'(q) \in Q_{A'}$. Algorithm 2^X sets $\nu_{A'}(h'(q)) = q$ and adds sink transitions on v (and uv if $u = v$). Thus, $\delta_C(h'(q), w) = \delta_C(q, w)$ for all $w \neq v$ (and $w \neq uv$ if $u = v$), and is otherwise undefined. Likewise, Algorithm 1 defines $\delta_{B'}(h(q), w) = \delta_{B'}(q, w)$ for all $w \neq v$ (and $w \neq uv$ if $u = v$) and as $\bot$ otherwise. Interestingly, if $\delta_B(\delta_B(q, u), v)$ is *also* defined, then B' maps $h(q)$ to a new state $h(\delta_B(q, u))$. Meanwhile, $\delta_{A'}(h'(q), u) = \bot$. Yet, $\delta_C(h'(q), u) = \delta_C(q, u) = h'(\delta_A(q, u))$. Again, we can easily verify that $\delta_{B'}(f(q), w) = f(\delta_C(q, w))$, for all new states in $Q_{A'}$.   □

Corollary 1. *For any string DFA A and proper dictionary D, applying Algorithm 2^X to merge all rules in D to A produces a D^2FA accepting the languages of correct tokenizations of $\mathcal{L}(A)$.*

Proof. Algorithm 1 produces the language of correct tokenizations of $\mathcal{L}(A)$ [2], and the D^2FA being equivalent can be seen by iterating Lemma 1.   □

Having established the correctness of Algorithm 2 with the optimizations $O2$ and $O3$ disabled, we now show that enabling these preserves correctness.

Lemma 2. *For any D^2FA A and rule $u \wr v$, applying Algorithm 2 with or without optimization O2 or O3 (i.e. keeping or removing the blue or red lines) results in equivalent automata.*

Proof. Although $O2$ and $O3$ never apply to the same state, they may apply to different states within the same automaton. The correctness of $O3$ is immediate: it reuses an existing state instead of introducing a redundant copy $fresh(s_2)$ with identical local behaviour. Since such reused states only have sink and default outgoing transitions, they are unaffected by later steps in Algorithm 2. Thus, we may safely focus solely on the effect of $O2$.

If all $s_2 \in S_2$ are initial or multi-use, then enabling or disabling $O2$ has no effect; the equivalence follows immediately. Otherwise, assume there exists a non-initial, single-use state s_2 on v—meaning that $O2$ applies when enabled. Then, for all $q \in P_{s_2,v}$, (1) either q has an incoming transition on u and $q \in S_2$, or (2) for any $p \in P_{q,v}$ such that p has an incoming transition, $p \in S_2$.

Without $O2$, Algorithm 2 creates a new state $fresh(q)$ for each $q \in S_2$, redirects all incoming u-transitions from q to $fresh(q)$, and gives $fresh(q)$ sink transitions on v (and uv if $u = v$), defaulting to q otherwise. Line 26 then renders s_2 unreachable on u (directly or via defaults), making its behaviour on v (and uv) irrelevant and safely modifiable.

When enabled, $O2$ skips creating $fresh(s_2)$, leaves s_2 reachable on u, and assigns it a sink transition on v (and uv if $u = v$), which preserves language equivalence.   □

With that, we can conclude that Algorithm 2 is correct.

Theorem 1. *For any string DFA A and proper dictionary D applying Algorithm 2 to merge all rules in D to A produces a D^2FA accepting the languages of correct tokenizations of $\mathcal{L}(A)$.*

Proof. Corollary 1 establishes the correctness of Algorithm 2 with respect to *O1*, when *O2* and *O3* are disabled. Lemma 2 shows that enabling *O2* and *O3* preserves correctness: the relation f in Lemma 1 may no longer be bijective, but it remains homomorphic, which suffices to establish equivalence.   □

5   Efficiently Encoding Sink Transitions

The majority of the remaining transitions after applying optimizations *O1–O3* originate from lines 2–3, both in practice (see Table 3) and in the theoretical worst case (see Example 3). Fortunately, these sink transitions have a predictable form; recall Fig. 1. This section introduces the final optimization, *O4*, which compresses these transitions into a single 'predicate'. To preview the substantial transition reduction, compare column 4 and 6 in Table 3.

We start by presenting a one-to-one relationship between tokens and dictionary rules when all rules are useful, followed by a definition of the data structure used to compactly encode transitions to the sink reject state.

Lemma 3. *Let $D = [u_1 \wr v_1, \ldots, u_n \wr v_n]$, with all rules useful. Then $u_i v_i = u_j v_j$ implies $i = j$.*

Proof. Follows directly from [3], Corollary 1.   □

Definition 5. *The* prefix forest V_D *for a dictionary $D = [u_1 \wr v_1, \ldots, u_n \wr v_n]$ is a forest of nodes labelled by token-index pairs (w, i), where $i \in \{1, \ldots, n\}$. We refer to a node only by its associated token when the index is not important. Let V_{D_0} be the empty forest. For $i \in [1, \ldots, n]$, construct the prefix forest V_{D_i} for the first i rules of D, from $V_{D_{i-1}}$, as follows:*

1. *If no node v_i exists in $V_{D_{i-1}}$, add node (v_i, i); and*
2. *if node u_i exists in $V_{D_{i-1}}$, or if we just added it in step 1, i.e. when $u_i = v_i$, add node $(u_i v_i, i)$ and make it a child of u_i.*

Observation 2. By Lemma 3, V_D is indeed a forest, given node (w, i) receives an incoming edge only when it is created by processing rule $u_i \wr v_i$, where $w = u_i v_i$ – which only happens once. In addition, Lemma 3 guarantees that referring to a node by its token is unambiguous.

Example 4. For $D = [a \wr a, \ldots, a^{2^i} \wr a^{2^i}, \ldots, a^{2^{n-1}} \wr a^{2^{n-1}}]$, the prefix forest V_D is a unary tree of height n, where the $n + 1$ nodes from root to leaf correspond to the tokens $a, \ldots, a^{2^n}$. A state with a sink transition on the token a^i, also has sink transitions on all descendant tokens of a^i in this tree, see Fig. 1.

Example 5. For $D = [a \wr b, ab \wr b, abb \wr b, \ldots]$, the prefix forest V_D is the single node b.

Proposition 1. *Let A be the token D^2FA resulting from applying Algorithm 2 to the rules in $D = [u_1 \wr v_1, \ldots, u_n \wr v_n]$, starting with an initial token DFA. Suppose state q in A receives an outgoing transition to the sink state on token v_i in Lines 16 or 22. while applying Algorithm 2 with rule $u_i \wr v_i$. Then q may also receive one additional sink transitions by applying Lines 18 or 24, or multiple additional sink transitions by applying Line 3. The tokens labelling these sink transitions are precisely all descendants of v_i in the prefix forest V_D whose associated indices j satisfy $j \geq i$.*

Proof (sketch). Once q receives a sink transition labelled by the token v_i in Lines 16 or 22, it will also receive a sink transition on $u_i v_i$ in Lines 18 or 24 if $u_i = v_i$. But if $u_i = v_i$, then $u_i v_i$ is a descendant of v_i in V_D with index i. Algorithm 2 ensures that any subsequent rule of the form $v_i \wr w$ adds a sink transition to q on $v_i w$ in Line 3. This process recurses and subsequent rules of the form $v_i w \wr w'$ lead to a sink transitions on $v_i w w'$, also added in Line 3, and so on. Thus, q accumulates sink transitions on all tokens formed by successively extending v_i through rules applied at or after step i. But these tokens are precisely all descendants of v_i in the prefix forest V_D whose associated indices j satisfy $j \geq i$.   □

Observation 3. To realize the compression, a copy of V_D is kept with the constructed D^2FA. Each sink transition on v_i introduced by Line 16 or 22 is replaced with a predicate matching the token v_i, and all of v_i's descendants in V_D with indices greater than or equal to i.

Note that the height of V_D determines the worst-case time complexity of deciding whether a given symbol matches a transition predicate by traversing the forest. Fortunately, V_D is in practice very shallow; see Table 4.

6   Theoretical Bounds

With all the optimizations in place, we can observe some straightforward bounds.

Proposition 2. *Starting from an n-state string DFA, and then repeatedly applying Algorithm 2 while using optimization O4 (see Observation 3) to merge all rules in dictionary D, yields a token D^2FA with $\mathcal{O}(n|D|)$ transitions.*

Proof. We require $|D|$ applications of Algorithm 2. With $O4$, line 3 no longer contributes additional transitions. In each iteration, the remaining loops add $\mathcal{O}(|S_2|)$ transitions. The complexity bounds in [2] guarantee that $|S_2| \leq n$ in every iteration, yielding the result.   □

Optimization $O4$ is required for this bound. Without it, Example 3 leads to a quadratic number of sink transitions. As we will see in the next section, the difference is significant in practice as well.

Next, we provide an upper bound on the default chain-length, i.e. the maximum number of default transitions that can be taken consecutively. This demonstrates that Algorithm 2 does not create cycles of default transitions, but the bound is also important for matching efficiency. First, we define subsets of D that might lead to default chains of transitions.

Definition 6. *For a fixed token u, let $l(u, D) := \{u' \wr v' \in D \mid u' = u\}$. Then the* common left token cardinality *of D is defined to be the maximum of $|l(u, D)|$ over all tokens u.*

Proposition 3. *Let A be a token D^2FA obtained by repeatedly applying Algorithm 2 to merge rules from a dictionary D, starting from some token DFA. Then the common left token cardinality of D is an upper bound on the default chain-length of A.*

Proof. Let u be a token maximizing $n = |l(u, D)|$, let $u \wr v_1, \ldots, u \wr v_n$ be the subsequence of rules in D, having u as left token. Then a chain can be formed by first merging $u \wr v_1$, where some state s_2 is reached on u in Algorithm 2, and a new state $fresh(s_2)$ being created that defaults to s_2. However, the only way of lengthening that chain is to again create a new state from this previous $fresh(s_2)$, but by construction it is reachable only on u, so a second rule $u \wr v_2$ is required. Then this argument is simply iterated. □

There are interesting interactions between *O2* and how chains grow. Table 4 shows that optimization *O2* indeed makes the chains much shorter in practice.

7 Testing Real-World Dictionaries

The algorithm was instrumented to enable and disable optimizations O1–O4 individually, aiming to evaluate their impact on BPE dictionaries from prominent large language systems primarily trained on English corpora, see Table 1.

Table 1. The four dictionaries used for testing.

Short name	System	Source	Rule count
GPT-3	OpenAI GPT-2 and GPT-3	[11]	50 000
SmolLM-2	HuggingFace SmolLM 2 language model	[6]	48 900
Whisper	OpenAI Whisper speech recognition system	[12]	50 000
Phi-4	Microsoft Phi 4 large language model	[10]	100 000

The number of states produced using the relevant combinations of optimizations is shown in Table 2. Clearly, *O2* accounts for most of the reductions, with *O3* offers marginal savings. The optimizations are not additive: applying both

Table 2. The number of states resulting when applying Algorithm 2, starting from a universal DFA for the various dictionaries and optimizations.

Dictionary	No opt.	$O2$ only	$O3$ only	$O2$ and $O3$
GPT-3	50 002	15 854	49 304	15 700
SmolLM-2	48 902	15 905	48 201	15 730
Whisper	50 002	16 732	49 035	16 561
Phi-4	100 002	30 557	98 495	30 159

$O2$ and $O3$ saves more than either alone, but less than their sum. The fraction of states saved is consistent across the four dictionaries.

To simplify transition counting for optimized cases, we count the overall number of normal transitions, default transitions, and predicate transitions – counting each predicate transition as one and thus not accounting for the encoding of V_D. For the non-optimized case, we count only transitions to non-sink states from Algorithm 1. The results are shown in Table 3. Not all combinations are considered, since $O1$ dominates transition reduction.

Table 3. The number of transitions resulting when applying Algorithm 2, starting from a universal DFA for the various dictionaries and optimizations.

Dictionary	No opt.	$O1$ only	$O1$–$O3$	$O1$ and $O4$	$O1$–$O4$
GPT-3	2 421 614 550	891 455	835 632	150 434	115 580
SmolLM-2	2 299 440 749	942 645	888 855	147 044	113 342
Whisper	2 453 532 446	704 658	652 783	150 367	116 109
Phi-4	9 577 741 705	2 599 616	2 493 209	300 502	229 425

The state-count optimizations $O2$ and $O3$ also significantly reduce the lengths of default transition chains by preventing the creation of unnecessary intermediate states. We analyze the distribution of maximal default chains – chains starting at states without incoming default transitions and ending in states without an outgoing default transition. For all dictionaries, the distribution is heavily skewed toward short chains, with a long tail of very rare, longer chains. Table 4 shows the effects of optimization $O2$ and $O3$ on the median and maximum lengths of default chains for the respective dictionary. An example histogram giving a sense of the distribution is shown in Fig. 3.

The last two columns of Table 4 show the maximum and median lengths of maximal paths in the prefix forest V_D (see Definition 5). These paths are consistently short, with maximum not exceeding 8 and medians of 1 or 2. The rare longer paths all stem from tokens such as "#####..." or "-----...", containing up to 80 repeated symbols – commonly used as separators in emails and code. This makes Example 3 relevant, as an A_8-like case is realized in practice. Since

Table 4. The maximum and median length of the maximal default chains (DC), as well as the maximum and median length of the maximal paths in the prefix forest V_D (see Definition 5) for the various dictionaries.

Dict.	Median DC (O1)	Max DC (O1)	Median DC (O1–O3)	Max DC (O1–O3)	Median V_D –	Max V_D –
GPT-3	12	330	7	114	1	6
SmolLM-2	13	301	7	134	2	7
Whisper	10	282	6	80	1	6
Phi-4	14	956	8	203	2	8

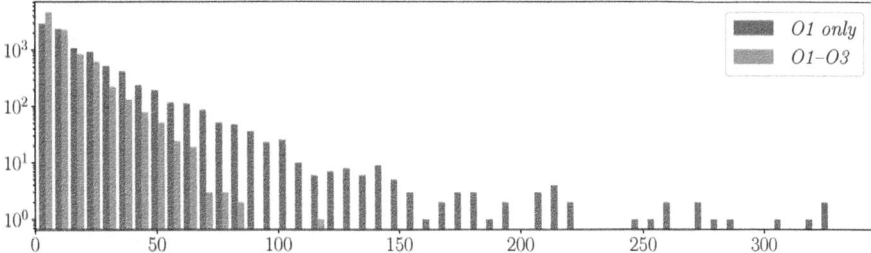

Fig. 3. The distribution of default transition chain lengths in the automaton built for the GPT-3 dictionary, shown as a histogram with 50 bins. A log scale is used as longer chains are very uncommon. Observe that optimizations $O2$ and $O3$ shorten the length of default chains substantially.

V_D solely depends on dictionary D, these patterns are not affected by any of our optimizations.

8   Conclusions

With the proposed optimizations, the algorithm produces automata with substantially fewer states, and vastly fewer transitions, making them suited for practical use. It should be observed that the state savings are more robust to transforming the automata (e.g. taking the intersection with another language), and in the future further state saving optimizations should be investigated.

Acknowledgments. This work was partially supported by the Wallenberg AI, Autonomous Systems and Software Program (WASP) funded by the Knut and Alice Wallenberg Foundation. Thanks to Viktor Lindmark who performed an initial exploration of the default transition optimization [9].

References

1. Aho, A.V., Corasick, M.J.: Efficient string matching: an aid to bibliographic search. Commun. ACM **18**(6), 333–340 (1975). https://doi.org/10.1145/360825.360855
2. Berglund, M., Martens, W., van der Merwe, B.: Constructing a BPE tokenization DFA. In: Fazekas, S.Z. (ed.) Implementation and Application of Automata, pp. 66–78. Springer, Cham (2024). https://doi.org/10.1007/978-3-031-71112-1_5
3. Berglund, M., van der Merwe, B.: Formalizing BPE tokenization. In: 13th International Workshop on Non-Classical Models of Automata and Applications, NCMA 2023, Famagusta, Cyprus, 18–19 September 2023, pp. 16–27. Open Publishing Association (2023)
4. Bille, P., Gørtz, I.L., Pedersen, M.R.: Fast practical compression of deterministic finite automata. In: Královič, R., Kůrková, V. (eds.) SOFSEM 2025: Theory and Practice of Computer Science, pp. 136–150. Springer, Cham (2025). https://doi.org/10.1007/978-3-031-82670-2_11
5. Björklund, H., Björklund, J., Zechner, N.: Compression of finite-state automata through failure transitions. Theor. Comput. Sci. **557**, 87–100 (2014)
6. HuggingFace: SmolLM 2 BPE dictionary (2024). https://huggingface.co/HuggingFaceTB/SmolLM2-1.7B/blob/main/merges.txt. Accessed 22 Apr 2025
7. Kourie, D.G., Watson, B.W., Cleophas, L., Venter, F.: Failure deterministic finite automata. In: Holub, J., Žďárek, J. (eds.) Proceedings of the Prague Stringology Conference 2012, pp. 28–41. Czech Technical University in Prague, Czech Republic (2012)
8. Kumar, S., Dharmapurikar, S., Yu, F., Crowley, P., Turner, J.: Algorithms to accelerate multiple regular expressions matching for deep packet inspection. SIGCOMM Comput. Commun. Rev. **36**(4), 339–350 (2006). https://doi.org/10.1145/1151659.1159952
9. Lindmark, V.: Analyzing the Effect of DFA Compression in Token DFAs. Master's thesis, Umeå University, Department of Computing Science (2024)
10. Microsoft: Phi 4 BPE dictionary (2024). https://huggingface.co/microsoft/phi-4/blob/main/merges.txt. Accessed 22 Apr 2025
11. OpenAI: GPT-2/GPT-3 BPE dictionary (2019). https://huggingface.co/openai-community/gpt2/blob/main/merges.txt. Accessed 22 Apr 2025
12. OpenAI: Whisper BPE dictionary (2024). https://huggingface.co/openai/whisper-large/blob/main/merges.txt. Accessed 22 Apr 2025
13. Sennrich, R., Haddow, B., Birch, A.: Neural machine translation of rare words with subword units. In: Proceedings of the 54th Annual Meeting of the Association for Computational Linguistics, vol. 1: Long Papers, pp. 1715–1725. Association for Computational Linguistics, Berlin (2016)

Epsilon Automata on Linear Orderings

Bernard Boigelot$^{(\boxtimes)}$ ⓘ, Thomas Braipson ⓘ, and Tom Clara ⓘ

Montefiore Institute, B28, University of Liège, Liège, Belgium
Bernard.Boigelot@uliege.be, {Thomas.Braipson,Tom.Clara}@student.uliege.be

Abstract. Automata on linear orderings are a form of finite-state automata introduced by Bruyère and Carton, that generalize the concepts of finite-word, infinite-word, and transfinite-word automata. They recognize words indexed by linear orderings, defined as mappings from the elements of such orderings to a finite alphabet. Some theoretical developments involving automata on linear orderings are hindered by the fact that their definition does not allow epsilon transitions, i.e., transitions with an empty label. In this paper, we define a variant of automata on linear orderings that allows such transitions in their transition relation, and show that this extension does not modify the expressive power of the automata. We motivate the usefulness of epsilon automata on linear orderings by using them for correcting an erroneous construction in the original proof of equivalence between automata on linear orderings and the corresponding notion of regular expressions.

Keywords: Finite automata · Linear orderings · Epsilon transitions

1 Introduction

Finite automata are mathematical objects that are highly expressive and yet easy to handle algorithmically. For those reasons, they provide powerful theoretical tools for establishing the decidability of various logics. For instance, a simple and elegant way of deciding Presburger arithmetic, i.e., the first-order additive theory of integer numbers, consists in building a finite-word automaton recognizing the models of a given formula, and then checking whether this automaton accepts a non-empty language [3]. In 1962, Büchi introduced infinite-word automata in order to extend this approach to logics such as the second-order monadic theory of one successor (S1S) [5]. These automata have then been generalized into automata on bi-infinite words [8] and automata on transfinite words [6,12].

Automata on linear orderings are a more general class of automata defined by Bruyère and Carton, aimed at unifying and extending finite-word, (bi-)infinite-word, and transfinite-word automata [4]. Those automata recognize words indexed by linear orderings, defined as mappings from the elements of an arbitrary totally ordered set to a finite alphabet. Their expressive power has been precisely characterized, by establishing that the languages that they accept are exactly those that can be described by an extension of the notion of regular expression [1,4]. It is however worth mentioning that the properties of automata on linear orderings do not straightforwardly generalize those of finite-word and infinite-word automata. For instance, language complementation is only feasible with such automata if some restrictions are imposed on the orderings that are considered [10].

Automata on linear orderings have mainly been introduced as theoretical tools, although there are projects to use them as actual data structures [2]. They have been useful in particular to design decision procedures for temporal logics over continuous time [7,9]. For theoretical developments, one missing feature of those automata is that their transition relation cannot contain epsilon transitions, i.e., transitions with an empty label. This needlessly complicates some constructions, in particular those required by the proof of equivalence between automata on linear orderings and regular expressions [1,4]. For instance, showing that the concatenation of two languages accepted by automata on linear orderings is itself recognizable by such automata needs to find a way of connecting every final state of the first automaton with every initial state of the second one by combinations of transitions. With epsilon transitions, this operation becomes immediate. The main motivation behind this work actually comes from the discovery of a (minor) error in [1] that, in our opinion, results from a construction made unnecessarily difficult by the lack of epsilon transitions.

The contribution of this work is to introduce a variant of automata on linear orderings that allows epsilon transitions, which we call epsilon automata on linear orderings. We show that, with the appropriate choice of semantics, this adaptation does not modify the expressive power of automata, and provide an algorithm that removes the epsilon transitions from a given input automaton without affecting its accepted language. We then demonstrate the usefulness of epsilon automata on linear orderings by correcting the erroneous construction of [1] with the help of epsilon transitions.

2   Basic Notions

2.1   Orderings and Cuts

We recall elementary notions on orderings and cuts, and refer to [11] for further details. A *linear ordering* is a set J associated with a binary relation $<_J$ that is complete (for every $j \neq k \in J$, either $j <_J k$ or $k <_J j$), irreflexive (for every $j \in J$, $j \not<_J j$), antisymmetric (for every j, $k \in J$, $j <_J k$ implies $k \not<_J j$) and transitive (for every $j, k, \ell \in J$, $j <_J k$ and $k <_J \ell$ imply $j <_J \ell$). When the ordering relation is clear from the context, we simply write $<$ in place of $<_J$. Two elements $j < k \in J$ are said to be *consecutive* if there does not exist $\ell \in J$ such that $j < \ell < k$. In such a case, k is the *successor* of j, and j the *predecessor* of k. An ordering J is *dense* if for every $j < k \in J$, there exists ℓ such that $j < \ell < k$.

A *cut* of a linear ordering J is a partition, written (K, L), of J into two sets K and L such that for every $k \in K$ and $\ell \in L$, one has $k < \ell$. The set of cuts of J is denoted by $\hat{J}$. This set contains a *first cut* $\hat{J}_{min} = (\emptyset, J)$ and a *last cut*

$\widehat{J}_{max} = (J, \emptyset)$. The set $\widehat{J} \setminus \{\widehat{J}_{min}, \widehat{J}_{max}\}$ is denoted by $\widehat{J}^*$. A cut $(K, L) \in \widehat{J}^*$ is called a *gap* if K does not have a greatest element and L does not have a least element. If J does not have any gap, then it is said to be *complete*.

The set $\widehat{J}$ can itself be turned into a linear ordering by equipping it with the relation $<_{\widehat{J}}$ such that $(K_1, L_1) <_{\widehat{J}} (K_2, L_2)$ iff $K_1 \subsetneq K_2$ (or equivalently $L_2 \subsetneq L_1$). Note that J and $\widehat{J}$ may have different cardinalities, for instance, the cuts of the set $\mathbb{Q}$ of rational numbers form a set that is uncountable. The set $J \cup \widehat{J}$ that combines both the elements of J and its cuts also forms a linear ordering, by considering that the subsets J and $\widehat{J}$ keep their internal ordering, i.e., for all $j_1, j_2 \in J$ and $c_1, c_2 \in \widehat{J}$, one has $j_1 < j_2$ iff $j_1 <_J j_2$ and $c_1 < c_2$ iff $c_1 <_{\widehat{J}} c_2$, and that for $j \in J$ and $(K, L) \in \widehat{J}$, one has $j < (K, L)$ iff $j \in K$, and $(K, L) < j$ iff $j \in L$.

2.2   Words and Automata on Linear Orderings

We now summarize some definitions borrowed from [4]. A *word indexed by a linear ordering* is a mapping $\rho : J \to \Sigma$ from a linear ordering J to a finite alphabet Σ. The ordering J is called the *length* of w; this is denoted by $|w| = J$. Note that choosing J to be equal to a finite set, the set of natural numbers $\mathbb{N}$, or the set of integer numbers $\mathbb{Z}$, leads to the familiar notions of finite, infinite, and bi-infinite words. The *empty word* ε is the only word indexed by the empty set, i.e., such that $|\varepsilon| = \emptyset$.

Definition 1. *An* automaton on linear orderings *is a tuple* $\mathcal{A} = (Q, \Sigma, \Delta, I, F)$ *where*

- Q *is a finite set of* states,
- Σ *is a finite* alphabet,
- $\Delta \subseteq (Q \times \Sigma \times Q) \cup (Q \times 2^Q) \cup (2^Q \times Q)$ *is a transition relation,*
- $I \subseteq Q$ *is a set of* initial *states, and*
- $F \subseteq Q$ *is a set of* final *states.*

The transition relation Δ contains three types of transitions. Those of the form (q_1, a, q_2), with $q_1, q_2 \in Q$ and $a \in \Sigma$ are called *successor transitions*. They are alternatively denoted by $q_1 \xrightarrow{a} q_2$. Transitions (q, P) with $q \in Q$ and $P \subseteq Q$ are *right-limit transitions*; they can be written $q \to P$. Similarly, transitions (P, q) with $P \subseteq Q$ and $q \in Q$ are *left-limit transitions*, written $P \to q$.

The semantics of automata on linear orderings is defined as follows.

Definition 2. *A* run *of $\mathcal{A}$ reading a word w is a mapping $\rho : \widehat{J} \to Q$, where $J = |w|$, that satisfies the following conditions:*

- $\rho(\widehat{J}_{min}) \in I$.
- $\rho(\widehat{J}_{max}) \in F$.
- *For every consecutive $c_1 = (K_1, L_1)$, $c_2 = (K_2, L_2) \in \widehat{J}$, i.e., such that $K_2 = K_1 \cup \{j\}$ for some $j \in J$, one has $(\rho(c_1), w(j), \rho(c_2)) \in \Delta$.*

- For every $c \neq \widehat{J}_{min} \in \widehat{J}$ that does not have a predecessor, one has $\lim_{c^-} \rho \rightarrow \rho(c) \in \Delta$, where $\lim_{c^-} \rho = \{q \in Q \mid (\forall c_1 < c)(\exists c_2)(c_1 < c_2 < c \wedge \rho(c_2) = q)\}$.
- For every $c \neq \widehat{J}_{max} \in \widehat{J}$ that does not have a successor, one has $\rho(c) \rightarrow \lim_{c^+} \rho \in \Delta$, where $\lim_{c^+} \rho = \{q \in Q \mid (\forall c_1 > c)(\exists c_2)(c < c_2 < c_1 \wedge \rho(c_2) = q)\}$.

Intuitively, a run ρ must start in an initial state and end in a final state. Pairs of states corresponding to consecutive cuts must be linked by an instance of a successor transition reading the appropriate symbol. The *left limit* $\lim_{c^-} \rho$ of ρ at the cut c is the set of states P that are visited by ρ before c, arbitrarily close to c. Following a left-limit transition $P \rightarrow \rho(c)$ moves from such a set P to the state mapped to c. Similarly, the *right limit* $\lim_{c^+} \rho$ is the set of states P visited by ρ after c, arbitrarily close to c. A right-limit transition $\rho(c) \rightarrow P$ moves from the state mapped to c to the set P.

A word w indexed by a linear ordering is *accepted* by $\mathcal{A}$ if this automaton admits a run reading w. The class of all such words forms the *language accepted* by $\mathcal{A}$.

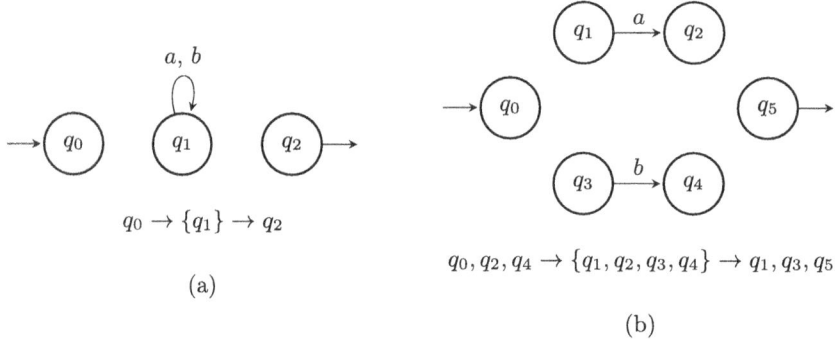

Fig. 1. Examples of automata on linear orderings.

We illustrate the concept of automata on linear orderings with the help of two examples. Initial and final states are respectively identified by incoming transitions without an origin, and outgoing transitions without a destination. For the sake of clarity, successor transitions sharing the same origin and destination are grouped together, as well as right-limit and left-limit transitions based on the same set of states. For instance, p, $q \rightarrow \{p, q\} \rightarrow p$ is shorthand for the transitions $p \rightarrow \{p, q\}$, $q \rightarrow \{p, q\}$ and $\{p, q\} \rightarrow p$. The automaton in Fig. 1a accepts all bi-infinite words on the alphabet $\Sigma = \{a, b\}$, in other words all mappings from $\mathbb{Z}$ to Σ. The automaton in Fig. 1b accepts all words $w : J \rightarrow \Sigma$ such that

- their length J is a non-empty complete set that does not contain any least or greatest element, and
- for every $j, k \in J$ such that $j < k$, there exist $\ell_a, \ell_b \in J$ such that $j < \ell_a < k$, $j < \ell_b < k$, $w(\ell_a) = a$, and $w(\ell_b) = b$.

An example of such a word can be constructed by setting $J = \mathbb{R}$, and for every $j \in \mathbb{R}$, $w(j) = a$ if $j \in \mathbb{Q}$ and $w(j) = b$ if $j \in \mathbb{R} \setminus \mathbb{Q}$. A run ρ reading this word is then obtained by defining

- $\rho((\emptyset, \mathbb{R})) = q_0$ and $\rho((\mathbb{R}, \emptyset)) = q_5$,
- for all $j \in \mathbb{Q}$, $\rho((\mathbb{R}_{<j}, \mathbb{R}_{\geq j})) = q_1$ and $\rho((\mathbb{R}_{\leq j}, \mathbb{R}_{>j})) = q_2$, where $\mathbb{R}_{\bowtie j}$ stands for $\{k \in \mathbb{R} \mid k \bowtie j\}$ with $\bowtie \in \{<, \leq, >, \geq\}$,
- for all $j \in \mathbb{R} \setminus \mathbb{Q}$, $\rho((\mathbb{R}_{<j}, \mathbb{R}_{\geq j})) = q_3$ and $\rho((\mathbb{R}_{\leq j}, \mathbb{R}_{>j})) = q_4$.

It is easy to see that this definition satisfies all the conditions of Definition 2.

2.3   Operators on Orderings and Words

Given two linear orderings J_1 and J_2, the ordering $J = J_1 + J_2$ is defined as the ordering obtained by placing the elements of J_1 before those of J_2, both sets keeping the internal order among their elements. Formally, J can be defined as the set $(J_1 \times \{1\}) \cup (J_2 \times \{2\})$ such that $(j, m) < (k, n)$ iff either $m <_{\mathbb{N}} n$, or $m = n$ and $j <_{J_m} k$. This leads to a natural definition for a concatenation operator on words: given two words w_1 and w_2 indexed by linear orderings, the word $w = w_1 \cdot w_2$ is such that $|w| = |w_1| + |w_2|$, and for every $j \in |w|$, one has $w(j) = w_1(j)$ if $j \in |w_1|$ and $w(j) = w_2(j)$ if $j \in |w_2|$. Note that we slightly abuse the notation by writing $|w_1|$ and $|w_2|$ for the two distinct parts of $|w|$ joined by the sum operator[1].

Concatenation can be seen as a particular case of a more general operator. Given linear orderings J and K_j for each $j \in J$, the ordering $J = \sum_{j \in J} K_j$ is obtained, schematically, by replacing every element j of J by the corresponding set K_j, keeping the internal order among the elements of this set. Formally, we have $\sum_{j \in J} K_j = \{(k, j) \mid j \in J \wedge k \in K_j\}$, with $(k_1, j_1) < (k_2, j_2)$ iff either $j_1 <_J j_2$, or $j_1 = j_2$ and $k_1 <_{K_{j_1}} k_2$. For a linear ordering J and words w_j for each $j \in J$, over the same alphabet, we define $\prod_{j \in J} w_j$ as the word w such that $|w| = \sum_{j \in J} |w_j|$, and for all $k \in |w|$, $w(k) = w_j(k)$ iff $k \in |w_j|$.

3   Epsilon Transitions

In this section, we motivate the need for epsilon transitions in automata on linear orderings, and introduce them with the help of an extended syntax and semantics.

---

[1] A more rigorous definition involves the notion of *order type*, that we do not introduce in this paper for the sake of simplicity.

3.1   Motivation

It has been established that the languages that can be accepted by automata on linear orderings correspond exactly to those that can be described by a generalization of the concept of regular expression. This result has first been proved in [4] with a restriction to *countable and scattered* words, i.e., words w for which $|w|$ is a countable ordering that does not contain a dense infinite sub-ordering. It is then extended to unrestricted words in [1].

In this paper, we do not need to introduce the full formalism of regular expressions for words indexed by linear orderings, and we focus instead on a specific operator, defined as follows.

Definition 3. *Let L_1, L_2 be languages of words indexed by linear orderings, sharing the same alphabet Σ. The language $L = L_1 \diamond L_2$ contains all the words of the form $w = \prod_{j \in J \cup \widehat{J}^*} w_j$, where J is any non-empty linear ordering, and one has $w_j \in L_1$ for each $j \in J$ and $w_j \in L_2$ for each $j \in \widehat{J}^*$.*

Intuitively, $L_1 \diamond L_2$ is obtained by considering all non-empty linear orderings J, and replacing all their elements by words from L_1, and all their non-extreme cuts by words from L_2.

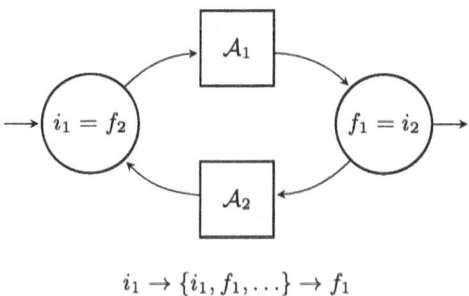

$$i_1 \rightarrow \{i_1, f_1, \ldots\} \rightarrow f_1$$

Fig. 2. Automaton accepting $L_1 \diamond L_2$.

A construction building an automaton $\mathcal{A}$ accepting $L_1 \diamond L_2$ from automata $\mathcal{A}_1$ and $\mathcal{A}_2$ accepting respectively L_1 and L_2 is outlined in [1,4]. It consists in first expressing each $\mathcal{A}_k$, for $k \in \{1, 2\}$, as a *normalized automaton*, which has a single initial state i_k and a single final state f_k such that $i_k \neq f_k$. In addition, i_k (resp. f_k) cannot be the destination (resp. the origin) of a successor or limit transition.

The next step is to combine $\mathcal{A}_1$ and $\mathcal{A}_2$ to form the automaton $\mathcal{A}$. The combination scheme is shown in Fig. 2, in the particular case where neither $\mathcal{A}_1$ nor $\mathcal{A}_2$ accepts the empty word. In this construction, the limit transitions of the resulting automaton are those of $\mathcal{A}_1$ and $\mathcal{A}_2$, with additional transitions $i_1 \rightarrow P$ and $P \rightarrow f_1$ for all supersets P of $\{i_1, f_1\}$.

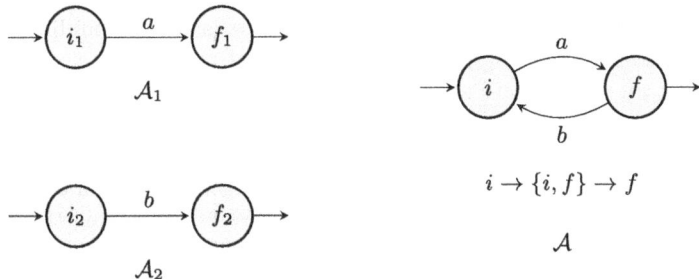

Fig. 3. Automata accepting $\{a\}$, $\{b\}$, and (erroneously) $\{a\} \diamond \{b\}$.

This construction is proved to be correct in [4], for the particular case of words restricted to be indexed by countable and scattered linear orderings. The claim made in [1] that it also applies to the general case is unfortunately invalid. Consider for instance the automata in Fig. 3, accepting the languages $L_1 = \{a\}$ and $L_2 = \{b\}$. Let us show that the resulting automaton $\mathcal{A}$ accepts a language that differs from $L_1 \diamond L_2$. Indeed, $\mathcal{A}$ accepts in particular the word $w = \prod_{j \in \mathbb{R}} b$, i.e., the word indexed by $\mathbb{R}$ that only contains occurrences of the symbol b: a run ρ reading w can be obtained by setting $\rho((\emptyset, \mathbb{R})) = i$, $\rho((\mathbb{R}, \emptyset)) = f$, and for every $j \in \mathbb{R}$, $\rho((\mathbb{R}_{<j}, \mathbb{R}_{\geq j})) = f$ and $\rho((\mathbb{R}_{\leq j}, \mathbb{R}_{>j})) = i$. It is clear however that w does not belong to $\{a\} \diamond \{b\}$, since a occurs in every word of this language.

The error is caused by the fact that the construction in Fig. 2 makes it possible to visit infinitely often all the states of $\mathcal{A}$ without ever following the successor transition labeled by a. This situation only happens when $\mathcal{A}_1$ admits a path from i_1 to f_1 that does not visit any other state.

A simple correction of this problem would thus consist in making sure that an internal state of $\mathcal{A}_1$ must necessarily be visited in order to enable the added limit transitions. However, this cannot be achieved with our current definition of automata on linear orderings, since it follows from Definitions 1 and 2 that any automaton accepting $\{a\}$ must contain a successor transition from an initial state to a final one, labeled by a. This motivates the need for a more general form of automaton on linear orderings, offering the possibility of defining internal states linked by transitions with an empty label. We introduce such automata in the next section. The correction of the construction of an automaton accepting $L_1 \diamond L_2$ is addressed in Sect. 4.

3.2   Epsilon Automata on Linear Orderings

We define our extended form of automata over linear orderings as follows.

Definition 4. *An* epsilon automaton on linear orderings *is a tuple* $\mathcal{A} = (Q, \Sigma, \Delta, I, F)$ *where the set of states* Q, *the alphabet* Σ, *the sets of initial and final states* I *and* F *are as in Definition 1, and the relation transition* Δ *is of the form* $\Delta \subseteq (Q \times (\Sigma \cup \{\varepsilon\}) \times Q) \cup (Q \times 2^Q) \cup (2^Q \times Q)$.

Compared to Definition 1, the only difference is the possibility of having successor transitions (q_1, ε, q_2) with an empty label ε.

Designing an appropriate semantics for epsilon automata is not immediate. Following a successor transition labeled by ε should naturally make a run move from a state to another without reading any symbol in its accepted word. The question is to decide whether following an infinite combination of epsilon transitions should enable limit transitions. In the positive case, adding the transition (q_1, ε, q_1) to the automaton in Fig. 1a would make it also accept all finite words over the alphabet $\{a, b\}$. In the negative case, its accepted language would not be affected.

Our choice is to opt for the latter approach, that conservatively minimizes the impact of epsilon transitions on the semantics of automata. This leads to a semantics in which the states visited by a run reading a word w are still associated to the cuts of $|w|$, but with the property that following epsilon transitions does not modify the current cut.

Formally, given an epsilon automaton $\mathcal{A} = (Q, \Sigma, \Delta, I, F)$, we define the set $S \subseteq Q \times 2^Q \times Q$ of all the triples (q_1, U, q_2) such that $\mathcal{A}$ admits a finite path from q_1 to q_2 visiting exactly the states in U, entirely composed of successor transitions labeled by ε. In other words, such a path must be of the form $s_0 \xrightarrow{\varepsilon} s_1 \xrightarrow{\varepsilon} \cdots \xrightarrow{\varepsilon} s_n$ with $n \geq 0$, $s_0 = q_1$, $s_n = q_2$, $U = \{s_0, s_1, \ldots, s_n\}$, and for every $0 \leq k < n$, $(s_k, \varepsilon, s_{k+1}) \in \Delta$. For each $\sigma = (q_1, U, q_2) \in S$, we define $first(\sigma) = q_1$, $last(\sigma) = q_2$, and $states(\sigma) = U$.

Definition 5. *A run of $\mathcal{A}$ reading a word w is a mapping $\rho : \widehat{J} \to S$, where $J = |w|$, that satisfies the following conditions:*

- *$first(\rho(\widehat{J}_{min})) \in I$.*
- *$last(\rho(\widehat{J}_{max})) \in F$.*
- *For every $c_1 = (K_1, L_1)$, $c_2 = (K_2, L_2) \in \widehat{J}$ such that $K_2 = K_1 \cup \{j\}$ for some $j \in J$, one has $(last(\rho(c_1)), w(j), first(\rho(c_2))) \in \Delta$.*
- *For every $c \neq \widehat{J}_{min} \in \widehat{J}$ that does not have a predecessor, one has $\lim_{c^-} \rho \to first(\rho(c)) \in \Delta$, where $\lim_{c^-} \rho = \{q \in Q \mid (\forall c_1 < c)(\exists c_2)(c_1 < c_2 < c \land q \in states(\rho(c_2)))\}$.*
- *For every $c \neq \widehat{J}_{max} \in \widehat{J}$ that does not have a successor, one has $last(\rho(c)) \to \lim_{c^+} \rho \in \Delta$, where $\lim_{c^+} \rho = \{q \in Q \mid (\forall c_1 > c)(\exists c_2)(c < c_2 < c_1 \land q \in states(\rho(c_2)))\}$.*

Intuitively, instead of mapping the cuts of $|w|$ onto states of the automaton as in Definition 2, those cuts are now associated with finite paths of epsilon successor transitions, in which the only relevant information is their origin and destination states, together with the set of all the states that they visit.

An example of epsilon automaton on linear orderings is given in Fig. 4. This automaton accepts the language $(ab)^\omega$, composed of a single word w mapping $\mathbb{N}$ onto $\{a, b\}$, such that $w(j) = a$ for even values of j, and $w(j) = b$ for odd j. Note that this automaton does not accept the empty word.

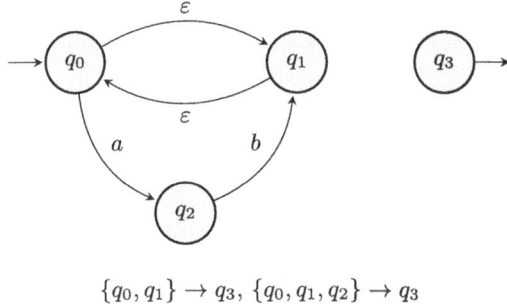

$$\{q_0, q_1\} \rightarrow q_3, \ \{q_0, q_1, q_2\} \rightarrow q_3$$

Fig. 4. Example of epsilon automaton on linear orderings.

3.3 Expressiveness

We now establish that, with the choices made in Sect. 3.2, adding epsilon transitions to automata on linear orderings does not increase their expressive power.

Theorem 1. *A language of words indexed by linear orderings can be accepted by an epsilon automaton on linear orderings iff it can be accepted by an automaton on linear orderings.*

Proof. One direction is immediate: by Definitions 1 and 4, an automaton on linear orderings $\mathcal{A}$ can also be seen as an epsilon automaton $\mathcal{A}_\varepsilon$ without any epsilon transition. Every run ρ of $\mathcal{A}$ can be turned into a run of $\mathcal{A}_\varepsilon$ reading the same word w, by replacing $\rho(c)$ by $(\rho(c), \{\rho(c)\}, \rho(c))$ for each $c \in |\widehat{w}|$.

For the other direction, we prove a stronger result by providing an algorithm that turns an epsilon automaton $\mathcal{A}_\varepsilon$ into an equivalent automaton $\mathcal{A}$ on linear orderings. Let $\mathcal{A}_\varepsilon = (Q_\varepsilon, \Sigma_\varepsilon, \Delta_\varepsilon, I_\varepsilon, F_\varepsilon)$. The automaton $\mathcal{A} = (Q, \Sigma, \Delta, I, F)$ is built as follows:

- $Q \subseteq Q_\varepsilon \times 2^{Q_\varepsilon} \times Q_\varepsilon$ is the set of all triples (q_1, U, q_2) such that $\mathcal{A}_\varepsilon$ admits a finite path of epsilon successor transitions from q_1 to q_2, visiting exactly the states that belong to U. (This corresponds to the definition of the set S in Sect. 3.2).
- $\Sigma = \Sigma_\varepsilon$.
- Δ contains
  - all $(\sigma_1, a, \sigma_2) \in Q \times \Sigma \times Q$ such that $(last(\sigma_1), a, first(\sigma_2)) \in \Delta_\varepsilon$,
  - all $(P, \sigma) \in 2^Q \times Q$ such that $\left(\bigcup_{\sigma' \in P} states(\sigma'), first(\sigma)\right) \in \Delta_\varepsilon$,
  - all $(\sigma, P) \in Q \times 2^Q$ such that $\left(last(\sigma), \bigcup_{\sigma' \in P} states(\sigma')\right) \in \Delta_\varepsilon$.
- $I = \{\sigma \in Q \mid first(\sigma) \in I_\varepsilon\}$.
- $F = \{\sigma \in Q \mid last(\sigma) \in F_\varepsilon\}$.

It follows from Definition 5 that every run of $\mathcal{A}_\varepsilon$ describes a run of $\mathcal{A}$ reading the same word. $\qquad\square$

The algorithm given in the proof of Theorem 1 can incur an exponential blowup in the number of states of the automaton, and a double exponential blowup in the number of limit transitions. This is not problematic, since we expect epsilon automata on linear orderings to be mainly useful for theoretical developments.

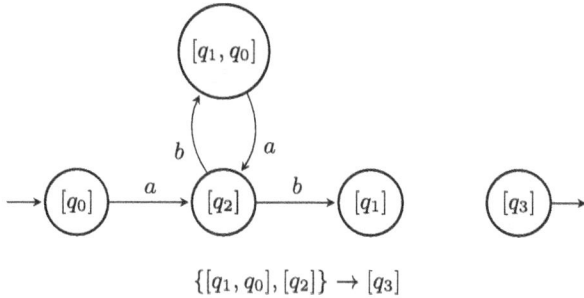

$$\{[q_1, q_0], [q_2]\} \rightarrow [q_3]$$

Fig. 5. Automaton after eliminating epsilon transitions.

We illustrate this result by showing in Fig. 5 the construction of an automaton on linear orderings equivalent to the one in Fig. 4. We use the notation $[q_i]$ as shorthand for $(q_i, \{q_i\}, q_i)$, for $i \in \{0, 1, 2, 3\}$, and write $[q_1, q_0]$ in place of $(q_1, \{q_0, q_1\}, q_0)$. To keep the picture simple, some states and transitions that cannot be visited or followed by any run are omitted.

4   Application

We are now ready to correct the construction of an automaton accepting $L_1 \diamond L_2$ with the help of epsilon automata. Let $\mathcal{A}_1$ and $\mathcal{A}_2$ be automata accepting respectively L_1 and L_2. Recall that automata on linear orderings as defined in [1,4] can be seen as particular cases of epsilon automata. We therefore assume w.l.o.g. that $\mathcal{A}_1$ and $\mathcal{A}_2$ are epsilon automata. Furthermore, we assume like in Sect. 3.1 that $\mathcal{A}_1$ and $\mathcal{A}_2$ are normalized, by extending to epsilon automata the same definition of normalization as for plain automata on linear orderings. Two remarks are in order. Firstly, epsilon transitions make it easy to normalize an arbitrary automaton: this operation amounts to replacing the initial states by a fresh state with outgoing epsilon transitions to them, and the final states by a fresh state with incoming epsilon transitions from them. Secondly, it is possible for a normalized epsilon automaton to accept the empty word: all that is needed is an epsilon transition linking its initial to its final states. This greatly simplifies the developments compared to [4], where languages that contain the empty word have to be handled as particular cases.

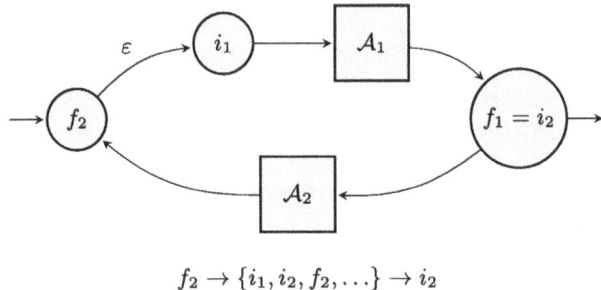

$$f_2 \rightarrow \{i_1, i_2, f_2, \ldots\} \rightarrow i_2$$

Fig. 6. Automaton accepting $L_1 \diamond L_2$.

The construction of an automaton $\mathcal{A}$ accepting $L = L_1 \diamond L_2$ is shown in Fig. 6. It consists in

- merging the initial state i_2 of $\mathcal{A}_2$ with the final state f_1 of $\mathcal{A}_1$ into a single state, which becomes the only final state of $\mathcal{A}$,
- making f_2 the only initial state of $\mathcal{A}$,
- adding an epsilon transition from f_2 to i_1,
- adding right-limit transitions from f_2 and left-limit transitions to i_2, for all sets of states containing i_1, i_2 and f_2.

Compared to the construction in Fig. 2, the difference is that the presence of the epsilon transition (f_2, ε, i_1) now forces the state i_1, and therefore the branch associated to $\mathcal{A}_1$, to be visited in order to enable one of the added limit transitions. Let us show that this new construction is correct.

Theorem 2. *The automaton in Fig. 6 accepts* $L = L_1 \diamond L_2$.

Proof. We first show that every word in $L_1 \diamond L_2$ is accepted by the constructed automaton $\mathcal{A}$. For any such word w, there exists a non-empty linear ordering J and words $w_j \in L_1$ for each $j \in J$ and $w_j \in L_2$ for each $j \in \widehat{J}^*$ such that $w = \prod_{j \in J \cup \widehat{J}^*} w_j$. For each $j \in J$ (resp. $j \in \widehat{J}^*$), let ρ_j be a run of $\mathcal{A}_1$ (resp. $\mathcal{A}_2$) that reads w_j. We build a run ρ of $\mathcal{A}$ reading w. This run maps every cut c of $|w|$ onto a triple $(q_1, U, q_2) \in Q \times 2^Q \times Q$.

As a first step, we introduce a concatenation operator on such triples. Let $\sigma_1, \sigma_2 \in Q \times 2^Q \times Q$ be two triples such that $last(\sigma_1) = first(\sigma_2)$ or $last(\sigma_1) \xrightarrow{\varepsilon} first(\sigma_2) \in \Delta$. Their concatenation $\sigma_1 \cdot \sigma_2$ is defined as $(first(\sigma_1), states(\sigma_1) \cup states(\sigma_2), last(\sigma_2))$.

We also define two operators $first(\cdot)$ and $last(\cdot)$ on runs. For a run ρ reading a word w, we have $first(\rho) = \rho(|w|_{min})$ and $last(\rho) = \rho(|w|_{max})$. Intuitively, these operators respectively identify the triples onto which the first and last cuts of $|w|$ are mapped.

Now, we are ready to build a run of $\mathcal{A}$ reading w from the words w_j, with $j \in J \cup \widehat{J}^*$. Each internal cut c of $|w_j|$ is mapped to $\rho_j(c)$. The first cut of $|w_j|$ is mapped to $last(\rho_k) \cdot first(\rho_j)$ if j has a predecessor k in $J \cup \widehat{J}^*$, to

$(f_2, \{f_2\}, f_2) \cdot \mathit{first}\,(\rho_j)$ if j is the first element of $j \in J \cup \widehat{J}^*$, and to $\mathit{first}\,(\rho_j)$ otherwise. The last cut of $|w_j|$ is mapped to $\mathit{last}\,(\rho_j) \cdot \mathit{first}\,(\rho_k)$ if j has a successor k in $J \cup \widehat{J}^*$, and to $\mathit{first}\,(\rho_j)$ otherwise. Note that if $|w_j| = \emptyset$, and k and ℓ are respectively the predecessor and the successor of j in $J \cup \widehat{J}^*$, the only cut c of $|w_j|$ is mapped onto $\mathit{last}\,(\rho_k) \cdot \rho_j\,(c) \cdot \mathit{first}\,(\rho_\ell)$.

The second part of the proof is to show that every word accepted by $\mathcal{A}$ belongs to the language $L_1 \diamond L_2$. Let ρ be a run of $\mathcal{A}$ reading w. The idea is to show that ρ is a succession of passages through $\mathcal{A}_1$ and $\mathcal{A}_2$. Formally, let us define the following subsets of $|\widehat{w}|$:

$$\Gamma_1 = \left\{ c \in |\widehat{w}| \text{ such that } i_1 \in \mathit{states}\,(\rho(c)) \right\}$$

and

$$\Gamma_2 = \left\{ c \in |\widehat{w}| \text{ such that } i_2 \in \mathit{states}\,(\rho(c)) \right\}.$$

Intuitively, a cut in Γ_1 represents a passage through $\mathcal{A}_1$, since after visiting the state i_1, a word accepted by $\mathcal{A}_1$ must necessarily be read. The set Γ_2 can be interpreted in a similar way in $\mathcal{A}_2$. The set $\Gamma = \Gamma_1 \cup \Gamma_2$ can be turned into a linear ordering by keeping the order relation from $|\widehat{w}|$. The intuition is that Γ represents the succession of passages through $\mathcal{A}_1$ and $\mathcal{A}_2$. However, Γ_1 and Γ_2 might have a non empty intersection, since a cut c can be mapped onto a triple $\rho(c)$ such that $\mathit{states}\,(\rho(c))$ contains both i_1 and i_2 (this is only possible if $\mathcal{A}_1$ or $\mathcal{A}_2$ accepts the empty word). In other words, there might exist cuts in Γ that represent a passage through $\mathcal{A}_1$ followed by a passage through $\mathcal{A}_2$, or conversely. We need to distinguish these two types of passages, hence we define $\bar{\Gamma}$ as the linear ordering Γ in which any cut c that belongs to $\Gamma_1 \cap \Gamma_2$ has been replaced by two distinct elements c_1, c_2 such that:

- c_2 is the successor of c_1 in $\bar{\Gamma}$ if c represents a passage through $\mathcal{A}_1$ followed by a passage through $\mathcal{A}_2$. This happens when $\mathit{first}(\rho(c)) \in \{i_1\} \cup Q_2 \setminus \{i_2\}$.
- c_1 is the successor of c_2 in $\bar{\Gamma}$ if c represents a passage through $\mathcal{A}_2$ followed by a passage through $\mathcal{A}_1$. This happens when $\mathit{first}(\rho(c)) \in \{i_2\} \cup Q_1 \setminus \{i_1\}$.

The linear ordering $\bar{\Gamma}$ can then be partitioned into $\bar{\Gamma}_1$ and $\bar{\Gamma}_2$, whose elements represent passages through $\mathcal{A}_1$ and $\mathcal{A}_2$, respectively.

We now enumerate three properties that come directly from the structure of $\mathcal{A}$. Let us denote by $\Delta^\diamond$ the set of limit transitions $f_2 \to \{i_1, i_2, f_2, \dots\} \to i_2$. Firstly, in the ordering $\bar{\Gamma}$, each element $c_1 \in \bar{\Gamma}_1$ that is not the first or last one has a successor and a predecessor in $\bar{\Gamma}_2$. Indeed, the limit transitions in $\Delta^\diamond$ do not make it possible to leave i_2 by a right-limit transition, or to arrive in f_2 or i_1 after following a left-limit transition. Secondly, the ordering $\bar{\Gamma}$ is complete. This comes from the fact that there does not exist any cut $c \in |\widehat{w}|$ that is mapped onto a triple σ such that $\mathit{first}(\sigma)$ is the destination of a limit transition in $\Delta^\diamond$ and $\mathit{last}(\sigma)$ is the source of a right-limit transition in $\Delta^\diamond$. Finally, for all $c_2 <_{\bar{\Gamma}} c_2' \in \bar{\Gamma}$, there exists $c_1 \in \bar{\Gamma}_1$ such that $c_2 <_{\bar{\Gamma}} c_1 <_{\bar{\Gamma}} c_2'$, since c_2 and c_2' cannot be consecutive in $\bar{\Gamma}$, and $\bar{\Gamma}$ does not admit an infinite dense subordering that only contains elements of $\bar{\Gamma}_2$. This last property is ensured by the presence of the state i_1 and the transition (f_2, ε, i_1).

We now show that those properties imply that the ordering $\bar{\Gamma}$ is isomorphic to $\bar{\Gamma}_1 \cup \widehat{\bar{\Gamma}_1}^{*}$, i.e., that there exists a one-to-one correspondence between these orderings that preserves order. This together with Definition 3 establishes that the word read by ρ belongs to $L_1 \diamond L_2$. Consider the mapping $\tau : \bar{\Gamma} \to \bar{\Gamma}_1 \cup \widehat{\bar{\Gamma}_1}^{*}$ defined as $\tau(c_1) = c_1$ for all $c_1 \in \bar{\Gamma}_1$ and $\tau(c_2) = (\{c_1 \in \bar{\Gamma}_1 : c_1 <_{\bar{\Gamma}} c_2\}, \{c_1 \in \bar{\Gamma}_1 : c_2 <_{\bar{\Gamma}} c_1\})$ for all $c_2 \in \bar{\Gamma}_2$. The fact that this mapping is an order-preserving bijection is a consequence of the following properties:

- Each element of $\bar{\Gamma}_1$ does not belong to $\bar{\Gamma}_2$, and is mapped by τ onto itself.
- For all elements $c_2, c_2' \in \bar{\Gamma}_2$ such that $c_2 <_{\bar{\Gamma}} c_2'$, one has $\tau(c_2) <_{\bar{\Gamma}} \tau(c_2')$. This follows from the definition of τ and the existence of an element $c_1 \in \bar{\Gamma}_1$ between any two elements of $\bar{\Gamma}_2$.
- For all $(K, L) \in \widehat{\bar{\Gamma}_1}^{*}$, there exists $c_2 \in \bar{\Gamma}_2$ such that $\tau(c_2) = (K, L)$ and $c_1 <_{\bar{\Gamma}} c_2 <_{\bar{\Gamma}} c_1'$ for all $c_1 \in K$ and $c_1' \in L$. Indeed, if K (resp. L) has a greatest element K_{max} (resp. a least element L_{min}), then c_2 can be chosen equal to the successor of K_{max} (resp. the predecessor of L_{min}) in $\bar{\Gamma}$. Otherwise, (K, L) is a gap in $\bar{\Gamma}_1$, and the completeness property of $\bar{\Gamma}$ ensures the existence of c_2. $\qquad\square$

5   Conclusions and Perspectives

The contribution of this paper is to introduce an extension of automata on linear orderings, in which the transition relation can contain successor transitions with an empty label. The semantics of those automata has been defined so as to minimize the impact of epsilon transitions; in particular, following infinite combinations of epsilon transitions does not suffice to enable limit transitions. We have established that moving from plain automata to epsilon automata on linear orderings does not increase their expressive power, by providing an algorithm that removes epsilon transitions without modifying the language accepted by the automaton. We have illustrated the usefulness of our extended form of automata on linear orderings by using them for correcting a minor error in [1].

We mainly expect epsilon automata on linear orderings to be useful for theoretical developments, for which the high cost of eliminating epsilon transitions is not prohibitive. There are projects however to exploit those automata to develop an effective decision procedure for some logics, in particular the first-order theory of order over $\mathbb{R}$ and $\mathbb{Q}$ with uninterpreted predicates [2]. For such applications, it will become necessary to find efficient strategies for restricting the set of limit transitions that are handled by the epsilon transition elimination procedure.

Acknowledgments. The authors wish to thank the anonymous reviewers for their insightful comments and constructive suggestions. This work is partially supported by the FNRS-DFG PDR Weave (SMT-ART) grant 40019202.

Disclosure of Interests. The authors do not have competing interests to declare that are relevant to the content of this article.

References

1. Bès, A., Carton, O.: A Kleene theorem for languages of words indexed by linear orderings. Int. J. Found. Comput. Sci. **17**(03), 519–541 (2006)
2. Boigelot, B., Fontaine, P., Vergain, B.: Non-emptiness test for automata over words indexed by the reals and rationals. In: Fazekas, S.Z. (ed.) CIAA 2024. LNCS, vol. 15015, pp. 94–108. Springer, Cham (2024). https://doi.org/10.1007/978-3-031-71112-1_7
3. Bruyère, V., Hansel, G., Michaux, C., Villemaire, R.: Logic and p-recognizable sets of integers. Bull. Belgian Math. Soc. **1**(2), 191–238 (1994)
4. Bruyère, V., Carton, O.: Automata on linear orderings. J. Comput. Syst. Sci. **73**(1), 1–24 (2007)
5. Büchi, J.R.: On a decision method in restricted second order arithmetic. In: Proceedings of the International Congress on Logic, Methodology and Philosophy of Science, pp. 1–12. Stanford University Press, Stanford (1962)
6. Choueka, Y.: Finite automata, definable sets, and regular expressions over ω^n-tapes. J. Comput. Syst. Sci. **17**(1), 81–97 (1978)
7. Cristau, J.: Automata and temporal logic over arbitrary linear time. In: Proceedings of the IARCS 29th Annual Conference on Foundations of Software Technology and Theoretical Computer Science (FSTTCS). LIPIcs, vol. 4, pp. 133–144 (2009)
8. Nivat, M., Perrin, D.: Ensembles reconnaissables de mots biinfinis. In: Proceedings of the 14th Annual ACM Symposium on Theory of Computing (STOC), pp. 47–59. ACM (1982)
9. Rabinovich, A.: Temporal logics over linear time domains are in PSPACE. Inf. Comput. **210**, 40–67 (2012)
10. Rispal, C., Carton, O.: Complementation of rational sets on countable scattered linear orderings. Int. J. Found. Comput. Sci. **16**(04), 767–786 (2005)
11. Rosenstein, J.G.: Linear Orderings. Academic Press (1982)
12. Wojciechowski, J.: Finite automata on transfinite sequences and regular expressions. Fund. Inform. **8**(3–4), 379–396 (1985)

Multi-entry DFA with Reduced Initial States to Speedup Parallel Recognition

Angelo Borsotti⑩, Luca Breveglieri$^{(\boxtimes)}$⑩, Stefano Crespi Reghizzi⑩, and Angelo Morzenti⑩

Politecnico di Milano, 20133 Milan, Italy
angelo.borsotti@mail.polimi.it,
{luca.breveglieri,stefano.crespireghizzi,angelo.morzenti}@polimi.it

Abstract. Parallel algorithms for regular language recognition have been widely experimented for various types of finite automata (FA), deterministic (DFA) and nondeterministic (NFA), often derived from regular expressions (RE). Such algorithms cut the input string into finitely many sections or chunks, independently recognize each one in parallel by means of identical FAs, and at last join the results to check overall consistency. Each FA, except the first, needs to speculatively start a run assuming any state as initial, though most runs will abort or be rejected when joining the computations of adjacent chunks. The ensuing overhead may nullify the gain over a serial algorithm. Existing parallel DFA-based recognizers suffer from the excessive number of starting states, while the NFA-based ones suffer from the number of nondeterministic transitions. Our algorithm is based on a new type of multi-entry DFA called reduced interface DFA (RI-DFA), which can reduce speculation by cutting the number of initial states down to the size of an equivalent NFA. The algorithm benefits from the smaller size of an NFA combined with the speed of deterministic transitions. We have also developed a new initial-state reduction method for RI-DFA. A quantitative comparison of the number of starting states of RI-DFA vs DFA for a large public benchmark of complex FAs confirms the effectiveness of our approach. Speed measurements of text recognition on a multi-core computer show that RI-DFA matches the DFA-based recognizer on some benchmarks and is much faster on some others. The moderate time increase to convert NFA to RI-DFA does not hinder practical use.

Keywords: Regular language · Parallel recognition · Finite automata · Multi-entry DFA · Speculation overhead · Multi-core computer architecture

---

A very short presentation of the main ideas and of the experimental setting is in [4], which does not include theorems and their proofs, and does not present the initial state reduction procedure of Sect. 4. A more complete account of our experimental results can be found in [3].

© The Author(s), under exclusive license to Springer Nature Switzerland AG 2026
G. Castiglione and S. Mantaci (Eds.): CIAA 2025, LNCS 15981, pp. 55–72, 2026.
https://doi.org/10.1007/978-3-032-02602-6_5

1   Introduction

Regular language recognition by a finite automaton (FA) is a widely used basic algorithm. It has been extensively investigated to take advantage of parallel computing architectures of different types, especially the multi-processor/multi-core ones of concern here. We present a novel algorithm that belongs to the so-called data-parallel type. We recall the classical speculative data-parallel algorithm (for short CSDPA), see, e.g., [11] for an early reference, [20] for theoretical and practical aspects, and [12] for code description. After splitting the *input* text into a number $c \geq 1$ of substrings called *chunks*, CSDPA operates in two phases. First, the *reach* phase processes in parallel each chunk by using the same FA called *chunk automaton* (CA), and answers two questions: is the chunk a legal substring for the input language, and which pairs of starting and ending FA states are involved in substring recognition. Then, the *join* phase checks the consistency of the sets of state pairs for any two adjacent chunks. The dominating execution time is in the reach phase, which depends on the input length. The number $|Q|$ of CA states affects the recognition time, because a CA (except for the first one), not knowing the last state reached by the upstream CA, is forced to start a run in each state, though many runs will abort or will be later rejected in the join phase. Thus in the worst case, the number of state transitions that are needed to recognize a string of length $n \geq 1$, segmented into c chunks, is $\mathcal{O}(n \times |Q|)$, instead of just $\mathcal{O}(n)$ as in serial recognition. Since the size $|Q|$ of the CA is often much larger than the number of available computing processors, a parallel recognizer may turn out to be slower than a serial one! The join phase is lighter and is typically executed serially, since its time complexity does not depend on n but only on c, with $c \ll n$.

A CA, viewed in isolation, belongs to a historical formal automaton type called *multi-entry finite automaton* (MEFA) [9], where the transitions are deterministic and all states are initial. Our theoretical contribution is a novel MEFA to be used as CA type, which is more efficient in theory and performs well on multi-core computers. The SW tool[1] is described in [4]. Experimental results are shown in [3] and summarized in this paper.

To contextualize our contribution, it helps to survey other variants of CSDPA. To reduce the very large speculation overhead of a DFA used as CA, researchers have also considered an NFA as CA, since its state size $|Q_N|$ is typically less than $|Q|$. Unfortunately, serial NFA simulation is expensive and parallel NFA simulation is not convenient on multi-processors: for instance, the time complexity of the parallel algorithm [16] based on prefix-sum is $\mathcal{O}(n^3 \times \log m)$, with m processors. In practice, the state reduction brought by an NFA does not pay in general. This finding of prior experimental studies that use an NFA as CA, e.g., [6], is also confirmed by our measurements [3]. For completeness, we add that in some on-line applications, such as regular expression (RE) matching, an NFA may be preferred to a DFA because the conversion from RE to DFA has an exponential time complexity and may be too slow.

---

[1] The tool is available on Zenodo at https://zenodo.org/records/14219357.

Many optimizations on CSDPA are known, and we exemplify with a popular one. The last character, denoted b, of chunk $i - 1$, is passed to the processor of chunk i to restrict the initial states to those entered by b-labeled transitions. Others, e.g., [25], pre-process the DFA in order to select as initial the states that are more likely to be successful, i.e., to have longer runs; the remaining states are left for serial execution in case of failure. The benefit, if any, relies on language statistics, while we aim at a good performance in general. Notice that many of the proposed optimizations are compatible with our approach, and it would be interesting to experiment how they jointly perform.

In this paper, we focus on the speculation cost induced by multiple starting states, but an extremely large size $|Q|$ may slow-down the hardware for different reasons, primarily for the increase in the misses of cache-memory accesses [20]. This affects the so-called Simultaneous Finite Automata (SFA) algorithm [23], which completely avoids speculation at the cost of state explosion. Given a deterministic CA where all the states are initial, the equivalent SFA is a much larger DFA, each state of which is characterized by a set of pairs (i, f), where i and f are respectively the starting and arrival states of a run on the chunk. Therefore, speculation disappears since the multiple parallel runs of the CA are mapped on the single run between two SFA states characterized by the corresponding sets of pairs (i, f). Though deterministic, such a CA may suffer from cache memory misses and, for an RE of moderate size, its construction can be thousand times slower than for a DFA [6]. In the sequel, the cost of a single state-transition is assumed to be independent of CA size.

Between Scylla and Charybdis, i.e., state explosion and nondeterminism, our novel type of CA, called *reduced-interface* DFA (RI-DFA), cuts down the number of initial states and avoids both dangers. The number of initial-states equals the NFA state size $|Q_N|$. As said, the RI-DFA is a special type of MEFA that, to our knowledge, is unknown in previous studies on multi-entry machines [7,9,13,14,21,24]. For each chunk, the RI-DFA starts in the NFA states and then operates as a deterministic FA. Since the number $|Q_N|$ is typically smaller than the state size $|Q|$ of the DFA, a significant reduction in speculation overhead is obtained without paying the price of NFA simulation. As for the cost of constructing the RI-DFA from an NFA, we have found that, for a large public collection of big NFAs, it is only moderately higher than the cost of the classical NFA-to-DFA transformation and remains practical for applications.

We call reduced-interface device (RID) the recognizer that uses an RI-DFA as CA. To complete the definition of RID, we reformulate the join phase and prove overall correctness, also after a subsequent formal optimization that reduces the number of initial states. Then we present a few experimental results to compare the performance of our recognizer with respect to the classical DFA-based and NFA-based ones, in terms of reduction in speculation cost and speedup of recognition on a multi-core computer.

Paper Organization. Section 2 recalls CSDPA with its DFA and NFA variants. Section 3 presents the RI-DFA and proves correctness. Section 4 exploits Nerode-equivalence to reduce the initial states. Section 5 summarizes experimental results. Section 6 outlines future developments. References to related work are distributed by a best fit criterion.

2   Preliminaries on Data-Parallel Recognition Algorithm

We formalize the classic speculative data-parallel algorithm CSDPA (see in particular [12]) in a sufficiently general description that fits our new developments. Let $L \subseteq \Sigma^+$ be a regular language and let $x \in L$ be an *input* string of length $|x| = n \geq 1$. A *segmentation* of x into $c \geq 2$ *chunks* is $x = y_1 y_2 \ldots y_c$, where the y_i have approximately equal length. We assume that L is defined by a finite automaton (FA) $A = (Q, \Sigma, \delta, q_0, F)$, either nondeterministic (NFA) or deterministic (DFA), with state set Q, alphabet Σ, state-transition graph δ, initial state $q_0 \in Q$ and set of final states $F \subseteq Q$.

CSDPA is abstractly modeled by a series $A_1 \ldots A_c$, $c \geq 2$, of identical *chunk automata* (CA), called A_i, which are obtained from the FA A as later explained. A CA is an FA with multiple initial and final states denoted by $A_i = (Q, \Sigma, \delta^A, I_i^A, F_i^A)$, where Q is the state set, and $I_i^A = F_i^A = Q$ are the initial and final state sets, respectively. Notice that all states are both initial and final, with two specializations: the first CA has $I_1^A = \{q_0\}$ and the last CA has $F_c^A = F$. The state-transition graph, $\delta^A = \delta$, is identical to the graph of A. Depending on δ^A being a relation or a function, we call a CA nondeterministic or deterministic; the latter with a misuse of language since the initial move is nondeterministic as in a multi-entry DFA. A chunk y_i is accepted by A_i if it is consumed by at least one run in $\delta^A (q, y_i)$, for some state $q \in I_i^A$, denoted for brevity as $\delta^A (I_i^A, y_i)$. In isolation, a generic CA A_i recognizes all the substrings of L.

The recognizer operates *serially* if at any time only one CA is active, all the upstream CAs have successfully finished, and all the downstream CAs are awaiting. In *serial operation*, CA A_1 processes chunk y_1 starting in state $I_1^A = \{q_0\}$, and passes to CA A_2 the set of *last active states* $\text{LAS}_1 = \delta^A (I_1^A, y_1)$ as those to start from. Similarly, each CA A_i, for $2 \leq i \leq c$, is initialized with the states $I_i^A = \text{LAS}_{i-1}$ and passes (except for the last CA A_c) the set $\text{LAS}_i = \delta^A (I_i^A, y_i)$ to its next downstream CA A_{i+1}. The device accepts if the last set LAS contains a final state, i.e., $\text{LAS}_c \cap F \neq \emptyset$.

We focus on the cost for recognizing string x, defined as the total number of state transitions executed. We disregard the cost for checking acceptance, as it does not depend on the input size. In serial operation, if the CA is deterministic, the overall number of transitions is $\sum_{i=1}^c |y_i| = |x| = n$, and is independent of language L and CA size. On the other hand, if the CA is nondeterministic, the number of transitions may exceed the length n and depends on the degree of nondeterminism of the CA and on the input.

In *parallel operation*, all CAs start in parallel in the initial states $I_i^A = Q$, with the optimization of I_1^A to $\{q_0\}$, and scan their chunks. Each CA A_i, for $1 \leq i \leq c$, defines a mapping $\lambda_i : Q \to 2^Q$ from the *possible initial states* (PIS) of A_i to the *possible last active states* (PLAS) of A_i, such that $q \xmapsto{\lambda_i} \{q' \mid q' \in \delta^A (q, y_i)\}$.

To complete recognition, the device *joins* the mappings λ_i for all chunks. Since join complexity does not depend on input length, we only describe serial join. For chunk one, the possible initial states are $\text{PIS}_1 = I_1^A = \{q_0\}$, the mapping is $q_0 \xmapsto{\lambda_1} \delta^A(q_0, y_1)$, and the possible last active states are $\text{PLAS}_1 = \lambda_1(\text{PIS}_1)$. For every chunk y_i, $2 \leq i \leq c$, set $\text{PIS}_i \subseteq I_i^A = Q$ contains all the states q such that $\delta^A(q, y_i)$ is defined, and is represented by the mapping λ_i s.t. for each $q \in \text{PIS}_i$ it holds $q \xmapsto{\lambda_i} \delta^A(q, y_i)$. The join operation intersects the possible initial states and the possible last states of the upstream CA, and applies the mapping to obtain the possible last states: $\text{PLAS}_i = \lambda_i(\text{PLAS}_{i-1} \cap \text{PIS}_i)$, where $\text{PIS}_i = Q$. The last CA accepts by condition $\text{PLAS}_c \cap F \neq \emptyset$.

Example 1 (**CSDPA**). An example with a deterministic CA is shown in Fig. 1. The values of the sets PIS and PLAS for the two-chunk input string $bab \cdot aaa$ are listed.   □

The language recognized by each CA A_i, for $1 < i < c$, is the set of substrings of $L(A)$, but it would be wrong to replace A_i with the minimal DFA. Just observe that A_i, to operate correctly in the join phase, has to maintain the distinction between its states.

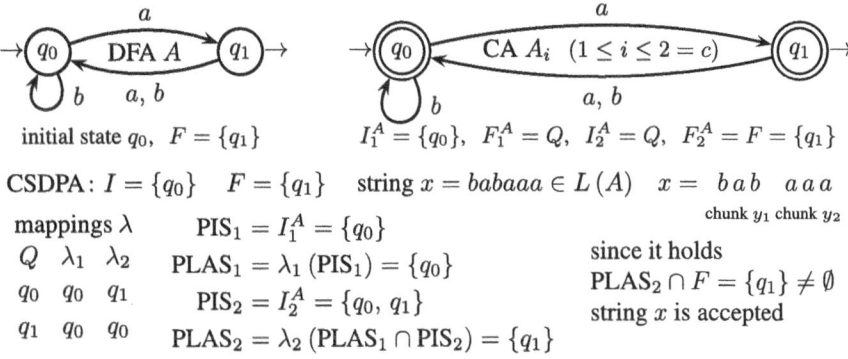

Fig. 1. CSDPA with a deterministic CA; all CA states are both initial and final (double lined)

In parallel *deterministic* operation, the overall number of transitions is bounded by $\sum_{i=1}^{c}(|y_i| \times |I_i^A|) \leq n \times |Q|$, since each CA is a multi-entry machine but some runs may prematurely terminate in error, e.g., the discussion in [6,15,20]. The factor $|Q|$ is the *speculation cost*, due to the need to speculate on the CA starting states. In the example, the worst possible speculation cost occurs, though this is false in general.

The CSDPA scheme continues to inspire many implementations, e.g., the references in [4], for various parallel computing architectures. Each CA is assigned a processing unit that serially executes the runs imposed by speculation. As said, non-deterministic CAs were sometimes chosen to reduce the number of starting states, so obtaining occasional benefits for restricted regular languages of practical interest. Yet, in general, the benefits are outbalanced by the cost of the multiple runs inherent to NFA operation. The next section presents a new technique for reducing speculation, by combining the size reduction of a non-deterministic machine with the speed of a deterministic one.

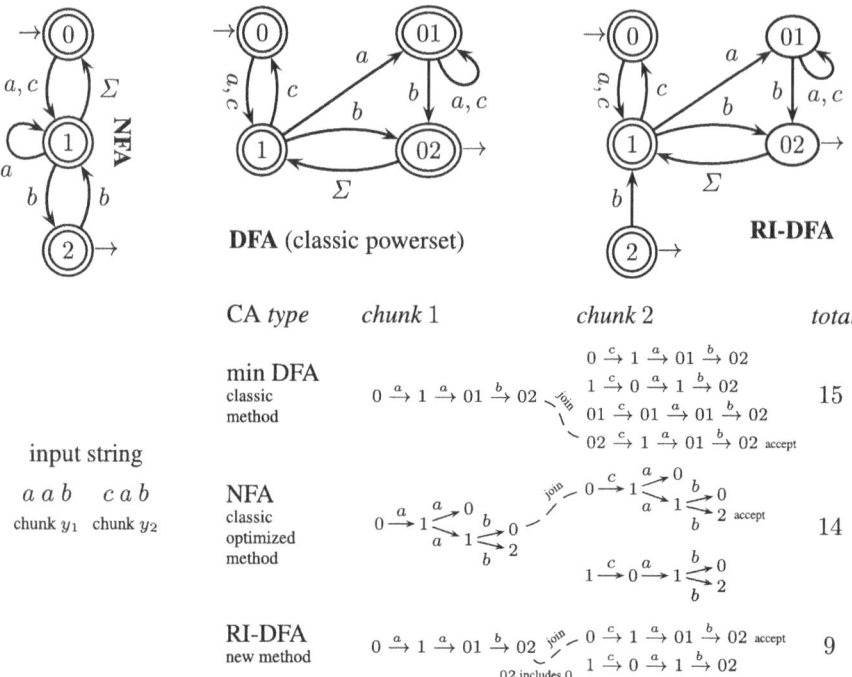

Fig. 2. Three models of CA and their speculation costs for the input string $aab \cdot cab$. Top: NFA with the equivalent (classic) powerset DFA and the new RI-DFA, over the alphabet $\Sigma = \{a, b, c\}$. States acting as initial in the CA are double lined. Bottom: runs executed by the reach phase, join of the PLAS and PIS states of adjacent chunks, and count of transitions

3   Reduced-Interface Recognizer

Our *reduced-interface device* (RID) is based on a series of deterministic CAs as the classic device, with one important difference. The CA initial states are a typically smaller subset of the entire state set, as their number is bounded by the state size of an NFA for language L. An efficient *interface function* maps the possible last active states reached by the $(i-1)$-th CA onto the possible initial states of the i-th CA. After all CAs have finished, the join phase accepts the input if all interface mappings are consistent, as in the classic case. We anticipate the formalization by a motivating example that compares the speculation costs of three CA types: deterministic, nondeterministic and RI-DFA.

*Example 2 (**CSDPA with various CAs**).* Suppose language L is defined by an NFA that we convert into the equivalent DFA by the well-known power-set construction, and then possibly minimize. The NFA and minimal DFA are shown in Fig. 2 (top). For use as CA, all the NFA and DFA states $\{0, 1, 2\}$ and $\{0, 1, 01, 02\}$ are taken as initial. As typical, the NFA has fewer states than the DFA: $3 = |Q_N| < |Q| = 4$. We count how many transitions the DFA and NFA in Fig. 2 (bottom) execute to recognize the two-chunk input $aab \cdot cab$: 15 (DFA) and 14 (NFA). Such numbers approximately measure the overall work done by the CSDPA recognizer. The serial DFA recognizer executes exactly $n = |aabcab| = 6$ transitions on the whole input, therefore, the exceeding transitions measure the speculation cost, which is lower when using the DFA as CA.

The RI-DFA in Fig. 2 (top) has five states $Q_{\text{RI-DFA}} = \{0, 1, 2, 01, 02\}$, but only $\{0, 1, 2\}$ are initial, i.e., exactly set Q_N. The RI-DFA graph may seem a superimposition of the NFA and DFA ones, but is defined more subtly. Intuitively, it is obtained by starting in any NFA state, namely 0, 1 or 2, and jumping into the existing DFA states. Since states 0 and 1 are already present in DFA, their subgraphs in NFA and RI-DFA are identical, and it remains to see what happens from state 2. Since there is a transition $2 \xrightarrow[\text{NFA}]{b} 1$, we add an edge $2 \xrightarrow[\text{RI-DFA}]{b} 1$. The other DFA states 01 and 02 are not initial for the CA, as they do not belong to NFA. Thus RI-DFA has fewer initial states than DFA and makes fewer speculative transitions. The count in Fig. 2 (bottom) is 9, the minimum. The results in [3] (summarized in Sect. 5) confirm and quantify the gain for representative benchmarks. The conversion from NFA to RI-DFA, formalized later, is moderately more complex than the one from NFA to DFA. □

The best saving in transition count for the RI-DFA is obtained when the NFA is state-wise minimal. Unfortunately, no polynomial-time algorithms are known for NFA minimization [10,19] and we could not rely on minimal NFAs in our experiments. Yet, we obtained a significant reduction in speculation cost (Sect. 4) by unconventionally applying to the initial states of RI-DFA the state-equivalence (undistinguishability or Nerode) relation between DFA states. Note that minimizing all DFA states would confound the PIS and PLAS states, thus making such a machine useless for the join phase.

Construction of the RI-DFA Chunk Automaton. RI-DFA is obtained from the given NFA through a novel iterative application of the classic powerset construction. The NFA is denoted $N = (Q_N, \Sigma, \rho, q_0, F)$, where $Q_N = \{q_0, \ldots, q_{\ell-1}\}$, $\ell \geq 1$, the transition relation is $\rho \subseteq Q_N \times \Sigma \times Q_N$, the initial state is $q_0 \in Q_N$ and $F \subseteq Q_N$. Let 2^{Q_N} denote the powerset. Each element $p \in 2^{Q_N}$ is called a *group* of NFA states. Relation ρ induces a transition function δ on the powerset, $\delta: 2^{Q_N} \times \Sigma \rightarrow 2^{Q_N}$, where $(p, a) \overset{\delta}{\mapsto} \{q' \mid \exists q \in p \ (q, a, q') \in \rho\}$. Together, 2^{Q_N} and δ define a deterministic state-transition graph $\mathcal{P}$. The classic powerset DFA A equivalent to NFA N consists of all the (group-)states and transitions of $\mathcal{P}$ that belong to an accepting path, which connects the initial singleton state $\{q_0\}$ to a final state $p \subset Q_N$, i.e., s.t. $p \cap F \neq \emptyset$.

Definition 1 (RI-DFA chunk automaton). *The reduced-interface DFA (RI-DFA) of the NFA N is a multi-entry DFA denoted $B = \left(P, \Sigma, \delta^B, I^B, F^B\right)$. The initial state set is $I^B = \left\{\, \{q\} \mid q \in Q_N \,\right\} = \left\{\, \{q_0\}, \ldots, \{q_{\ell-1}\} \,\right\}$, the state set $P \subseteq 2^{Q_N}$ and the transition function δ^B consist of all the states and transitions reachable (on the graph δ above) from any state in I^B. The final state set is $F^B = P$, i.e., all states are final.* □

Said differently, the RI-DFA is obtained by *cleaning* graph $\mathcal{P}$ of all the states and transitions that are *unreachable* from any initial singleton state $\{q\}$. It contains all the states and transitions of the DFA A, since the latter includes only the accepting paths in $\mathcal{P}$. However, automaton B may have some additional states (and their transitions): all the singletons $\{q\}$, some states needed to connect them to A, and some that do not reach a final state of A. Thus B has at least as many states as A, but usually not many more, as it will be shown in Sect. 5. Figure 2 shows an RI-DFA (top right), which includes the DFA (top middle); both derive from the NFA (top left).

The initial states of B are in one-to-one correspondence with the states of N, and all the states of B are final. Thus B recognizes language $L = L\,(N)$ united with the set of all the substrings of L. To practically compute the RI-DFA, instead of building the large graph $\mathcal{P}$ we use a faster incremental algorithm, which places all initial singletons and then adds only the (group-)states and transitions reachable from the singletons:

Input: NFA N, i.e., state set Q_N and transition relation ρ (see above)
Output: RI-DFA B, i.e., group-state set P and transition function δ^B
         (Def. 1)
Variables: q and q' states of N, p and p' group-states of B, a input symbol

```
P := { {q₀},…,{qₗ₋₁} }    δᴮ := ∅              // initialize P and δᴮ
OUTER:  foreach p ∈ P do                      // scan group-states
   INNER:  foreach a ∈ Σ do                   // scan input symbols
      p' := {q' |  ∃q ∈ p  (q, a, q') ∈ ρ}    // new group-state
      if p' ≠ ∅ then                          // group-state not empty
         if p' ∉ P then P := P ∪ {p'};        // add group-state to P
         δᴮ := δᴮ ∪ {p →ᵃ p'}                 // add transition to δᴮ
```

Cycle OUTER scans set P for states p and cycle INNER progressively puts new reachable states in P. If cycle OUTER starts from $p = \{q_0\}$ and scans the new elements of P in LIFO order, the algorithm builds first the DFA and then the additional states and transitions. Anyway, the order does not affect the resulting RI-DFA. Set P is larger than set Q_N, but the set of initial states will be reduced in Sect. 4. The construction NFA $\rightarrow$ RI-DFA is quite efficient, as witnessed by the measurements in Sect. 5.

Parallel Operation and Interface Function. The RID essentially conforms to CSDPA, and we focus on their differences. The initial and final state(s) are $I^{\mathrm{RID}} = \{ \{q_0\} \}$ and $F^{\mathrm{RID}} = \{p \in P \mid p \cap F \neq \emptyset\}$. All CAs B_i are launched in parallel and each of them scans chunk y_i. The CA B_1 starts from I^{RID} and the others from I^B. Upon termination, each B_i returns a mapping λ_i that for each *possible initial state* (PIS) gives the *possible last active states* (PLAS): it holds $p \xmapsto{\lambda_i} \delta^B (p, y_i)$.

All the consecutive mappings λ_i and λ_{i+1} are joined to check that at least one sequence of chunk runs is consistent. The join operation differs from the CSDPA one in the way it maps the set PLAS of a CA to the set PIS of the downstream CA, by using the *interface function if*, next defined. The argument of *if* is a subset S of group-states of an RI-DFA and its result $if(S)$ is a set of initial singleton states $\{q\}$, namely $if: 2^P \to 2^{I^B}$ with $S \xmapsto{if} \bigcup_{p \in S} \{ \{q\} \mid q \in p \}$, where $S \subseteq P$, $p \in S$ and $q \in p$. The reason for introducing this interface function will become clear in Sect. 4. Then we have:

– For the first chunk, $\mathrm{PIS}_1 = \{p_0\}$, with $p_0 = \{q_0\}$, the mapping λ_1 is $p_0 \xmapsto{\lambda_1} \delta^B (p_0, y_1)$, and for B_1 it holds $\mathrm{PLAS}_1 = \lambda_1 (\mathrm{PIS}_1)$.
– For every other chunk with $2 \leq i \leq c$, set $\mathrm{PIS}_i \subseteq I^B$ contains the states $p \in I^B$ such that $\delta^B (p, y_i)$ is defined. It is represented by a mapping λ_i such that for each $p \in \mathrm{PIS}_i$ it holds $p \xmapsto{\lambda_i} \delta^B (p, y_i)$. Set PLAS_i of B_i is the intersection of PIS_i with the upstream PLAS_{i-1} through function if, i.e., $\mathrm{PLAS}_i = \lambda_i (if(\mathrm{PLAS}_{i-1}) \cap \mathrm{PIS}_i)$.
– RID recognizes the input if the last CA passes the acceptance condition $\mathrm{PLAS}_c \cap F^{\mathrm{RID}} \neq \emptyset$, i.e., at least one of the possible last active states in PLAS_c is final for RID.

The crucial differences between RID and the DFA-based device are that (i) the RID, through the interface function if, remaps the possible last active states PLAS of the upstream CA B_{i-1} onto the possible initial states PIS of the downstream CA B_i, and that (ii) set PIS is a subset of the set of the initial states I^B, which in turn is a (potentially much smaller) subset of the whole state set P of the CA.

Figure 3 for Example 2 (Fig. 2) shows a case with input $aab \cdot cab$. Processing aab yields $\mathrm{PLAS}_1 = \{ \{0, 2\} \}$, thus $if(\mathrm{PLAS}_1) = \{ \{0\}, \{2\} \}$. The set of possible initial states for cab is $\mathrm{PIS}_2 = \{ \{0\}, \{1\} \}$, thus $if(\mathrm{PLAS}_1) \cap \mathrm{PIS}_2 = \{ \{0\} \}$. Hence $\mathrm{PLAS}_2 = \lambda_2 (if(\mathrm{PLAS}_1) \cap \mathrm{PIS}_2) = \{ \{0, 2\} \}$, which accepts the input in state $\{0, 2\}$.

Correctness of the RID. We define some functions for proving Theorem 1. For the NFA $N = (Q_N, \Sigma, \rho, q_0, F)$, we formulate the transition relation ρ as a function $\rho: Q_N \times \Sigma \to 2^{Q_N}$ and we extend to all strings: $\rho(q, \varepsilon) = \{q\}$ and $\rho(q, xa) = \bigcup_{q' \in \rho(q, x)} \rho(q', a)$. Hence $\rho(q, x)$ is the state set reached by N from state q after reading string x. For any group-state subset $S \subseteq P$ and group-state $p \in P$, functions *Nst* and *sgl* flatten S and p into a state set and a set of singletons: $Nst(S) = \bigcup_{p \in S} p$ and $sgl(p) = \{ \{q\} \mid q \in p \}$.

run / mapping of chunk 1:

$\{0\} \xrightarrow{a} \{1\} \xrightarrow{a} \{0,1\} \xrightarrow{b} \{0,2\}$

$\{0\} \xmapsto{\lambda_1} \{0,2\}$

$I^{\mathrm{RID}} = \{\, \{0\} \,\}$

runs / mapping of chunk 2:

$\{0\} \xrightarrow{c} \{1\} \xrightarrow{a} \{0,1\} \xrightarrow{b} \{0,2\}$

$\{1\} \xrightarrow{c} \{0\} \xrightarrow{a} \{1\} \xrightarrow{b} \{0,2\}$

$\{0\} \xmapsto{\lambda_2} \{0,2\} \quad \{1\} \xmapsto{\lambda_2} \{0,2\}$

$F^{\mathrm{RID}} = \{\, \{2\}, \{0,2\} \,\}$

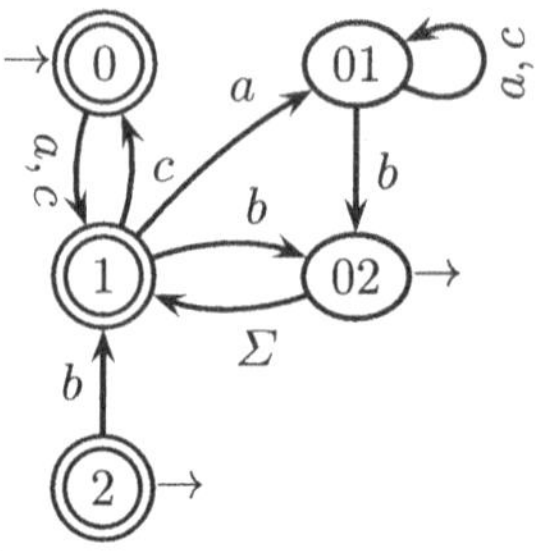

chunk 1: $\mathrm{PLAS}_1 = \{\, \{0,2\} \,\}$ and $if(\{\, \{0,2\} \,\}) = \{\, \{0\}, \{2\} \,\}$

chunk 2: $\mathrm{PIS}_2 = \{\, \{0\}, \{1\} \,\}$ and $\mathrm{PLAS}_2 = \{\, \{0,2\} \,\}$, thus $\mathrm{PLAS}_2 \cap F^{\mathrm{RID}} = \{\, \{0,2\} \,\} \cap \{\, \{2\}, \{0,2\} \,\} = \{\, \{0,2\} \,\} \neq \emptyset$

Fig. 3. CA (right), runs of CAs B_1 and B_2 (left), and use of the interface function if(bottom).

Lemma 1. *The set of the states of the NFA N that are included in the elements of each set PLAS_i, is equal to the set of the states reached by N after reading string $y_1 \ldots y_i$, i.e., for all i with $1 \leq i \leq c$ it holds $Nst(\mathrm{PLAS}_i) = \rho(q_0, y_1 \ldots y_i)$.* □

Proof. By induction on index i. For chunk y_1, $\mathrm{PLAS}_1 = \delta^B(p_0, y_1) = \delta^B(\{q_0\}, y_1)$. From Definition 1 it follows $Nst(\mathrm{PLAS}_1) = \rho(q_0, y_1)$. For chunk y_i, $2 \leq i \leq c$, inductively assume $Nst(\mathrm{PLAS}_{i-1}) = \rho(q_0, y_1 \ldots y_{i-1})$. Then $\mathrm{PIS}_i = \{\, \{q\} \mid \delta^B(\{q\}, y_i) \text{ is def.} \,\} = \{\, \{q\} \mid \rho(q, y_i) \neq \emptyset \,\}$. From the inductive assumption $if(\mathrm{PLAS}_{i-1}) = \bigcup_{p \in \mathrm{PLAS}_{i-1}} \{\, \{q\} \mid q \in p \,\} = sgl(Nst(\mathrm{PLAS}_{i-1})) = sgl(\rho(q_0, y_1 \ldots y_{i-1}))$. So $if(\mathrm{PLAS}_{i-1}) \cap \mathrm{PIS}_i = sgl(\rho(q_0, y_1 \ldots y_{i-1})) \cap \{\, \{q\} \mid \rho(q, y_i) \neq \emptyset \,\}$, thus $\mathrm{PLAS}_i = \lambda_i(if(\mathrm{PLAS}_{i-1}) \cap \mathrm{PIS}_i) = \{\, \delta^B(\{q\}, y_i) \mid \{q\} \in (if(\mathrm{PLAS}_{i-1}) \cap \mathrm{PIS}_i) \,\}$, hence $Nst(\mathrm{PLAS}_i) = \{\, q' \mid \exists q \, (q \in \rho(q_0, y_1 \ldots y_{i-1}) \wedge q' \in \rho(q, y_i)) \,\} = \rho(q_0, y_1 \ldots y_i)$. □

Theorem 1 (correctness). *The RID accepts the same language as the NFA N.* □

Proof. The theorem will follow from the two implications $\mathrm{PLAS}_c \cap F^{\mathrm{RID}} \neq \emptyset \iff \rho(q_0, y_1 \ldots y_c) \cap F \neq \emptyset$, which we prove by using Lemma 1:

- $\mathrm{PLAS}_c \cap F^{\mathrm{RID}} \neq \emptyset \Rightarrow \rho(q_0, y_1 \ldots y_c) \cap F \neq \emptyset$. If $\mathrm{PLAS}_c \cap F^{\mathrm{RID}} \neq \emptyset$, some element of PLAS_c includes a final state of N, and so does $Nst(\mathrm{PLAS}_c)$, thus $Nst(\mathrm{PLAS}_c) \cap F \neq \emptyset$. From Lemma 1, $Nst(\mathrm{PLAS}_c) = \rho(q_0, y_1 \ldots y_c)$, hence $\rho(q_0, y_1 \ldots y_c) \cap F \neq \emptyset$.
- $\rho(q_0, y_1 \ldots y_c) \cap F \neq \emptyset \Rightarrow \mathrm{PLAS}_c \cap F^{\mathrm{RID}} \neq \emptyset$. From Lemma 1, $\rho(q_0, y_1 \ldots y_c) = Nst(\mathrm{PLAS}_c)$, thus if $\rho(q_0, y_1 \ldots y_c)$ includes a final state of N, so does PLAS_c, i.e., it is an element of F^{RID}, hence $\mathrm{PLAS}_c \cap F^{\mathrm{RID}} \neq \emptyset$.

□

4   Further Reduction of the Starting States

The program used in our experimentation (Sect. 5) realizes the RID by using an RI-DFA as CA, with an optimization to further reduce the interface size. The crucial observation is that the (Nerode) language-equivalence relation between DFA states can be extended to RI-DFA, since the transitions from each state, including the initial ones, are deterministic. Thus we apply to RI-DFA the standard (Myhill-Nerode) state-partition algorithm for converting a DFA into the equivalent state-minimal DFA, but we restrict its application to the RI-DFA initial states only. Then, we downgrade to non-initial those initial states the role of which as initial can be taken by another initial state.

Figure 4 shows how to reduce the number of initial states. The states of RI-DFA B (right) are labeled with the NFA (left) states, e.g., $p_4 = \{0, 3\}$, for short 03; states p_0, p_1, p_2 and p_3 (double lined) are initial. States p_1 and p_3 are equivalent, and p_3 (arbitrarily chosen) is downgraded to non-initial, thus obtaining an RI-DFA B_{min} with only three initial states p_0, p_1 and p_2 (dashed box). State p_1 is relabeled from 1 to 13 to record that, if the upstream CA admits state 3 as last possible, then the current CA should start (also) in state 3. We say that state p_3 *delegates* to state p_1 its role as initial. States p_4, p_5 and p_7 are also equivalent, but we do not touch them, since we only want to reduce the size of the RI-DFA interface. The state-transition graph is unchanged.

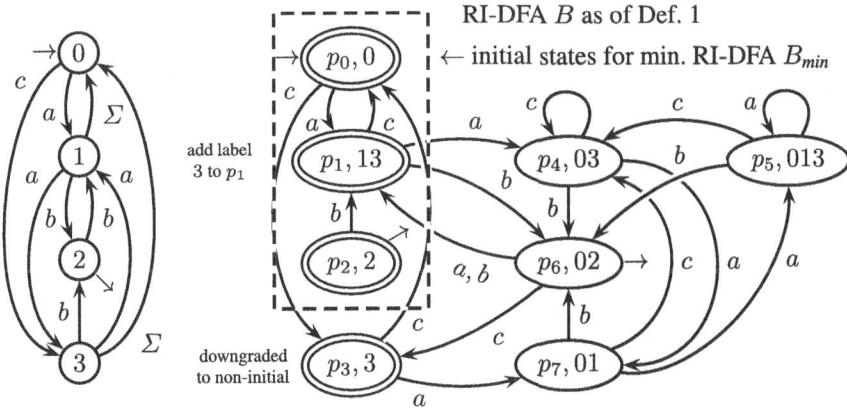

Fig. 4. Minimization of the initial state interface of an RI-DFA (right) from an NFA (left)

We argue why delegation is preferable to merging equivalent initial states. Figures 5a and 5b schematize the RI-DFAs B and B_{min}, where state $\{q_2\}$ delegates to state $\{q_1\}$, while the (equivalent) machine in Fig. 5c loses determinism after merging the two initial states, with ensuing inconveniences like a loss of efficiency. Say B_{min} is an RI-DFA as of Definition 1, with an initial state set $I^{B_{min}}$ containing only the delegated states. Say device RID_{min} is an

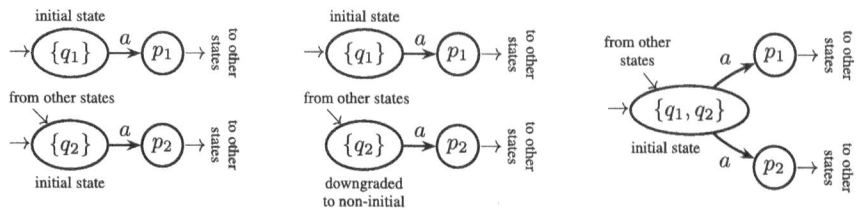

(a) equivalence $\{q_1\} \equiv \{q_2\}$ (b) $\{q_2\}$ delegates to $\{q_1\}$ (c) merging $\Rightarrow$ nondeterminism

Fig. 5. Reduction of the size of the RI-DFA interface by delegation of the initial role

RID using B_{min} as CA and this new interface function: $S \xrightarrow{if_{min}}$ $\{\, p \in I^{B_{min}} \mid \exists \, \{q\} \in if(S) \text{ such that } \{q\} = p \text{ or } \{q\} \text{ delegates to } p \,\}$.

Theorem 2. (interface minimization). *Devices* RID$_{min}$ *and* RID *are equivalent.* □

Proof. The speculative runs of B_{min} do not include those that start from the downgraded states $I^B \setminus I^{B_{min}}$. The interface function if_{min} maps all such downgraded states, possibly included in set PLAS, to the initial states $I^{B_{min}}$ that are their delegates and, thus, are language-equivalent. Therefore, no accepting RID run is lost in RID$_{min}$ and no new accepting run is introduced, hence devices RID$_{min}$ and RID are equivalent. □

For instance, with chunks $y_1 = caa$ and $y_2 = aab$, the RID of Fig. 4 speculates for y_2 from states p_0, p_1 (p_2 aborts soon) and p_3, while RID$_{min}$ only from p_0 and p_1, since p_3 delegates to p_1. Notice it holds PLAS$_1 = \{p_5\} = \{\{0,1,3\}\}$ and if_{min} (PLAS$_1$) $= \{\{0\}, \{1\}\}$. A natural question is then how the CA constructed from an NFA compares with the CA obtained by the same procedure, but starting from a state-minimal NFA.

Theorem 3 (minimality). *Let N be an NFA and N_{min} be one of the state-minimal equivalent machines. Let B and C be the RI-DFAs from N and N_{min}, respectively (Sect. 3), and let B_{min} be the RI-DFA from B (see Theorem 2). For any NFA N, the number of initial states in C is less than or equal to the number of initial states in B_{min}.* □

Proof. By refutation. For any two initial states $\{q_1\}$ and $\{q_2\}$ of B such that in B_{min} it happens that $\{q_2\}$ delegates to $\{q_1\}$, merge states q_1 and q_2 in N, and so obtain an equivalent smaller NFA N'. But then, supposing B_{min} has fewer initial states than C, it follows that N' has even fewer states than N_{min}, which is a contradiction. □

From Theorem 3, applying initial-state reduction to the RI-DFA of a minimal NFA is useless.

5   Summary of Experimental Results

We summarize our findings from [3], which experimentally compares the performance of RID vs classic CSDPA. The measures obtained with our tool (on Zenodo) concern: (*i*) reduction of CA initial states achieved by RI-DFA vs DFA and NFA variants, (*ii*) conversion times from NFA to RI-DFA vs from NFA to DFA, (*iii*) reduction of the number of transitions executed by RI-DFA CAs vs the same variants, and (*iv*) speedup of text recognition by RID vs the same two variants on a multi-processor. Brevity prevents describing the data-sets (benchmarks) and other relevant features.

Results on Interface Minimality and CA Construction Time. The RI-DFA in Fig. 2 has fewer initial states than the minimal DFA. We experimentally addressed two questions: how often this happens and how large the state reduction is. The tables below compare the numbers of NFA states and RI-DFA initial states against the number of states of the minimal DFA, for all FAs in the public collection Ondrik of $1,084$ large FAs (avg. 2490 states per FA) from various applications, e.g., system modeling and formal verification:

interval < 1	NFA		RI-DFA	
$0.5 - 0.6$	110		636	
$0.6 - 0.7$	677		355	
$0.7 - 0.8$	173		34	
$0.8 - 0.9$	60		40	
$0.9 - 1.0$	25		19	
subtotal	1045	(96.4%)	1084	(100%)

interval > 1	NFA		RI-DFA
$1.0 - 1.1$	19		none
$1.1 - 1.2$	16		none
$1.2 - 1.3$	3		none
$1.3 - 1.4$	1		none
subtotal	39	(3.6%)	none

The two tables list the distribution of Ondrik machines with respect to the values $N = \frac{number\ of\ NFA\ states}{number\ of\ DFA\ states}$ and $R = \frac{no.\ of\ RI\text{-}DFA\ initial\ states}{number\ of\ DFA\ states}$. In the left and right tables there are the values < 1 and > 1, respectively. For instance, in row one, left table, it holds $0.5 \le N \le 0.6$ for 110 NFAs, and $0.5 \le R \le 0.6$ for 636 RI-DFAs, i.e., more than half collection. The subtotal 1084 shows that all RI-DFAs have fewer initial states than their equivalent DFAs, thus the frequency of cases that may benefit from our approach is very high. Column 2 (NFA), left table, shows that for 96.4% of the collection the FAs are smaller than their minimal DFAs, but we do not know whether they are minimal.

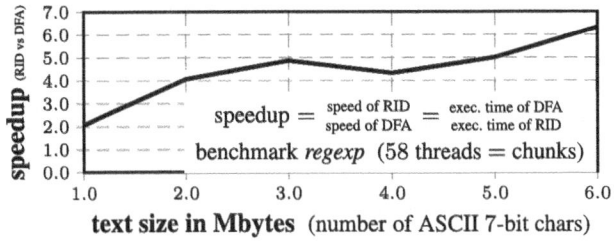

Fig. 6. Speedup RI-DFA / DFA vs text length for regexp $(a \mid b)^* a (a \mid b)^k$ with $k = 6$

The conversion NFA $\to$ RI-DFA is more complex than NFA $\to$ DFA: on a simple laptop, the conversion of Ondrik 1.084 FAs for a total of 2.699.411 states, respectively takes 2.994 s (seconds) vs 146 s, with ratio about 20. The total numbers of states constructed are respectively 1.485.483 for the DFAs and 6.753.792 for the RI-DFAs.

We argue that the above value of the ratio is much less than what should be expected from a worst-case theoretical analysis. The average number of states of Ondrik FAs is $|Q|_{avg} = \frac{2,699.411}{1,084} = 2,490.2$. In principle, the conversion NFA $\to$ RI-DFA computes one powerset per NFA state, while NFA $\to$ DFA computes just one, say at a cost C. But in reality each successive powerset computation in NFA $\to$ RI-DFA creates fewer and fewer states, so that the overall cost is much less than the product $|Q|_{avg} \times C$. As said, the measured time ratio is just 20, so that the actual individual (i.e., per NFA) cost is $20 \times C$. This is much lower than the theoretical worst-case cost estimation $2.490 \times C$ obtained by taking the average NFA size $|Q|_{avg}$. We may disregard the cost of state minimization, which is done only once at the end of both constructions.

Results on Speculation Overhead and Parallel Recognition Speed. We have compared the speculation costs by counting the state-transitions for recognizing the input texts that accompany certain collections of regular expressions or FAs (see [3] for details). Transition counts depend also on the number of chunks; here we refer to 32 chunks, which is midway between one (serial execution) and the 64 processors available on our computer. The ratio $\frac{no.\ of\ transitions\ of\ DFA}{no.\ of\ transitions\ of\ RI\text{-}DFA}$ can be as high as 130 for the NFAs defining the language series $(a \mid b)^* a (a \mid b)^k$, with parameter $k \geq 0$, a classic case of DFA vs NFA state explosion as k grows. For other benchmarks the ratios are lower, down to $1 \pm 10\%$. Such measurements are moderately sensitive to text length, presumably because on the input data most runs are short-lived and stop after just a few transitions.

The number of transitions executed by each CA is obviously correlated with the actual time for text recognition on a 64-core machine, which is the final result we discuss. For each text, the *reach* phase launches from 2 to 66 concurrent threads, one per chunk. Since the number of cores is 64, we can reasonably assume that each thread runs on the same core until termination. To exclude I/O time, the texts are loaded into memory before processing. For all benchmarks, we measured the recognition time as a function of two independent variables, text length and number of threads; here we report just one instance. In all cases, the reach phase takes the longer time, while the join phase accounts for less than 1% of the total time.

The speedup of RID vs the DFA variant of CSDPA strongly depends on the benchmarks considered. For some the speedup is > 1, and for some others it is barely < 1. The RID always dominates by far the NFA variant. In the winning group, the speedup of RID vs the DFA variant is from 3 to over 6 for the language series $(a \mid b)^* a (a \mid b)^k$, with parameter $k = 6$. For the latter, in Fig. 6 we show a plot that typifies the ideal conditions for a top RID performance:

availability of a state-minimal NFA such that the equivalent minimal DFA has an exponential blow-up of states.

Concerning performance sensitivity to text length and chunk number, the speedup over DFA may decrease when the text of fixed length is cut into more (shorter) chunks, because of the growing weight of chunk management overhead. The opposite effect emerges: the speedup over DFA increases with text length, for a fixed number of chunks.

6   Conclusion and Future Work

The new type RI-DFA of multi-entry DFA can improve the performance of classic data-parallel recognition algorithms on multiprocessors, because it reduces the number of initial states in the chunk automata with their useless runs. The optimal speedup occurs when the language is such that its NFA is much smaller than its minimal DFA. In our experimentation it is not uncommon that for very long texts, or in the case of DFA state explosion, the classic DFA-based variant of the algorithm fails or takes too long to run to completion, while the RI-DFA variant succeeds. Since the RI-DFA approach is compatible with most existing optimization such as state-convergence, state speculation exploiting look-back, and even higher-order speculation, as described in [22], it should be valuable for the implementation of parallel finite-state machines.

Of course, an issue for a successful application is the availability of small and possibly minimal NFAs that define the languages to be recognized. In the following, NFA denotes the given automaton, DFA_{min} the minimal equivalent DFA, and NFA_{min} arbitrarily one of the minimal-state FAs equivalent to NFA (transition minimization is not relevant for speculation reduction). We know from Theorem 3 that the minimal initial state set is obtained if RI-DFA is generated from NFA_{min}, which may be quite rare in practice. Since the simpler languages are typically defined by hand as REs (without any guarantee of minimality) and are converted to NFA by standard algorithms, such as GMY [18], the result may be much larger than NFA_{min}. Yet improvements are possible: by using more sophisticated RE $\rightarrow$ NFA converters or by optimizing the RE prior to its use in a converter, e.g., see [8] and its references. For more complex languages, such as those that occur in model checking or formal verification, the NFAs are typically generated by a program and their level of minimization is often unknown.

Unfortunately, NFA minimization is PSPACE-hard [19], and it cannot even be approximated within a factor of $\mathcal{O}(n)$ unless P $\equiv$ PSPACE [10]. Many papers [1,2,5,17] have focused on efficient heuristic algorithms for state reduction that sometimes produce a minimal NFA. It would be interesting to see how such heuristics perform on collections of practically relevant NFAs, both in terms of construction time and approximation to the minimum, and whether they pay off in terms of RI-DFA performance.

At last, we observe that the RI-DFA fits in between two extremes: the multientry DFA having maximal degree of initial nondeterminism, and the *simultaneous* FA [23] having one initial state but a giant number of states. The former suffers from speculation overhead, while the latter is impractical for most languages due to state explosion and the long conversion time from NFA. We wonder whether, in between the two extremes, one could find some other type of CA that would improve over RI-DFA for the situations where the sizes of the DFA and minimal NFA do not significantly differ.

Acknowledgments. We warmly thank the anonymous reviewers for their valuable advices and suggestions.

References

1. Bianchini, C., Policriti, A., Riccardi, B., Romanello, R.: Incremental NFA minimization. Theor. Comput. Sci. **1004**, 114621 (2024). https://doi.org/10.1016/J. TCS.2024.114621
2. Björklund, J., Cleophas, L.: Aggregation-based minimization of finite-state automata. Acta Informatica **58**(3), 177–194 (2021). https://doi.org/10.1007/ S00236-019-00363-5
3. Borsotti, A., Breveglieri, L., Crespi Reghizzi, S., Morzenti, A.: Minimizing speculation overhead in a parallel recognizer for regular texts. CoRR abs/2412.14975 (2024). https://doi.org/10.48550/arXiv.2412.14975
4. Borsotti, A., Breveglieri, L., Crespi Reghizzi, S., Morzenti, A.: Minimizing speculation overhead in a parallel recognizer for regular texts. In: 30th Annual Symposium on Principles and Practice of Parallel Programming, PPoPP 2025, Las Vegas, NV, USA, 1–5 March 2025, pp. 569–572. ACM (2025). https://doi.org/10.1145/ 3710848.3710866
5. Brzozowski, J.A., Tamm, H.: Minimal nondeterministic finite automata and atoms of regular languages. CoRR abs/1301.5585 (2013). https://doi.org/10.48550/ arXiv.1301.5585
6. Fu, Z., Liu, Z., Li, J.: Efficient parallelization of regular expression matching for deep inspection. In: 26th International Conference on Computer Communication and Networks, ICCCN 2017, Vancouver, BC, Canada, 31 July–3 August 2017, pp. 1–9. IEEE Computer Society (2017). https://doi.org/10.1109/ICCCN.2017. 8038377
7. Galil, Z., Simon, J.: A note on multiple-entry finite automata. J. Comput. Syst. Sci. **12**(3), 350–351 (1976). https://doi.org/10.1016/S0022-0000(76)80006-2
8. García, P., López, D., Ruiz, J., Alvarez, G.I.: From regular expressions to smaller NFAs. Theor. Comput. Sci. **412**(41), 5802–5807 (2011). https://doi.org/10.1016/ J.TCS.2011.05.058
9. Gill, A., Kou, L.T.: Multiple-entry finite automata. J. Comput. Syst. Sci. **9**(1), 1–19 (1974). https://doi.org/10.1016/S0022-0000(74)80034-6

10. Gramlich, G., Schnitger, G.: Minimizing NFAs and regular expressions. J. Comput. Syst. Sci. **73**(6), 908–923 (2007). https://doi.org/10.1016/J.JCSS.2006.11.002
11. Hillis, W.D., Steele, Jr., G.L.: Data parallel algorithms. Commun. ACM **29**(12), 1170–1183 (1986). https://doi.org/10.1145/7902.7903
12. Holub, J., Štekr, S.: On parallel implementations of deterministic finite automata. In: Maneth, S. (ed.) CIAA 2009. LNCS, vol. 5642, pp. 54–64. Springer, Heidelberg (2009). https://doi.org/10.1007/978-3-642-02979-0_9
13. Holzer, M., Salomaa, K., Yu, S.: On the state complexity of k-entry deterministic finite automata. J. Autom. Lang. Comb. **6**(4), 453–466 (2001). https://doi.org/10.25596/JALC-2001-453
14. Kappes, M.: Descriptional complexity of deterministic finite automata with multiple initial states. J. Autom. Lang. Comb. **5**(3), 269–278 (2000). https://doi.org/10.25596/JALC-2000-269
15. Ko, Y., Jung, M., Han, Y.-S., Burgstaller, B.: A speculative parallel DFA membership test for multicore, SIMD and cloud computing environments. Int. J. Parallel Prog. **42**(3), 456–489 (2013). https://doi.org/10.1007/s10766-013-0258-5
16. Ladner, R.E., Fischer, M.J.: Parallel prefix computation. J. ACM **27**(4), 831–838 (1980). https://doi.org/10.1145/322217.322232
17. Lombardy, S., Sakarovitch, J.: Morphisms and minimisation of weighted automata. Fundam. Informaticae **186**(1–4), 195–218 (2022). https://doi.org/10.3233/FI-222126
18. McNaughton, R., Yamada, H.: Regular expressions and state graphs for automata. IRE Trans. Electron. Comput. **EC-9**(1), 39–47 (1960). https://doi.org/10.1109/TEC.1960.5221603
19. Meyer, A.R., Stockmeyer, L.J.: The equivalence problem for regular expressions with squaring requires exponential space. In: 13th Annual Symposium on Switching and Automata Theory, SWAT 1972, College Park, MD, USA, 25–27 October 1972, pp. 125–129. IEEE Computer Society (1972). https://doi.org/10.1109/SWAT.1972.29
20. Mytkowicz, T., Musuvathi, M., Schulte, W.: Data-parallel finite-state machines. In: Balasubramonian, R., Davis, A., Adve, S.V. (eds.) Architectural Support for Programming Languages and Operating Systems – 19th International Conference, ASPLOS 2014, Salt Lake City, UT, USA, 1–5 March 2014, pp. 529–542. ACM (2014). https://doi.org/10.1145/2541940.2541988
21. Palioudakis, A., Salomaa, K., Akl, S.G.: Finite nondeterminism vs. DFAs with multiple initial states. In: Jurgensen, H., Reis, R. (eds.) DCFS 2013. LNCS, vol. 8031, pp. 229–240. Springer, Heidelberg (2013). https://doi.org/10.1007/978-3-642-39310-5_22
22. Qiu, J., Sun, X., Sabet, A.H.N., Zhao, Z.: Scalable FSM parallelization via path fusion and higher-order speculation. In: Sherwood, T., Berger, E.D., Kozyrakis, C. (eds.) Architectural Support for Programming Languages and Operating Systems – 26th International Conference, ASPLOS 2021, Virtual Event, USA, 19–23 April 2021, pp. 887–901. ACM (2021). https://doi.org/10.1145/3445814.3446705
23. Sin'ya, R., Matsuzaki, K., Sassa, M.: Simultaneous finite automata: an efficient data-parallel model for regular expression matching. In: 42nd International Conference on Parallel Processing, ICPP 2013, Lyon, France, 1–4 October 2013, pp. 220–229. IEEE Computer Society (2013). https://doi.org/10.1109/ICPP.2013.31

24. Veloso, P.A.S., Gill, A.: Some remarks on multiple-entry finite automata. J. Comput. Syst. Sci. **18**(3), 304–306 (1979). https://doi.org/10.1016/0022-0000(79)90038-2

25. Yang, Y.E., Prasanna, V.K.: Space-time tradeoff in regular expression matching with semi-deterministic finite automata. In: 30th International Conference on Computer Communications – Joint Conference of the IEEE Computer and Communications Societies, INFOCOM 2011, Shanghai, China, 10–15 April 2011, pp. 1853–1861. IEEE Computer Society (2011). https://doi.org/10.1109/INFCOM.2011.5934986

Two-Way Automata and Bounded Languages

Alessandro Clerici Lorenzini[1] , Giovanni Pighizzini[1] ,
and Luca Prigioniero[2]([✉])

[1] Dipartimento di Informatica, Università degli Studi di Milano, Via Celoria, 18,
20133 Milan, Italy
research@aclerici.me, pighizzini@di.unimi.it
[2] Department of Computer Science, Loughborough University, Epinal Way,
Loughborough LE11 3TU, UK
l.prigioniero@lboro.ac.uk

Abstract. We prove that, at the cost of a polynomial increase of the number of states, each two-way nondeterministic automaton accepting a letter-bounded language can be simulated by an equivalent machine of the same type, where nondeterministic transitions and inversions of head movement are possible only when the head visits one of the tape end-markers. This result has several consequences. Two of them are related to the open question posed by Sakoda and Sipser concerning the state cost of the conversions of one-way and two-way nondeterministic automata into equivalent two-way deterministic finite automata: in the letter-bounded case, the cost of the latter conversion is bounded by a subexponential function, while that of the former is polynomial in the number of states of the given machine. Other consequences are related to the cost of complementing two-way automata, to make them halting, self-verifying or unambiguous. Connections with the question of the power of the nondeterminism in logarithmic space bounded complexity classes are also stated.

1 Introduction

It is well known that the elimination of nondeterminism from one-way finite automata requires, in the worst case, an exponential increase in the number of states [15]. In the case of two-way automata, the situation is not so clear. Two-way nondeterministic and deterministic finite automata (2NFAs and 2DFAs, for short) can be simulated by one-way deterministic finite automata (1DFAs). Furthermore, the number of states of the machines resulting from such simulations is bounded by a function which is exponential in the number of states of the given two-way automata [17,19]. But *what can be said for the state cost of the simulation of* 2NFA*s by* 2DFA*s?*

More generally, in 1978 Sakoda and Sipser raised the question of whether the use of two-way motion can help to reduce the number of states in the elimination

G. Castiglione and S. Mantaci (Eds.): CIAA 2025, LNCS 15981, pp. 73–85, 2026.
https://doi.org/10.1007/978-3-032-02602-6_6

of nondeterminism. They asked whether it is possible to obtain polynomial simulations of 2NFAs and 1NFAs (namely, one-way nondeterministic finite automata) by 2DFAs [18]. They conjectured a negative answer for both simulations.

In spite of all efforts, after almost 50 years both problems are still open. Progress has been achieved for some weaker formulations. In particular, the question has been solved by considering restrictions or extensions of the class of simulating machines (e.g., negative answers have been proved for the *sweeping* [20], *oblivious* [10], and *few-reversals* [12] 2DFA restrictions – for a survey and further references see [16]; positive answers have been obtained extending the class of simulating machines to some variants of *linear-time one-tape deterministic Turing machines* that are able to accept regular languages only [7]).

Another restriction of the Sakoda and Sipser questions that deserves a special mention is to languages defined over a one-letter alphabet, also called *unary languages*. In this case, the simulation of 1NFAs by 2DFAs has a quadratic cost in the number of states [1],[1] while for the simulation of 2NFAs by 2DFAs a subexponential but superpolynomial upper bound has been obtained [4]. In [6], the latter question has been related to space complexity, by proving that if $L = NL$ (the classical logarithmic space bounded complexity classes), then the state cost of the simulation of unary 2NFAs by 2DFAs is polynomial. Hence, proving that the above-mentioned simulation of unary 2NFAs by 2DFAs is optimal would imply $L \neq NL$. (For further connection with the L vs NL question, we address the reader to [13,14]). In [3], these results have been extended to general alphabets in the case of a restricted class of 2NFAs, in which nondeterministic decisions can be taken only when the input head is scanning one of the two end-markers of the tape. These machines are called *outer-nondeterministic automata* (2ONFAs, for short).

In this paper, we continue this line of research by considering *letter-bounded languages* (shortly called *bounded* in the following), namely subsets of $a_1^* a_2^* \cdots a_k^*$, for some symbols $a_1, a_2, \ldots, a_k$. Our main result is a construction that allows to transform each 2NFA accepting a bounded language into an equivalent sweeping 2ONFA at the cost of a polynomial increase in the number of states. According to [3], this allows to extend to the bounded case the two above-mentioned results for the unary case, and to derive some other interesting results. In particular, for each n-state 2NFA A accepting a bounded language, we obtain the following:

– The 2NFA A can be made halting with a polynomial increase in the number of states.
– A 2NFA with a number of states polynomial in n can be built which accepts the complement of the language accepted by A (this extends a result for the unary case [5]).
– The 2NFA A can be simulated by a 2DFA with a number of states subexponential in n. This cost would reduce to a polynomial in case $L = NL$.
– The 2NFA A can be made *unambiguous* with a polynomial increase in the number of states (this extends a result for the unary case [6]).

---

[1] A subtle error in [1] has been discovered and fixed by To [21].

As a byproduct of our transformation of 2NFAs into 2ONFAs, we are also able to prove that the state cost of the simulation of 1NFAs accepting bounded languages by 2DFAs is polynomial in the size of the original machine, as in the unary case, thus answering the Sakoda and Sipser question concerning 1NFAs vs 2DFAs in the bounded case. For the simulation of 1NFAs accepting bounded languages by 1DFAs we address the reader to [8].

2   Preliminaries

Let us recall some basic definitions useful in the paper. We assume the reader familiar with notions from formal languages and automata theory. For any unfamiliar terminology see, e.g., [9].

Given a set S, $\#S$ denotes its cardinality and 2^S the family of all its subsets. Given an alphabet Σ and a string $w \in \Sigma^*$, $|w|$ denotes the length of w. Languages defined over a one-letter alphabet (and the devices accepting them) are called *unary*. In this paper we are mainly interested in bounded languages. A language L over Σ is said to be *letter-bounded* (or simply *bounded* in the following[2]) if $L \subseteq a_1^* a_2^* \cdots a_k^*$, for $k \geq 1$ and $a_i \in \Sigma$ for $i = 1, \ldots, k$. Given a word $x \in a_1^* a_2^* \cdots a_k^*$, an a_i-*block* (or simply *block*, when a_i can be understood from the context) is a maximal factor of x belonging to a_i^*.

A *two-way nondeterministic finite automaton* (2NFA, for short) is a quintuple $\mathcal{A} := \langle Q, \Sigma, \delta, q_0, F \rangle$, where Q is a finite set of states, Σ is a finite input alphabet, $\delta : Q \times (\Sigma \cup \{\triangleright, \triangleleft\}) \to 2^{Q \times \{-1, 0, +1\}}$ is a transition function, with the two special symbols $\triangleright, \triangleleft \notin \Sigma$ called the left and the right end-markers, respectively, $q_0 \in Q$ is an initial state, and $F \subseteq Q$ is a set of final states. The input is stored onto the tape surrounded by the two end-markers, the left end-marker being at the position zero. Hence, on input w, the right end-marker is on the cell in position $|w| + 1$. In one move, $\mathcal{A}$ reads an input symbol, changes its state, and moves the input head one position to the right or left depending on whether δ returns, as second component, $+1$ or -1 respectively, or keeps it stationary if δ returns 0. Furthermore, the head cannot pass the end-markers, except at the end of computation, to accept the input, as explained below. The machine accepts the input if there exists a computation path from the initial state q_0 with the head on the left end-marker, ending in a final state $q \in F$ after passing the right end-marker. The language accepted by $\mathcal{A}$ is denoted by $\mathcal{L}(\mathcal{A})$. A 2NFA $\mathcal{A}$ is said to be *deterministic* (2DFA), whenever $\#\delta(q, \sigma) \leq 1$, for each $q \in Q$ and $\sigma \in \Sigma \cup \{\triangleright, \triangleleft\}$.

By 1NFAs and 1DFAs we denote *one-way* nondeterministic and deterministic finite automata, respectively.

A two-way finite automaton is said to be *sweeping* if the direction of the head is changed only at the end-markers [20]. It is said to be *outer-nondeterministic*

---

[2] In the literature, *word-bounded* languages are also defined, which are subsets of $w_1^* w_2^* \cdots w_k^*$, for some fixed words $w_1, \ldots, w_k$. Furthermore, a letter-bounded language is called *strongly bounded* if all symbols $a_1, \ldots, a_k$ are pairwise distinct. All of the results in this paper hold for general letter-bounded languages.

(or simply *outer*, 2ONFA for short) if the nondeterministic choices are made only at the end-markers [3].

In this paper we are interested in comparing the size of finite automata. The *size* of a device is given by the total number of symbols used to write down its description. Therefore, the size of finite automata is linear in the number of "instructions", which is bounded by a polynomial in the number of states and in the number of input symbols.

3   From 2NFAs for Bounded Languages to 2ONFAs

This section is devoted to our main construction. Indeed, we show how to restrict nondeterminism and two-way motion in 2NFA accepting bounded languages solely to the end-markers, while keeping the number of states polynomial in the original number of states. In other words, we show how to transform a 2NFA accepting a bounded language into an equivalent sweeping 2ONFA that is polynomially larger.

We first show that every 2NFA accepting a bounded language can be simulated by an equivalent *block-sweeping* 2NFA of polynomial size. The accepting computations of such a machine can be divided into traversals from left to right and from right to left. During each traversal, the head can visit each *block* (composed by a maximal sequence of the same letter) only once. More precisely, the computation inside each block is still two-way, but the head cannot be moved back to a block that has already been visited or to the starting end-marker in the same traversal, until the other end-marker is reached. Then, such a machine is transformed into a sweeping 2NFA that is allowed to perform nondeterministic moves only on the first entered cell of each block. This transformation has a polynomial cost in the size of the original machine, and it derives from results obtained for the unary case [4]. Finally, we move the nondeterministic choices performed on the first cell of each block to the left and right end-markers only, thus obtaining a sweeping 2ONFAs that is polynomially larger.

3.1   From 2NFAs to Block-Sweeping 2NFAs

As a first step to obtain our main results, we start by describing how to convert a 2NFA accepting a bounded language into an equivalent block-sweeping 2NFA with a polynomial-size increase.

Let $\mathcal{A}$ be a two-way finite automaton accepting a bounded language $L \subseteq a_1^* a_2^* \cdots a_k^*$. A *right- (left-) segment* in a computation of $\mathcal{A}$ is a computation path that starts on the left (right) end-marker, ends on the right (left) end-marker, and does not visit any end-markers in between. A *U-turn* is a computation path that enters a block from one side, ends by leaving it to the same side, and does not visit any other blocks in between. $\mathcal{A}$ is *block-sweeping* if each accepting computation can be divided into a sequence of left- and right-segments such that no segment contains a U-turn. Therefore, in a right-segment all blocks are entered from the left and exited to the right (vice versa in left-segments), and each block is traversed exactly once per segment.

As long as the original machine does not perform a U-turn, it can be simulated directly. When a move to the previous block is detected, the simulating machine intercepts it, saves the information useful to recover the computation in the previous block (i.e., the state that the simulated machine would have entered), and moves the head in the opposite direction instead, until it reaches the end-marker and thus completes a segment. It then initiates an opposite segment and moves the head up to the cell where the original transition would have moved, it restores the saved information, and resumes the direct simulation.

In order to intercept a transition to the previous block without performing it, the machine uses a counter to keep track of the position of the head inside the current block. If the position is 0, namely the head is at the beginning of the block, and the transition that is chosen to simulate is towards the previous block, then the afore-mentioned technique is executed.

In order to limit the maximum value of the counter, we state the following lemma:[3]

Lemma 1 (Corollary of [2, Lemma 3]**).** *For each U-turn that starts in state p and ends in state q, there exists a U-turn on the same block that starts in state p, ends in state q, and does not move farther than n^2 cells from the beginning of the block.*

Therefore, if the simulating machine detects that a move to the previous block has to be simulated and the position counter reached n^2 within the block, it can safely reject. To implement this, once the counter reaches n^2, it is never decremented.

Theorem 1 *Each n-state 2NFA accepting a bounded language can be simulated by an equivalent block-sweeping 2NFA with $O(n^3)$ states. Furthermore, the simulation preserves determinism.*

3.2 Reducing Nondeterminism in Block-Sweeping 2NFAs

By leveraging some tools from [4], we now describe a construction for simulating an n-state block-sweeping 2NFA $\mathcal{A}$ with an equivalent sweeping 2NFA $\mathcal{A}'$ that, on each segment, performs nondeterministic moves only on the first-entered cell of each block, and is otherwise deterministic and one-way.

Before giving an outline of the simulation, let us state a lemma that was originally proven by Geffert for space-bounded Turing machines over a unary input alphabet [2] and then reformulated in the case of unary 2NFAs [4]. The formulation we give is adapted to consider input blocks in the case of a 2NFA $\mathcal{A}$ with n states accepting a bounded language:

---

[3] Our definition of U-turns is actually a particular case of the one used in [2], where the role of the position i where the U-turns starts (and ends) is played by the last cell of the block that the machine leaves (and comes back to).

Lemma 2 ([4, Lemma 3]). *Given two states q_1, q_2 of $\mathcal{A}$ and a block a^m, with $m > n$, for each traversal of the block from q_1 to q_2, there is a traversal from q_1 to q_2 where $\mathcal{A}$: having traversed s_1 cells of the block, gets into a loop (called* dominant loop*) of length ℓ, which starts from a state p and is repeated λ times, then it traverses the remaining s_2 block cells, and finally reaches the next block, for some p, s_1, s_2, satisfying $0 \le \lambda$, $1 \le \ell \le n$, and $s_1 + s_2 \le 3n^2$.*

Roughly, Lemma 2 states that if a block is long enough, then a traversal of the block from a state q_1 to a state q_2 can be simulated by a traversal from q_1 to q_2 in which almost all of the computation is performed by repeating a loop. In the case of unary 2NFAs, such a traversal of the entire input is replaced by a traversal that repeats the loop on the same string [4]. In our case, we adopt such a strategy to process each input block. More precisely, a block is called *long* if its length exceeds $5n^2$ cells, so that it guarantees the existence of a dominant loop. (The threshold of $5n^2$ instead of the $3n^2$ that appears in Lemma 2 will be clear later, see Lemma 4.) Otherwise, the block is called *short*.

To process a new block, $\mathcal{A}'$ starts by simulating a state of $\mathcal{A}$. Here, it non-deterministically guesses whether the block is long or short. In the first case, $\mathcal{A}'$ enters a loop which serves the purpose of simulating the dominant loop. It then follows the loop until a new block is detected, when the machine nondeterministically selects and enters a state that $\mathcal{A}$ could have entered upon leaving the loop and reaching the new block (we shall present the details later).

In the second case, $\mathcal{A}'$ traverses the block measuring its length, until it finds a new block. It then selects a state of the original machine to enter, among the ones that are possible in $\mathcal{A}$ after computation paths on words of the found length on the block symbol, which were precomputed.

While performing the above steps, the machine verifies that the initial guess of whether the block is short or long is correct, and rejects otherwise. The process then starts over for the new block.

We now give some details about the simulation.

The Case of Long Blocks. Let us first address the case of long blocks. Fixed a symbol $a \in \Sigma$, consider the weighted digraph representing the transition diagram of $\mathcal{A}$, projected on the sole transitions on a, and in which we set weights $+1$, -1, or 0 to arcs depending on whether they represent transitions where the input head is moved right, left, or kept stationary, respectively. Every cycle of weight ℓ in such a graph represents a computation loop of length ℓ in the automaton $\mathcal{A}$, taking place in an a-block and far enough from the other blocks so that none is visited. Let us denote with $\mathscr{C}_1^{(a)}, \ldots, \mathscr{C}_{r^{(a)}}^{(a)}$ the strongly connected components of this digraph, and with $\ell_i^{(a)} > 0$ the greatest common divisor of the absolute values of the cycle weights in the component $\mathscr{C}_i^{(a)}$, for $i = 1, \ldots, r^{(a)}$. Because $\mathscr{C}_1^{(a)}, \ldots, \mathscr{C}_{r^{(a)}}^{(a)}$ are pairwise disjoint sets of states of $\mathcal{A}$, it holds that

$$\ell_1^{(a)} + \cdots + \ell_{r^{(a)}}^{(a)} \le n. \tag{1}$$

We now define a set of states Q_L of $\mathcal{A}'$, which, along with a copy of the original state set Q, will be used to simulate block-traversals over long blocks. For each segment direction d and strongly connected component $\mathscr{C}_i^{(a)}$, we add the new states $t_{i,0}^{(a,d)}, \ldots, t_{i,\ell_i^{(a)}-1}^{(a,d)}$ forming a loop of length $\ell_i^{(a)}$. More precisely, for $j = 0, \ldots, \ell_i^{(a)} - 1$, we have the following deterministic one-way transitions:

$$\delta'(t_{i,j}^{(a,d)}, a) := \left\{ (t_{i,(j+1) \bmod \ell_i^{(a)}}^{(a,d)}, d) \right\}.$$

In what follows, we analyze the case of right-segments and block-traversals. All of the following notations and results can be analogously stated for left-segments and block-traversals.

Let $p \overset{a^m}{\underset{+1}{\rightsquigarrow}} q$ denote that there is a computation path in an a-block of $\mathcal{A}$ that starts in state p and ends m cells to the right in state q, and only visiting the cells in between. By Lemma 1, such a path does not depend on the initial or final position as long as they are at least n^2 cells away from the neighboring blocks. A component $\mathscr{C}_i^{(a)}$ is said to be *positive* (resp. *negative* for left-segments) if it contains at least one cycle of positive (negative) weight. All the states used in the dominant loop of a left-to-right block-traversal belong to the same positive component. Given a positive component $\mathscr{C}_i^{(a)}$, we designate some state $q_{(i)}^{(a)}$ as the *center* of the component. To map the states of the original state set to the ones of the new loops, given $p \in \mathscr{C}_i^{(a)}$ we define

$$\kappa^{(a)}(p) := \min\left\{ x \in \mathbb{N} \mid q_{(i)}^{(a)} \overset{a^x}{\underset{+1}{\rightsquigarrow}} p \right\} \quad \text{and} \quad \varphi^{(a,+1)}(p) := t_{i,\kappa^{(a)}(p) \bmod \ell_i^{(a)}}^{(a,+1)}.$$

States in $\mathscr{C}_i^{(a)}$ with the same image under $\varphi^{(a,d)}(p)$ have the same *"distance"* modulo $\ell_i^{(a)}$ from the center, and for sufficiently large blocks they can be considered equivalent.

As an extension of [4, Lemma 5], we can prove that, once entered one of the loops, $\mathcal{A}'$ correctly simulates a dominant loop of the original machine, provided that the block is long enough:

Lemma 3. *Given $a \in \Sigma$, let $q, p \in Q$ be in the same positive component $\mathscr{C}_i^{(a)}$. Then $p \overset{a^m}{\underset{+1}{\rightsquigarrow}} q$ if and only if $\varphi^{(a,+1)}(p) \overset{a^m}{\underset{+1}{\rightsquigarrow}} \varphi^{(a,+1)}(q)$, for each $m \geq 2n^2 + n$.*

To prove the correctness of the simulation of block-traversals, we now need to define how the new loops are entered and exited. Intuitively, the machine, rather than directly simulating the computation path performed by $\mathcal{A}$ before entering the dominant loop in a state q, calculates the state of the new loop that can be entered at the beginning of the block, so that the state entered when reaching the same cell is in fact q. The opposite operation happens when leaving the block.

Because the possible dominant loop and entering state are not unique, this step has to be performed by a nondeterministic transition, happening on the first cell of the traversed block for entering the loop, and on the first cell of the following block for exiting it.

To understand how these possible entry states are calculated, let us define

$$\rho^{(a)} := \mathrm{lcm}_{i \in \{1,\dots,r^{(a)}\}} \ell_i^{(a)} \qquad \text{and} \qquad M^{(a)} := \rho^{(a)} \cdot \left\lceil \frac{3n^2+1}{\rho^{(a)}} \right\rceil,$$

where $\rho^{(a)}$ is the least common multiple of the loop lengths of all strongly connected components with respect to a and $M^{(a)}$ is the smallest multiple of $\rho^{(a)}$ that is greater than $3n^2$.

Now, consider a block $a^{2M^{(a)}}$. After reading the first $M^{(a)}$ symbols, $\mathcal{A}$ reaches a state p. Since $M^{(a)} > 3n^2$, p belongs to the dominant loop, and hence also to some positive component $\mathscr{C}_i^{(a)}$. In order to have $\mathcal{A}'$ reach state $\tilde{p} := \varphi^{(a,+1)}(p)$ after reading the first $M^{(a)}$ symbols, it is sufficient to have it enter state $\tilde{p}$ at the beginning of the block (with a stationary move), as $M^{(a)}$ is a multiple of the loop length $\ell_i^{(a)}$. Similarly, we want $\mathcal{A}'$, starting in state $\tilde{p}$, to read the remaining $M^{(a)}$ symbols of the current block, and enter the state q that $\mathcal{A}$ enters when started from p and having traversed the same cells. Because the simulated path is the same as in the previous case, a single stationary move, performed nondeterministically on the first cell of the next block, is again sufficient.

Precisely, for each $a \in \Sigma$, $d \in \{+1, -1\}$, and $q \in Q$, we have the following transitions:

$$\delta'(q, a) \supseteq \{(\varphi^{(a,d)}(p), 0) \mid q \overset{a^{M^{(a)}}}{\underset{d}{\leadsto}} p\}.$$

Moreover, for $b \neq a$, $i = 1, \dots, r^{(a)}$, and $j = 0, \dots, \ell_i^{(a)} - 1$, we have

$$\delta'(t_{i,j}^{(a,d)}, b) := \{(p, 0) \mid p' \in \varphi^{(a,d)-1}(t_{i,j}^{(a,d)}), p' \overset{a^{M^{(a)}}}{\underset{d}{\leadsto}} p\}.$$

The following lemma, again adapted from [4, Theorem 1], proves that the simulation of block-traversals is correct for long blocks:

Lemma 4. *Let $m > 5n^2$ and $q_1, q_2 \in Q$. There exists a left-to-right traversal of $\mathcal{A}$ of a block a^m from q_1 to q_2 if and only if there exists a left-to-right traversal of $\mathcal{A}'$ of the same block that starts in q_1 and is followed by a stationary move to reach q_2. An analogous statement holds for traversals from right to left.*

Note that the required block length does not need to reach $M^{(a)}$.

Finally, for simplicity, we omitted checking that the block length exceeds $5n^2$. This can be done by counting the length in parallel with the previous steps, up to $5n^2 + 1$ and only performing the stationary move on the next block if the counter has hit its limit (thus rejecting otherwise).

The Case of Short Blocks. Blocks of length at most $5n^2$ are easier to handle, as the computation of $\mathcal{A}$ over them can be precomputed. In particular, when $\mathcal{A}'$ guesses to be in a short block, it saves the current state $q \in Q$ and performs a stationary move to a state $q_0^{(a,d)}$, where d is the direction of the current segment. Here, d is the direction of every first non-stationary transition of $\mathcal{A}$ on a starting in state q, as otherwise the automaton would not be block-sweeping. For the sake of simplicity we shall write $\delta(q,a) \ni (\cdot, d)$. The states $q_0^{(a,d)}, \ldots, q_{5n^2}^{(a,d)}$ simply count the length l of the current block. If the length exceeds $5n^2$, the machine rejects. If the first cell of a new block is entered, then the machine enters, with a stationary move, one of the possible states reached by $\mathcal{A}$ after a computation starting in state q on the first cell of an a-block of length l in a d-segment. Note that a nondeterministic transition can be performed here as the head is on the first cell of a block. Because the set of such counter states has polynomial size in n, their exiting transitions can be precomputed.

Therefore we have, for $q \in Q$, $a \in \Sigma$, and $d \in \{+1, -1\}$ such that $\delta(q,a) \ni (\cdot, d)$:

$$\delta'(q,a) \ni (q_0^{(a,d)}, 0),$$

for $i = 0, \ldots, 5n^2 - 1$:

$$\delta'(q_i^{(a,d)}, a) := \{(q_{i+1}^{(a,d)}, d)\},$$

and, given $b \neq a$:

$$\delta'(q_i^{(a,d)}, b) := \{(p, 0) \mid q \overset{a^i}{\underset{d}{\rightsquigarrow}} p\}.$$

Note that there are no outgoing transitions from the states $q_{5n^2}^{(a,d)}$, namely, for each $q \in Q$, $a \in \Sigma$, and $d \in \{+1, -1\}$, we have $\delta'(q_{5n^2}^{(a,d)}, a) := \emptyset$.

By combining the two cases, we obtain the following:

Theorem 2. *Every n-state block-sweeping 2NFA $\mathcal{A}$ can be simulated by an equivalent sweeping 2NFA $\mathcal{A}'$ having the following properties:*

- *in each segment of its computation, $\mathcal{A}'$ performs nondeterministic moves only on the first entered cell of each block;*
- *in the rest of each segment, $\mathcal{A}'$ performs only deterministic one-way moves;*
- *$\mathcal{A}'$ has $O(n^3)$ states.*

3.3   Moving Nondeterminism to the End-Markers

We now complete our construction by presenting a way to simulate an n-state automaton in the form described in Theorem 2 using a sweeping 2ONFA, i.e., moving the nondeterministic guesses to the endmarkers.

To do so, first notice that on every block in every segment the stationary moves can be combined into one non-stationary nondeterministic move, which combines leaving the state processing the previous block, guessing whether the block is long or short, and performing the first non-stationary move. Note that such an operation maintains all the properties given in Theorem 2, while making the states in Q unreachable, allowing us to remove them from the state set. The same pruning of stationary moves can be done for every automaton that satisfies the properties given in the theorem statement. Thus, in this new refined form, every segment contains at most one nondeterministic move per block, which is performed at the beginning of the block.

To move the nondeterministic transitions to the end-markers, an easy way is simply guessing k states with the first transition of each segment, and verifying that, when entering a new block, the guessed state can in fact be reached in the original machine from the current state. This construction yields a sweeping 2ONFA, at the cost of an exponential size in k, while still being polynomial in n.

To produce a smaller automaton, which is also polynomial in k, we implement a different strategy. The new automaton will only guess the starting state of a single block at a time: when a not yet visited block is reached, the head is moved back to the end-marker, after completing the traversal, to perform the nondeterministic choice, before returning to the beginning of the block and resuming the simulation. More in detail:

1. Within each block, the automaton simulates the states of the original machine, while also keeping track of the index i, $1 \le i \le k$, of the current block's symbol a_i in $a_1^* \cdots a_i^* \cdots a_k^*$.[4] This process requires kn states.
2. Upon reaching a new block, which is detected when the read symbol a_j is such that $j \ne i$, the current state and the value j are saved.
3. The head is moved in the same direction until reaching the end-marker, then brought back to the opposite end-marker (which started the previous segment). Note that there is no need to keep track of the segment's direction, as it is given by the direction of every move from the saved state.[5] This phase can be split into two parts: one to bring the head from the block containing a_i to the end-marker and back (which requires $2kn$ states), and one to reach the opposite end-marker (which requires kn states).
4. The first state of the new block is nondeterministically guessed while the head is at the opposite end-marker, replacing the one saved previously.
5. The beginning of the new block is again reached, by using the saved value j. This phase requires, once again, kn states.
6. The process starts over from the saved state.

By using the above strategy we obtain the following:

Theorem 3. *Every n-state block-sweeping* 2NFA *$\mathcal{A}$ such that*

– *in each segment of its computation, $\mathcal{A}$ performs nondeterministic moves only on the first entered cell of each block, and*

---

[4] In general, it is not sufficient to save the symbols themselves, as they might repeat in different blocks.

[5] However, this might be necessary if we relax the condition about the original machine performing only one-way moves within blocks.

– *in the rest of each segment, $\mathcal{A}$ performs only deterministic one-way moves,*
can be simulated by an equivalent sweeping 2ONFA *with $5kn$ states.*

Summing up, from Theorem 1, 2 and 3 we obtain our main simulation result:

Theorem 4. *Each n-state* 2NFA *accepting a bounded language can be simulated by an equivalent sweeping* 2ONFA *with a number of states polynomial in n.*

4   Consequences

In this section we present some important consequences of the polynomial simulation in Theorem 4, that derive from general results for 2ONFAs proved in [3].

The first result we present is related to the elimination of infinite computations from 2NFAs. A 2NFA $\mathcal{A}$ is said to be *halting* if each of its computation paths ends within a finite number of steps. It is *self-verifying* [11] if $\mathcal{A}$ is equipped, besides the set of accepting states F, also a set of rejecting states disjoint from F. Each string should have either at least one computation ending in an accepting state (and in this case it is accepted) or at least one computation ending in a rejecting state (and in this case it is rejected) but it cannot have both (note that some computation paths may end with a state that is neither accepting nor rejecting, or may enter an infinite loop). As a consequence of constructions and simulations in [3] (in particular Theorem 5.1), we obtain the following results:

Theorem 5. *Each n-state* 2NFA *accepting a bounded language can be simulated by an equivalent halting and self-verifying* 2ONFA *with a number of states polynomial in n.*

From a self-verifying automaton, we can immediately obtain an automaton accepting the set of rejected strings, by choosing as set of final states the set of rejecting states. Hence, from Theorem 5 we obtain:

Corollary 1. *For each n-state* 2NFA *accepting a bounded language L there exists a halting* 2ONFA *accepting the complement of L and having a number of states polynomial in n.*

Other consequences of Theorem 4 are related to the Sakoda and Sipser question, concerning the cost of the conversion of 2NFAs into equivalent 2DFAs. The following result generalizes to the bounded case a result obtained in [4] for the unary case. It derives from Theorem 7.1 in [3]:

Theorem 6. *Each n-state* 2NFA *accepting a bounded language can be simulated by an equivalent* 2DFA *with $n^{O(\log n)}$ states.*

It is natural to ask if the subexponential and superpolynomial upper bound in Theorem 6 is optimal. This question is related to the longstanding open problem of whether deterministic and nondeterministic logarithmic space are equivalent. From the next result, which is derived from Theorem 8.1 in [3] and generalizes a result obtained in [6] for unary languages, a superpolynomial lower bound for the conversion in Theorem 6 would imply the separation of deterministic from nondeterministic logarithmic space.

Theorem 7. *If* $\mathrm{L} = \mathrm{NL}$ *then each* n*-state* 2NFA *accepting a bounded language can be simulated by an equivalent* 2DFA *with a number of states polynomial in* n.

Finally, we remind the reader that a machine is *unambiguous* whenever there exists at most one accepting computation path for each input. The next result, derived from Theorem 8.2 in [3], states a polynomial cost in size for the elimination of ambiguity in 2NFAs accepting bounded languages:

Theorem 8. *Each* n*-state* 2NFA *accepting a bounded language can be simulated by an equivalent unambiguous* 2DFA *with a number of states polynomial in* n.

5   The Case of 1NFAs

The simulation to obtain a 2DFA (Theorem 6) can be improved in the case the given machine is a 1NFA. In this case, as 1NFAs are already block-sweeping, we can apply our simulations to move the nondeterministic choices to the beginning of the k blocks. The obtained machine performs only k nondeterministic moves, one per block, along a unique traversal. Notice that the number of blocks k is constant with respect to the size of the automaton. Therefore, we can simulate the obtained machine with a 2DFA of polynomial size that enumerates all the n^k possible combinations of the nondeterministic choices at the beginning of each block and iteratively performs all of them, one per left-to-right traversal.

Theorem 9. *Each* n*-state* 1NFA *accepting a subset of* $a_1^* a_2^* \cdots a_k^*$, $k > 0$, *can be simulated by an equivalent* 2DFA *with* $n^{O(k)}$ *states.*

It is possible to observe that, for every fixed $k > 0$, this result answers the Sakoda and Sipser question on 1NFAs in the restricted case of bounded languages, extending to this class the polynomial-size conversion that was previously only known for unary alphabets.

References

1. Chrobak, M.: Finite automata and unary languages. Theor. Comput. Sci. **47**(3), 149–158 (1986). https://doi.org/10.1016/0304-3975(86)90142-8. Errata: Theor. Comput. Sci. **302**(1–3), 497–498 (2003)
2. Geffert, V.: Nondeterministic computations in sublogarithmic space and space constructibility. SIAM J. Comput. **20**(3), 484–498 (1991). https://doi.org/10.1137/0220031
3. Geffert, V., Guillon, B., Pighizzini, G.: Two-way automata making choices only at the endmarkers. Inf. Comput. **239**, 71–86 (2014). https://doi.org/10.1016/J.IC.2014.08.009
4. Geffert, V., Mereghetti, C., Pighizzini, G.: Converting two-way nondeterministic unary automata into simpler automata. Theor. Comput. Sci. **295**, 189–203 (2003). https://doi.org/10.1016/S0304-3975(02)00403-6
5. Geffert, V., Mereghetti, C., Pighizzini, G.: Complementing two-way finite automata. Inf. Comput. **205**(8), 1173–1187 (2007). https://doi.org/10.1016/j.ic.2007.01.008

6. Geffert, V., Pighizzini, G.: Two-way unary automata versus logarithmic space. Inf. Comput. **209**(7), 1016–1025 (2011). https://doi.org/10.1016/J.IC.2011.03.003

7. Guillon, B., Pighizzini, G., Prigioniero, L., Průša, D.: Converting nondeterministic two-way automata into small deterministic linear-time machines. Inf. Comput. **289**, Part A, 104938 (2022). https://doi.org/10.1016/j.ic.2022.104938

8. Herrmann, A., Kutrib, M., Malcher, A., Wendlandt, M.: Descriptional complexity of bounded regular languages. J. Autom. Lang. Comb. **22**(1-3), 93–121 (2017). https://doi.org/10.25596/jalc-2017-093

9. Hopcroft, J.E., Ullman, J.D.: Introduction to Automata Theory, Languages and Computation. Addison-Wesley (1979)

10. Hromkovič, J., Schnitger, G.: Nondeterminism versus Determinism for two-way finite automata: generalizations of Sipser's separation. In: Baeten, J.C.M., Lenstra, J.K., Parrow, J., Woeginger, G.J. (eds.) ICALP 2003. LNCS, vol. 2719, pp. 439–451. Springer, Heidelberg (2003). https://doi.org/10.1007/3-540-45061-0_36

11. Jirásková, G., Pighizzini, G.: Optimal simulation of self-verifying automata by deterministic automata. Inf. Comput. **209**(3), 528–535 (2011). https://doi.org/10.1016/j.ic.2010.11.017

12. Kapoutsis, C.A.: Nondeterminism is essential in small two-way finite automata with few reversals. Inf. Comput. **222**, 208–227 (2013). https://doi.org/10.1016/J.IC.2012.11.001

13. Kapoutsis, C.A.: Two-way automata versus logarithmic space. Theory Comput. Syst. **55**(2), 421–447 (2014). https://doi.org/10.1007/S00224-013-9465-0

14. Kapoutsis, C.A., Pighizzini, G.: Two-way automata characterizations of L/poly versus NL. Theory Comput. Syst. **56**(4), 662–685 (2015). https://doi.org/10.1007/S00224-014-9560-X

15. Meyer, A.R., Fischer, M.J.: Economy of description by automata, grammars, and formal systems. In: SWAT 1971, pp. 188–191. IEEE Computer Society (1971). https://doi.org/10.1109/SWAT.1971.11

16. Pighizzini, G.: Two-way finite automata: old and recent results. Fundam. Informaticae **126**(2–3), 225–246 (2013). https://doi.org/10.3233/FI-2013-879

17. Rabin, M.O., Scott, D.S.: Finite automata and their decision problems. IBM J. Res. Dev. **3**(2), 114–125 (1959). https://doi.org/10.1147/RD.32.0114

18. Sakoda, W.J., Sipser, M.: Nondeterminism and the size of two way finite automata. In: STOC 1978, pp. 275–286. ACM (1978). https://doi.org/10.1145/800133.804357

19. Shepherdson, J.C.: The reduction of two-way automata to one-way automata. IBM J. Res. Dev. **3**(2), 198–200 (1959). https://doi.org/10.1147/RD.32.0198

20. Sipser, M.: Lower bounds on the size of sweeping automata. J. Comput. Syst. Sci. **21**(2), 195–202 (1980). https://doi.org/10.1016/0022-0000(80)90034-3

21. To, A.W.: Unary finite automata vs. arithmetic progressions. Inf. Process. Lett. **109**(17), 1010–1014 (2009)

An Algebraic Approach
to the Equivalence Checking
of Deterministic Top-down Tree
Transducers

Deng Zhibo⊙, Tang Tianxiang⊙, and Vladimir A. Zakharov$^{(\boxtimes)}$⊙

Shenzhen MSU-BIT University, Shenzhen, China
{dengzb,tangtx}@smbu.edu.cn, zakh@cs.msu.su

Abstract. For many classes of automata, denotational semantics provides suitable means for specifying functions or relations $[\![s]\!]_\pi$, computed in all states s of an automaton π, through the system of algebraic equations. By adding to this system an equation $[\![s_1]\!]_\pi = [\![s_2]\!]_\pi$ which declares that π computes in some states s_1 and s_2 the same function or relation we reduce the equivalence checking problem to that of checking the solvability of the system of equations obtained thus. In general, this problem is intractable, but when dealing only with deterministic automata the solvability checking can be performed efficiently by means of a technique very much similar to Gaussian variable elimination used in linear algebra. In this paper we show that equivalence of total deterministic tree transducers can be resolved just as efficiently using this fairly general method. Moreover, this method can be very easy adapted for computing the most general unifiers of parameterized deterministic tree transducers.

Keywords: tree transducer · equivalence checking · unification · denotational semantics · polynomial time complexity · proof by consistency

The main goal of the paper is to draw attention to some algebraic approach to designing efficient equivalence checking algorithms in various classes of automata. The principles, advantages and limitations of this approach are illustrated using top-down tree transducers as a test site. Previously, this approach was successfully applied to construct polynomial-time algorithms for checking the equivalence of deterministic string transducers, two-tape automata, biautomata and push-down (stateless) automata (see [13,14]).

Typically, the introduction of a class of automata begins with an exposition of the operational semantics, and it is almost always defined in terms of transitions, runs, etc. Such a viewpoint provides a good intuitive understanding of the model and the potential of its application for the formalization and analysis of dynamic

---

This study is supported by the Guangdong Provincial Talented Scientist Promotion Program (Project 2023CX10X178) and MSU-BIT University in Shenzhen.

G. Castiglione and S. Mantaci (Eds.): CIAA 2025, LNCS 15981, pp. 86–98, 2026.
https://doi.org/10.1007/978-3-032-02602-6_7

processes in technology, nature and social life. It is therefore not surprising that algorithmic problems in automata theory are almost always treated exclusively on the basis of operational semantics. Very often, this attitude makes it possible to find suitable mathematical structures and problems, like, say, the reachability problem in graphs, and by reducing to them to obtain a ready-made effective solution. Meanwhile, an alternative interpretation of the behavior of automata, which is given by denotational semantics, turns out to be advantageous as well. Denotational semantics assigns a meaning to every state s of an automaton π by associating s with a certain function or relation $[\![s]\!]_\pi$ which is usually defined implicitly by means of a system of equations $\mathcal{E}_\pi$ constructed according to the transition rules of π. By adding to $\mathcal{E}_\pi$ an equation (constraint) $[\![s_1]\!]_\pi = [\![s_2]\!]_\pi$ which declares that π computes the same function or relation in the states s_1 and s_2 we reduce the equivalence checking problem to that of checking the solvability of the system of equations obtained thus.

The similar approach is known in computer-aided theorem proving under the names "proof by consistency" or "inductionless induction" (see [2] for the survey). This technique also adds the target equation to a set of equality axioms and then applies the saturation-based Knuth-Bendix completion algorithm to show consistency of the extended collection of equalities. In the 1980s, great hopes were associated with this method in the field of automated reasoning, which, as usual, were only partially fulfilled. Perhaps our paper will highlight new possibilities for applying this method to solving the equivalence checking problems in automata theory.

Dealing with functional equations on words or trees is not always easy but for some classes of automata the variable elimination techniques provides an effective means for checking the solvability of systems of equations $\mathcal{E}_\pi \cup \{[\![s_1]\!]_\pi = [\![s_2]\!]_\pi\}$. Using certain features of the equations specifying the behavior of automata, one can express some variables through others and, by performing substitutions, exclude variables one by one from equations of the system while preserving its canonical form. Eventually, either all constraints will be eliminated and the preserved canonical form of the system of equations will indicate its solvability, or, upon the next restoration of the canonical form of the system, an inconsistent equation will be discovered. If the recovery of the canonical form does not require a radical increase in the size of the system of equations, then such a process of solvability checking can be performed in polynomial time.

We show in details how to apply this algebraic technique to some classes of top-down tree transducers. The Tree Transducer is the brainchild of two respected parents, the Tree Automaton and the String Transducer, and it inherits the talents of both: the ability to inspect trees and the aptitude for recording observations in words or trees. Tree transducers have found application in many areas such as XML databases, computational linguistics, image processing, information security, and program translation. In order for a mathematical model to become not just a subject of academic research but also a framework for practical implementation, it must be supplied with efficient algorithms for solving the most important problems that inevitably arise when working with this model.

One of such problems is the equivalence checking of states in automata. Unfortunately, the Tree Transducer borrowed from its parents not only computing power, but also some "hereditary diseases". The equivalence problem is undecidable for non-deterministic tree transducers since it is the same for nondeterministic string transducers [6], and it is *EXPTIME*-hard for deterministic tree transducers, since the emptiness problem for alternating tree automata can be reduced to it [9] (see also [3]).

A survey on the decidable cases of the equivalence problem for tree transducers is presented in [7]. If we restrict ourselves to the basic model of deterministic top-down tree transducers (DTDTs), two results are decisive. In 1980, Zoltán Ésik [15] proved the decidability of the equivalence checking problem for DTDTs, following the approach suggested by Leslie Valiant in [12]. If two automata are equivalent, then neither will get very far ahead of the other during their runs on the same input. This "bounded balance" allows simulating parallel runs of two DTDTs by a single automaton of the same type, equipped with additional bounded memory to keep track of the difference in outputs. We thus reduce the equivalence checking problem to the emptiness problem for tree automata. The main drawback of this fairly universal approach to equivalence checking is that the amount of additional memory required for such simulation typically depends exponentially on the size of the machines being verified.

The authors of [4] looked at this problem from a different angle. Mehryar Mohri noticed in [10] that one can check the equivalence of a pair of deterministic string transducers by bringing them to the common canonical form. To this end, for each state of the machine, it is necessary to find the longest common prefix of all output words that the machine produces in the runs from this state and then make changes to the transitions so that the transducer outputs this prefix as early as possible. As shown in [4], any DTDT has the canonical form of this kind, but to construct it one has to trace the parallel operation of several copies of the automaton on the same input tree. This does not make much difficulty for those DTDTs that are applicable to any input, and thus a polynomial-time algorithm for checking the equivalence of total DTDTs was obtained in [4]. But in the general case, it is necessary to monitor the domains of functions that are computable in different states of the transducer; such an inspection can be done by a special tree automaton, which may, however, have an exponential size.

We propose an alternative approach aimed at proving that the equation $[\![s_1]\!]_\pi = [\![s_2]\!]_\pi$, which requires that the states s_1 and s_2 of the DTDT π be equivalent, is consistent with the equations describing the functions computed by the machine. This idea was inspired by the unification algorithm for first-order terms suggested in 1982 by Alberto Martelli and Ugo Montanari [8]. Firstly, we define a denotational semantics for DTDTs and specify the functions $[\![s]\!]_\pi$ computed by a DTDT π in every state s by a system of equations $\mathcal{E}_\pi$. Next, we show how to check the solvability of the system of equations $\mathcal{E}_\pi \cup \{[\![s_1]\!]_\pi = [\![s_2]\!]_\pi\}$ using simple algebraic rules when a DTDT π is total. Finally, we show how to adapt this method for solving the unification problem for parameterized DTDTs that have unknown parameters instead of some output letters in their transitions.

1   Preliminaries

Let $\mathcal{F}$ be a finite set of *functional symbols*. A ranked alphabet is a couple $(\mathcal{F}, Arity)$, where $Arity$ is a mapping from $\mathcal{F}$ into $\mathbb{N}$. The set of symbols of arity k is denoted by $\mathcal{F}^{(k)}$. Let $\mathcal{X}$ be a set of variables, which is disjoint from the functional symbols in $\mathcal{F}$. The variables from $\mathcal{X}$ are used in $\mathcal{F}$-terms only as placeholders, and, to distinguish them from functional variables that are introduced late, we will call them *arguments*. The set of $\mathcal{F}$-*terms* $\mathcal{T}(\mathcal{F}, \mathcal{X})$ is the smallest set satisfying the following conditions:

- all arguments in $\mathcal{X}$ and all functional symbols in $\mathcal{F}^{(0)}$ are $\mathcal{F}$-terms,
- if $t_1, \ldots, t_k$ are $\mathcal{F}$-terms and $f \in \mathcal{F}^{(k)}$, then $f(t_1, \ldots, t_k)$ is $\mathcal{F}$-term.

If a term contains only functional symbols then it is called *ground term*. The set of ground $\mathcal{F}$-terms is denoted as $\mathcal{T}(\mathcal{F})$.

The arguments are assigned values by means of *substitutions* of the type $\theta : \mathcal{X} \to \mathcal{T}(\mathcal{F})$. Finite substitutions are represented by lists of bindings of the form $\theta = \{x_1/t_1, \ldots, x_n/t_n\}$, where $x_1, \ldots, x_n$ are arguments and $t_1, \ldots, t_n$ are ground $\mathcal{F}$-terms. The application $t\theta$ of a substitution θ to an $\mathcal{F}$-term t replaces all occurrences of arguments $x_1, \ldots, x_n$ with ground $\mathcal{F}$-terms $t_1, \ldots, t_n$.

Every $\mathcal{F}$-term t has a *meaning* $[\![t]\!]$. The meaning $[\![t_0]\!]$ of a ground $\mathcal{F}$-term t_0 is t_0 itself. The meaning of a term t whose arguments are $x_1, \ldots, x_n$ is a function $[\![t]\!]$ of the type $\mathcal{T}(\mathcal{F})^n \to \mathcal{T}(\mathcal{F})$ such that $[\![t]\!](t_1, \ldots, t_n) = t\{x_1/t_1, \ldots, x_n/t_n\}$ for every tuple of ground $\mathcal{F}$-terms $t_1, \ldots, t_n$. The meaning $[\![f]\!]$ of every functional symbol $f \in \mathcal{F}^{(n)}$ is understood as $[\![f(x_1, \ldots, x_n)]\!]$.

Clearly, for any functional symbol $f \in \mathcal{F}^{(n)}$ and terms $t_1, \ldots, t_n$ an equality $[\![f]\!]([\![t_1]\!], \ldots, [\![t_n]\!]) = [\![f(t_1, \ldots, t_n)]\!]$ holds. An algebra defined thus is called *free term algebra*. This algebra satisfies the laws of congruence and decomposition

$$[\![t'_1]\!] = [\![t''_1]\!], \ldots, [\![t'_n]\!] = [\![t''_n]\!] \quad \Rightarrow \quad [\![f(t'_1, \ldots, t'_n)]\!] = [\![f(t''_1, \ldots, t''_n)]\!],$$
$$[\![f(t'_1, \ldots, t'_n)]\!] = [\![g(t''_1, \ldots, t''_n)]\!] \Rightarrow f = g, [\![t'_1]\!] = [\![t''_1]\!], \ldots, [\![t'_n]\!] = [\![t''_n]\!]$$

that are crucial for the development of efficient decision procedures for tree automata and transducers.

Let Σ and Δ be two finite sets of ranked functional symbols. Tree transducers take at their inputs ground Σ-terms and compute transduction functions or relations by converting these terms into ground Δ-terms. Since in what follows we will focus only on deterministic transducers, all subsequent definitions are adapted for this class of machines. Let $\mathcal{C}$ be a countable set of unary functional variables which are called *transduction variables*. The range of values of each transduction variable C is a set of functions (in general, partial ones) of type $\mathcal{T}(\Sigma) \to \mathcal{T}(\Delta)$. These values will be specified implicitly by means of systems of equations where transduction variables are included in Σ-terms and Δ-terms. A $\Delta\mathcal{C}$-term is an expression of the form $t(C_1(x_{i_1}), \ldots, C_n(x_{i_n}))$, where $t(y_1, \ldots, y_n)$ is a Δ-term, and $C_1, \ldots, C_n \in \mathcal{C}$. When the meanings $[\![C]\!]$ of transduction variables are fixed, the meanings of $\Delta\mathcal{C}$-terms are the compositions of the functions corresponding to Δ-terms and transduction variables.

A *Deterministic Top-Down Transducer* (DTDT) over an input alphabet Σ and an output alphabet Δ is a pair $\pi = (\mathcal{C}_\pi, R_\pi)$, where $\mathcal{C}_\pi$ is a finite set of transduction variables, and R_π is a finite set of *transduction rules* of the form

$$S(f(x_1, \ldots, x_n)) \;\rightarrow\; t(S_1(x_{i_1}), \ldots, S_k(x_{i_k})) , \tag{1}$$

which satisfy the following requirements:

- $S, S_1, \ldots, S_k$ are transduction variables in $\mathcal{C}_\pi$,
- $f \in \Sigma^{(n)}$, and $t(S_1(x_{i_1}), \ldots, S_k(x_{i_k}))$ is a $\Delta \mathcal{C}$-term such that $x_{i_j} \in \{x_1, \ldots, x_n\}$ for every $j, 1 \leq j \leq k$,
- there no two rules in R_π with the same left-hand side.

The variables $S \in \mathcal{C}_\pi$ can be considered as states of π, which is what we will sometimes call them in the following. A DTDT π is called *total* if for every its state S and each $f \in \Sigma$ the set R_π has a rule with a left-hand side $S(f(x_1, \ldots, x_n))$.

The rules of a DTDT π define a monotonic operator

$$F_\pi : (\mathcal{T}(\Sigma) \rightarrow \mathcal{T}(\Delta))^N \rightarrow (\mathcal{T}(\Sigma) \rightarrow \mathcal{T}(\Delta))^N ,$$

where $|\mathcal{C}_\pi| = N$, on the lattice of (partial) transduction functions. For every tuple of transductions $(\llbracket S_1 \rrbracket', \ldots, \llbracket S_N \rrbracket')$ assigned to the transduction variables (states) of π this operator computes transductions $(\llbracket S_1 \rrbracket'', \ldots, \llbracket S_N \rrbracket'')$ such that for every transduction rule (1) and for every tuple of ground Σ-terms $t_1, \ldots, t_n$ the equality $\llbracket S_1 \rrbracket''(f(t_1, \ldots, t_n)) = t(\llbracket S_1 \rrbracket'(t_{i_1}), \ldots, \llbracket S_N \rrbracket'(t_{i_k}))$ holds. The least fixed point $(\llbracket S_1 \rrbracket_\pi, \ldots, \llbracket S_N \rrbracket_\pi)$ of F_π, which exists due to Kleene fixed-point theorem, defines the tuple of transduction functions computed in all states of π.

Fixed-point denotational semantics may be quite suitable for reasoning about fundamental properties of tree transduction functions, but it is not so much good for designing decision procedures. Fortunately, total DTDTs allow another, more practical way of specifying their denotational semantics. We associate with every transduction rule (1) of a total DTDT $\pi = (\mathcal{C}_\pi, R_\pi)$ an algebraic equation

$$S(f(x_1, \ldots, x_n)) = t(S_1(x_{i_1}), \ldots, S_k(x_{i_k}))$$

over the set of transduction variables $\mathcal{C}_\pi$ and consider the system $\mathcal{E}_\pi$ of such equations for all rules in R_π.

Proposition 1. *For every total DTDT π the system of equations $\mathcal{E}_\pi$ over a set of transduction variables $\mathcal{C}_\pi$ has the unique solution $(\llbracket S_1 \rrbracket_\pi, \ldots, \llbracket S_N \rrbracket_\pi)$.*

Proof. Since $(\llbracket S_1 \rrbracket_\pi, \ldots, \llbracket S_N \rrbracket_\pi)$ is the least fixed point of F_π, this tuple of transduction functions is a solution to $\mathcal{E}_\pi$. Since π is a total DTDT, it follows from Kleene fixed point theorem that all transduction functions $\llbracket S_N \rrbracket_\pi, 1 \leq i \leq N$, are total and, therefore, this is the only fixed point of F_π. $\square$

We say, that S_1 and S_2 are *equivalent states* of DTDT π iff $\llbracket S_1 \rrbracket_\pi = \llbracket S_2 \rrbracket_\pi$.

Proposition 2. *States S_1 and S_2 of a total DTDT π are equivalent iff the system of equations $\mathcal{E}_{\pi,S_1,S_2} = \mathcal{E}_\pi \cup \{S_1(x) = S_2(x)\}$ has a solution.*

Thus, the equivalence checking problem for total DTDTs is reduced to the solvability checking problem for systems of equations of the form $\mathcal{E}_{\pi,S_1,S_2} = \mathcal{E}_\pi \cup \{S_1(x) = S_2(x)\}$ in the free term algebra, and this problem can be solved by pure algebraic means.

2 Equivalence Checking Total DTDTs

An equivalence checking algorithm presented below operates with systems of equations $\mathcal{E}$ over transduction variables. Every such system is composed of three subsystems $\mathcal{E}^\Sigma, \mathcal{E}^\Delta$, and $\mathcal{E}^C$. The subsystem $\mathcal{E}^\Sigma$ is a total collection of basic equations over some set of transduction variables $\mathcal{C}_0$, i.e. for every transduction variable $S \in \mathcal{C}_0$ and for each symbol $f \in \Sigma$ a subsystem $\mathcal{E}^\Sigma$ includes an equation of the form $S(f(x_1, \ldots, x_n)) = t(S_1(x_{i_1}), \ldots, S_k(x_{i_k}))$, where $x_{i_j} \in \{x_1, \ldots, x_n\}$ for every $j, 1 \le j \le k$. The subsystem $\mathcal{E}^\Delta$ has only equations of the form $t' = t''$, where both t' and t'' are ΔC-terms over the set of transduction variables $\mathcal{C}_0$. These equations are regarded as constraints on the solution of $\mathcal{E}^\Sigma$ which, by Proposition 1, always exists. The subsystem $\mathcal{E}^C$ has only equations of the form $S(x) = t$ such that t is a ΔC-term over the set of transduction variables $\mathcal{C}_0$, and $S(x)$ is a transduction variable that occurs only in the left-hand side of this equation and does not appear anywhere else. These equations are viewed as definitions of the left-hand side variables, and these variables are called *resolved*. Note that the system of equations $\mathcal{E}_{\pi,S_1,S_2}$ falls under this form: $\mathcal{E}^\Sigma_{\pi,S_1,S_2} = \mathcal{E}_\pi$, $\mathcal{E}^\Delta_{\pi,S_1,S_2} = \{S_1(x) = S_2(x)\}$, and $\mathcal{E}^C_{\pi,S_1,S_2} = \emptyset$. Our aim is to check the solvability of $\mathcal{E}_{\pi,S_1,S_2}$, and we know for sure that some systems of the kind defined above have a solution. We say that a system of equations $\mathcal{E}$ is *reduced* if $\mathcal{E}^\Delta = \emptyset$.

Proposition 3. *Any reduced system of equations $\mathcal{E}$ is solvable.*

Proof. By Proposition 1, the total subsystem of basic equations $\mathcal{E}^\Sigma$ has the unique solution. The resolved variables are evaluated through their definitions in $\mathcal{E}^C$. Since there are no constraints in $\mathcal{E}$, the solution to the subsystem $\mathcal{E}^\Sigma$ provides the solution to the whole system $\mathcal{E}$. □

We propose a nondeterministic algorithm which has weak Church–Rosser property (i.e. gives the same result regardless of the order in which the rules are applied) and checks the solvability of a system of equations $\mathcal{E}_{\pi,S_1,S_2}$ by bringing it to reduced form. Beginning with $\mathcal{E}_0 = \mathcal{E}_{\pi,S_1,S_2}$, the equivalent transformation rules defined below should be applied in any order to the systems of equations $\mathcal{E}$ as long as it is possible. The correctness of every rule, unless it is obvious, is explained.

1) *Removing identities.* Equations of the form $t = t$ are removed.
2) *Decomposition of equations.* An equation of the form $f(t'_1, \ldots, t'_k) = f(t''_1, \ldots, t''_k)$ in the subsystem $\mathcal{E}^\Delta$ is replaced with a set of equations $t'_i = t''_i, 1 \le i \le k$.

The correctness of this rule follows from the law of decomposition, which is valid in the free algebra of terms.

3) *Detecting conflict in $\mathcal{E}^{\Delta}$.* If $\mathcal{E}^{\Delta}$ includes an equation of the form $f(t'_1, \ldots, t'_k) = g(t''_1, \ldots, t''_m)$, where f and g are different symbols in Δ, then terminate and announce the unsolvability of the system $\mathcal{E}$.

This decision is correct, since in free term algebra if $f \neq g$ then the functions $[\![f]\!]$ and $[\![g]\!]$ in this algebra do not have common values.

4) *Swapping equations.* An equation of the form $f(t'_1, \ldots, t'_k) = S(x)$ in the subsystem $\mathcal{E}^{\Delta}$ is replaced with the equation $S(x) = f(t'_1, \ldots, t'_k)$.

5) *Self-reference detection.* If $\mathcal{E}^{\Delta}$ includes such an equation $S(x) = t$ that it is not an identity and the transduction variable $S(x)$ occurs in Δ-term t, then terminate and announce the unsolvability of the system $\mathcal{E}$.

Since the subsystem of basic equations $\mathcal{E}^{\Sigma}$ is total, it has only total solutions. But for a total transduction function $[\![S]\!]$ the equality $[\![S]\!](\alpha) = t(\ldots, [\![S]\!](\alpha), \ldots)$ is impossible for any input term $\alpha \in \mathcal{T}(\Sigma)$.

6) *Substitution of ground terms.* If $\mathcal{E}^{\Delta}$ includes an equation $S(x) = t$, where the term t contains such a transduction variable $S'(y)$ that $S \neq S'$ and $x \neq y$, then

- choose an arbitrary symbol $f_0 \in \Sigma^{(0)}$ and the corresponding basic equation $S'(f_0) = t_0$ in $\mathcal{E}^{\Sigma}$ (note that in this case t_0 is a ground Δ-term);
- in every equation of the system $\mathcal{E}$ replace all occurrences of terms of the form $S'(t')$ headed by the transduction variable S' with the term t_0;
- add the equation $S'(y) = t_0$ to the subsystem $\mathcal{E}^{\mathcal{C}}$.

For example, basic equations $S'(f_0) = t_0$ and $S'(f(x_1, x_2)) = h(S(x_1), S'(x_2))$ in $\mathcal{E}^{\Sigma}$, and a constraint $S(x) = g(S'(y), S''(x))$ in $\mathcal{E}^{\Delta}$ are replaced with equations $t_0 = t_0$, $t_0 = h(S(x_1), t_0)$, and $S(x) = g(t_0, S''(x))$ which are moved to $\mathcal{E}^{\Delta}$.

To see that this rule is correct assume that the system of equations $\mathcal{E}$ has a solution. Then the equality $[\![S]\!]_{\mathcal{E}}(x) = t(\ldots, [\![S']\!]_{\mathcal{E}}(y), \ldots)$ holds for all possible values of arguments x and y. Hence, the function $[\![S']\!]_{\mathcal{E}}(y)$ does not depend on y. Since $\mathcal{E}^{\Sigma}$ has a basic equation $S'(f_0) = t_0$, this means that $[\![S']\!]_{\mathcal{E}} \equiv t_0$.

7) *Reducing variables.* If $\mathcal{E}^{\Delta}$ includes an equation $S(x) = t(S_{i_1}(x), \ldots, S_{i_k}(x))$, where all transduction variables have the same argument, and $S(x)$ does not occur in the right-hand side of the equation, then

- in every basic equation of the form $S(f(x_1, \ldots, x_n)) = t'$ in $\mathcal{E}^{\Sigma}$ replace its left-hand side with the term $t(t_{i_1}, \ldots, t_{i_k})$, where every subterm $t_{i_j}, 1 \leq j \leq k$, is the right-hand side of the basic equation $S_{i_j}(f(x_1, \ldots, x_n)) = t_{i_j}$ in $\mathcal{E}^{\Sigma}$, and move the resulting equation to the subsystem $\mathcal{E}^{\Delta}$;
- in every equation of the system replace all occurrences of terms $S(y)$ for any argument y with the term $t(S_{i_1}(y), \ldots, S_{i_k}(y))$;
- add the equation $S(x) = t(S_{i_1}(x), \ldots, S_{i_k}(x))$ to the subsystem $\mathcal{E}^{\mathcal{C}}$.

For example, if the system $\mathcal{E}^{\Sigma}$ includes basic equations

$$S_1(f(x_1, x_2)) = f_1(S_2(x_1)),$$
$$S_2(f(x_1, x_2)) = f_2(S_1(x_1), S(x_2)),$$

and $\mathcal{E}^{\Delta}$ includes a constraint $S(x) = h(S_1(x), S_2(x))$, then a basic equation

$$S(f(x_1, x_2)) = g(S(x_1), S_3(x_2))$$

is replaced with the constraint

$$h(f_1(S_2(x_1)), f_2(S_1(x_1), h(S_1(x_2), S_2(x_2)))) = g(h(S_1(x_1), S_2(x_1)), S_3(x_2)),$$

which is moved to the subsystem $\mathcal{E}^\Delta$. As for the equation $S(x) = h(S_1(x), S_2(x))$, it is moved to $\mathcal{E}^C$, and the transduction variable $S(x)$ becomes resolved.

This is an equivalent transformation due to the law of congruence.

If none of the rules 1) – 7) can be applied to $\mathcal{E}$ then the algorithm terminates and announces the solvability of the system $\mathcal{E}$.

Theorem 1. *For every total DTDT π and every pair of its states S_1 and S_2 the algorithm defined above always terminates and correctly verifies the equivalence of these states.*

Proof. For a system of equations $\mathcal{E}$ let (n_1, n_2) be a pair of integers, where

- n_1 is the number of basic equations in $\mathcal{E}$,
- n_2 is the total number of symbols in the left-hand sides of equations in $\mathcal{E}$.

As can be seen from the definition of rules 1), 2), 4), 6), and 7), if the system $\mathcal{E}'$ is obtained from $\mathcal{E}$ after using any of these rules, then the pair (n_1', n_2') corresponding to $\mathcal{E}'$ is lexicographically smaller than the pair (n_1, n_2). Therefore, the rules of the algorithm can not be invoked infinitely long.

As it was noticed, the rules 1), 2), 4), 6), and 7), preserve the solvability of the systems of equations, and the rules 3) and 5) correctly detect their unsolvability. Hence, if the algorithm terminates at applying the rules 3) or 5) then the obtained system of equations $\mathcal{E}$ as well as the initial system $\mathcal{E}_{\pi, S_1, S_2}$ are unsolvable. By Proposition 2, this means that the states S_1 and S_2 are not equivalent.

If none of the rules 1) – 7) can be activated then the subsystem $\mathcal{E}^\Delta$ is empty. This means that $\mathcal{E}$ is a reduced system of equations, and, by Proposition 3, it has a solution. Therefore, the system of equations $\mathcal{E}_{\pi, S_1, S_2}$ has a solution as well, and, by Proposition 2, the states S_1 and S_2 are equivalent. $\qquad\square$

The naive non-deterministic algorithm presented above can be made much more efficient. The set of terms in the definition of a DTDT π can be represented by a labeled directed acyclic graph (DAG) G_π to avoid duplication of subterms. Then by the size of π we mean the size of G_π. Like any DAG, G_π can be represented as a reference data structure. This representation allows all manipulations specified in rules 1) – 7) (such as substitution of terms, checking self-reference, generation of new equations) to be performed in linear time without increasing the number of nodes in G_π. Every equation $t' = t''$ in the system $\mathcal{E}$ becomes associated with a pair of nodes in G_π corresponding to the terms t' and t''. Following the principles of dynamic programming, we will keep track of the equations in subsystem $\mathcal{E}^\Delta$ to avoid repetitive processing of the same equation. Thus, the number of equations the algorithm deals with does not exceed the number of pairs of nodes of G_π. Thus, we arrive at

Theorem 2. *The equivalence checking for a total DTDT π can be performed in time $O(n^3)$, where n is the size of π.*

3   Unifying Parameterized DTDTs

The use of systems of equations for describing the functioning of automata has certain advantages when solving some other algorithmic problems. Usually, in the early stages of system design, developers do not know the exact values of some data or a detailed description of the system's behavior. Then, the using of parameters in the project sketch makes it possible to inspect and study the operation of the entire system as a whole and clarify the values and the structure of the unknown components gradually, as the project develops. Naturally, the question arises: is it possible, given some additional information about the expected functionality of the designed system, to automatically determine the values of some of its parameters?

Suppose, for example, that several teams are involved in the design of the same information processing system, with each team responsible for its own part of the functionality of the entire system. Each team can present its own version of the project (say, as a DTDT), in which fragments outside the team's area of competence are replaced by parameters that act as placeholders. Is it possible to achieve consistency among the proposed drafts by choosing appropriate parameter values and thus combining several parameterized models together? It is from these practical considerations that the unification problem for parameterized models of computation arises.

Parameterization is a costly option in terms of the complexity of the computational model analysis. In [1] it was discovered that the introduction of parameters even into Rabin-Scott finite automata and regular expressions leads to a high jump in the complexity of decision problems. However, in [11] we were able to find out that for a parameterized model of deterministic string transducers many important problems still can be solved efficiently. We believe that the same is true for parameterized tree transducers, and we will demonstrate the validity of this assumption using the example of the unification problem.

Let $\mathcal{P}$ be a set of ranked functional symbols different from the symbols in Σ and Δ; they are called *parameters*. We assume that $\Delta^{(n)} = \emptyset$ implies $\mathcal{P}^{(n)} = \emptyset$ for any $n \geq 0$. Terms in $\mathcal{T}(\mathcal{P} \cup \Delta, \mathcal{X})$ will be called *templates*. Parameters are instantiated by means of arity-preserving substitutions $\eta : \mathcal{P} \to \mathcal{P} \cup \Delta$, i.e. $p \in \mathcal{P}^{(n)}$ implies $\eta(p) \in \mathcal{P}^{(n)} \cup \Delta^{(n)}$. If $\eta(p) \in \Delta$ for every $p \in \mathcal{P}$ then η is called *ground substitution*. The set of substitutions (ground substitutions) of this type is denoted by $Subst(\mathcal{P}, \Delta)$ (respectively $GSubst(\mathcal{P}, \Delta)$). Applying a substitution η to a template t results in a template $t\eta$ where all occurrences of each parameter p in t are replaced with the functional symbol $\eta(p)$. Given a pair of substitutions $\eta_1, \eta_2 \in Subst(\mathcal{P}, \Delta)$ we denote by $\eta_1\eta_2$ their composition.

A *Parameterized Deterministic Top-Down Transducer* (PDTDT) is a triple $\Pi = (\mathcal{C}_\Pi, S_0, R_\Pi)$, where $\mathcal{C}_\Pi$ is a finite set of transduction variables, S_0 is an initial state (main transduction variable), and R_Π is a finite set of transduction rules of the form (1) with one single difference: a term $T(y_1, \ldots, y_k)$ is a template in $\mathcal{T}(\Delta \cup \mathcal{P}, \mathcal{X})$. We denote by $\mathcal{P}(\Pi)$ the set of all parameters occurred in the rules of R_Π. Substitutions are applicable to PDTDTs in the same way as to templates: a PDTDT $\Pi\eta$ is obtained by applying a substitution η to all templates in the transduction rules of Π. A PDTDT $\Pi\eta$ is called an *instance* of PDTDT Π. If η is a ground substitution then $\Pi\eta$ is a *ground instance* of Π.

The semantics of a PDTDT $\Pi = (\mathcal{C}_\Pi, S_0, R_\Pi)$ is characterized by the set of transduction functions $[\![S_0]\!]_{\Pi\eta}$ computed in the initial states of all ground instances $\Pi\eta$ of Π. Two PDTDTs Π_1 and Π_2 are called equivalent ($\Pi_1 \sim \Pi_2$ in symbols) if $[\![S_{10}]\!]_{\Pi_1\eta} = [\![S_{20}]\!]_{\Pi_2\eta}$ holds for every ground substitution η.

PDTDT Π can be viewed as a rough draft of a control system, some elements of which remain undefined. This initial design can be specialized by applying substitutions to Π. As soon as a ground instance $\pi = \Pi\eta$ is obtained we have a complete project of a control system as a DTDT model. One way of such specialization is to compare several alternative control system designs in an attempt to unify their behavior and obtain thus a well coordinated system project that is consistent with all its drafts. This approach to system design leads to the Unification Problem for PDTDTs, which is as follows.

Let Π_1 and Π_2 be PDTDTs such that $\mathcal{P}(\Pi_1) \cap \mathcal{P}(\Pi_2) = \emptyset$. A substitution η is called a *unifier* of Π_1 and Π_2 if $\Pi_1\eta \sim \Pi_2\eta$, i.e. η reduces two different control system designs to such instances that have the same behavior. A unifier η of Π_1 and Π_2 is called *the most general unifier* of these PDTDTs if for any ground substitution η' which is a unifier of Π_1 and Π_2 there exists such a substitution $\rho \in GSubst(\mathcal{P}, \Delta)$ that $\eta' = \eta\rho$. Informally speaking, the most general unifier instantiates as loosely as possible parameters to combine several rough designs of the same control system into a single overall project. This explains the practical importance of the *Unification Problem* for PDTDTs: given a pair of PDTDTs Π_1 and Π_2, compute their most general unifier.

To solve the Unification Problem, we adapt the algebraic method that was used to check the equivalence of states of total DTDTs.

Proposition 4. *Let Π_1 and Π_2 be two total PDTDTs such that $\mathcal{C}_{\Pi_1} \cap \mathcal{C}_{\Pi_2} = \emptyset$, and let S_{10} and S_{20} be their initial states. Then a substitution η is a unifier of Π_1 and Π_2 iff for every ground substitution ρ the system of equations*

$$\mathcal{E}_{\Pi_1\eta\rho, \Pi_2\eta\rho} = \mathcal{E}_{\Pi_1\eta\rho} \cup \mathcal{E}_{\Pi_2\eta\rho} \cup \{S_{10}(x) = S_{20}(x)\}$$

has a solution.

Proof. By definition, a substitution η is a unifier of Π_1 and Π_2 iff $[\![S_{10}]\!]_{\Pi_1\eta\rho} = [\![S_{20}]\!]_{\Pi_1\eta\rho}$ holds for every ground substitution ρ. By Proposition 1, the subsystems of equations $\mathcal{E}_{\Pi_1\eta\rho}$ and $\mathcal{E}_{\Pi_2\eta\rho}$ have the unique solution and, therefore, this is also a solution to the system of equations $\mathcal{E}_{\Pi_1\eta\rho, \Pi_2\eta\rho}$. $\square$

Thus, to compute the most general unifier of Π_1 and Π_2, it suffices to find the most general substitution η that guarantees the solvability of the system of equations $\mathcal{E}_{\Pi_1\eta\rho, \Pi_2\eta\rho}$ for every ground substitution ρ. This can be achieved by applying the same rules of equivalent transformations that have been used for the equivalence checking of total DTDTs.

A unification algorithm operates with systems of equations $\mathcal{E}$ partitioned into 4 subsystems $\mathcal{E}^\Sigma$, $\mathcal{E}^\Delta$, $\mathcal{E}^C$, and $\mathcal{E}^P$. The first three subsystems have the same structure as in the equivalence checking algorithm. Every equation of a subsystem $\mathcal{E}^{\mathcal{P}}$ is of the form $p = r$, where $p \in \mathcal{P}$, $r \in \Delta \cup \mathcal{P}$, and all left-hand sides of these equations are pairwise distinct. If a subsystem $\mathcal{E}^{\mathcal{P}}$ includes only equations $p_{i_1} = z_1, \ldots, p_{i_m} = z_m$ then denote by $\eta_{\mathcal{E}}$ the substitution $\{p_{i_1}/z_1, \ldots, p_{i_m}/z_m\}$.

The same as before, a system of equations $\mathcal{E}$ is *reduced* if $\mathcal{E}^{\Delta} = \emptyset$. Given a system of equation $\mathcal{E}$ and a substitution ρ we denote by $\mathcal{E}\rho$ a system of equation obtained from $\mathcal{E}$ by applying to all its terms the substitution ρ.

Proposition 5. *Suppose that a system of equations $\mathcal{E}$ is reduced. Then for every ground substitution ρ the system $\mathcal{E}\rho$ has a solution iff $\rho = \eta_{\mathcal{E}}\rho'$ for some $\rho' \in GSubst(\mathcal{P}, \Delta)$.*

Proof. As it follows from Proposition 1, for every reduced system of equations $\mathcal{E}$ the subsystem $\mathcal{E}^{\Sigma}\rho \cup \mathcal{E}^{\mathcal{C}}\rho$ has a solution for any ground substitution ρ. In [8] it was shown that if all equations in $\mathcal{E}^{\mathcal{P}}$ are of the form $p = z$ and the parameter p does not occur anywhere else then for every ground substitution ρ the equations of $\mathcal{E}^{\mathcal{P}}\rho$ become identities iff $\rho = \eta_{\mathcal{E}}\rho'$. $\qquad\square$

Given two total PDTDTs Π_1 and Π_2, the unification algorithm starts with a system of equations $\mathcal{E}_{\Pi_1, \Pi_2}$ and applies the rules of equivalent transformations 1) – 7) naturally extended to work with templates and the rules 8) and 9) described below until it either constructs a reduced system of equations or detects the unsolvability of the system.

8) *Parameter elimination.* If for some $p \in \mathcal{P}$ a subsystem $\mathcal{E}^{\Delta}$ has an equation of the form $p(t'_1, \ldots, t'_k) = r(t''_1, \ldots, t''_k)$ or $r(t'_1, \ldots, t'_k) = p(t''_1, \ldots, t''_k)$, then

- replace this equation with the equations $t'_i = t''_i, 1 \leq i \leq k$,
- apply the substitution $\eta = \{p/r\}$ to all equations in the system $\mathcal{E}$,
- add the equation $p = r$ to the subsystem $\mathcal{E}^{\mathcal{P}}$, and
- remove all identities of the form $z = z$ from the subsystem $\mathcal{E}^{\mathcal{P}}$, and replace every equation of the form $z = q$, where $q \in \mathcal{P}$, with the equation $q = z$.

9) *Detecting conflict in $\mathcal{E}^{\mathcal{P}}$.* If a subsystem $\mathcal{E}^{\mathcal{P}}$ has an equation $f = g$, where f and g are different functional symbols in Δ, then terminate and announce the unsolvability of the system $\mathcal{E}$.

If none of the rules above is applicable to a system of equations $\mathcal{E}$ then the algorithm terminates and outputs the substitution $\eta_{\mathcal{E}}$.

Theorem 3. *Let Π_1 and Π_2 be two total PDTDTs. Then the Unification Algorithm defined above, when applied to the system of equations $\mathcal{E}_{\Pi_1, \Pi_2}$ always terminates in $O(n^3)$ time. The algorithm outputs a substitution η iff Π_1 and Π_2 are unifiable; then η is one of the most general unifiers of Π_1 and Π_2.*

Proof. Termination and time complexity of the algorithm are established by the same reasonings as in Theorem 1 and Corollary 2. After analyzing the rules 1), 2), 4), 6) – 8), one can easily see that if a system of equations $\mathcal{E}'$ is obtained from the system $\mathcal{E}$ by applying any of these rules, then for each ground substitution ρ the systems $\mathcal{E}\rho$ and $\mathcal{E}'\rho$ are equivalent.

Suppose that the Unification Algorithm terminates and builds at the end the system of equations $\mathcal{E}$. If one of the rules 3), 5), or 9) can be applied to $\mathcal{E}$ then the system is, obviously, inconsistent, and for every ground substitution ρ the system $\mathcal{E}\rho$ does not have a solution. Since the systems of equations $\mathcal{E}\rho$ and $\mathcal{E}_{\Pi_1\rho,\Pi_2\rho}$ are equivalent, the latter is unsolvable as well. Hence, by Proposition 4, the PDTDTs Π_1 and Π_2 can not be unified.

If none of the rules 1) – 9) can be applied to $\mathcal{E}$ then $\mathcal{E}^\Delta = \emptyset$. As it can be seen from the definition of the rule 8), it always retains the canonical form of the system $\mathcal{E}^{\mathcal{P}}$. Therefore, $\mathcal{E}$ is a reduced system of equations. Hence, by Proposition 5, all ground substitutions ρ for which the system $\mathcal{E}\rho$ has a solution are of the form $\rho = \eta_{\mathcal{E}}\rho'$, where $\rho' \in GSubst(\mathcal{P}, \Delta)$. Since the systems of equations $\mathcal{E}$ and $\mathcal{E}_{\Pi_1,\Pi_2}$ are equivalent, by Proposition 4, the substitution $\eta_{\mathcal{E}}$ is the most general unifier of PDTDTs Π_1 and Π_2. □

4   Conclusion

Theorems 2 and 3 show that the algebraic method of checking equivalence developed in [13,14] is applicable to analyze the behaviour of machines working not only with strings, but also with trees. However, extending it further to other classes of tree transducers is not so easy, and there are at least two reasons for this. Firstly, for an arbitrary DTDT π, the system of equations $\mathcal{E}_\pi$ can have several solutions, of which only the least one is semantically meaningful. Secondly, the values $[\![S]\!]_\pi$ of transduction variables C of an arbitrary DTDT π are, in general, partial functions. To deal with an indefinite value, one has to add a constant $\bot$ to the free term algebra, together with identities of the form $f(\ldots, \bot, \ldots) \equiv \bot$. In this algebra the simple decomposition law on which some rules of our algorithms are based, is not valid. These rules can be improved by taking into account the domains of terms, but this will require introducing new variables, equations, and, possibly, new algebraic operations that specify these domains. The difficulties in manipulating with such equations are serious, but we believe that they are quite surmountable.

We considered the unification problem in a rather simplified setting: substitutions allow replacing parameters only with functional symbols. In a more general case, instead of n-place parameters, one can substitute arbitrary terms having n arguments. However, as was shown in [5], the unification problem in such a formulation (second-order unification) is undecidable. Nevertheless, Theorem 3 gives an example of a successful solution to the unification problem for total PDTDTs in a simpler setting of the problem. It shows that, although parameterization on data can lead to a significant increase in the complexity of decision problems (see [1,11]), it does not rule out the possibility of effectively solution to some of them. We think that the study of complexity issues for parameterized top-down and bottom-up tree transducers worth attention and will be the topic of our further research.

When the equivalence checking of the states of automata is concerned, one more issue should be taken into account: generating counterexamples in the case when the states are not equivalent. We believe that our algorithm can be adapted to solve this problem if each equation in the system is supplied with a history—a description of the input term that contributes to the emergence of this equation as a result of applying rules 6) and 7).

Acknowledgments. The authors express their deep gratitude to anonymous reviewers for their valuable comments and interesting ideas suggested by them.

References

1. Barceló, P., Reutter, J., Libkin, L.: Parameterized regular expressions and their languages. Theor. Comput. Sci. **474**, 21–45 (2013)
2. Comon, H.: Inductionless induction. In: Robinson, J.A., Voronkov, A., (eds.) Handbook of Automated Reasoning (in 2 volumes), pp. 913–962. Elsevier and MIT Press, Cambridge (2001)
3. Comon, H., et al.: Tree automata techniques and applications, Christof Löding (2008)
4. Engelfriet, J., Maneth, S., Seidl, H.: Deciding equivalence of top–down xml transformations in polynomial time. J. Comput. Syst. Sci. **75**(5), 271–286 (2009)
5. Goldfarb, W.D.: The undecidability of the second-order unification problem. Theoret. Comput. Sci. **13**(2), 225–230 (1981)
6. Griffiths, T.V.: The unsolvability of the equivalence problem for lambda-free nondeterministic generalized machines. J. ACM **15**(3), 409–413 (1968)
7. Maneth, S.: A survey on decidable equivalence problems for tree transducers. Int. J. Found. Comput. Sci. **26**(8), 1069–1100 (2015)
8. Martelli, A., Montanari, U.: An efficient unification algorithm. ACM Trans. Program. Lang. Syst. **4**(2), 258–282 (1982)
9. Martens, W., Neven, F.: On the complexity of typechecking top-down xml transformations. Theoret. Comput. Sci. **336**(1), 153–180 (2005)
10. Mohri, M.: Minimization algorithms for sequential transducers. Theoret. Comput. Sci. **234**(1), 177–201 (2000)
11. Tang, T., Zakharov, V.A.: On the complexity of decision problems for parameterized finite state synchronous transducers. In: Fazekas, S.Z. (ed.) Implementation and Application of Automata. CIAA 2024. LNCS, vol. 15015, pp. 332–346. Springer, Cham (2024). https://doi.org/10.1007/978-3-031-71112-1_24
12. Valiant, L.G.: The equivalence problem for deterministic finite-turn pushdown automata. Inf. Control **25**(2), 123–133 (1974)
13. Zakharov, V.A.: Equivalence checking of prefix-free transducers and deterministic two-tape automata. In: Martín-Vide, C., Okhotin, A., Shapira, D. (eds.) LATA 2019. LNCS, vol. 11417, pp. 146–158. Springer, Cham (2019). https://doi.org/10.1007/978-3-030-13435-8_11
14. Zakharov, V.A.: Efficient equivalence checking technique for some classes of finite-state machines. Autom. Control Comput. Sci. **55**(7), 670–701 (2021)
15. Ésik, Z.: Decidability results concerning tree transducers. Acta Cybernet. **5**(1), 1–20 (1980)

An Active Learning Algorithm for Bidirectional Deterministic Finite Automata

Simon Dieck$^{(\boxtimes)}$ and Sicco Verwer

Department of Software Technology, TU Delft, Delft, Netherlands
{S.Dieck,S.E.Verwer}@tudelft.nl

Abstract. In this paper, we present an L^*-style algorithm for actively learning a bidirectional deterministic finite automaton (biDFA) in polynomial time using three types of oracles. We show how the W-method for the equivalence oracle can be adapted to our algorithm and present a novel heuristic for choosing the orientation of states. With this algorithm, one can identify automata for a subset of the linear languages that includes but is not limited to the regular languages. Since the equivalence oracle is an important part of the algorithm, we also discuss complexity bounds for different versions of the language equivalence problem for biDFAs. These results, together with our algorithm, also prove complexity bounds for the biDFA minimisation problem. Finally, we provide an implementation of the algorithm and experimentally show its performance with different approximation heuristics.

1 Introduction

A deterministic finite automaton (DFA) is a versatile model that is used across a variety of different fields ranging from applications in biology for parsing DNA [18], in compilers as scanners [12], to cyber security for fingerprinting [9]. While there are several ways of obtaining DFAs, for many fields, it is useful to infer them by interacting with a system whose behaviour the DFA should replicate [7]. This process, normally referred to as active learning, ideally produces a DFA whose behaviour is equivalent to the system from which it was inferred. For example, one can use DFAs inferred from programs that were infected by malicious software to identify security vulnerabilities [2,5,6]. Similarly, one can analyse the behaviour of opaque models, like neural networks, by analysing a DFA surrogate model inferred from them instead [17].

One issue that all of these applications face is that DFAs are limited to recognising the regular languages. And while learning procedures for more expressive models exist, often based on a notion of substitutability, these algorithms are computationally significantly more expensive [4,23].

In this context bidirectional deterministic automata (biDFAs) are interesting. One can think of these models as DFAs with two reading heads. At the beginning, one reading head is placed at the start of the string, and the other one is placed at the end. In each step, the model decides on which reading head will read the next character. This way, a string is parsed from both ends by heads moving inwards. The process terminates if both heads meet. One can think of this meeting point as a "centre" of a string. For the deterministic version of these models, the decision of which reading head to move is also deterministic, and as such, every word has a fixed centre where the heads meet. Nondeterministic biDFAs can recognise all linear languages, while the deterministic version can recognise a subset of the linear languages that strictly includes the regular languages. Importantly, they can do this without requiring additional memory structures like, for example, pushdown automata [13, 15].

While not significantly more expressive than DFAs, the additional languages biDFA can recognise are such that normally require counting, like $a^n b^n$. They can also recognise languages like palindromes, which notably even deterministic pushdown automata cannot. Further, Dieck and Verwer have shown that on palindrome-like regular languages biDFAs can be exponentially smaller than DFAs [8]. Especially in the field of inferring models from software, one can expect such patterns. If software produces some output at the beginning and end of a recursive function, the output will have a palindrome-like structure. Accordingly, an inference algorithm for biDFAs has potential uses in many fields but is of special interest for software inference.

In this paper, we show that the L^* inference algorithm for DFA can be adapted to infer a biDFA. We show that if an oracle for state orientation, that is which reading head is used in a state, is given, this algorithm has the same runtime guarantees as L^* for DFAs. We also prove that the algorithm produces a biDFA that is correct and minimal with regard to a centre function defined by the oracle. Since L^* requires an equivalence oracle, we analyse the complexity of this problem, which is an open question for biDFAs, and prove results for a special case. The algorithm introduced in this work also proves that the biDFA minimisation problem is NP-complete if the biDFA recognises a regular language and in P if the centre function for the minimised model is equivalent to the centre function of the input model.

In practice, especially when inferring software, one has only access to membership queries. Accordingly, model-checking techniques such as the W-method [3] are often used to approximate an equivalence query heuristically [14]. In this work, we also prove that an adaptation of the W-method heuristic [3] works for biDFAs, giving similar guarantees. We also provide a heuristic that can be used for the state orientation oracle to decide which end of the string to read from in a given state.

Finally, we test the algorithm and heuristics experimentally. Here, we show that with the right setup for the heuristics, the algorithm consistently correctly identifies non-regular languages. This includes a palindrome language and even some that were chosen to be difficult for the heuristic. We also show that it often finds the globally minimal model on those languages.

The full version of this paper which includes all proofs is available together with the code used for the experiments at https://github.com/SimonDieck/biDFA_L_star.

1.1   Related Work

Bidirectional automata, then called biautomata, were first introduced by Klíma and Polák [16]. While these models also read a string from both ends moving inwards and are almost equivalent to biDFAs, Klíma and Polák defined the transition function in such a way that these models could only recognise regular languages. This transition function was then changed to allow non-deterministic transitions by Holzer and Jakobi [13]. They analysed the language theoretical properties of these biautomata and showed that they could recognise all linear languages [13].

Jirásková and Klíma then adjusted the transition function of Holzer and Jakobi once again to be deterministic both in allowing only one transition per character and state and enforcing that all transitions from a single state need to read a character from the same end [15]. These models are equivalent to biDFAs since they implicitly introduce an orientation of the state and otherwise are fully deterministic. Jiráskova and Klíma prove most language theoretical properties of these models in their work and show that the class of languages they can recognise is a proper subset of the linear languages and a proper superset of the regular languages [15]. Dieck and Verwer provide a Myhill-Nerode style characterisation of the class of languages recognised by these models [8].

The contributions of these works regarding the language equivalence and minimisation problem, together with our novel contributions are summarised in Table 1.

Table 1. Summary of some problems relating to biDFA. Results in bold are novel results from this paper. NP-hardness of the minimisation problem was proven by Dieck and Verwer [8] and the other known results by Jirásková and Klíma [15].

Given B_1 and B_2, recognising L_1 and L_2, defining centres c_1 and c_2	L_1 or L_2 is regular	$c_1 = c_2$	No restrictions
Emptiness of intersection	NL-complete	**NL-complete**	Undecideable
Language equivalence problem	NL-complete	**In NL**	Unknown
Minimisation problem	**NP-complete**	**In P**	NP-hard

Independently by Nagy very similar bidirectional models have been proposed in the context of DNA processing [18,19]. Whether these models are equivalent has not been proven yet.

For the active learning algorithm, much of our work is based on the L^* algorithm introduced by Angluin [1]. However, for the counterexample processing we use the method first introduced by Rivest and Schapire [20,21].

2   Background

In this section, we will introduce some notation and results from other works that are used in this work. For notation, we will mostly refer to what was used by Dieck and Verwer [8] since we are focused on the deterministic version of these bidirectional models and will heavily use their characterisation. We assume familiarity with standard definitions and notations regarding formal languages and DFA.

For biDFAs we need to introduce the notion of a centre function c.

Definition 1 (Centre function). *A centre function $c : \Sigma^* \to \mathbb{N}$ maps a word $w \in \Sigma^*$ to the range $[0, |w|]$, where $|w|$ denotes the length of a word which is equal to the number of characters it contains.*

Given such a centre function, every word can be split into a left and a right half $w = w_{c,l} \cdot w_{c,r}$ s.t. if $w = \sigma_1 \sigma_2 \ldots \sigma_{|w|}$ then $w_{c,l} = \sigma_1 \ldots \sigma_{c(w)}$ and $w_{c,r} = \sigma_{c(w)+1} \ldots \sigma_{|w|}$.

For the definition of a biDFA we use the one given by Dieck and Verwer [8]:

Definition 2 (biDFA). *A bidirectional deterministic finite automaton (biDFA) is a six-tuple $B = (Q_l, Q_r, \Sigma, \delta, q_\lambda, F)$, where $Q = Q_l \cup Q_r$, with $Q_l \cap Q_r = \emptyset$, forms a finite set of states, Σ is an alphabet, $\delta : Q \times \Sigma \to Q$ a transition function, $q_\lambda \in Q$ the initial state and $F \subseteq Q$ the set of accepting states.*

For this definition to work on words and not just characters, δ is extended to $\delta^* : Q \times \Sigma^* \to Q$ with the following definition:

$$\delta^*(q, (\sigma_1 \ldots \sigma_n)) = \begin{cases} q & \text{if } \sigma_1 \ldots \sigma_n = \lambda \\ \delta(q, (\sigma_1 \ldots \sigma_n)) & \text{if } n = 1 \\ \delta^*(\delta(q, \sigma_1), (\sigma_2 \ldots \sigma_n)) & \text{if } n > 1 \text{ and } q \in Q_l \\ \delta^*(\delta(q, \sigma_n), (\sigma_1 \ldots \sigma_{n-1})) & \text{if } n > 1 \text{ and } q \in Q_r \end{cases}$$

The language L accepted by B can then be defined as $\{w \in \Sigma^* | \delta^*(q_\lambda, w) \in F\}$. One can think of a biDFA as parsing a word by parsing a left string from left to right and a right string from right to left. Whenever B is in a state in Q_l, δ^* will process the next character from the left. When in a state of Q_r, it will process the next character from the right. With this, one can use a biDFA B to define a centre function c_B. To obtain $c_B(w)$ one can count the number of states in Q_l that were encountered when parsing w with B, except for the final state in which the process terminates. This is exactly the number of characters that were read from the left end of the string.

Dieck and Verwer have shown, given a centre function c and a language L one can define an equivalence relation on Σ^* as follows [8]:

$u \equiv_{c,L} v$ iff for all $m \in \Sigma^* : u_{c,l} \cdot m \cdot u_{c,r} \in L \iff v_{c,l} \cdot m \cdot v_{c,r} \in L$

In words, two words are considered inter-equivalent w.r.t. c and L if and only if all words that can be produced from them by inserting a word at their centre behave the same with regards to membership in L. As this interjection comes up often, we introduce a shorthand $u \cdot_c m = u_{c,l} \cdot m \cdot u_{c,r}$, where "$\cdot_c$" is a left-associative operator. We will denote the equivalence classes for this by $[w]_{c,L}$ where $[w]_{c,L} = \{w' \in \Sigma^* | w' \equiv_{c,L} w\}$.

Finally, there is the important restriction for a centre function of "strict stability" which was introduced by Dieck and Verwer [8]:

Definition 3 (Strict stability). *A centre function c is considered stable if for all words $w \in \Sigma^*$ and all characters $s \in \Sigma$ $c(w) \leq c(w \cdot_c s) \leq c(w) + 1$.*

A stable centre function c is considered "strictly stable" for a language L if for all $u, v \in \Sigma^*$ $u \equiv_{c,L} v$ implies that for all $s, s' \in \Sigma : c(u) = c(u \cdot_c s) \iff c(v) = c(v \cdot_c s')$

In words, a strictly stable centre function shifts the centre by either 0 or 1 if a single character is inserted and two inter-equivalent words shift by the same amount. Since every word is inter-equivalent to itself, that also means that for a single word, the centre shifts in the same direction independent of which character is inserted. Also note that every centre function defined by a biDFA is strictly stable.

Given this, Dieck and Verwer prove a Myhill-Nerode style theorem whose corollary is important for this work [8]:

Corollary 1. *For a given strictly stable centre function c and a language L the number $|Q|$ of states of the minimal biDFA defining the same centre function c and recognising L is equal to the number of equivalence classes of the inter-equivalence w.r.t. c and L. There exists an isomorphism between such a minimal biDFA and the "canonical" minimal biDFA w.r.t. c and L.*

This corollary means that we can identify a biDFA that is minimal w.r.t. a given centre function if we can identify the corresponding equivalence classes.

3   The Algorithm

In this section, we will introduce the active learning algorithm for biDFA. The algorithm is in large parts an adaptation of the L^* algorithm introduced by Angluin [1] but uses a counterexample processing based on Rivest and Schapire [20,21]. For the description of the algorithm, we will assume the existence of a teacher that acts as a membership and equivalence oracle, like the original L^* algorithm, in addition to a centre function oracle. For a teacher of a language $L \subseteq \Sigma^*$ the oracles are expected to have the following capabilities:

The membership oracle can answer whether a given word $w \in \Sigma^*$ is in L or not. The equivalence oracle can answer for a given biDFA B whether the language recognised by B is equivalent to L. If it is not equivalent the oracle returns a counterexample. The centre function oracle for a fixed strictly stable centre function c will return for a given word $w \in \Sigma^*$ its centre $c(w)$.

We will discuss how each of the oracles can be realised in Sect. 3.2.

The algorithm at all times maintains two sets $A, E \subset \Sigma^*$, the set A of access words and the set E of extensions. Both are initialised with $A, E = \{\lambda\}$. In addition to these two sets it maintains two tables, D, the distinguishing table, with $|A|$ rows and $|E|$ columns and X, the extended table, with $|A||\Sigma|$ rows and $|E|$ columns. For $a \in A$ and $e \in E$, the distinguishing table has the entry $d_{a,e} = 1$ if $a \cdot_c e \in L$ and 0 otherwise. Similarly for $a \in A$, $s \in \Sigma$ and $e \in E$ the extended table has the entry $x_{a,s,e} = 1$ if $a \cdot_c s \cdot_c e \in L$ and 0 otherwise.

We consider A to be **complete** if for all $a' \in A$ and $s \in \Sigma$ there exists an $a \in A$ s.t. $x_{a',s,e} = d_{a,e}$ for all $e \in E$. If A is not **complete** we will add one $a' \cdot_c s$ to A for which for all $a \in A$ there exists an $e \in E$ s.t. $x_{a',s,e} \neq d_{a,e}$. One can think of A as the set of unique equivalence classes and, therefore, states that have already been identified. We know for all $a, a' \in A$ that $[a]_{c,L} \neq [a']_{c,L}$ since there exists an $e \in E \subset \Sigma^*$ s.t. $a \cdot_c e \in L$ and $a' \cdot_c e \notin L$ or vice versa, since $d_{a,e} \neq d_{a',e}$ for at least one $e \in E$. Since we want to maintain this property we will not add new access words outside of this routine to make A complete. This differs from the original L^* algorithm, which also adds new access words when obtaining a counterexample [1].

An important observation here is that we can make the queries to the membership oracle, that are necessary to fill the tables, even with only partial information on the centre function. If we know $c(a)$, we can form $a \cdot_c e$ for any $e \in E$. The same is true for $a \cdot_c s \cdot_c e$ if we know $c(a)$ and $c(a \cdot_c s)$. Further, since c is strictly stable, it suffices to know $c(a \cdot_c s)$ for one $s \in \Sigma$. This way, we can limit the number of times we need to call the centre function oracle. We can save all the information the algorithm needs to function by having a label $o(a)$ for all $a \in A$. This label will be "l" if we believe the state associated with a is in Q_l, and the centre function will shift forward by 1 upon inserting a single character and "r" otherwise. We can obtain this label by querying the centre function oracle for the centre of $a \cdot_c s$ for any $s \in \Sigma$ when a is added to A. Since $c(\lambda) = 0$ and all other access words are one-letter extensions of other access words, this is sufficient to build o starting from the initialisation.

Given a **complete** A, we can now construct a biDFA by creating a state $q_a \in Q_l$ for all $a \in A$ with $o(a) = $ "l" and a state $q_{a'} \in Q_r$ for all $a' \in A$ with $o(a') = $ "r". Since A is **complete** for every $a' \in A$ and $s \in \Sigma$ there exists a unique $a \in A$ s.t. $d_{a,e} = x_{a',s,e}$ for all $e \in E$. We will call this a $x(a', s)$. With this we can define $\delta(q_a, s) = q_{x(a,s)}$. Further, we will set q_λ to be the state created from $\lambda \in A$ and $F = \{q_a | d_{a,\lambda} = 1\}$. This fully defines a biDFA since we assume Σ is given.

Once a biDFA B is built from a **complete** set A we can use the equivalence oracle. If the oracle certifies that B recognises L, we terminate returning B. Otherwise, we process the counterexample w given by the oracle. Here, we use a method based on the work of Rivest and Schapire [20,21]. Since w is a counterexample, meaning either B accepts it, while it is not in L or vice versa, we can use it to identify a new extension that distinguishes a word $a' \cdot_c s$ from every row in the distinguishing table. There must exist a pre- and inter- and suffix of w, say w_p, w_i and w_s with $w_p \cdot w_i \cdot w_s = w$ and $c(w_p w_s) = |w_p|$ s.t. $\delta^*(q_\lambda, w_p w_s) = q_a$ but $[w_p w_s]_{c,L} \neq [a]_{c,L}$ and w.l.o.g. $w_p w_s \cdot_c w_i \notin L$ and $a \cdot_c w_i \in L$. If this was not the case, B would also reject w. Let $w_p w_s$ be the shortest of such pre- and suffixes. So if w.l.o.g. $w'_p \cdot \sigma \cdot w_s = w_p w_s$ for some $\sigma \in \Sigma$, and $\delta^*(q_\lambda, w'_p w_s) = q_{a'}$, then $x_{a',\sigma,w_i} \neq d_{a,w_i}$. But since the row associated with a is unique and otherwise equal with the row associated with a', $\sigma \ w_i$ distinguishes $a' \cdot \sigma$ from every row in the distinguishing table, which was our claim. Finally, observe that we can identify w_i since for some pre- and suffix $w_p w_s$ with $\delta(q_\lambda, w_p w_s) = q_a$ if $[w_p w_s]_{c,L} = [a]_{c,L}$ then $w_p w_s \cdot_c w_i \in L \iff a \cdot_c w_i \in L$. It, therefore, suffices to process w character by character and substituting the pre- and suffix already processed with the access trace of the state that was reached. A membership query is asked before and after the substitution. The first time the queries differ before and after substitution, the shortest pre- and suffixes are identified. This can also be used to improve the query complexity of this search by using binary search to find the shortest pre- and suffix from which predictions begin to differ, which is the result obtained by Rivest and Schapire [20,21].

3.1   Correctness and Running Time

While many of the key arguments have already been made in the previous section, we formalise the correctness of the presented algorithm with the following theorem:

Theorem 1. *Given a teacher T for a language L and a strictly stable centre function c, the presented algorithm will return a biDFA that is isomorphic to the canonical biDFA given c and L.*

Theorem 1 shows that if a biDFA is returned by the algorithm, then it will be the minimal one w.r.t. c and L. However, it gives no guarantee for the algorithm terminating and how long it will take for the algorithm to terminate. For the L^*-style algorithms, they are often evaluated on how many calls to the membership and equivalence oracle they make [1,7]. We will do the same but also include the number of calls to the centre function oracle.

Theorem 2. *Given a teacher T for a language L and a centre function c the presented algorithm will identify the minimal biDFA $B = (Q_l, Q_r, \Sigma, \delta, q_\lambda, F)$ after asking at most $O(|Q|)$ equivalence queries, $O(|Q|)$ centre function queries and $O(|Q|^2|\Sigma| + |Q| \log_2(|x|))$ membership queries, where $Q = Q_l \cup Q_r$ and x is the longest counterexample.*

Essentially, we retain the same running time guarantees as modern L^*-style algorithms used for learning a DFA while learning a biDFA.

3.2 Heuristics

While the version of the algorithm relying on oracles is interesting from a theoretical perspective for many applications where DFA active learning is normally used, namely learning surrogate models for more complex models, these oracles are normally not available. In this section, we will discuss how each of the oracles can be implemented.

First, note that the membership oracle does not interact with a biDFA and, therefore, is independent of the model we are trying to learn. The implementation of the membership oracle just relies on what model is used as a teacher. When learning from a more complex model, this normally involves running the complex model with the queried input to obtain the output [14].

While in Sect. 4 we will discuss how the language equivalence problem, and therefore oracle, can be solved if the teacher is a biDFA or DFA, this problem cannot be solved exactly for arbitrary models. There exist methods that heuristically predict equivalence while using only membership queries. The two most popular ones are the W-method [3] and random walks [14]. The W-method generates the set of all words of length less than a given parameter m, $W = \{u \in \Sigma^* | |u| \leq m\}$ and then asks membership queries of both the teacher and the hypothesized model for all words $a \cdot u \cdot e$ for all $a \in A$, $u \in W$ and $e \in E$. It compares the output on both queries. When learning DFA, if all query pairs give the same output, it guarantees that the models either recognise the same language or that the DFA would require at least m more states to recognise the same language [3]. We can easily adjust this method to work with biDFAs by querying for all words $a \cdot_c u \cdot_c e$ for all $a \in A, u \in W, e \in E$ instead. One can prove that such an adaptation retains the same guarantees by showing that the shortest distinguishing sequence being longer than m implies that at least m states need to be split into at least two states.

Lastly, we will discuss how the centre function oracle can be implemented. Unless P=NP, a centre function oracle, that runs in polynomial time but results in the globally minimal biDFA will not be possible even if a biDFA is given as a teacher since that would result in a P-time algorithm for the minimisation problem with a regular language, which we show in Sect. 4 to be an NP-complete problem. In the more realistic setting where the teacher is some opaque model, which we can only interact with through membership queries and the centre function is no longer given, it is, therefore, unrealistic to expect a heuristic that leads to a globally good solution. Accordingly, we propose a greedy heuristic that attempts to minimise the immediate number of new states that would appear as a result of fixing the centre function. This heuristic is shown in Algorithm 1.

A simple summary of the heuristic is that it chooses the direction in such a way that the new rows that will be added to the extended table have the maximum overlap with already existing rows in the distinguishing table. If a newly added row in the extended table overlaps with an existing one the distinguishing table one can think of this as an indication that likely no new state will be introduced by this row. The heuristic therefore choses the direction of a state in such a way that locally the number of states is greedily minimised. If no direction is better, it flips a coin. One can think of this as looking one step into the future since only single-character extensions are considered.

---

input : A word $a^* = a \cdot_c s'$, a teacher T, a set of access words A, a set of distinguishing words E, a partial centre function o and a distinguishing table d

output: An orientation $o(a^*)$

1 $l \leftarrow 0$;   // An overlap counter for assuming $o(a^*) = ``\mathrm{l}"$
2 $r \leftarrow 0$;   // An overlap counter for assuming $o(a^*) = ``\mathrm{r}"$
3 **for** *all $s \in \Sigma$* **do**
4   Assume $o(a^*) = "\mathrm{l}"$;
5      $x^l_{a^*,s,e} = $ MembershipQuery$(a^* \cdot_c s \cdot_c e)$;
6   **if** $\exists a \in A : \forall e \in E : d_{a,e} = x^l_{a^*,s,e}$ **then**
7        $\lfloor \; l \leftarrow l+1$;
8   Assume $o(a^*) = "\mathrm{r}"$;
9      $x^r_{a^*,s,e} = $ MembershipQuery$(a^* \cdot_c s \cdot_c e)$;
10   **if** $\exists a \in A : \forall e \in E : d_{a,e} = x^r_{a^*,s,e}$ **then**
11       $\lfloor \; r \leftarrow r+1$;

12 **if** $l > r$ **then**
13     $\lfloor$ **return** $o(a^*) = "\mathit{l}"$
14 **else if** $r > l$ **then**
15     $\lfloor$ **return** $o(a^*) = "\mathit{r}"$
16 **else**
17     $\lfloor$ **return** $o(a^*) = "\mathit{l}"$ *or* $o(a^*) = "\mathit{r}"$ *with probability* 0.5 *each*

Algorithm 1: Heuristic for choosing state orientation

One could also extend the heuristic by running the algorithm for n steps and choosing the direction that minimised the number of states after those steps. But since each step requires choosing a direction again, this would require making $2^n|A||E||\Sigma|$ membership queries. This quickly becomes infeasible but might be worth considering for very small values of n.

As given, there are instances where the heuristic of Algorithm 1 will result in the algorithm taking exponential time in expectation if we assume E always contains a distinguishing sequence if it exists. For example $L = \{a, b\}^n (c\{a, b\}^{10})^n$ over the alphabet $\Sigma = \{a, b, c\}$ is problematic for the heuristic. Since every time the algorithm chooses a state to read from the left without choosing right-oriented states for the next 11 successors, the number of minimal states of a model which respect this centre function will increase by 12.

However, while this example gives a good idea that the heuristic performs worse the longer and more generic the pumpable sequences are, it is not very useful for the actual application of the algorithm. This is due to the assumption that E contains a distinguishing sequence if it exists. Generally, the heuristic approach relies on an accurate prediction of inter-equivalence. This is also the main motivation for using the counterexample processing of Rivest and Schapire [20,21]. If we did not have the invariant that entries in A represent unique states then in a heuristic setting we would need to guarantee that entries in A that cannot be distinguished by E get assigned the same orientation. However, after adding new entries to E one might discover that those states are not equivalent but it would be no longer possible to assign a new orientation to one of the states. By setting up the algorithm as we did we delay orientation assignments as long as possible. This should lead to more accurate assessments of overlap by the heuristic since E will contain more entries. For our experiments, we wanted to further improve this overlap assessment by initialising E to contain several words. We achieved this in two different ways. One was to simply add all words up to a certain length to E. The other method was to run the algorithm for some iterations and then restart it while retaining E between restarts. We call this second method **Reset and Remember**.

3.3   Experiments

Setup. For experiments, we implemented the algorithm in C++ and ran it on several simple languages as a proof of concept. For each language, the teacher was given in the form of a biDFA. However, the algorithm only had access to it in the form of membership queries to simulate the more common use case. For the equivalence query, we implemented the W-method, and for the centre function oracle, we implemented Algorithm 1. Additionally, we used three methods to initialise the algorithm, one **standard** method where $A = E = \{\lambda\}$ and one **μ-warm start** method, where E is initialised as $\{s \in \Sigma^* || s | \leq \mu\}$. Finally, we implemented the **Reset and Remember** method described at the end of Sect. 3.2. For this, we reset the algorithm for the kth time if a model with at least 2^{1+k} states was discovered. The algorithm terminated if it predicted the smallest so far known model twice in a row. For all methods, we also set a limit to forcibly terminate once the distinguishing table A contained more than 100 rows. The algorithm was used 10 times on each language for each experiment, and we recorded the average runtime, the number of times the forced termination occurred, as well as the maximum, minimum, and average number of states of the final model, excluding the ones that were forcibly terminated.

The languages we tested on where $L_1 = a^n b^n$, $L_2 = a^n ccb^n$, $L_3 = a^n (bb)^n$ as some simple linear languages that give an idea of how the algorithm performs on these simple variations of a two-sided pump pattern. We also ran experiments on a regular language $L_4 = (abbb)^n$, which illustrates some of the pitfalls of the algorithm. To showcase the strength of the algorithm, we used the palindrome language over $\Sigma = \{a, b, c\}$, which is $L_5 = ww^{-1}$ a string followed by its reversed string. Finally, we chose two languages designed to challenge our proposed heuristic of Algorithm 1. The chosen languages try to avoid revealing local information immediately. These languages are $L_6 = ((\{a, b\})^2)^n (b\{a, b\}^5)^n$ and $L_7 = (\{a, b\}^2 a\{a, b\}^2)^n (\{a, b\}^2 b\{a, b\}^2)^n$.

Experimental Results. An informative subset of the experimental results can be found in Table 2 while the full experimental results are available in the extended version of the paper[1].

Table 2. Experimental results for different initialisation methods.

		$m = 3, 7, 8$ $\mu = 0$ "RR"=No	$m = 1, 5, 6$ $\mu = 2$ "R and R"=No	$m = 3, 7, 8$ $\mu = 0$ "R and R"=Yes	$m = 1, 5, 6$ $\mu = 2$ "R and R"=Yes
L_1	#MinDisc	6/10	10/10	10/10	10/10
	#ForcedTerm	0/10	0/10	0/10	0/10
	Average #states	5.8	3	3	3
L_2	#MinDisc	5/10	10/10	10/10	10/10
	#ForcedTerm	0/10	0/10	0/10	0/10
	Average #states	7.1	5	5	5
L_3	#MinDisc	6/10	5/10	10/10	8/10
	#ForcedTerm	0/10	0/10	0/10	0/10
	Average #states	7.3	5.1	4	4.5
L_4	#MinDisc	1/10	1/10	1/10	5/10
	#ForcedTerm	0/10	0/10	0/10	0/10
	Average #states	8.8	7.4	9.0	6.3
L_5	#MinDisc	2/10	10/10	10/10	10/10
	#ForcedTerm	8/10	0/10	0/10	0/10
	Average #states	5	5	5	5
L_6	#MinDisc	0/10	1/10	0/10	2/10
	#ForcedTerm	10/10	3/10	8/10	5/10
	Average #states	_*	14.4	18	10.8
L_7	#MinDisc	2/10	4/10	6/10	7/10
	#ForcedTerm	0/10	0/10	0/10	0/10
	Average #states	15.8	12.9	14.2	12.6

A initial observation of the experiments was that one can reduce the value of m by the value of μ since the W method compares outputs on all strings $a \cdot_c u \cdot_c e$ for all $a \in A, u \in W$ and $e \in E$. Normally E initially contains only λ. If strings of length μ are added, the W-method can find μ longer counterexamples during the first equivalence query.

In the experiments, both the μ-**warm start** method and **Reset and Remember** method were successful in making the heuristic notably more performant. Without it, the algorithm essentially always failed on languages L_5 and

---

[1] https://github.com/SimonDieck/biDFA_L_star/blob/main/
CIAA_2025_biDFA_active_learning_camera_ready__full_version_.pdf.

L_6 and struggled with L_2. Combining the two methods showed some improvement in the average number of states. For running time, the μ-**warm start** was fastest, followed by the combined method with **Reset and Remember** being the slowest.

When run with one of the initialisation methods, the algorithm almost always finds a minimal size model for our simple languages L_1 and L_2. At the same time, only **Reset and Remember** was very consistent on L_3. It also often finds these minimal size models for these simple languages when run without adding extra words to E. On the palindrome language L_5 however, it seems the initialisation methods are necessary. Without them, the model struggled to find the minimal size model and even often encountered the forced termination. Nevertheless, if one of the initialisation methods was used, the algorithm consistently found a minimal-size model for the palindrome language. For L_4, the regular language, one can however observe that the algorithm rarely finds the minimal size model with only the combined initialisation performing well. The minimal size model orients all states in the same direction, and as soon as one non-sink state is oriented differently the number of required states is almost doubled. However, the heuristic cannot obtain this information by looking only one step into the future and thus struggles to find a minimal-size model. Surprisingly though the algorithm performed well on L_6 and L_7 which were designed to exploit the same problems of the centre function heuristic. While the algorithm, even with the initialisation methods, still sometimes hit forced termination on those languages it often found a small model that recognised those languages and sometimes even a minimal-size model.

4   Language Theoretical Properties

Language Equivalence. The algorithm that is presented in this work needs access to an equivalence oracle that produces a counterexample if languages are not equivalent. However, the complexity of the problem of deciding whether two arbitrary biDFA recognise the same language is currently not known. We however conjecture that the problem is at least in co-NP. Jirásková and Klíma have shown that the problem is in NL if one of the languages is regular [15]. For the non-regular case we believe that for a biDFA to correctly identify a two-sided pump-pattern like $uv^n wx^n y$, where both v and x are non-empty it will have to place the centre in between the copies of v and x. If this placement can be limited to at most some constant it would force two biDFAs whose languages have a significant overlap to define similar centre functions. If one can prove and combine these results it should be possible to limit the size of the smallest counterexample.

A simpler version of the language equality problem we can consider, even if the defined languages are non-regular, is if two biDFA define the same centre function.

Lemma 1. *Given two biDFAs* $B_1 = (Q_l, Q_r, \Sigma, \delta, q_\lambda, F)$ *and* $B_2 = (Q_l', Q_r', \Sigma, \delta', q_\lambda', F')$ *which recognise* L_1 *and* L_2 *and define the centre functions* c_1 *and* c_2 *respectively. If* $c_1 = c_2$ *determining whether* $L_1 \cap L_2 = \emptyset$ *is NL-complete.*

The core idea of the proof is that once the centre function is fixed one can construct an intersection biDFA which respects the same centre function.

Corollary 2. *The language equality problem for two biDFAs B_1 and B_2 defining L_1 and L_2 respectively, that both define the same centre function c is in NL.*

To prove the corollary one can use the standard construction of $L_1 \cap \overline{L_2} = \emptyset$ and $\overline{L_1} \cap L_2 = \emptyset$ implies $L_1 = L_2$.

Minimisation. The algorithm shown in the previous section also answers some questions about the complexity of the biDFA minimisation problem. One can define this problem as given a biDFA B which recognises L and a constant k to decide whether there exists a biDFA B^* that also recognises L with $|Q^*| \leq k$.

Theorem 3 (Minimisation complexity). *Given a membership and equivalence oracle for L there exists a nondeterministic polynomial time algorithm that solves the biDFA minimisation problem which requires a polynomial number of calls to the membership oracle and a linear number of calls to the equivalence oracle.*

Proof. Note that there are only two possible outcomes whenever the algorithm calls the centre function oracle. A nondeterministic algorithm can guess the outcome and otherwise run the algorithm as described. The algorithm will return "yes" if the final biDFA has at most k states and "no" otherwise. Theorems 1 and 2 guarantee that the returned algorithm recognises L and that this procedure had a nondeterministic polynomial running time.

Since the input biDFA to the minimisation problem can be given as the teacher, and therefore the membership query can be solved in linear time, the complexity of the minimisation problem depends on the complexity of the language equivalence problem. This leads us to the following corollary:

Corollary 3. *Given a biDFA B recognising the language L and defining the centre function c and a constant k the minimisation problem is:*

1. *In P when restricted to finding a biDFA that also defines c.*
2. *NP-complete if L is regular.*

Proof. 1. follows from Theorem 3 and Corollary 2 as well as the observation that a biDFA can be used as an oracle for its own centre function.
2. follows from Theorem 3 and Dieck and Verwer's proof that the problem is NP-hard [8] as well as Jiráskova and Klíma's proof that the language equivalence problem is NL-complete if L is regular [15].

5   Discussion

In this work, we have introduced an active learning algorithm for biDFAs and some proofs about the complexity of the language equivalence problem for biD-FAs. Together this also answered some open questions about the complexity of the biDFA minimisation problem. When combined with the heuristics we provided for the oracle required by the algorithm, we believe this work to be an important step to move bidirectional automata from a theoretical novelty towards something that can be used in practice. Indeed for DFA applications where the memory footprint is of significant importance, one could use the NP algorithm for minimisation introduced in this work. For the more common reverse engineering applications in security where DFA are currently learned one could experiment with the algorithm introduced in this work. For this, we expect that there is also significant room for improving the heuristics we provided. Especially the centre function heuristic we proposed, while working well on toy examples, provides no theoretical guarantees. Since beyond their superior expressiveness biDFA are also interesting because they can be potentially exponentially smaller than DFA when recognising regular languages [8], it would be desirable to find a centre function heuristic that works well on regular languages.

Another issue that keeps these results from being applicable to many real-life scenarios is that often not DFA are learned but Mealy machines [7,14]. Nevertheless, with the theoretical foundations provided in this work, we expect the same algorithm can used to learn bidirectional Mealy machines, similarly to how L^* is often used to learn them [14].

If one could show that the language equivalence problem is indeed in co-NP it would likely also imply that biDFAs are learnable in the limit [10]. If that is the case it would also be interesting if passive learning algorithms [22] or exact algorithms [11], that learn from a fixed sample could be adapted to learn biDFAs instead.

From a theoretical perspective, there are also some open questions left. Most importantly, whether the language equality problem is indeed as we conjectured in co-NP for the general case. Together with Theorem 3 this would imply that minimisation is also in co-NP.

For two biDFA that define the same centre function we have proven membership in NL for the equivalence problem and membership in P for the minimisation problem but completeness remains an open question.

In conclusion, we believe that bidirectional automata can be useful in real-life applications and to that end, we have in this work provided a foundation upon which such methods can be built and with which initial experiments can be conducted.

References

1. Angluin, D.: Learning regular sets from queries and counterexamples. Inf. Comput. **75**(2), 87–106 (1987)

2. Cho, C.Y., Babić, D., Poosankam, P., Chen, K.Z., Wu, E.X.J., Song, D.: Mace: model-inference-assisted concolic exploration for protocol and vulnerability discovery. In: 20th USENIX Security Symposium (USENIX Security 2011) (2011)
3. Chow, T.S.: Testing software design modeled by finite-state machines. IEEE Trans. Softw. Eng. **4**(3), 178–187 (1978)
4. Clark, A., Eyraud, R.: Polynomial identification in the limit of substitutable context-free languages. J. Mach. Learn. Res. **8**(8), 1–21 (2007)
5. Cook, J.E., Wolf, A.L.: Discovering models of software processes from event-based data. ACM Trans. Softw. Eng. Methodol. (TOSEM) **7**(3), 215–249 (1998)
6. Cui, W., Kannan, J., Wang, H.J.: Discoverer: automatic protocol reverse engineering from network traces. In: USENIX Security Symposium, pp. 1–14 (2007)
7. De la Higuera, C.: Grammatical Inference: Learning Automata and Grammars. Cambridge University Press, Cambridge (2010)
8. Simon Dieck and Sicco Verwer. On bidirectional deterministic finite automata. In: Fazekas, S.Z. (ed) Implementation and Application of Automata. CIAA 2024. LNCS, vol. 15015, pp. 109–123. Springer, Cham (2024). https://doi.org/10.1007/978-3-031-71112-1_8
9. Ficara, D., Giordano, S., Procissi, G., Vitucci, F., Antichi, G., Di Pietro, A.: An improved DFA for fast regular expression matching. ACM SIGCOMM Comput. Commun. Rev. **38**(5), 29–40 (2008)
10. Mark Gold, E.: Language identification in the limit. Inf. Control **10**(5), 447–474 (1967)
11. Heule, M.J.H., Verwer, S.: Exact DFA identification using SAT solvers. In: Sempere, J.M., García, P. (eds.) ICGI 2010. LNCS (LNAI), vol. 6339, pp. 66–79. Springer, Heidelberg (2010). https://doi.org/10.1007/978-3-642-15488-1_7
12. Hoe, A.V., Sethi, R., Ullman, J.D.: Compilers—Principles, Techniques, and Tools. Pearson Addison Wesley Longman (1986)
13. Holzer, M., Jakobi, S.: Minimization and characterizations for biautomata. Fund. Inform. **136**(1–2), 113–137 (2015)
14. Isberner, M., Howar, F., Steffen, B.: The open-source LearnLib. In: Kroening, D., Păsăreanu, C.S. (eds.) CAV 2015. LNCS, vol. 9206, pp. 487–495. Springer, Cham (2015). https://doi.org/10.1007/978-3-319-21690-4_32
15. Jirásková, G., Klíma, O.: On linear languages recognized by deterministic biautomata. Inf. Comput. **286**, 104778 (2022)
16. Klíma, O., Polák, L.: On biautomata. RAIRO-Theoret. Inform. App. **46**(4), 573–592 (2012)
17. Muškardin, E., Aichernig, B.K., Pill, I., Tappler, M.: Learning finite state models from recurrent neural networks. In: ter Beek, M.H., Monahan, R. (eds.) integrated formal methods. IFM 2022. LNCS, vol. 13274, pp. 229–248. Springer, Cham (2022). https://doi.org/10.1007/978-3-031-07727-2_13
18. Nagy, B.: On a hierarchy of $5' \to 3'$ sensing Watson-crick finite automata languages. J. Log. Comput. **23**(4), 855–872 (2013)
19. Nagy, B., Parchami, S.: On deterministic sensing $5' \to 3'$ Watson-crick finite automata: a full hierarchy in 2detlin. Acta Informatica **58**, 153–175 (2021)
20. Rivest, R.L., Schapire, R.E.: Inference of finite automata using homing sequences. Inf. Comput. **103**(2), 299–347 (1993). https://www.sciencedirect.com/science/article/pii/S0890540183710217, https://doi.org/10.1006/inco.1993.1021
21. Rivest, R.L., Schapire, R.E.: Inference of finite automata using homing sequences. In: Proceedings of the Twenty-first Annual ACM Symposium on Theory of Computing, pp. 411–420 (1989)

22. Verwer, S., Hammerschmidt, C.A.: Flexfringe: a passive automaton learning package. In: 2017 IEEE International Conference on Software Maintenance and Evolution (ICSME), pp. 638–642. IEEE (2017)
23. Yoshinaka, R.: Efficient learning of multiple context-free languages with multi-dimensional substitutability from positive data. Theoret. Comput. Sci. **412**(19), 1821–1831 (2011)

Dynamically Weighted Tree Transducers

Frank Drewes[1] , Marco Kuhlmann[2] , and Olle Torstensson[2]([⊠])

[1] Umeå University, Umeå, Sweden
drewes@cs.umu.se
[2] Linköping University, Linköping, Sweden
{marco.kuhlmann,olle.torstensson}@liu.se

Abstract. We introduce *dynamically weighted tree transducers (dyn-wtts)*, a weighted generalization of top-down tree transducers with regular look-ahead in which rule weights are determined by external *tree weighters* mapping input trees to values in a commutative semiring. The general framework allows for any kind of device defining a weighted tree language to serve as a tree weighter. In this paper, we focus on weighters implemented by different classes of tree automata and show how the resulting classes of weighted tree transformations relate to one another and to known classes. In particular, we show how conventional top-down weighted tree transducers (with and without regular look-ahead) can be expressed as dyn-wtts, also in the linear and non-deleting cases.

Keywords: Weighted tree transducers · Weighted tree automata · Regular look-ahead

1 Introduction

Tree transducers [4] are a fundamental tool in theoretical computer science and its applications, providing a formal mechanism for transforming structured data between different representations. In weighted extensions of these models, the transformation process is enriched with quantitative information, often representing probabilities, costs, or preferences. Weighted tree transducers (wtts, [6,9,10]), in particular, integrate the structural aspects of tree rewriting with a weight algebra such as a semiring. Traditionally, this integration is tightly coupled: the total weight assigned to a derivation is computed compositionally, in lockstep with the application of transduction rules, using static rule weights defined in the syntactic specification of the transducer.

While this tight coupling enables elegant algebraic formulations and tractable algorithmic properties, it also imposes limitations. In many practical applications of tree automata, especially those involving machine learning or probabilistic modeling, it is useful to compute weights in a more flexible and context-sensitive manner. For example, in natural language processing tasks like machine translation or semantic parsing, the desirability of an output structure may depend not only on local rule applications but also on global properties of the input [16]. Static rule weights cannot capture such dependencies.

G. Castiglione and S. Mantaci (Eds.): CIAA 2025, LNCS 15981, pp. 115–128, 2026.
https://doi.org/10.1007/978-3-032-02602-6_9

In this paper, we propose a novel framework of *dynamically weighted tree transducers (dyn-wtts)* that partly decouples the structural and quantitative components of tree transduction by replacing static rule weights with *weighters*—computational devices that assign weights to rule applications dynamically, based on the context of the input node. Dynamically weighted tree transducers of the type studied here are related to (unweighted) top-down tree transducers with regular look-ahead [5]. These were invented to overcome a weakness of standard top-down tree transducers: they cannot check properties of input subtrees which are potentially deleted. Look-ahead solves this problem by letting the automata inspect the subtrees before a rule is applied. Intuitively, our weighters play a similar role as look-ahead, but they compute the weight of a rule application rather than acting as a filter that checks for rule applicability, thereby compensating for missing weights due to deletion. In particular, a top-down dyn-wtt over the Boolean semiring whose weighters are ordinary weighted tree automata is a top-down tree transducer with regular look-ahead.

To illustrate our proposed framework, we may consider a simple dyn-wtt over the semiring of natural numbers (with the usual multiplication and addition) and the ranked alphabet consisting of a binary symbol σ and a nullary symbol α. Assuming access to a weighter W mapping each tree t to its height[1] $\mathrm{ht}(t)$, we define the dyn-wtt via the rules

$$q[\sigma[x_1, x_2]] \xrightarrow[W]{} q[x_1] \qquad \text{and} \qquad q[\alpha] \xrightarrow[W]{} \alpha. \tag{1}$$

At each rule application, the weighter calculates the height of the input subtree and uses that as the weight of the rule application. The resulting weighted tree transformation maps each pair (t, α) to $\prod_{t' \in L_t} \mathrm{ht}(t')$, where L_t is the set of subtrees in t that are rooted along the left-most branch. This transformation is not definable by any ordinary top-down wtt over the above semiring for essentially three reasons: (i) ordinary wtts over this semiring are not flexible enough to capture and utilize the height function in their weight computation; (ii) the deletion of each right subtree, which is necessary to implement the structural aspect of the transformation, denies an ordinary top-down wtt access to use it in the weight computation; and (iii) the total weight computed grows too fast, relative to the size n of input trees, for a top-down wtt over the considered semiring to produce it. (Its growth exceeds any exponential function k^n, where k is some constant, being bounded only by $n!$.)

Without any restriction on the type of weighters used, dyn-wtts can implement arbitrarily complex functions; hence, to obtain meaningful results, we need to consider reasonably restricted weighters. In this paper, we take a first step by looking at weighters which are tree automata of various descriptions. Our aim is to relate the resulting classes of weighted tree transductions to classes known from the literature, thus showing how the proposed model fits into the theoretical landscape and creating a baseline for further studies. We compare top-down dyn-wtts over arbitrary commutative semirings with standard top-down

---

[1] Here we consider trees consisting of a single node to have height 1.

weighted tree transducers in terms of their expressive power, considering also the linear and non-deleting cases. We organize the results according to the type of weighters used: constant weighters (Sect. 3.1), tree automata over the Boolean semiring (Sect. 3.2), tree automata over arbitrary semirings (Sect. 3.3), and a generalization of these that we call *weighted copying tree automata* (Sect. 3.4). Our main result (Propositions 6 and 7) is that top-down dyn-wtts over the latter class are equivalent to Maletti's generalized notion of *top-down weighted tree transducers with regular look-ahead* [14,15]. Interestingly, the two frameworks differ when we restrict our attention to the linear case (Proposition 5).

2   Preliminaries

We write $\mathbb{N}$ for the set of non-negative natural numbers. For $k \in \mathbb{N}$, we let $[k]$ denote the set $\{1, \dots, k\}$.

Trees. A *ranked set* is a set Σ together with a mapping $\mathrm{rank}_\Sigma \colon \Sigma \to \mathbb{N}$. For $k \geq 0$, we write $\Sigma^{(k)}$ to denote the subset $\{\sigma \in \Sigma \mid \mathrm{rank}_\Sigma(\sigma) = k\}$ and use the notation $\sigma^{(k)}$ to clarify that $\sigma \in \Sigma^{(k)}$. A *ranked alphabet* is a finite ranked set. The *set of trees over* Σ, denoted by T_Σ, is the smallest set $T \subseteq \Sigma^*$ such that if $k \geq 0$, $\sigma \in \Sigma^{(k)}$, and $t_1, \dots, t_k \in T$, then $t = \sigma t_1 \cdots t_k \in T$. To enhance readability, we write trees using the familiar bracket notation, i.e. $t = \sigma[t_1, \dots, t_k]$. For an arbitrary set A, the *set of trees over* Σ *indexed by* A is the set $T_\Sigma(A) \subseteq T_{\Sigma \cup A}$ where elements $a \in A$ are treated as having rank zero. Moreover, for any set A, we define $\Sigma(A) = \{\sigma[t_1, \dots, t_k] \mid k \geq 0, \sigma \in \Sigma^{(k)}, t_1, \dots t_k \in A\} \subseteq T_\Sigma(A)$. Throughout, we fix the *set of variables* $X = \{x_1, x_2, \dots\}$, and for $k \geq 0$, we let $X_k = \{x_1, \dots, x_k\}$. For a tree $t \in T_\Sigma(X)$, $\ell \geq 0$, and trees $t_1, \dots, t_\ell \in T_\Delta$ over some ranked set Δ, we define the *substitution* $t[\![t_1, \dots, t_\ell]\!]$ as the tree obtained from t by replacing each occurrence of x_i by t_i, for all $i \in [\ell]$. A tree t is *linear* (respectively, *non-deleting*) with respect to X_ℓ, $\ell \geq 0$, if every x_i occurs in t at most once (respectively, at least once). For any $\ell \geq 0$, we denote by $\widehat{T_\Sigma}(X_\ell)$ the set of all trees t in $T_\Sigma(X_\ell)$ which are linear and non-deleting with respect to X_ℓ and where the variables in t occur in the order $x_1, \dots, x_\ell$.

Semirings. A *semiring* [2] is a 5-tuple $\mathbb{K} = (K, +, \cdot, 0, 1)$ consisting of a carrier set K, two binary operations $+$ and $\cdot$, called *addition* and *multiplication*, and two distinct elements $0, 1 \in K$, such that: $(K, +, 0)$ is a commutative monoid, i.e., $+$ is associative and commutative and 0 is its identity element; $(K, \cdot, 1)$ is a monoid, i.e., $\cdot$ is associative and and 1 is its identity element; 0 is an annihilator for $\cdot$, i.e., $0 \cdot a = a \cdot 0 = 0$ for all $a \in K$; and $\cdot$ distributes over $+$, i.e., $a \cdot (b + c) = a \cdot b + a \cdot c$ and $(a + b) \cdot c = a \cdot c + b \cdot c$, for all $a, b, c \in K$. We overload the symbol $\mathbb{K}$ to also denote the carrier set. A semiring is *commutative* if $\cdot$ is commutative, and *locally finite* if, for any finite $A \subseteq K$, the set generated by A via iterated application of $+$ and $\cdot$ is also finite. An important semiring with both these properties is the *Boolean semiring* $\mathbb{B} = (\{0, 1\}, \vee, \wedge, 0, 1)$. The *natural semiring* $(\mathbb{N}, +, \cdot, 0, 1)$ is commutative but not locally finite. *For the rest of the paper, we always let* $\mathbb{K}$ *denote a commutative semiring.* We will use the familiar symbols $\sum$ and $\prod$ for the extension of addition and multiplication to finitely indexed sets, and also allow infinitely indexed sets if the number of non-neutral elements is still finite.

Weighted Tree Languages and Automata. Given a ranked alphabet Σ and a semiring $\mathbb{K}$, a *weighted tree language over Σ and $\mathbb{K}$* is any function $\mathcal{L}\colon \mathrm{T}_\Sigma \to \mathbb{K}$ that assigns semiring values to trees. The set $\mathrm{supp}(\mathcal{L}) = \{t \in \mathrm{T}_\Sigma \mid \mathcal{L}(t) \neq 0\}$ of trees with non-zero weight is called the *support* of $\mathcal{L}$. In the special case where $\mathbb{K} = \mathbb{B}$, we identify $\mathcal{L}$ with its support and view it as an unweighted *tree language* $\mathcal{L} \subseteq \mathrm{T}_\Sigma$. If there is an $a \in \mathbb{K}$ such that $\mathcal{L}(t) = a$ for all $t \in \mathrm{T}_\Sigma$, we denote $\mathcal{L}$ by $\tilde{a}$.

A *weighted tree automaton (wta)* [10] is a tuple $\mathcal{A} = (\Sigma, Q, Q_d, R, \mathbb{K}, c)$ where: Σ is a ranked alphabet of *input symbols*; $Q = Q^{(1)}$ is a ranked alphabet of *states* such that $Q \cap \Sigma = \emptyset$; $Q_d \subseteq Q$ is a set of *designated states*; R is a ranked alphabet of *transition rules* such that, for all $k \geq 0$, $R^{(k)} \subseteq Q \times \Sigma^{(k)} \times Q^k$; $\mathbb{K}$ is a semiring; and $c\colon R \to \mathbb{K}$ is a total function mapping rules to weights. A weighted tree automaton $\mathcal{A}$ defines a weighted tree language $\mathcal{L}_\mathcal{A}\colon \mathrm{T}_\Sigma \to \mathbb{K}$ as follows. First, for every $q \in Q$, $k \geq 0$ and $s = \sigma[s_1, \ldots, s_k] \in \mathrm{T}_\Sigma$, we let

$$\mathcal{L}_{\mathcal{A};q}(s) = \sum_{\rho=(q,\sigma,q_1\cdots q_k)\in R} c(\rho) \cdot \prod_{i\in[k]} \mathcal{L}_{\mathcal{A};q_i}(s_i) \, . \tag{2}$$

Second, for every $s \in \mathrm{T}_\Sigma$, we let $\mathcal{L}_\mathcal{A}(s) = \sum_{q\in Q_d} \mathcal{L}_{\mathcal{A};q}(s)$. A weighted tree language defined by a wta is called *recognizable*, and the class of all such languages for a given semiring $\mathbb{K}$ is denoted by $\mathrm{Rec}(\mathbb{K})$ or, for a specific Σ, by $\mathrm{Rec}(\Sigma, \mathbb{K})$.

Weighted Tree Transformations and Transducers. Given ranked alphabets Σ, Δ and a semiring $\mathbb{K}$, a $\mathbb{K}$-*weighted tree transformation* (or simply *weighted tree transformation* if $\mathbb{K}$ is clear from the context) is a function $\tau\colon \mathrm{T}_\Sigma \times \mathrm{T}_\Delta \to \mathbb{K}$.

A *top-down tree transducer (td-tt)* [4] is a tuple $\mathcal{M} = (\Sigma, \Delta, Q, Q_d, R)$, where: Σ and Δ are ranked alphabets of *input symbols* and *output symbols*, respectively; $Q = Q^{(1)}$ is a ranked alphabet of *states* such that $Q \cap (\Sigma \cup \Delta) = \emptyset$; $Q_d \subseteq Q$ is a set of *initial states*; and R is a ranked alphabet of *transformation rules* such that, for all $k \geq 0$, $R^{(k)} \subseteq Q(\Sigma^{(k)}) \times \mathrm{T}_\Delta(Q(\mathrm{X}_k))$. We write rules $\rho \in R^{(k)}$ as

$$q[\sigma[x_1, \ldots, x_k]] \to t'[\![q_1[x_{i_1}], \ldots, q_\ell[x_{i_\ell}]]\!] \, , \tag{3}$$

where $\ell \geq 0$, $t' \in \widehat{\mathrm{T}}_\Delta(\mathrm{X}_\ell)$, and $i_j \in [k]$ for every $j \in [\ell]$. The td-tt is called *linear* (respectively, *non-deleting*) if in each rule $\rho \in R^{(k)}$, the tree on the right-hand side of ρ is linear (non-deleting) with respect to X_k. A *top-down weighted tree transducer (td-wtt)* [10,12] is a tuple $\mathcal{M} = (\Sigma, \Delta, Q, Q_d, R, \mathbb{K}, c)$, where $(\Sigma, \Delta, Q, Q_d, R)$ is a td-tt, $\mathbb{K}$ is a semiring, and $c\colon R \to \mathbb{K}$ is a function assigning weights to rules. A td-wtt inherits linearity and the non-deleting property from its underlying td-tt, and we sometimes express a rule $\rho = l \to r$ and its weight $a = c(\rho)$ simultaneously by writing $l \xrightarrow{a} r$. A td-wtt defines a weighted tree transformation $\tau_\mathcal{M}\colon \mathrm{T}_\Sigma \times \mathrm{T}_\Delta \to \mathbb{K}$ as follows. First, for every $q \in Q$, $k \geq 0$, $s = \sigma[s_1, \ldots, s_k] \in \mathrm{T}_\Sigma$, and $t = t'[\![t_1, \ldots, t_\ell]\!] \in \mathrm{T}_\Delta$, we let

$$\tau_{\mathcal{M};q}(s, t) = \sum_{\rho=(q[\sigma[x_1,\ldots,x_k]]\to t'[\![q_1[x_{i_1}],\ldots,q_\ell[x_{i_\ell}]]\!])\in R} c(\rho) \cdot \prod_{j\in[\ell]} \tau_{\mathcal{M};q_j}(s_{i_j}, t_j) \, . \tag{4}$$

Then, for every $s \in T_\Sigma$ and $t \in T_\Delta$, we define $\tau_\mathcal{M}(s, t) = \sum_{q \in Q_d} \tau_{\mathcal{M};q}(s, t)$. Note that $\tau_{\mathcal{M};q}$ is the weighted tree transformation defined by the transducer $(\Sigma, \Delta, Q, \{q\}, R, \mathbb{K}, c)$, and that by letting $\mathbb{K} = \mathbb{B}$, we get the semantics of td-tts.

We let $\mathrm{TOP}(\mathbb{K})$ denote the class of weighted tree transformations defined by td-wtts employing the semiring $\mathbb{K}$. By prepending the class name by l and/or n we restrict the class to transformations defined by *linear* and/or *non-deleting* transducers. For example, $\mathrm{ln}\text{-}\mathrm{TOP}(\mathbb{K})$ is the class of weighted tree transformations defined by linear and non-deleting td-wtts over $\mathbb{K}$. In formal statements, we put "l" or "n" in square brackets (such as in [l]n-TOP($\mathbb{K}$)) if the statement holds both with and without the corresponding restriction for all its instances.

3   Top-Down Dynamically Weighted Tree Transducers

We now introduce our proposed model of top-down dynamically weighted tree transducers. These are directly inspired by (unweighted) top-down tree transducers with regular look-ahead [5]. In that framework, each transducer rule is associated with a set of tree automata that determine whether the rule can be applied at a given node of the input tree, based on the immediate subtrees at that node. In our model, we extend this idea by using weighted look-ahead automata, which allows them to be used as more than just filters. We also generalize the mechanism by permitting computational devices beyond traditional tree automata.

More formally, a *(tree) weighter* is any computational device (such as a wta) that specifies a weighted tree language $W : T_\Sigma \to \mathbb{K}$. We overload notation and write W both for the language and a weighter that specifies it, and we write $\mathcal{W}_{\Sigma \to \mathbb{K}}$ to denote a class of weighters of the specified type.

Definition 1 (top-down dynamically weighted tree transducer). *Given a class $\mathcal{W}_{\Sigma \to \mathbb{K}}$ of weighters, a top-down dynamically weighted tree transducer (td-dyn-wtt) with weighters in $\mathcal{W}$ is a tuple $\mathcal{M} = (\Sigma, \Delta, Q, Q_d, R, \mathbb{K}, w)$, where $(\Sigma, \Delta, Q, Q_d, R)$ is a td-tt, $\mathbb{K}$ is a semiring, and $w : R \to \mathcal{W}$ is a function assigning a weighter from $\mathcal{W}$ to every rule ρ in R.*

A td-dyn-wtt $\mathcal{M}$ defines a weighted tree transformation $\tau_\mathcal{M} : T_\Sigma \times T_\Delta \to \mathbb{K}$ given by $\tau_\mathcal{M}(s, t) = \sum_{q \in Q_d} \tau_{\mathcal{M};q}(s, t)$ for all $s \in T_\Sigma$ and $t \in T_\Delta$. Here, for all $q \in Q$, $k \geq 0$, $s = \sigma[s_1, \ldots, s_k] \in T_\Sigma$, and $t = t'[\![t_1, \ldots, t_\ell]\!] \in T_\Delta$,

$$\tau_{\mathcal{M};q}(s, t) = \sum_{\rho = (q[\sigma[x_1, \ldots, x_k]] \to t'[\![q_1[x_{i_1}], \ldots, q_\ell[x_{i_\ell}]]\!]) \in R} w(\rho)(s) \cdot \prod_{j \in [\ell]} \tau_{\mathcal{M};q_j}(s_{i_j}, t_j). \quad (5)$$

A class $\mathcal{W}_{\Sigma \to \mathbb{K}}$ of weighters is called *compositional*, if, for every weighter $W \in \mathcal{W}$, $k \geq 0$, and $\sigma \in \Sigma^{(k)}$, there exist a constant $w_0 \in \mathbb{K}$ and weighters $W_1, \ldots, W_k \in \mathcal{W}$ such that $W(\sigma[s_1, \ldots, s_k]) = w_0 \cdot \prod_{i \in [k]} W_i(s_i)$, for all trees $s_1, \ldots, s_k \in T_\Sigma$. In the remainder of this paper, we will only consider compositional classes of weighters. This means that in a td-dyn-wtt, the weighter $w(\rho)$ of every rule $\rho \in R^{(k)}$ can be assumed to be such that

$$w(\rho)(\sigma[s_1, \ldots, s_k]) = w(\rho)_0 \cdot \prod_{i \in [k]} w(\rho)_i(s_i), \quad (6)$$

for appropriate $w(\rho)_0 \in \mathbb{K}$ and weighters $w(\rho)_1, \ldots, w(\rho)_k \in \mathcal{W}$. If convenient, we specify such a rule ρ together with its weighter as $l \xrightarrow{W} r$ or $l \xrightarrow[W_1,\ldots,W_k]{w_0} r$, where $w_0 = w(\rho)_0$, $W_i = w(\rho)_i$ for all $i \in [k]$, and W is the weighter composed of w_0 and $W_1, \ldots, W_k$.

Consider a td-dyn-tt $\mathcal{M}$. It is clear from Definition 1 that $\tau_{\mathcal{M}}$ does not change if the weighters $w(\rho)$ of the rules of $\mathcal{M}$ are replaced with weighters $w'(\rho)$ such that $w'(\rho)(s) = w(\rho)(s)$ for all $s \in T_\Sigma$ and rules ρ of $\mathcal{M}$. In other words, $\tau_{\mathcal{M}}$ depends only on the weighted tree languages specified by the weighters but not on the particular weighters themselves. Therefore, the following notation makes sense: For a class $\mathcal{W}$ of weighted tree languages, we let $\mathrm{TOP}_{\mathcal{W}}(\mathbb{K})$ denote the class of all $\mathbb{K}$-weighted tree transformations defined by top-down dynamically weighted tree transducers with weighters specifying languages in $\mathcal{W}$.[2] As for td-wtts, linearity and non-deletion are properties a td-dyn-wtt inherits from its underlying tree transducer, and we likewise prepend "l", "n", and "ln" to the class name to denote the linear, non-deleting, and linear non-deleting subclasses, respectively.

3.1 Constant Weighters

Clearly, the class of td-wtts can be identified with the class of td-dyn-wtts whose weighters assign a constant weight to every tree; i.e., each rule $l \xrightarrow{a} r$ of a td-wtt is identified with the rule $l \xrightarrow{\mathcal{A}} r$, where $\mathcal{L}_{\mathcal{A}} = \tilde{a}$. This yields the following result, where $\mathrm{const}(\mathbb{K})$ denotes the class of weighters defining constant functions into $\mathbb{K}$.

Proposition 1. $[l][n]\text{-}\mathrm{TOP}_{\mathrm{const}}(\mathbb{K}) = [l][n]\text{-}\mathrm{TOP}(\mathbb{K})$.

3.2 Boolean Recognizable Weighters

Next, we investigate wta over the Boolean semiring $\mathbb{B}$ as weighters, thereby augmenting our transducers with exactly the same mechanism that adds a regular look-ahead to unweighted top-down tree transducers [5]. To make this compatible with our definition (in which weighters define functions into an arbitrary semiring $\mathbb{K}$), we define the class $\mathrm{RecB}(\mathbb{K})$ that, for all $\mathcal{L} \in \mathrm{Rec}(\mathbb{B})$ and $a \in \mathbb{K}$, contains the weighted tree language that maps a tree t to a if $t \in \mathcal{L}$ and to 0 otherwise. We use the notation $\{t \in \mathcal{L}\}_a$ to denote such a language.

The original idea behind regular look-ahead was to extend top-down tree transducers by the capability to inspect subtrees before potentially deleting them. We illustrate the additional power of our top-down dyn-wtts via an example.

Example 1. The well-known example showing the limitations of top-down tree transducers without regular look-ahead [1] can be transferred to our setting as follows. Let $\Omega = \{\alpha^{(0)}, \beta^{(2)}\}$ and $\mathbb{K}$ be arbitrary. Then let $\tau \colon T_\Omega \times T_\Omega \to \mathbb{K}$ be

---

[2] For the classes of weighted tree languages we cover, the $\mathbb{K}$ specified in the name used to denote the class is dropped in this notation—e.g., $\mathrm{Rec}(\mathbb{K})$ becomes Rec.

the weighted tree transformation that assigns 0 to all pairs of trees in $T_\Omega \times T_\Omega$ except for $(\beta[\alpha, \alpha], \alpha)$, which is given some weight $a \in \mathbb{K}\backslash\{0\}$. Then $\tau \notin \mathrm{TOP}(\mathbb{K})$. However, τ is induced by the top-down dyn-wtt with the single (initial) state q and the single rule $q[\beta[x_1, x_2]] \xrightarrow[\{\beta[\alpha,\alpha]\}_a]{} \alpha$. Thus $\tau \in \mathrm{TOP}_{\mathrm{RecB}}(\mathbb{K})$.

As expected, we get the following inclusion (cf. [5], Corollary 2.3).

Proposition 2. $[l]\text{-}\mathrm{TOP}(\mathbb{K}) \subsetneq [l]\text{-}\mathrm{TOP}_{\mathrm{RecB}}(\mathbb{K})$.

Proof. The inclusion follows from Proposition 1 and the additional fact that $\mathrm{const}(\mathbb{K}) \subseteq \mathrm{RecB}(\mathbb{K})$; strictness is demonstrated in Example 1. □

Note 1. Just as in the unweighted case, the essential use of our weighters is on the parts of the input tree that get deleted by a rule. In fact, when using weighters from $\mathrm{RecB}(\mathbb{K})$, for any subtree s that does not get deleted by a rule, we can have the part of the automaton responsible for checking s integrated into the overall transduction. Thus, for a td-dyn-wtt $\mathcal{M}$ with $\tau_\mathcal{M} \in \mathrm{TOP}_{\mathrm{RecB}}(\mathbb{K})$, we can assume that, for each of its rules $q[\sigma[x_1, \ldots, x_k]] \xrightarrow[\mathcal{A}_1, \ldots, \mathcal{A}_k]{a} t'[\![q_1[x_{i_1}], \ldots, q_\ell[x_{i_\ell}]]\!]$, $\mathcal{L}_{\mathcal{A}_i} = \tilde{1}$ for each $i \in \{i_1, \ldots, i_\ell\}$. In particular, $[l]\mathrm{n}\text{-}\mathrm{TOP}_{\mathrm{RecB}}(\mathbb{K}) = [l]\mathrm{n}\text{-}\mathrm{TOP}(\mathbb{K})$.

3.3 Recognizable Weighters

In the previous section we saw how applying essentially unweighted tree automata as weighters in a top-down dyn-wtt increases expressivity compared to standard top-down wtts. This added power is in its essence *structural*. Next, when we consider more general wta as weighters, we will see that these add power also in terms of achievable weights of transductions, provided that our semiring is sufficiently rich. We start with the case in which this last condition is not satisfied.

Proposition 3. *For every locally finite semiring* $\mathbb{K}$*, we have* $[l][n]\text{-}\mathrm{TOP}_{\mathrm{Rec}}(\mathbb{K}) = [l][n]\text{-}\mathrm{TOP}_{\mathrm{RecB}}(\mathbb{K})$.

Proof. We only need to prove $[l][n]\text{-}\mathrm{TOP}_{\mathrm{Rec}}(\mathbb{K}) \subseteq [l][n]\text{-}\mathrm{TOP}_{\mathrm{RecB}}(\mathbb{K})$. Let $\mathbb{K}$ be locally finite. Given a $\tau_\mathcal{M} \in \mathrm{TOP}_{\mathrm{Rec}}(\mathbb{K})$ defined by the td-dyn-wtt $\mathcal{M}$, we construct a td-dyn-wtt $\mathcal{M}'$ with $\tau_{\mathcal{M}'} \in \mathrm{TOP}_{\mathrm{RecB}}(\mathbb{K})$ and $\tau_\mathcal{M} = \tau_{\mathcal{M}'}$. We do this by replacing each rule $l \xrightarrow{\mathcal{A}} r$ of $\mathcal{M}$ with the set of all rules $l \xrightarrow[\mathcal{A}'_a]{} r$ such that $a \in \mathrm{Im}(\tau_\mathcal{M})$, where $\mathcal{L}_{\mathcal{A}'_a}(s) = a$ if $s \in \mathcal{L}_\mathcal{A}^{-1}(a)$ and $\mathcal{L}_{\mathcal{A}'_a}(s) = 0$ otherwise. Since $\mathbb{K}$ is locally finite, $\mathrm{Im}(\tau_\mathcal{M})$ is finite, and each $\mathcal{L}_\mathcal{A}^{-1}(a)$ is recognizable ([3], Lemma 6.1). □

In contrast to Proposition 3, the next example illustrates the kind of weighting power added when using a less restrictive semiring.

Example 2. Let $\Sigma_{\mathrm{m}} = \{\alpha^{(0)}, \gamma^{(1)}\}$. Consider a wta $\mathcal{A}$ such that $\mathcal{L}_{\mathcal{A}}$ is the weighted tree language in $\mathrm{Rec}(\Sigma_{\mathrm{m}}, \mathbb{N})$ that, for each $n \geq 0$, maps $\gamma^n \alpha$ to 2^n. (It requires one state such that the (unique) rule processing γ has the weight 2.) Now, letting $\mathcal{M}_{\mathrm{id}} = (\Sigma_{\mathrm{m}}, \Sigma_{\mathrm{m}}, \{*\}, \{*\}, R, \mathbb{N}, w)$ consist of the rules $*[\gamma[x_1]] \xrightarrow[\mathcal{A}]{} \gamma[*[x_1]]$ and $*[\alpha] \xrightarrow[\mathcal{A}]{} \alpha$, we get that $\tau_{\mathcal{M}_{\mathrm{id}}}(\gamma^n \alpha, \gamma^n \alpha) = 2^{\frac{n(n+1)}{2}}$ for all $n \in \mathbb{N}$. In contrast, for any top-down dyn-wtt $\mathcal{M}$ with weighters from $\mathrm{RecB}(\mathbb{N})$, it is well known that $\tau_{\mathcal{M}_{\mathrm{id}}}(\gamma^n \alpha, \gamma^n \alpha)$ is bounded from above by b^n for some constant $b \in \mathbb{N}$. Hence, $\tau_{\mathcal{M}_{\mathrm{id}}}$ is not in $\mathrm{TOP}_{\mathrm{RecB}}(\mathbb{N})$.

Together with the obvious inclusion $[\mathrm{l}][\mathrm{n}]\text{-}\mathrm{TOP}_{\mathrm{RecB}}(\mathbb{K}) \subseteq [\mathrm{l}][\mathrm{n}]\text{-}\mathrm{TOP}_{\mathrm{Rec}}(\mathbb{K})$, the previous example yields the next result.

Proposition 4. *The inclusion* $[\mathrm{l}][\mathrm{n}]\text{-}\mathrm{TOP}_{\mathrm{RecB}}(\mathbb{K}) \subseteq [\mathrm{l}][\mathrm{n}]\text{-}\mathrm{TOP}_{\mathrm{Rec}}(\mathbb{K})$ *holds for all semirings* $\mathbb{K}$, *and there are semirings* $\mathbb{K}$ *for which the inclusion is strict. In particular,* $[\mathrm{l}][\mathrm{n}]\text{-}\mathrm{TOP}_{\mathrm{RecB}}(\mathbb{N}) \subsetneq [\mathrm{l}][\mathrm{n}]\text{-}\mathrm{TOP}_{\mathrm{Rec}}(\mathbb{N})$.

The increased expressiveness of $\mathrm{TOP}_{\mathrm{Rec}}(\mathbb{K})$ for $\mathbb{K}$ not restricted to being locally finite comes at a price. For example, given a weighted tree transformation $\tau \colon \mathrm{T}_\Sigma \times \mathrm{T}_\Delta \to \mathbb{K}$, and a weighted tree language $\mathcal{L} \colon \mathrm{T}_\Delta \to \mathbb{K}$, we define (a) the *domain of* τ, $\mathrm{Dom}(\tau) \colon \mathrm{T}_\Sigma \to \mathbb{K}$, as $\mathrm{Dom}(\tau)(s) = \sum_{t \in \mathrm{T}_\Delta} \tau(s, t)$ for each $s \in \mathrm{T}_\Sigma$; and (b) the *pre-image of* $\mathcal{L}$ *under* τ, $\tau^{-1}(\mathcal{L}) \colon \mathrm{T}_\Sigma \to \mathbb{K}$, as $\tau^{-1}(\mathcal{L})(s) = \sum_{t \in \mathrm{T}_\Delta} \mathcal{L}(t) \cdot \tau(s, t)$ for each $s \in \mathrm{T}_\Sigma$. Note that for $\tau^{-1}(\mathcal{L})$ to be well defined, $\mathbb{K}$ needs to be *(countably) complete*[3]. Whereas both $\mathrm{Dom}(\tau)$ and $\tau^{-1}(\mathcal{L})$ for $\tau \in \mathrm{l\text{-}TOP}_{\mathrm{RecB}}(\mathbb{K})$ and $\mathcal{L} \in \mathrm{Rec}(\mathbb{K})$ can be shown to be recognizable (see Proposition 5 together with [8], Thms. 5.1 and 5.2), $\mathrm{Dom}(\mathcal{M}_{\mathrm{id}})$ is not recognizable for the td-dyn-wtt $\mathcal{M}_{\mathrm{id}}$ from Example 2, and the same holds for $\mathcal{M}_{\mathrm{id}}^{-1}(\widetilde{1})$, despite the fact that $\mathcal{M}_{\mathrm{id}}$ is even linear and non-deleting. It may therefore be an interesting aim for future research to find conditions under which $\mathrm{Dom}(\tau)$ and $\tau^{-1}(\mathcal{L})$ are recognizable in the case of not locally finite semirings.

3.4 Weighters Beyond Recognizability

In this section we show that, by defining a suitable class of weighters, we obtain td-dyn-wtts equivalent to the top-down weighted tree transducers with regular look-ahead first outlined by Engelfriet et al. [6] and later formally studied by Maletti [14]. The idea behind the concept is to let the right-hand side of top-down wtt rules be of the form $t'[\![q_1[x_{i_1}], \ldots, q_{\ell'}[x_{i_{\ell'}}]]\!]$ where, in contrast to the traditional definition, t' does not need to be non-deleting in $X_{\ell'}$.[4] As a consequence, some of the ℓ' output subtrees may be discarded, but their computation still contributes to the weight of the output tree. By sorting the list $q_1[x_{i_1}], \ldots, q_{\ell'}[x_{i_{\ell'}}]$ and renaming variables in t', it can be assumed that there is some $\ell \leq \ell'$ such

---

[3] Roughly, infinite sums are defined and they are associative, commutative, and satisfy the distributivity laws. For the precise definition see, e.g., [2].

[4] It does not have to be linear either, but linearity can be assumed without loss of generality.

that $x_1, \ldots, x_\ell$ are exactly the variables occurring in t'. Hence, the first ℓ output subtrees computed contribute to both the structure and the weight of the produced output tree (as usual) whereas the remaining ones only add weight.

This formalism is the canonical generalization of top-down tree transducers with regular look-ahead [5] in the sense that it allows the lifting of results relating classes of transductions to the weighted case. A definition geared towards formal tree series can be found in [15]. Here we give a slightly modified, but equivalent, definition that is closer in style to that of td-dyn-wtts.

A *top-down weighted tree transducer with regular look-ahead* (td-wtt^r)[5] is a tuple $\mathcal{M} = (\Sigma, \Delta, Q, Q_d, R, \mathbb{K}, c)$, where Σ, Δ, Q, Q_d, $\mathbb{K}$, and c are as in td-wtts and R is a ranked alphabet of transformation rules such that, for all $k \geq 0$, $R^{(k)} \subseteq Q(\Sigma^{(k)}) \times (\mathrm{T}_\Delta(Q(\mathrm{X}_k)) \times Q(\mathrm{X}_k)^*)$.

The, in comparison to ordinary top-down rules, added component $Q(\mathrm{X}_k)^*$ is the second part of the partition described earlier, and can be viewed as the look-ahead of the rule. We write rules $\rho \in R^{(k)}$ as

$$q[\sigma[x_1, \ldots, x_k]] \xrightarrow{a} \left(t'[\![q_1[x_{i_1}], \ldots, q_\ell[x_{i_\ell}]]\!], \langle q_{\ell+1}[x_{i_{\ell+1}}], \ldots, q_{\ell+m}[x_{i_{\ell+m}}]\rangle\right) , \quad (7)$$

where $\ell, m \geq 0$, $t' \in \widehat{\mathrm{T}_\Delta}(\mathrm{X}_\ell)$, $i_j \in [k]$ for every $j \in [\ell + m]$, and $a = c(\rho)$.

We define the semantics of a td-wtt^r $\mathcal{M} = (\Sigma, \Delta, Q, Q_d, R, \mathbb{K}, c)$ as a function $\tau_\mathcal{M} \colon \mathrm{T}_\Sigma \times \mathrm{T}_\Delta \to \mathbb{K}$ as follows. First, for every $q \in Q$, $k \geq 0$, $s = \sigma[s_1, \ldots, s_k] \in \mathrm{T}_\Sigma$, and $t = t'[\![t_1, \ldots, t_\ell]\!] \in \mathrm{T}_\Delta$, we let

$$\tau_{\mathcal{M};q}(s, t) = \sum_\rho c(\rho) \cdot \prod_{j \in [\ell]} \tau_{\mathcal{M};q_j}(s_{i_j}, t_j) \cdot \prod_{j \in [m]} \left(\sum_{s' \in \mathrm{T}_\Delta} \tau_{\mathcal{M};q_{\ell+j}}(s_{i_{\ell+j}}, s')\right) , \quad (8)$$

where the outer sum is taken over all $\rho \in R$ of the form

$$\rho = (q[\sigma[x_1, \ldots, x_k]] \to (t'[\![q_1[x_{i_1}], \ldots, q_\ell[x_{i_\ell}]]\!], \langle q_{\ell+1}[x_{i_{\ell+1}}], \ldots, q_{\ell+m}[x_{i_{\ell+m}}]\rangle)). \quad (9)$$

Then, for every $s \in \mathrm{T}_\Sigma$ and $t \in \mathrm{T}_\Delta$, we define $\tau_\mathcal{M}(s, t) = \sum_{q \in Q_d} \tau_{\mathcal{M};q}(s, t)$.

A td-wtt^r is *linear* if, in all its rules $\rho = (l, (r, v)) \in R^{(k)}$, each $x_i \in \mathrm{X}_k$ occurs at most once in r and v together; it is *non-deleting* if, in all its rules $\rho = (l, (r, v)) \in R^{(k)}$, r is non-deleting with respect to X_k and $v = \langle\rangle$. By [l][n]-TOP$^R(\mathbb{K})$ we denote the class of all weighted tree transformations defined by a (linear and/or non-deleting) td-wtt^r over the semiring $\mathbb{K}$.

Note that any non-deleting td-wtt^r is a non-deleting top-down wtt by definition, so [l]n-TOP$^R(\mathbb{K})$ = [l]n-TOP$(\mathbb{K})$. Considering Note 1, it is easy to see that l[n]-TOP$_{\mathrm{RecB}}(\mathbb{K}) \subseteq$ l[n]-TOP$^R(\mathbb{K})$. When instead considering weighters from Rec$(\mathbb{K})$, the opposite relation l[n]-TOP$^R(\mathbb{K}) \subseteq$ l[n]-TOP$_{\mathrm{Rec}}(\mathbb{K})$ clearly holds. We collect the relation between the linear and non-deleting restrictions of the three classes in the following result.

---

[5] Also known as a *(polynomial) tree series transducer of type II* and a *(polynomial) top-down tree series transducer with regular look-ahead* [14, 15].

Proposition 5. *For all semirings $\mathbb{K}$ it holds that*

$$\text{l[n]-TOP}_{\text{RecB}}(\mathbb{K}) \subseteq \text{l[n]-TOP}^{\text{R}}(\mathbb{K}) \subseteq \text{l[n]-TOP}_{\text{Rec}}(\mathbb{K}). \tag{10}$$

These inclusions are equalities if $\mathbb{K}$ is locally finite. Furthermore,

$$\text{[l]n-TOP}_{\text{RecB}}(\mathbb{K}) = \text{[l]n-TOP}^{\text{R}}(\mathbb{K}) \subseteq \text{[l]n-TOP}_{\text{Rec}}(\mathbb{K}) \tag{11}$$

and

$$\text{l-TOP}_{\text{RecB}}(\mathbb{K}) \subseteq \text{l-TOP}^{\text{R}}(\mathbb{K}) \subseteq \text{l-TOP}_{\text{Rec}}(\mathbb{K}), \tag{12}$$

and there exist semirings $\mathbb{K}$ for which the inclusions in (11) and (12) are strict.

Proof. The inclusions follow straightforwardly from the definitions; see also the preceding remarks. If $\mathbb{K}$ is locally finite, the inclusions in (10) are equations using a similar reasoning as in the proof of Proposition 3. The equality in (11) follows from both sides being equal to [l]n-TOP($\mathbb{K}$). Strictness of the inclusions in (11) and (12) can be shown for $\mathbb{K} = \mathbb{N}$ following the idea of Example 2. Indeed, strictness of the inclusion in (11) and the second inclusion in (12) are demonstrated by $\tau_{\mathcal{M}_{\text{id}}}$; strictness of the first inclusion of (12) is demonstrated by the following transformation. Let $\Omega = \{\alpha^{(0)}, \gamma^{(1)}, \beta^{(2)}\}$, and let $\tau \colon \text{T}_\Omega \times \text{T}_\Omega \to \mathbb{K}$ such that $\tau(\beta[\gamma^m\alpha, \gamma^n\alpha], \alpha) = 2^{m+n}$ for all $m, n \geq 0$ and $\tau(s, t) = 0$ for any other $s, t \in \text{T}_\Omega$. Then it can easily be checked that $\tau \in \text{l-TOP}^{\text{R}}(\mathbb{K})$ but $\tau \notin \text{l-TOP}_{\text{RecB}}(\mathbb{K})$. $\qquad\square$

Moving on from the linear and non-deleting restrictions of td-wtt^{r}s, we can see that their ability to recursively copy subtrees in the look-ahead of rules, is not something that can be captured by any of the weighter classes we have explored so far. To illustrate this, we look at an analogue of Example 2. Recall $\Sigma_{\text{m}} = \{\alpha^{(0)}, \gamma^{(1)}\}$. We define a td-wtt^{r} $\mathcal{M}$ with input and output alphabet Σ_{m} over $\mathbb{N}$ via the rules $q[\gamma[x_1]] \xrightarrow{2} (\gamma[q[x_1]], \langle q[x_1] \rangle)$ and $q[\alpha] \xrightarrow{1} (\alpha, \langle \rangle)$. Then $\tau_{\mathcal{M}}(\gamma^n\alpha, \gamma^n\alpha) = 2^{2^n - 1}$ for any $n \in \mathbb{N}$—a weight that, for large enough n, is unattainable by any top-down dyn-wtt with weighters from Rec($\mathbb{N}$), since its computed weight will be bounded above by b^{n^2} for some constant $b \in \mathbb{N}$. To express unrestricted td-wtt^{r}s as dyn-wtts, we therefore introduce a—to our knowledge—new type of weighted tree automaton model.

Definition 2. *A weighted copying tree automaton (wcta) is a 6-tuple $\mathcal{A} = (\Sigma, Q, Q_d, R, \mathbb{K}, c)$, where Σ, Q, Q_d, $\mathbb{K}$, and c is as for a wta and R is a ranked alphabet of transition rules such that, for all $k \geq 0$, $R^{(k)} \subseteq Q \times \Sigma^{(k)} \times \mathbb{N}^{[k] \times Q}$.*

The semantics of $\mathcal{A}$ is a weighted tree language $\mathcal{L}_{\mathcal{A}} \colon \text{T}_\Sigma \to \mathbb{K}$ defined for all $s \in \text{T}_\Sigma$ by $\mathcal{L}_{\mathcal{A}}(s) = \sum_{q \in Q_d} \mathcal{L}_{\mathcal{A};q}$, where, for every $q \in Q$, $k \geq 0$, and $s = \sigma[s_1, \ldots, s_k]$,

$$\mathcal{L}_{\mathcal{A};q}(s) = \sum_{\rho = (q, \sigma, \mu) \in R} c(\rho) \cdot \prod_{(i,q') \in [k] \times Q} (\mathcal{L}_{\mathcal{A};q'}(s_i))^{\mu_{i,q'}}. \tag{13}$$

The matrix μ of a rule (q, σ, μ) encodes the exponents in the monomial that defines the cost of applying the rule, i.e., how many times a certain subtree is sent to a certain state. Thus, if we for each rule (q, σ, μ) in a wcta $\mathcal{A}$ have that each row in μ sums to 1, $\mathcal{A}$ is equivalent to a conventional wta. We denote by $\mathrm{RecC}(\mathbb{K})$ the class of all weighted tree languages defined by a wcta over the semiring $\mathbb{K}$.

Lemma 1. *For any Σ, $\mathrm{RecC}(\Sigma, \mathbb{K})$ is closed under sum and Hadamard product, i.e., if $\mathcal{L}_1, \mathcal{L}_2 \in \mathrm{RecC}(\Sigma, \mathbb{K})$, then*

1. *$\mathcal{L}_1 \oplus \mathcal{L}_2 \in \mathrm{RecC}(\Sigma, \mathbb{K})$, where, for all $t \in T_\Sigma$, $(\mathcal{L}_1 \oplus \mathcal{L}_2)(t) = \mathcal{L}_1(t) + \mathcal{L}_2(t)$*
2. *$\mathcal{L}_1 \otimes \mathcal{L}_2 \in \mathrm{RecC}(\Sigma, \mathbb{K})$, where, for all $t \in T_\Sigma$, $(\mathcal{L}_1 \otimes \mathcal{L}_2)(t) = \mathcal{L}_1(t) \cdot \mathcal{L}_2(t)$.*

Proof. Given wcta $\mathcal{A}_1 = (\Sigma, Q_1, D_1, R_1, \mathbb{K}, c_1)$ and $\mathcal{A}_2 = (\Sigma, Q_2, D_2, R_2, \mathbb{K}, c_2)$, where Q_1 and Q_2 are assumed to be distinct, we can define a new wcta $\mathcal{A} = (\Sigma, Q, Q_d, R, \mathbb{K}, c)$ such that

1. $\mathcal{L}_\mathcal{A} = \mathcal{L}_{\mathcal{A}_1} \oplus \mathcal{L}_{\mathcal{A}_2}$, simply by letting $Q = Q_1 \cup Q_2$, $Q_d = D_1 \cup D_2$, $R = R_1 \cup R_2$ and by letting $c(\rho) = c_j(\rho)$ for $\rho \in R_j$, $j \in \{1, 2\}$.
2. $\mathcal{L}_\mathcal{A} = \mathcal{L}_{\mathcal{A}_1} \otimes \mathcal{L}_{\mathcal{A}_2}$ as follows. First, let $Q = Q_1 \cup Q_2 \cup (D_1 \times D_2)$ and $Q_d = D_1 \times D_2$. Then,
   (a) For any $k \geq 0$, $\sigma \in \Sigma^{(k)}$, $d_1 \in D_1$, $d_2 \in D_2$, and any two rules $\rho_1 = (d_1, \sigma, \mu_1) \in R_1$ and $\rho_2 = (d_2, \sigma, \mu_2) \in R_2$, we let R contain the rule $\rho = (\langle d_1, d_2 \rangle, \sigma, \mu)$, where, for any $i \in [k]$ and $q \in Q$, $\mu_{i,q} = \mu_{j_{i,q}}$ if $q \in Q_j$, for $j \in \{1, 2\}$ and $\mu_{i,q} = 0$ otherwise, and set $c(\rho) = c_1(\rho_1) \cdot c_2(\rho_2)$.
   (b) For any rule $\rho = (q, \sigma, \mu) \in R_j$, $j \in \{1, 2\}$, with $\sigma \in \Sigma^{(k)}$, we let R contain the rule $\rho' = (q, \sigma, \mu')$, where, for any $i \in [k]$ and $q' \in Q$, $\mu'_{i,q'} = \mu_{i,q'}$ if $q' \in Q_j$ and $\mu'_{i,q'} = 0$ otherwise, and set $c(\rho') = c_j(\rho)$. $\square$

The rest of this section is devoted to showing $\mathrm{TOP}^\mathrm{R}(\mathbb{K}) = \mathrm{TOP}_{\mathrm{RecC}}(\mathbb{K})$. The proof centers around the fact that wcta and td-wtt's are equally powerful in terms of weighting capabilities; the following lemma makes this relation explicit.

Lemma 2. *For any td-wtt^r $\mathcal{M} = (\Sigma, \Delta, Q, Q_d, R, \mathbb{K}, c)$, there is a wcta $\mathcal{A} = (\Sigma, Q, Q_d, R', \mathbb{K}, c')$ such that for all $q \in Q$ and $s \in T_\Sigma$, $\mathcal{L}_{\mathcal{A};q}(s) = \sum_{t \in T_\Delta} \tau_{\mathcal{M};q}(s, t)$.*

Proof. Let $\mathcal{M} = (\Sigma, \Delta, Q, Q_d, R, \mathbb{K}, c)$ be a td-wtt^r. We define the wcta $\mathcal{A} = (\Sigma, Q, Q_d, R', \mathbb{K}, c')$ as follows. For any $q \in Q$, $k \geq 0$, and $\sigma \in \Sigma^{(k)}$, let $R_{q;\sigma}$ denote the set of all rules in R having $q[\sigma[x_1, \ldots, x_k]]$ as a left hand side and, for $k \geq 1$, let $f_{q,\sigma} \colon R_{q;\sigma} \to \mathbb{N}^{[k] \times Q}$ be the function that maps each rule $\rho = (l \to (r, v))$ to the matrix μ, where each entry $\mu_{i,q'}$ is equal to the number of occurrences of $q'[x_i]$ in r and v together. For $\sigma \in \Sigma^{(0)}$, any rule in $R_{q;\sigma}$ is mapped by $f_{q;\sigma}$ to the empty matrix $[]$. Now, for any $\sigma \in \Sigma$, $q \in Q$, and $\mu \in \mathrm{Im}(f_{q,\sigma})$, we let the rule $\rho = (q, \sigma, \mu)$ be in R', and set $c'(\rho) = \sum_{\rho' \in f_{q,\sigma}^{-1}(\mu)} c(\rho')$.

Due to lack of space, we omit the straightforward remainder of the proof, which shows that $\mathcal{L}_{\mathcal{A};q}(s) = \sum_{t \in T_\Delta} \tau_{\mathcal{M};q}(s, t)$ for all $s \in T_\Sigma$ and $q \in Q$, by structural induction on s. $\square$

With Lemmas 1 and 2 in place, the proof of the first inclusion is straightforward.

Proposition 6. $\mathrm{TOP}^{\mathrm{R}}(\mathbb{K}) \subseteq \mathrm{TOP}_{\mathrm{RecC}}(\mathbb{K})$.

Proof. Let $\mathcal{M} = (\Sigma, \Delta, Q, Q_d, R, \mathbb{K}, c)$ be a td-wtt^{r} and $\mathcal{A}$ the wcta that exists by Lemma 2. We define a dyn-wtt $\mathcal{M}' = (\Sigma, \Delta, Q, Q_d, R', \mathbb{K}, w)$ as follows. For any $l \in Q(\Sigma)$ and $r \in \mathrm{T}_\Delta(Q(X))$, let $R_{(l,r)}$ be the set of all rules in R of the form $l \rightarrow (r, v)$. Then, for each $k \geq 0$ and $R_{(l,r)} \subseteq R^{(k)}$, let R' contain the rule $l \xrightarrow[\mathcal{A}_1, \ldots, \mathcal{A}_k]{a} r$, where $a = \sum \rho \in R_{(l,r)} c(\rho)$ if $k = 0$ and $a = 1$ otherwise, and for each $i \in [k]$, $\mathcal{A}_i$ is the automaton for which $\mathcal{L}_{\mathcal{A}_i} = \bigoplus_{\rho = (l \rightarrow (r,v)) \in R_{(l,r)}} \widetilde{c(\rho)} \bigotimes_{q'[x_i] \in v} \mathcal{L}_{\mathcal{A};q'}$, the existence of which is ensured by Lemma 1. It should now be clear that $\tau_{\mathcal{M}'} = \tau_{\mathcal{M}}$. $\qquad \square$

We now turn to the opposite inclusion.

Proposition 7. $\mathrm{TOP}_{\mathrm{RecC}}(\mathbb{K}) \subseteq \mathrm{TOP}^{\mathrm{R}}(\mathbb{K})$.

Proof. Let $\mathcal{M} = (\Sigma, \Delta, Q, Q_d, \mathbb{K}, R, w)$ and let $\{\mathcal{A}_j = (\Sigma, Q_j, D_j, R_j, \mathbb{K}, c_j)\}_{j \in [n]}$ be the family of wcta appearing as weighters in $\mathcal{M}$. Without loss of generality, we assume that $\bigcap_{j \in [n]} Q_j \cap Q = \emptyset$. We will now define a td-wtt^{r} $\mathcal{M}' = (\Sigma, \Delta, Q', Q_d', \mathbb{K}, R', c')$ such that $\tau_{\mathcal{M}'} = \tau_{\mathcal{M}}$. The idea is to emulate the rules of every $\mathcal{A}_j$ via the look-ahead of the rules in R'. To do this, we first let $Q' = \bigcup_{j \in [n]} Q_j \cup Q$ and $Q_d' = Q_d$. Then we build R' and define c' as follows.

1. For each $j \in [n]$, $k \geq 0$, and rule $\rho = (q, \sigma, \mu) \in R_j^{(k)}$, we add to R' the rule $q[\sigma[x_1, \ldots, x_k]] \xrightarrow{c_j(\rho)} (\delta, v)$, where $\delta \in \Delta^{(0)}$ is arbitrary but fixed, and v is made up of exactly $\mu_{i,q}$ copies of $q[x_i]$ for each $i \in [k]$ and $q \in Q_j$, in some arbitrary order.
2. For each $k \geq 0$ and rule $\rho = (l \xrightarrow[\mathcal{A}_{j_1}, \ldots, \mathcal{A}_{j_k}]{a} r)$ in $R^{(k)}$, we extend R' with all rules in the set $F_\rho = \{l \xrightarrow{a} (r, \langle p_{j_1}[x_1], \ldots, p_{j_k}[x_k] \rangle) \mid (\forall i \in [k])(p_{j_i} \in D_{j_i})\}$.

The rules constructed in item 1 encode the weighters of $\mathcal{M}$ in terms of (isolated) sets of transducer rules, to which input trees are referred by the look-ahead of the rules constructed in item 2, which are also in charge of performing the same "top-level" transduction as the rules in R.

We omit the straightforward inductive proof of the fact that $\tau_{\mathcal{M}'} = \tau_{\mathcal{M}}$. $\qquad \square$

Corollary 1. $\mathrm{TOP}^{\mathrm{R}}(\mathbb{K}) = \mathrm{TOP}_{\mathrm{RecC}}(\mathbb{K})$.

4   Conclusion

We have introduced the framework of dynamically weighted tree transducers, where the task of determining the weights of rule applications is outsourced to external devices (called *weighters*) that operate on the input subtrees. We have shown how conventional models of top-down weighted tree transducers, both with and without look-ahead, fit into this framework, and seen that certain classes of weighted tree transformations that are cumbersome to specify in a classical setting have natural definitions in our framework.

In this paper, we have restricted dyn-wtts to the top-down case; nevertheless, it is quite possible to also define bottom-up dyn-wtts. In such dyn-wtts, instead of pre-inspecting pieces of the input tree to determine a rule weight, weighters would "re-inspect" the already constructed pieces of the output tree. As such, their contribution is of more limited use; for example, $BOT_{RecB}(\mathbb{K}) = BOT(\mathbb{K})$. However, for sufficiently powerful weighter classes, transformations will move beyond $BOT(\mathbb{K})$, making the bottom-up case more interesting.

Whereas we in this paper have only explored the effects of having wta-like devices as weighters, our framework in principle allows *any* device defining a function from trees to values to act as a weighter. A natural step forward would be to investigate the result of letting the weighters employed be other automata-theoretic formalisms such as (weighted) pushdown tree automata [13] and weighted tree automata with storage [7,11]. For practical applications, we are especially interested in the potential of a neuro-symbolic integration, by letting neural networks take on the responsibility of weighting. For example, in natural language processing, neural weighters such as models based on the Transformer architecture [17] can use the global context of the input to help resolve the ambiguity of human language. The flexibility of integrating such powerful weighters underscores the broader potential of our framework as a foundation for practical, learning-driven applications of tree transduction.

Acknowledgments. This work was partially supported by the Wallenberg AI, Autonomous Systems and Software Program (WASP) funded by the Knut and Alice Wallenberg Foundation, and by the Swedish Research Council under grant no. 2024-05318.

Disclosure of Interests. The authors have no competing interests to declare that are relevant to the content of this article.

References

1. Dauchet, M.: Transductions inversibles de forêts. Ph.D. thesis, Univ. de Lille, France (1975)
2. Droste, M., Kuich, W.: Semirings and formal power series. In: Droste, M., Kuich, W., Vogler, H. (eds.) Handbook of Weighted Automata, pp. 3–28. Springer, Heidelberg (2009). https://doi.org/10.1007/978-3-642-01492-5_1
3. Droste, M., Vogler, H.: Weighted tree automata and weighted logics. Theor. Comput. Sci. **366**(3), 228–247 (2006). https://doi.org/10.1016/j.tcs.2006.08.025
4. Engelfriet, J.: Bottom-up and top-down tree transformations - a comparison. Math. Syst. Theory **9**(3), 198–231 (1975). https://doi.org/10.1007/BF01704020
5. Engelfriet, J.: Top-down tree transducers with regular look-ahead. Math. Syst. Theory **10**, 289–303 (1977)
6. Engelfriet, J., Fülöp, Z., Vogler, H.: Bottom-up and top-down tree series transformations. J. Automata Lang. Comb. **7**(1), 11–70 (2002). https://doi.org/10.25596/jalc-2002-011

7. Fülöp, Z., Herrmann, L., Vogler, H.: Weighted regular tree grammars with storage. Disc. Math. Theor. Comput. Sci. **20**(1) (2018). https://doi.org/10.23638/DMTCS-20-1-26

8. Fülöp, Z., Maletti, A., Vogler, H.: Weighted extended tree transducers. Fund. Inf. **111**, 163–202 (2011). https://doi.org/10.3233/FI-2011-559

9. Fülöp, Z., Vogler, H.: Weighted tree transducers. J. Autom. Lang. Comb. **9**(1), 31–54 (2004). https://doi.org/10.25596/JALC-2004-031

10. Fülöp, Z., Vogler, H.: Weighted tree automata and tree transducers. In: Droste, M., Kuich, W., Vogler, H. (eds.) Handbook of Weighted Automata, pp. 313–403. Springer, Heidelberg (2009). https://doi.org/10.1007/978-3-642-01492-5_9

11. Herrmann, L.: Weighted Automata with Storage. Ph.D. thesis, Technische Universität Dresden (2020)

12. Kuich, W.: Tree transducers and formal tree series. Acta Cybern. **14**(1), 135–149 (1999)

13. Kuich, W.: Pushdown tree automata, algebraic tree systems, and algebraic tree series. Inf. Comput. **165**(1), 69–99 (2001). https://doi.org/10.1006/inco.2000.2908

14. Maletti, A.: The power of tree series transducers of type I and II. In: De Felice, C., Restivo, A. (eds.) DLT 2005. LNCS, vol. 3572, pp. 338–349. Springer, Heidelberg (2005). https://doi.org/10.1007/11505877_30

15. Maletti, A.: Hierarchies of tree series transformations revisited. In: Ibarra, O.H., Dang, Z. (eds.) DLT 2006. LNCS, vol. 4036, pp. 215–225. Springer, Heidelberg (2006). https://doi.org/10.1007/11779148_20

16. Maletti, A.: Survey: weighted extended top-down tree transducers part II–application in machine translation. Fund. Inf. **112**(2–3), 239–261 (2011). https://doi.org/10.3233/FI-2011-589

17. Vaswani, A., et al.: Attention is all you need. In: Guyon, I., Luxburg, U.V., Bengio, S., Wallach, H., Fergus, R., Vishwanathan, S., Garnett, R. (eds.) Advances in Neural Information Processing Systems, vol. 30 (2017)

Engineering an LTLf Synthesis Tool

Alexandre Duret-Lutz[1(✉)], Shufang Zhu[2], Nir Piterman[3],
Giuseppe De Giacomo[4], and Moshe Y. Vardi[5]

[1] LRE, EPITA, Le Kremlin-Bicêtre, France
adl@lrde.epita.fr
[2] University of Liverpool, Liverpool, UK
[3] University of Gothenburg and Chalmers University of Technology,
Gothenburg, Sweden
[4] Sapienza University of Rome, Rome, Italy
[5] Rice University, Houston, TX, USA

Abstract. The problem of LTLf reactive synthesis is to build a transducer, whose output is based on a history of inputs, such that, for every infinite sequence of inputs, the conjoint evolution of the inputs and outputs has a prefix that satisfies a given LTLf specification.

We describe the implementation of an LTLf synthesizer that outperforms existing tools on our benchmark suite. This is based on a new, direct translation from LTLf to a DFA represented as an array of Binary Decision Diagrams (MTBDDs) sharing their nodes. This MTBDD-based representation can be interpreted directly as a reachability game that is solved on-the-fly during its construction.

1 Introduction

Reactive synthesis is concerned with synthesizing programs (a.k.a. strategies) for reactive computations (e.g., processes, protocols, controllers, robots) in active environments [26,30,45], typically, from temporal logic specifications. In AI, Reactive Synthesis, which is related to (strong) planning for temporally extended goals in fully observable nondeterministic domains [3,4,6,14–16,21,34], has been studied with a focus on logics on finite traces such as LTLf [7,22,23,33]. In fact, LTLf synthesis [23] is one of the two main success stories of reactive synthesis so far (the other being the GR(1) fragment of LTL [44]), and has brought about impressive advances in scalability [8,18,20,53].

Reactive synthesis for LTLf involves the following steps [23]: (1) distinguishing uncontrollable input ($\mathcal{I}$) and controllable output ($\mathcal{O}$) variables in an LTLf specification φ of the desired system behavior; (2) constructing a DFA accepting the behaviors satisfying φ; (3) interpreting this DFA as a two-player reachability game, and finding a controller winning strategy. Step (2) has two main bottlenecks: the DFA is worst-case doubly-exponential and its propositional alphabet $\Sigma = 2^{\mathcal{I} \cup \mathcal{O}}$ is exponential. The first only happens in the worst case, while the second blow-up – which we call *alphabet explosion* – always happens.

G. Castiglione and S. Mantaci (Eds.): CIAA 2025, LNCS 15981, pp. 129–147, 2026.
https://doi.org/10.1007/978-3-032-02602-6_10

Mona [39] addresses the alphabet-explosion problem, which happens also in MSO, by representing a DFA with Multi-Terminal Binary Decision Diagrams (MTBDDs) [36]. MTBDDs are a variant of BDDs [12] with arbitrary terminal values. If terminal values encode destination states, an MTBDD can compactly represent all outgoing transitions of a single DFA state. A DFA is represented, through its transition function, as an array of MTBDDs sharing their nodes.

The first $\mathsf{LTL_f}$ synthesizer, Syft [53], converted $\mathsf{LTL_f}$ into first-order logic in order to build a MTBDD-encoded DFA with Mona. Syft then converted this DFA into a BDD representation to solve the reachability game using a symbolic fixpoint computation. Syft demonstrated that DFA construction is the main bottleneck in $\mathsf{LTL_f}$ synthesis, motivating several follow-up efforts.

One approach to effective DFA construction uses compositional techniques, decomposing the input $\mathsf{LTL_f}$ formula into smaller subformulas whose DFAs can be minimized before being recombined. Lisa [8] decomposes top-level conjunctions, while Lydia [19] and LydiaSyft [29] decompose every operator.

Compositional methods construct the full DFA before synthesis can proceed, limiting their scalability. On-the-fly approaches [50] construct the DFA incrementally, while simultaneously solving the game, allowing strategies to be found before the complete DFA is built. The DFA construction may use various techniques. Cynthia [20] uses Sentential Decision Diagrams (SDDs) [17] to generate all outgoing transitions of a state at once. Alternatively, Nike [28] and MoGuSer [51] use a SAT-based method to construct one successor at a time. The game is solved by forward exploration with suitable backpropagation.

Contributions and Outline. In Sect. 3, we propose a direct and efficient translation from $\mathsf{LTL_f}$ to MTBDD-encoded DFA (henceforth called MTDFA). In Sect. 4, we show that given an appropriate ordering of BDD variables, $\mathsf{LTL_f}$ realizability can be solved by interpreting the MTBDD nodes of the MTDFA as the vertices of a reachability game, known to be solvable in linear time by backpropagation of the vertices that are winning for the *output* player. We give a linear-time implementation for solving the game on-the-fly while it is constructed. For more opportunities to abort the on-the-fly construction earlier, we additionally backpropagate vertices that are known to be winning by the *input* player. We implemented these techniques in two tools (`ltlf2dfa`☑ and `ltlfsynt`☑) that compare favorably with other existing tools in benchmarks from the $\mathsf{LTL_f}$-Synthesis Competition. To meet space limits, Sect. 5 only reports on the $\mathsf{LTL_f}$ realizability benchmark, and we refer readers to our artifact for the other results [24].

2   Preliminaries

2.1   Words over Assignments

A *word over* σ of length n over an alphabet Σ is a function $\sigma : \{0, 1, \ldots, n-1\} \to \Sigma$. We use Σ^n (resp. $\Sigma^\star$ and Σ^+) to denote the set of words of length n (resp. any length $n \geq 0$ and $n > 0$). We use $|\sigma|$ to represent the length of a word σ. For $\sigma \in \Sigma^n$ and $0 \leq i < n$, $\sigma(..i)$ denotes the prefix of σ of length $i + 1$.

Let $\mathcal{P}$ be a finite set of Boolean variables (a.k.a. *atomic propositions*). We use $\mathbb{B}^{\mathcal{P}}$ to denote the set of all assignments, i.e., functions $\mathcal{P} \to \mathbb{B}$ mapping variables to values in $\mathbb{B} = \{\bot, \top\}$.

Given two disjoint sets of variables $\mathcal{P}_1$ and $\mathcal{P}_2$, and two assignments $w_1 \in \mathbb{B}^{\mathcal{P}_1}$ and $w_2 \in \mathbb{B}^{\mathcal{P}_2}$, we use $w_1 \sqcup w_2 : (\mathcal{P}_1 \cup \mathcal{P}_2) \to \mathbb{B}$ to denote their combination.

In a system modeled using discrete Boolean signals that evolve synchronously, we assign a variable to each signal, and use a word $\sigma \in (\mathbb{B}^{\mathcal{P}})^+$ over assignments of $\mathcal{P}$ to represent the conjoint evolution of all signals over time.

We extend $\sqcup$ to such words. For two words $\sigma_1 \in (\mathbb{B}^{\mathcal{P}_1})^n$, $\sigma_2 \in (\mathbb{B}^{\mathcal{P}_2})^n$ of length n over assignments that use disjoint sets of variables, we use $\sigma_1 \sqcup \sigma_2 \in (\mathbb{B}^{\mathcal{P}_1 \cup \mathcal{P}_2})^n$ to denote a word such that $(\sigma_1 \sqcup \sigma_2)(i) = \sigma_1(i) \sqcup \sigma_2(i)$ for $0 \leq i < n$.

2.2   Linear Temporal Logic over Finite, Nonempty Words

We use classical LTL_f semantics over nonempty finite words [22].

Definition 1 (LTL_f formulas). *An LTL_f formula φ is built from a set $\mathcal{P}$ of variables, using the following grammar where $p \in \mathcal{P}$, and $\odot \in \{\wedge, \vee, \to, \leftrightarrow, ...\}$ is any Boolean operator: $\varphi ::= tt \mid ff \mid p \mid \neg\varphi \mid \varphi \odot \varphi \mid \mathsf{X}\varphi \mid \mathsf{X}^!\varphi \mid \varphi\, \mathsf{U}\, \varphi \mid \varphi\, \mathsf{R}\, \varphi \mid \mathsf{G}\varphi \mid \mathsf{F}\varphi$.*

Symbols tt and ff represent the true *and* false *LTL_f formulas. Temporal operators are X (weak next), $\mathsf{X}^!$ (strong next), U (until), R (release), G (globally), and F (finally). LTL_f$(\mathcal{P})$ denotes the set of formulas produced by the above grammar. We use $\mathsf{sf}(\varphi)$ to denote the set of subformulas for φ. A maximal temporal subformula of φ is a subformula whose primary operator is temporal and that is not strictly contained within any other temporal subformula of φ.*

The satisfaction of a formula $\varphi \in$ LTL_f$(\mathcal{P})$ by word $\sigma \in (\mathbb{B}^{\mathcal{P}})^+$ of length $n > 0$ at position $0 \leq i < n$, denoted $\sigma, i \models \varphi$, is defined as follows.

$$\sigma, i \models tt \iff i < n \qquad\qquad \sigma, i \models \mathsf{X}\varphi \iff (i+1=n) \vee (\sigma, i+1 \models \varphi)$$

$$\sigma, i \models ff \iff i = n \qquad\qquad \sigma, i \models \mathsf{X}^!\varphi \iff (i+1<n) \wedge (\sigma, i+1 \models \varphi)$$

$$\sigma, i \models p \iff p \in \sigma(i) \qquad\qquad \sigma, i \models \mathsf{F}\varphi \iff \exists j \in [i,n),\, \sigma, j \models \varphi$$

$$\sigma, i \models \neg\varphi \iff \neg(\sigma, i \models \varphi) \qquad \sigma, i \models \mathsf{G}\varphi \iff \forall j \in [i,n),\, \sigma, j \models \varphi$$

$$\sigma, i \models \varphi_1 \odot \varphi_2 \iff (\sigma, i \models \varphi_1) \odot (\sigma, i \models \varphi_2)$$

$$\sigma, i \models \varphi_1\, \mathsf{U}\, \varphi_2 \iff \exists j \in [i,n),\, (\sigma, j \models \varphi_2) \wedge (\forall k \in [i,j),\, \sigma, k \models \varphi_1)$$

$$\sigma, i \models \varphi_1\, \mathsf{R}\, \varphi_2 \iff \forall j \in [i,n),\, (\sigma, j \models \varphi_2) \vee (\exists k \in [i,j),\, \sigma, k \models \varphi_1)$$

The set of words that satisfy $\varphi \in$ LTL_f$(\mathcal{P})$ is $\mathscr{L}(\varphi) = \{\sigma \in (\mathbb{B}^{\mathcal{P}})^+ \mid \sigma, 0 \models \varphi\}$.

Example 1. Consider the following LTL_f formulas over $\mathcal{P} = \{i_0, i_1, i_2, o_1, o_2\}$: $\Psi_1 = \mathsf{G}((i_0 \to (o_1 \leftrightarrow i_1)) \wedge ((\neg i_0) \to (o_1 \leftrightarrow i_2)))$, and $\Psi_2 = (\mathsf{GF}o_2) \leftrightarrow (\mathsf{F}i_0)$. If we interpret i_0, i_1, i_2 as input signals, and o_1, o_2 as output signals, formula Ψ_1 specifies a 1-bit multiplexer: the value of the signal o_1 should be equal to the value of either i_1 or i_2 depending on the setting of i_0. Formula Ψ_2 specifies that the last value of o_2 should be $\top$ if and only if i_0 was $\top$ at some instant.

Definition 2 (Propositional Equivalence *[27]). For $\varphi \in \mathsf{LTL_f}(\mathcal{P})$, let φ_P be the Boolean formula obtained from φ by replacing every maximal temporal subformula ψ by a Boolean variable x_ψ. Two formulas $\alpha, \beta \in \mathsf{LTL_f}(\mathcal{P})$ are propositionally equivalent, denoted $\alpha \equiv \beta$, if α_P and β_P are equivalent Boolean formulas.*

Example 2. Formulas $\alpha = (\mathsf{G}b) \vee ((\mathsf{F}a) \wedge (\mathsf{G}b))$ and $\beta = \mathsf{G}b$ are propositionally equivalent. Indeed, $\alpha_P = x_{\mathsf{G}b} \vee (x_{\mathsf{F}a} \wedge x_{\mathsf{G}b}) = x_{\mathsf{G}b} = \beta_P$.

Note that $\alpha \equiv \beta$ implies $\mathscr{L}(\alpha) = \mathscr{L}(\beta)$, but the converse is not true in general. Since $\equiv$ is an equivalence relation, we use $[\alpha]_\equiv \in \mathsf{LTL_f}(\mathcal{P})$ to denote some unique representative of the equivalence class of α with respect to $\equiv$.

2.3 LTL$_f$ Realizability

Our goal is to build a tool that decides whether an LTL$_f$ formula is *realizable*.

Definition 3 (*[23,37]). Given two disjoint sets of variables $\mathcal{I}$ (inputs) and $\mathcal{O}$ (outputs), a controller is a function $\rho : \mathcal{I}^* \to \mathcal{O}$, that produces an assignment of output variables given a history of assignments of input variables.*

Given a word of n input assignments $\sigma \in (\mathbb{B}^\mathcal{I})^n$, the controller can be used to generate a word of n output assignments $\sigma_\rho \in (\mathbb{B}^\mathcal{I})^n$. The definition of σ_ρ may use two semantics depending on whether we want to the controller to have access to the current input assignment to decide the output assignment:

Mealy semantics: $\sigma_\rho(i) = \rho(\sigma(..i))$ for all $0 \le i < n$.
Moore semantics: $\sigma_\rho(i) = \rho(\sigma(..i-1))$ for all $0 \le i < n$.

A formula $\varphi \in \mathsf{LTL_f}(\mathcal{I} \cup \mathcal{O})$ is said to be Mealy-realizable or Moore-realizable if there exists a controller ρ such that for any word $\sigma \in (\mathbb{B}^\mathcal{I})^\omega$ there exists a position k such that $(\sigma \sqcup \sigma_\rho)(..k) \in \mathscr{L}(\varphi)$ using the desired semantics.

Example 3. Formula Ψ_1 (from Example 1) is Mealy-realizable but not Moore-realizable. Formula Ψ_2 is both Mealy and Moore-realizable.

2.4 Multi-terminal BDDs

Let $\mathcal{S}$ be a finite set. Given a finite set of variables $\mathcal{P} = \{p_0, p_1, \ldots, p_{n-1}\}$ (that are implicitly ordered by their index) we use $f : \mathbb{B}^\mathcal{P} \to \mathcal{S}$ to denote a function that maps an assignment of all those variables to an element of $\mathcal{S}$. Given a variable $p \in \mathcal{P}$ and a Boolean $b \in \mathbb{B}$, the function $f_{p=b} : \mathbb{B}^{\mathcal{P} \setminus \{p\}} \to \mathcal{S}$ represents a generalized co-factor obtained by replacing p by b in f. When $\mathcal{S} = \mathbb{B}$, a function $f : \mathbb{B}^\mathcal{P} \to \mathbb{B}$ can be encoded into a Binary Decision Diagram (BDD) [11]. Multi-Terminal Binary Decision Diagrams (MTBDDs) [32,39,42, 43], also called Algebraic Decision Diagrams (ADDs) [5,49], generalize BDDs by allowing arbitrary values on the leaves of the graph.

A *Multi-Terminal BDD* encodes any function $f : \mathbb{B}^\mathcal{P} \to \mathcal{S}$ as a rooted, directed acyclic graph. We use the term *nodes* to refer to the vertices of this graph. All nodes in an MTBDD are represented by triples of the form (p, ℓ, h).

In an internal node, $p \in \mathcal{P}$ and ℓ, h point to successors MTBDD nodes called the low and high links. The intent is that if (p, ℓ, h) is the root of the MTBDD representing the function f, then ℓ and h are the roots of the MTBDDs representing the functions $f_{p=\bot}$ and $f_{p=\top}$, respectively. Leaves of the graph, called *terminals*, hold values in $\mathcal{S}$. For consistency with internal nodes, we represent terminals with a triple of the form (∞, s, ∞) where $s \in \mathcal{S}$. When comparing the first elements of different triplets, we assume that ∞ is greater than all variables. We use MTBDD$(\mathcal{P}, \mathcal{S})$ to denote the set of MTBDD nodes that can appear in the representation of an arbitrary function $\mathbb{B}^{\mathcal{P}} \to \mathcal{S}$.

Following the classical implementations of BDD packages [1,11], we assume that MTBDDs are *ordered* (variables of $\mathcal{P}$ are ordered and visited in increasing order by all branches of the MTBDD) and *reduced* (isomorphic subgraphs are merged by representing each triplet only once, and internal nodes with identical low and high links are skipped over). Doing so ensures that each function $f : \mathbb{B}^{\mathcal{P}} \to \mathcal{S}$ has a unique MTBDD representation for a given order of variables.

Given $m \in$ MTBDD$(\mathcal{P}, \mathcal{S})$ and an assignment $w \in \mathbb{B}^{\mathcal{P}}$, we note $m(w)$ the element of $\mathcal{S}$ stored on the terminal of m that is reached after following the assignment w in the structure of m. We use $|m|$ to denote the number of MTBDD nodes that can be reached from m.

Let $m_1 \in$ MTBDD$(\mathcal{P}, \mathcal{S}_1)$ and $m_2 \in$ MTBDD$(\mathcal{P}, \mathcal{S}_2)$ be two MTBDD nodes representing functions $f_i : \mathbb{B}^{\mathcal{P}} \to \mathcal{S}_i$, and let $\odot : \mathcal{S}_1 \times \mathcal{S}_2 \to \mathcal{S}_3$, be a binary operation. One can easily construct $m_3 \in$ MTBDD$(\mathcal{P}, \mathcal{S}_3)$ representing the function $f_3(p_0, \ldots, p_{n-1}) = f_1(p_0, \ldots, p_{n-1}) \odot f_2(p_0, \ldots, p_{n-1})$, by generalizing the `apply2` function typically found in BDD libraries [32]. We use $m_1 \odot m_2$ to denote the MTBDD that results from this construction.

For $m \in$ MTBDD$(\mathcal{P}, \mathcal{S})$ we use $\texttt{leaves}(m) \subseteq \mathcal{S}$ to denote the elements of $\mathcal{S}$ that label terminals reachable from m. This set can be computed in $\Theta(|m|)$.

2.5   MTBDD-Based Deterministic Finite Automata

We now define an MTBDD-based representation of a DFA with a propositional alphabet, inspired by Mona's DFA representation [36,39].

Definition 4 (MTDFA). *An MTDFA is a tuple $\mathcal{A} = \langle \mathcal{Q}, \mathcal{P}, \iota, \Delta \rangle$, where $\mathcal{Q}$ is a finite set of states, $\mathcal{P}$ is a finite (and ordered) set of variables, $\iota \in \mathcal{Q}$ is the initial state, $\Delta : \mathcal{Q} \to$ MTBDD$(\mathcal{P}, \mathcal{Q} \times \mathbb{B})$ represents the set of outgoing transitions of each state. For a word $\sigma \in (\mathbb{B}^{\mathcal{P}})^{\star}$ of length n, let $(q_i, b_i)_{0 \le i \le n}$ be a sequence of pairs defined recursively as follows: $(q_0, b_0) = (\iota, \bot)$, and for $0 < i \le |\sigma|$, $(q_i, b_i) = \Delta(q_{i-1})(\sigma(i-1))$ is the pair reached by evaluating assignment $\sigma(i-1)$ on $\Delta(q_{i-1})$. The word σ is accepted by $\mathcal{A}$ iff $b_n = \top$. The language of $\mathcal{A}$, denoted $\mathscr{L}(A)$, is the set of words accepted by $\mathcal{A}$.*

Example 4. Figure 1 shows an MTDFA where $\mathcal{Q} \subseteq$ LTL$_f(\{i_0, i_1, i_2, o_1, o_2\})$. The set of states $\mathcal{Q}$ are the dashed rectangles on the left. For each such a state $q \in \mathcal{Q}$,

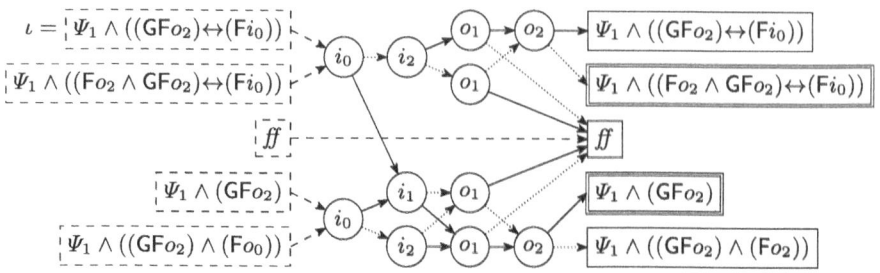

Fig. 1. An MTDFA where $\mathcal{P} = \{i_0, i_1, i_2, o_0, o_1\}$ and $\mathcal{Q} \subseteq \mathsf{LTL}_f(\mathcal{P})$. Following classical BDD representations a BDD node (p, ℓ, h) is represented by (p) with a link ℓ to h. A terminal $(\infty, (\alpha, b), \infty)$ is represented by $\boxed{\alpha}$ if $b = \bot$, or $\boxed{\boxed{\alpha}}$ if $b = \top$. Finally, MTBDD $m = \Delta(\alpha)$ representing the successors of state α is indicated with $\alpha \dashrightarrow m$. Subformula Ψ_1 abbreviates $\mathsf{G}((i_0 \rightarrow (o_1 \leftrightarrow i_1)) \wedge ((\neg i_0) \rightarrow (o_1 \leftrightarrow i_2)))$.

the dashed arrow points to the MTBDD node representing $\Delta(q)$. The MTBDD nodes are shared between all states. If, starting from the initial state ι at the top-left, we read the assignment $w = (i_0 \rightarrow \top, i_1 \rightarrow \top, i_2 \rightarrow \top, o_1 \rightarrow \top, o_2 \rightarrow \top)$, we should follow only the high links (plain arrows) and we reach the $\Psi_1 \wedge (\mathsf{GF}o_2)$ accepting terminal. If we read this assignment a second-time, starting this time from state $\Psi_1 \wedge (\mathsf{GF}o_2)$ on the left, we reach the same accepting terminal. Therefore, non-empty words of the form $www \ldots w$ are accepted by this automaton.

An MTDFA can be regarded as a semi-symbolic representation of a DFA over propositional alphabet. From a state q and reading the assignment w, the automaton jumps to the state q' that is the result of computing $(q', b) = \Delta(q)(w)$. The value of b indicates whether that assignment is allowed to be the last one of the word being read. By definition, an MTDFA cannot accept the empty word.

MTDFAs are compact representations of DFAs, because the MTBDD representation of the successors of each state can share their common nodes. Boolean operations can be implemented over MTDFAs, with the expected semantics, i.e., $\mathscr{L}(\mathcal{A}_1 \odot \mathcal{A}_2) = \{\sigma \in (\mathbb{B}^{\mathcal{P}})^+ \mid (\sigma \in \mathscr{L}(\mathcal{A}_1)) \odot (\sigma \in \mathscr{L}(\mathcal{A}_2))\}$.

3   Translating LTL_f to MTBDD and MTDFA

This section shows how to directly transform a formula $\varphi \in \mathsf{LTL}_f(\mathcal{P})$ into an MTDFA $\mathcal{A}_\varphi = \langle \mathcal{Q}, \mathcal{P}, \varphi, \Delta \rangle$ such that $\mathscr{L}(\varphi) = \mathscr{L}(\mathcal{A}_\varphi)$. The translation is reminiscent of other translations of LTL_f to DFA [20,22], but it leverages the fact that MTBBDs can provide a normal form for LTL_f formulas.

The construction maps states to LTL_f formulas, i.e., $\mathcal{Q} \subseteq \mathsf{LTL}_f(\mathcal{P})$. Terminals appearing in the MTBDDs of $\mathcal{A}_\varphi$ will be labeled by pairs $(\alpha, b) \in \mathsf{LTL}_f(\mathcal{P}) \times \mathbb{B}$, so we use $\mathsf{term}(\alpha, b) = (\infty, (\alpha, b), \infty)$ to shorten the notation from Sect. 2.4.

The conversion from φ to $\mathcal{A}_\varphi$ is based on the function $\mathsf{tr} : \mathsf{LTL}_f(\mathcal{P}) \to \mathsf{MTBDD}(\mathcal{P}, \mathsf{LTL}_f(\mathcal{P}) \times \mathbb{B})$ defined inductively as follows:

$$\mathsf{tr}(\mathit{ff}) = \mathsf{term}(\mathit{ff}, \bot) \qquad\qquad\qquad \mathsf{tr}(\mathsf{X}\alpha) = \mathsf{term}(\alpha, \top)$$

$$\mathsf{tr}(\mathit{tt}) = \mathsf{term}(\mathit{tt}, \top) \qquad\qquad\qquad \mathsf{tr}(\mathsf{X}^!\alpha) = \mathsf{term}(\alpha, \bot)$$

$$\mathsf{tr}(p) = (p, \mathsf{term}(\mathit{ff}, \bot), \mathsf{term}(\mathit{tt}, \top)) \text{ for } p \in \mathcal{P} \quad \mathsf{tr}(\neg\alpha) = \neg\mathsf{tr}(\alpha)$$

$$\mathsf{tr}(\alpha \odot \beta) = \mathsf{tr}(\alpha) \odot \mathsf{tr}(\beta) \text{ for any } \odot \in \{\wedge, \vee, \to, \leftrightarrow, \oplus\}$$

$$\mathsf{tr}(\alpha \, \mathsf{U} \, \beta) = \mathsf{tr}(\beta) \vee (\mathsf{tr}(\alpha) \wedge \mathsf{term}(\alpha \, \mathsf{U} \, \beta, \bot)) \qquad \mathsf{tr}(\mathsf{F}\alpha) = \mathsf{tr}(\alpha) \vee \mathsf{term}(\mathsf{F}\alpha, \bot).$$

$$\mathsf{tr}(\alpha \, \mathsf{R} \, \beta) = \mathsf{tr}(\beta) \wedge (\mathsf{tr}(\alpha) \vee \mathsf{term}(\alpha \, \mathsf{R} \, \beta, \top)) \qquad \mathsf{tr}(\mathsf{G}\alpha) = \mathsf{tr}(\alpha) \wedge \mathsf{term}(\mathsf{G}\alpha, \top)$$

Boolean operators that appear to the right of the equal sign are applied on MTBDDs as discussed in Sect. 2.4. Terminals in $\mathsf{LTL}_f(\mathcal{P}) \times \mathbb{B}$ are combined with: $(\alpha_1, b_1) \odot (\alpha_2, b_2) = ([\alpha_1 \odot \alpha_2]_\equiv, b_1 \odot b_2)$ and $\neg(\alpha, b) = ([\neg\alpha]_\equiv, \neg b)$.

Theorem 1. *For $\varphi \in \mathsf{LTL}_f(\mathcal{P})$, let $\mathcal{A}_\varphi = \langle \mathcal{Q}, \mathcal{P}, \iota, \Delta \rangle$ be the MTDFA obtained by setting $\iota = [\varphi]_\equiv$, $\Delta = \mathsf{tr}$, and letting $\mathcal{Q}$ be the smallest subset of $\mathsf{LTL}_f(\mathcal{P})$ such that $\iota \in \mathcal{Q}$, and such that for any $q \in \mathcal{Q}$ and for any $(\alpha, b) \in \mathit{leaves}(\Delta(q))$, then $\alpha \in \mathcal{Q}$. With this construction, $|\mathcal{Q}|$ is finite and $\mathscr{L}(\varphi) = \mathscr{L}(\mathcal{A}_\varphi)$.*

Proof. (sketch) By definition of tr, $\mathcal{Q}$ contains only Boolean combinations of subformulas of φ. Propositional equivalence implies that the number of such combinations is finite: $|\mathcal{Q}| \leq 2^{2^{|\mathsf{sf}(\varphi)|}}$. The language equivalence follows from the definition of LTL_f, and from some classical LTL_f equivalences. For instance the rule for $\mathsf{tr}(\alpha \, \mathsf{U} \, \beta)$ is based on the equivalence $\mathscr{L}(\alpha \, \mathsf{U} \, \beta) = \mathscr{L}(\beta \vee (\alpha \wedge \mathsf{X}^!(\alpha \, \mathsf{U} \, \beta)))$.

Example 5. Figure 1 is the MTDFA for formula $\Psi_1 \wedge \Psi_2$, presented in Example 1. Many more examples can be found in the associated artifact [24].

The definition of $\mathsf{tr}(\cdot)$ as an MTBDD representation of the set of successors of a state can be thought as a symbolic representation of Antimirov's linear forms [2] for DFA with propositional alphabets. Antimirov presented linear forms as an efficient way to construct all (partial) derivatives at once, without having to iterate over the alphabet. For LTL_f, *formula progressions* [20] are the equivalent of Brozozowski derivatives [13]. Here, $\mathsf{tr}(\cdot)$ computes all formulas progressions at once, without having to iterate over an exponential number of assignments.

Finally, note that while this construction works with any order for $\mathcal{P}$, different orders might produce a different number of MTBDD nodes.

Optimizations. The previous definitions can be improved in several ways.

Our implementation of MTBDD actually supports terminals that are the Boolean terminals of standard BDDs as well as the terminals used so far. So we are actually using $\mathsf{MTBDD}(\mathcal{P}, (\mathsf{LTL}_f(\mathcal{P}) \times \mathbb{B}) \cup \mathbb{B})$, and we encode $\mathsf{term}(\mathit{ff}, \bot)$ and $\mathsf{term}(\mathit{tt}, \top)$ directly as $\bot$ and $\top$ respectively. With those changes, `apply2` may be modified to shortcut the recursion depending on the values of m_1, m_2, and $\odot$.

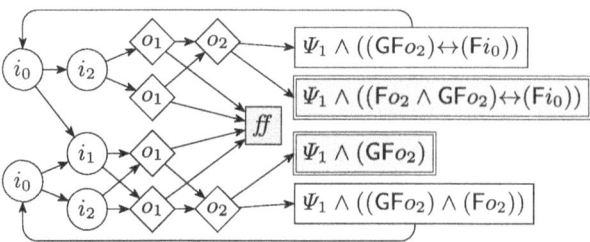

Fig. 2. Interpretation of the MTDFA of Fig. 1 as a game with $\mathcal{I} = \{i_0, i_1, i_2\}$, $\mathcal{O} = \{o_1, o_2\}$. Each MTBDD *node* of the MTDFA is viewed as a *vertex* of the game, with terminal of the form $(\alpha, \bot)$ looping back to $\Delta(\alpha)$. Player o decides where to go from diamond and rectangular vertices and wants to reach the green vertices corresponding to accepting terminals. Player i decides where to go from round vertices and wants to reach *ff* or avoid green vertices (Color figure online).

For instance if $\odot = \wedge$ and $m_1 = \top$, then m_2 can be returned immediately. Such shortcuts may be implemented for $\mathsf{MTBDD}(\mathcal{P}, \mathcal{S} \cup \mathbb{B})$ regardless of the nature of $\mathcal{S}$, so our implementation of MTBDD operations is independent of $\mathsf{LTL_f}$.

When combining terminals during the computation of tr, one has to compute the representative formula $[\alpha_1 \odot \alpha_2]_\equiv$. This can be done by converting α_{1P} and α_{2P} into BDDs, keeping track of such conversions in a hash table. Two propositionally equivalent formulas will have the same BDD representation. While we are looking for a representative formula, we can also use the opportunity to simplify the formula at hand. We use the following very simple rewritings, for patterns that occur naturally in the output of tr:

$(\alpha \mathbin{\mathsf{U}} \beta) \vee \beta \rightsquigarrow \alpha \mathbin{\mathsf{U}} \beta$, $(\alpha \mathbin{\mathsf{R}} \beta) \wedge \beta \rightsquigarrow \alpha \mathbin{\mathsf{R}} \beta$, $(\mathsf{F}\beta) \vee \beta \rightsquigarrow \mathsf{F}\beta$, $(\mathsf{G}\beta) \wedge \beta \rightsquigarrow \mathsf{G}\beta$.

Once $\mathcal{A}_\varphi$ has been built, two states $q, q' \in \mathcal{Q}$ such that $\Delta(q) = \Delta(q')$ can be merged by replacing all occurrences of q' by q in the leaves of Δ.

Example 6. The automaton from Fig. 1 has two pairs of states that can be merged. However, if the rule $(\mathsf{G}\beta) \wedge \beta \rightsquigarrow \mathsf{G}\beta$ is applied during the construction, then the occurrence of $(\mathsf{G}\mathsf{F}o_2) \wedge (\mathsf{F}o_2)$ will already be replaced by $\mathsf{G}\mathsf{F}o_2$, producing the simplified automaton without requiring any merging.

4   Deciding $\mathsf{LTL_f}$ Realizability

$\mathsf{LTL_f}$ realizability (Definition 3) is solved by reducing the problem to a two-player reachability game where one player decides the input assignments and the other player decides the output assignments [23]. Section 4.1 presents reachability games and how to interpret the MTDFA as a reachability game, and Sect. 4.2 shows how we can solve the game on-the-fly while constructing it.

4.1   Reachability Games and Backpropagation

Definition 5 (Rechability Game). *A Reachability Game is* $\mathcal{G} = \langle \mathcal{V} = \mathcal{V}_o \uplus \mathcal{V}_i, \mathcal{E}, \mathcal{F}_o \rangle$, *where* $\mathcal{V}$ *is a finite set of* vertices *partitioned to player* output *(abbreviated* O*) and player* input *(abbreviated* I*),* $\mathcal{E} \subseteq \mathcal{V} \times \mathcal{V}$ *is a finite set of edges, and* $\mathcal{F}_o \subseteq \mathcal{V}$ *is the set of target states. Let* $\mathcal{E}(v) = \{(v, v') \mid (v, v') \in \mathcal{E}\}$*. This graph is also referred to as the game* arena*.*

A strategy for player O *is a cycle-free subgraph* $\langle W, \sigma \rangle \subseteq \langle \mathcal{V}, \mathcal{E} \rangle$ *such that (a) for every* $v \in W$ *we have* $v \in \mathcal{F}_o$ *or* $\mathcal{E}(v) \cap \sigma \neq \emptyset$ *and (b) if* $v \in W \cap \mathcal{V}_I$ *then* $\mathcal{E}(v) \subseteq \sigma$*. A vertex* v *is winning for* O *if* $v \in W$ *for some strategy* $\langle W, \sigma \rangle$*.*

Such a reachability game can be solved by backpropagation identifying the maximal set W in a strategy. Namely, start from $W = \mathcal{F}_o$. Then W is iteratively augmented with every vertex in $\mathcal{V}_o$ that has some edge to W, and every (non dead-end) vertex in $\mathcal{V}_I$ whose edges all lead to W. At the end of this backpropagation, which can be performed in linear time [35, Theorem 3.1.2], every vertex in W is winning for O, and every vertex outside W is losing for O. Notice that every dead-end that is not in $\mathcal{F}_o$ cannot be winning. It follows that we can identify some (but not necessarily all) vertices that are losing by setting L as the set of all dead-ends and adding to L every $\mathcal{V}_I$ vertex that has some edge to L and every $\mathcal{V}_o$ vertex whose edges all lead to L.

Let $\mathcal{A}_\varphi = \langle \mathcal{Q}, \mathcal{I} \uplus \mathcal{O}, \iota, \Delta \rangle$ be a translation of $\varphi \in \mathsf{LTL}_f(\mathcal{I} \uplus \mathcal{O})$ (per Theorem 1) such that variables of $\mathcal{I}$ appear before $\mathcal{O}$ in the MTBDD encoding of Δ.

Definition 6 (Realizability Game). *We define the reachability game* $\mathcal{G}_\varphi = \langle \mathcal{V} = \mathcal{V}_I \uplus \mathcal{V}_o, \mathcal{E}, \mathcal{F}_o \rangle$ *in which* $\mathcal{V} \subseteq \mathsf{MTBDD}(\mathcal{I} \uplus \mathcal{O})$ *corresponds the set of nodes that appear in the MTBDD encoding of* Δ*.* $\mathcal{V}_o$ *contains all nodes* (p, ℓ, h) *such that* $p \in \mathcal{O}$ *or* $p = \infty$ *(terminals), and* $\mathcal{V}_I$ *contains those with* $p \in \mathcal{I}$*. The edges* $\mathcal{E}$ *follows the structure of* Δ*, i.e., if* $\mathcal{A}_\varphi$ *has a node* $r = (p, \ell, h)$*, then* $\{(r, \ell), (r, h)\} \subseteq \mathcal{E}$*. Additionally, for any terminal* $t = (\infty, (\alpha, \bot), \infty)$ *such that* $\alpha \neq \textit{ff}$*,* $\mathcal{E}$ *contains the edge* $(t, \Delta(\alpha))$*. Finally,* $\mathcal{F}_o$ *is the set of accepting terminals, i.e., nodes of the form* $(\infty, (\alpha, \top), \infty)$*.*

Theorem 2. *Vertex* $\Delta(\iota)$ *is winning for* O *in* $\mathcal{G}_\varphi$ *iff* φ *is Mealy-realizable.*

Moore realizability can be checked similarly by changing the order of $\mathcal{I}$ and $\mathcal{O}$ in the MTBDD encoding of Δ.

Example 7. Figure 2 shows how to interpret the MTDFA of Fig. 1 as a game, by turning each MTBDD node into a game vertex. The player owning each vertex is chosen according to the variable that labels it. Vertices corresponding to accepting terminals become winning targets for the output player, so the game stops once they are reached. Solving this game will find every internal node as winning for O, so the corresponding formula is Mealy-realizable.

The difference with DFA games [20, 23, 28, 51] is that instead of having player I select all input signals at once, and then player O select all output signals at once, our game proceeds by selecting one signal at a time. Sharing nodes that represent identical partial assignments contributes to the scalability of our approach.

4.2   Solving Realizability On-the-Fly

We now show how to construct and solve $\mathcal{G}_\varphi$ on-the-fly, for better efficiency. The construction is easier to study in two parts: (1) the on-the-fly solving of reachability games, based on backpropagation, and (2) the incremental construction of $\mathcal{G}_\varphi$, done with a forward exploration of a subset of the MTDFA for φ.

Algorithm 1 presents the first part: a set of functions for constructing a game arena incrementally, while performing the linear-time backpropagation algorithm on-the-fly. At all points during this construction, the winning status of a vertex ($winner[x]$) will be one of O (player O can force the play to reach $\mathcal{F}_0$, i.e., the vertex belongs to W), I (player I can force the play to avoid $\mathcal{F}_0$, i.e., the vertex belongs to L), or U (undetermined yet), and the algorithm will backpropagate both O and I. At the end of the construction, all vertices with status U will be considered as winning for I. Like in the standard algorithm for solving reachability games [35, Theorem 3.1.2] each state uses a counter (*count*, lines 7,15) to track the number of its undeterminated successors. When a vertex x is marked as winning for player w by calling set_winnerx, w, an undeterminated predecessor p has its counter decreased (line 15), and p can be marked as winning for w (line 16) if either vertex p is owned by w (player w can choose to go to x) or the counter dropped to 0 (meaning that all choices at p were winning for w).

To solve the game while it is constructed, we *freeze* vertices. A vertex should be frozen after all its successors have been introduced with new_edge. The counter dropping to 0 is only checked on frozen vertices (lines 11, 16) since it is only meaningful if all successors of a vertex are known.

Algorithm 2 is the second part. It shows how to build $\mathcal{G}_\varphi$ incrementally. It translates the states α of the corresponding MTDFA one at a time, and uses the functions of Algorithm 1 to turn each node of $\mathsf{tr}(\alpha)$ into a vertex of the game. Since the functions of Algorithm 1 update the winning status of the states as soon as possible, Algorithm 2 can use that to cut parts of the exploration.

Instead of using $\Delta(\varphi) = \mathsf{tr}(\varphi)$ as initial vertex of the game, as in Theorem 2, we consider $init = \mathsf{term}(\varphi, \bot)$ as initial vertex (line 3): this makes no theoretical difference, since $\mathsf{term}(\varphi, \bot)$ has $\mathsf{tr}(\varphi)$ as unique successor. Lines 5,9–11,13, and 32 implements the exploration of all the $\mathsf{LTL_f}$ formulas α that would label the states of the MTDFA for φ (as needed to implement Theorem 1). The actual order in which formulas are removed from *todo* on line 11 is free. (We found out that handling *todo* as a queue to implement a BFS exploration worked marginally better than using it as a stack to do a DFS-like exploration, so we use a BFS in practice.)

var: *owner*[];   `// map each vertex to one of {0,1}`

var: *pred*[];   `// map vertices to sets of predecessor vertices`

var: *count*[];   `// map vertices to # of undetermined successors`

var: *winner*[];   `// map vertices to one of {0,1,U}`

var: *frozen*[];   `// map vertices to their frozen status (a Boolean)`

1   **Function** `new_vertex`$(x \in \mathcal{V}, own \in \{0,1\})$ `// new vertex owned by` *own*

2     $owner[x] \leftarrow own;\ pred[x] \leftarrow \emptyset;\ count[x] \leftarrow 0;\ frozen[x] \leftarrow \bot;$

3     $winner[x] \leftarrow$ U; `// undeterminated winner`

4   **Function** `new_edge`$(src \in \mathcal{V}, dst \in \mathcal{V})$

5   **assert** $(frozen[src] = \bot)$;

6   **if** $winner[dst] =$ U **then**

7 | $count[src] \leftarrow count[src] + 1;\ pred[dst] \leftarrow pred[dst] \cup \{src\};$

8   **else if** $winner[dst] = owner[src]$ **then** `set_winner`$(src, owner[src])$;

     `// ignore the edge otherwise, it will never be used`

9   **Function** `freeze_vertex`$(x \in \mathcal{V})$ `// promise not to add more successors`

10    $frozen[x] \leftarrow \top;$ `// next line, we assume` $\neg$`1 = 0 and` $\neg$`0 = 1`

11   **if** $winner[x] =$ U $\land count[n] = 0$ **then** `set_winner`$(x, \neg owner[x])$;

12   **Function** `set_winner`$(x \in \mathcal{V}, w \in \{0,1\})$ `// with linear backprop.`

13   **assert** $(winner[x] =$ U$);\ winner[x] \leftarrow w;$

14   **foreach** $p \in pred[x]$ **such that** $winner[p] =$ U **do**

15       $count[p] \leftarrow count[p] - 1;$

16   **if** $owner[p] = w \lor (count[p] = 0 \land frozen[p])$ **then** `set_winner`(p, w);

Algorithm 1: API for solving a reachability game on-the-fly. Construct the game arena with **new_vertex** and **new_edge**. Once all successors of a vertex have been connected, call **freeze_vertex**. Call **set_winner** at any point to designate vertices winning for one player.

Each α is translated into an MTBDD $\mathsf{tr}(\alpha)$ representing its possible successors. The constructed game should have one vertex per MTBDD node in $\mathsf{tr}(\alpha)$. Those vertices are created in the inner **while** loop (lines 19–31). Function `declare_vertex` is used to assign the correct owner to each new node according to its decision variable (as in Definition 6) as well as adding those nodes to the *to_encode* set processed by this inner loop. Terminal nodes are either marked as winning for one of the players (lines 24–25) or stored in *leaves* (line 26).

Since connecting game vertices may backpropagate their winning status, the encoding loop can terminate early whenever the vertex associated to $\mathsf{term}(\alpha, \bot)$ becomes determined (lines 18 and 31). If that vertex is not determined, the *leaves* of α are added to *todo* (line 32) for further exploration.

The entire construction can also stop as soon as the initial vertex is determined (line 10). However, if the algorithm terminates with $winning[init] =$ U, it still means that 0 cannot reach its targets. Therefore, as tested by line 33, formula φ is realizable iff $winning[init] = 0$ in the end.

Theorem 3. *Algorithm 2 returns tt iff φ is Mealy-realizable.*

```
 1  Function realizability(φ ∈ LTL_f(I ⊎ O))
 2  |   configure the MTBDD library to put variables in I before those in O;
 3  |   init ← term(φ, ⊥); new_vertex(init, I);
 4  |   V ← {init}; // nodes created as game vertices
 5  |   Q ← ∅; // LTL_f formulas processed by main loop on line 10
 6  |   Function declare_vertex(r ∈ MTBDD(I ⊎ O, LTL_f(I ⊎ O) × B))
 7  |   |   (p, ℓ, h) ← r; if p = ∞ ∨ p ∈ I then own ← I else own ← O;
 8  |   |_  new_vertex(r, own); V ← V ∪ {r}; to_encode ← to_encode ∪ {r};
 9  |   todo ← {φ};
10  |   while todo ≠ ∅ ∧ winner[init] = U do
11  |   |   α ← todo.pop_any(); Q ← Q ∪ {α};
12  |   |   [optional: add one-step (un)realizability check here, see Sec. 5];
13  |   |   a ← term(α, ⊥); m ← tr(α);
14  |   |   if m ∈ V then // m has already been encoded
15  |   |   |_  new_edge(a, m); freeze_vertex(a); continue to line 10;
16  |   |   to_encode ← ∅; leaves ← ∅;
17  |   |   declare_vertex(m); new_edge(a, m); freeze_vertex(a);
18  |   |   if winner[a] ≠ U then continue to line 10;
19  |   |   while to_encode ≠ ∅ do
20  |   |   |   r ← to_encode.pop_any();
21  |   |   |   (p, ℓ, h) ← r;
22  |   |   |   if p = ∞ then // this is a terminal labeled by ℓ
23  |   |   |   |   (β, b) ← ℓ;
24  |   |   |   |   if b then set_winner(r, O);
25  |   |   |   |   else if β = ff then set_winner(r, I);
26  |   |   |   |   else if β ∉ Q then leaves ← leaves ∪ {β};
27  |   |   |   else
28  |   |   |   |   if ℓ ∉ V then declare_vertex(ℓ);
29  |   |   |   |   if h ∉ V then declare_vertex(h);
30  |   |   |   |_  new_edge(r, ℓ); new_edge(r, h); freeze_vertex(r);
31  |   |   |_  if winner[a] ≠ U then continue to line 10;
32  |   |_  todo ← todo ∪ leaves;
33  |   return winner[init] = O;
```

Algorithm 2: On-the-fly realizability check with Mealy semantics (for Moore semantics, swap the order of I and O on the first line).

5   Implementation and Evaluation

Our algorithms have been implemented in Spot [25], after extending its fork of BuDDy [40] to support MTBDDs. The release of Spot 2.14 distributes two new command-line tools: ltlf2dfa🔗 and ltlfsynt🔗, implementing translation from LTL_f to MTDFA, and solving LTL_f synthesis. We describe and evaluate ltlfsynt in the following.

Preprocessing. Before executing Algorithm 2, we use a few preprocessing techniques to simplify the problem. We remove variables that always have the same polarity in the specification (a simplification used also by Strix [48]), and we decompose the specifications into output-disjoint sub-specifications that can be solved independently [31]. A specification such as $\Psi_1 \wedge \Psi_2$, from Example 1, is not solved directly as demonstrated here, but split into two output-disjoint specifications Ψ_1 and Ψ_2 that are solved separately. Finally, we also simplify LTL$_f$ formulas using very simple rewriting rules such as $X(\alpha) \wedge X(\beta) \rightsquigarrow X(\alpha \wedge \beta)$ that reduce the number of MTBDD operations required during translation.

One-step (Un)realizability Checks. An additional optimization consists in performing one-step realizability and one-step unrealizability checks in Algorithm 2. The principle is to transform the formula α into two smaller Boolean formulas α_r and α_u, such that if α_r is realizable it implies that α is realizable, and if α_u is unrealizable it implies that α is unrealizable [50, Theorems 2–3]. Those Boolean formulas can be translated to BDDs for which realizability can be checked by quantification. On success, it avoids the translation of the larger formula α. The simple formula $\Psi_1 \wedge \Psi_2$ of our running example is actually one-step realizable.

Synthesis. After deciding realizability, ltlfsynt is able to extract a strategy from the solved game in the form of a Mealy machine, and encode that into an And-Invert Graph (AIG) [10]: the expected output of the Synthesis Competition for the LTL$_f$ synthesis tracks. The conversion from Mealy to AIG reuses prior work [46,47] developed for Spot's LTL (not LTL$_f$) synthesis tool. We do not detail nor evaluate these extra steps here due to lack of space.

Evaluation. We evaluated the task of deciding LTL$_f$ reachability over specifications from the Synthesis Competition [38]. We took all tlsf-fin specifications from SyntComp's repository 🔗, excluded some duplicate specifications as well as some specifications that were too large to be solved by any tool, and converted the specifications from TLSF v1.2 [37] to LTL$_f$ using syfco [37].

We used BenchExec 3.22 [9] to track time and memory usage of each tool. Tasks were run on a Core i7-3770 with *Turbo Boost* disabled, and frequency scaled down to 1.6 GHz to prevent CPU throttling. The computer has 4 physical cores and 16 GB of memory. BenchExec was configured to run up to 3 tasks in parallel with a memory limit of 4GB per task, and a time limit of 15 min.

Figure 3 compares five configurations of ltlfsynt against seven other tools. We verified that all tools were in agreement. Lydia 0.1.3 [19], SyftMax (or Syft 2.0) [52] and LydiaSyft 0.1.0-alpha [29] are all using Mona to construct a DFA by composition; they then solve the resulting game symbolically after encoding it using BDDs. Lisa [8] uses a hybrid compositional construction, mixing explicit compositions (using Spot), with symbolic compositions (using BuDDy), solving the game symbolically in the end. Cynthia 0.1.0 [20], Nike 0.1.0 [28], and MoGuSer [51] all use an on-the-fly construction of a DFA game that they solve via forward exploration with backpropagation, but they do not implement backpropagation in linear time, as we do. Yet, the costly part of synthesis is game generation, not solving. Cynthia uses SDDs [17] to compute successors and represent states, while Nike and MoGuSer use SAT-based techniques to compute

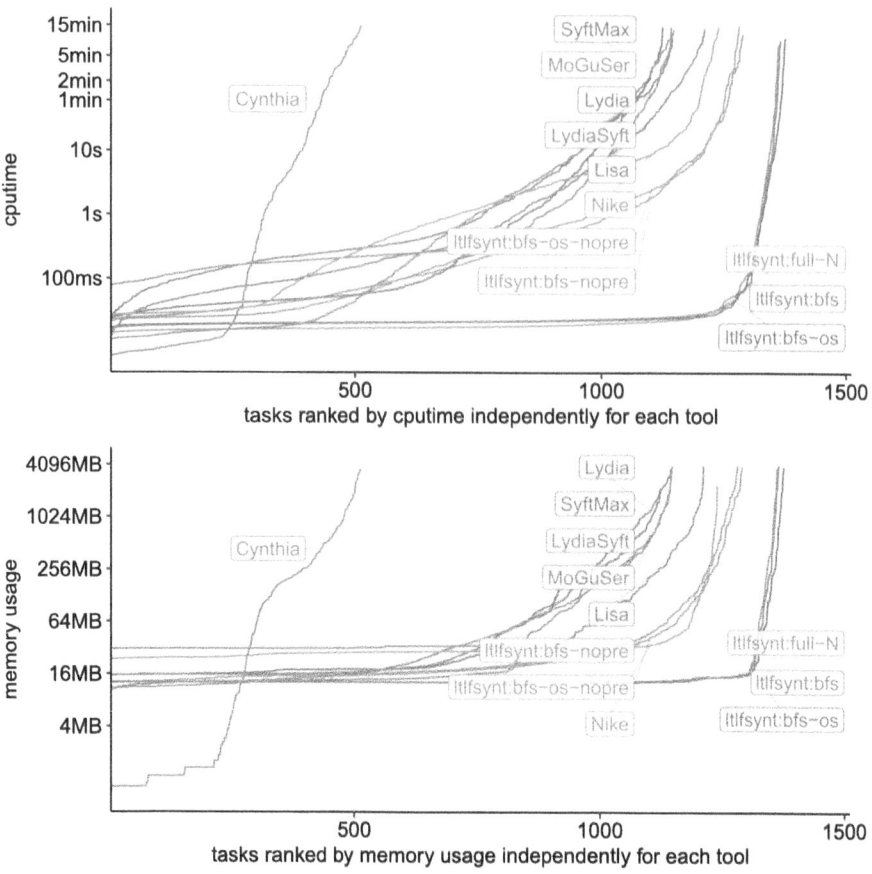

Fig. 3. Cactus plots comparing time and memory usage of different configurations.

successors and BDDs to represent states. Nike, Lisa, and LydiaSyft were the
top-3 contenders of the LTL$_f$ track of SyntComp in 2023 and 2024.

Configuration `ltlfsynt:bfs-nopre` corresponds to Algorithm 2 were *todo* is
a queue: it already solves more cases than all other tested tools. Suffix `-nopre`
indicates that preprocessings of the specification are disabled (this makes comparison fairer, since other tools have no such preprocessings). The version with
preprocessings enabled is simply called `ltlfsynt:bfs`. Variants with "-os" adds
the one-step (un)realizability checks that LydiaSyft, Cynthia, and Nike also perform. We also include a configuration `ltlfsynt:full-N` that corresponds to first
translating the specification into a MTDFA using Theorem 1, and then solving
the game by linear propagation. The difference between `ltlfsynt:full` and
`ltlfsynt:bfs` shows the gain obtained with the on-the-fly translation: although
that look small in the cactus plot, it is important in some specifications.

6   Conclusion

We have presented the implementation of ltlfsynt, and evaluated it to be faster at deciding LTL$_f$ realizability than seven existing tools, including the winners of SyntComp'24. The implementation uses a direct and efficient translation from LTL$_f$ to DFA represented by MTBDDs, which can then be solved as a game played directly on the structure of the MTBDDs. The two constructions (translation and game solving) are performed together on-the-fly, to allow early termination.

Although ltlsynt also includes a preliminary implementation of LTL$_f$ synthesis of And-Inverter graphs, we leave it as future work to document it and ensure its correctness.

Finally, the need for solving a reachability game while it is discovered also occurs in other equivalent contexts such as HornSAT, where linear algorithms that do not use "counters" and "predecessors" (unlike ours) have been developed [41]. Using such algorithms might improve our solution by saving memory.

Data Availability Statement. Implementation, supporting scripts, detailed analysis of this benchmark, and additional examples are archived on Zenodo [24].

References

1. Andersen, H.R.: An introduction to binary decision diagrams. Lecture notes for Efficient Algorithms and Programs, Fall 1999 (1999). https://web.archive.org/web/20090530154634/http://www.itu.dk:80/people/hra/bdd-eap.pdf
2. Antimirov, V.: Partial derivatives of regular expressions and finite automaton constructions. Theoret. Comput. Sci. **155**(2), 291–319 (1996). https://doi.org/10.1016/0304-3975(95)00182-4
3. Bacchus, F., Kabanza, F.: Planning for temporally extended goals. Ann. Math. Artif. Intell. **22**, 5–27 (1998). https://doi.org/10.1023/A:1018985923441
4. Bacchus, F., Kabanza, F.: Using temporal logics to express search control knowledge for planning. Artif. Intell. **116**(1–2), 123–191 (2000). https://doi.org/10.1016/S0004-3702(99)00071-5
5. Bahar, R.I., et al.: Algebraic decision diagrams and their applications. In: Proceedings of 1993 International Conference on Computer Aided Design (ICCAD 1993), pp. 188–191. IEEE Computer Society Press (1993). https://doi.org/10.1109/ICCAD.1993.580054
6. Baier, J.A., Fritz, C., McIlraith, S.A.: Exploiting procedural domain control knowledge in state-of-the-art planners. In: Proceedings of the International Conference on Automated Planning and Scheduling (ICAPS 2007), pp. 26–33. AAAI (2007). https://aaai.org/papers/icaps-07-004
7. Baier, J.A., McIlraith, S.A.: Planning with first-order temporally extended goals using heuristic search. In: Proceedings of the 21st International Conference on Artificial intelligence (AAAI 2006), pp. 788–795. AAAI Press (2006).https://doi.org/10.5555/1597538.1597664
8. Bansal, S., Li, Y., Tabajara, L.M., Vardi, M.Y.: Hybrid compositional reasoning for reactive synthesis from finite-horizon specifications. In: Proceedings of the 34th national conference on Artificial intelligence (AAAI 2020), pp. 9766–9774. AAAI Press (2020). https://doi.org/10.1609/AAAI.V34I06.6528

9. Beyer, D., Löwe, S., Wendler, P.: Reliable benchmarking: requirements and solutions. Int. J. Softw. Tools Technol. Transfer **21**, 1–29 (2019). https://doi.org/10.1007/s10009-017-0469-y

10. Biere, A., Heljanko, K., Wieringa, S.: AIGER 1.9 and beyond. Tech. Rep. 11/2, Institute for Formal Models and Verification, Johannes Kepler University, Altenbergerstr. 69, 4040 Linz, Austria (2011). https://fmv.jku.at/aiger/

11. Bryant, R.E.: Graph-based algorithms for Boolean function manipulation. IEEE Trans. Comput. **35**(8), 677–691 (1986). https://doi.org/10.1109/TC.1986.1676819

12. Bryant, R.E.: Symbolic Boolean manipulation with ordered binary-decision diagrams. ACM Comput. Surv. **24**(3), 293–318 (1992). https://doi.org/10.1145/136035.136043

13. Brzozowski, J.A.: Derivatives of regular expressions. J. ACM **11**(4), 481–494 (1964). https://doi.org/10.1145/321239.321249

14. Calvanese, D., De Giacomo, G., Vardi, M.Y.: Reasoning about actions and planning in LTL action theories. In: Proceedings of the Eights International Conference on Principles of Knowledge Representation and Reasoning (KR 2002), pp. 593–602. Morgan Kaufmann (2002). https://doi.org/10.5555/3087093.3087142

15. Camacho, A., Bienvenu, M., McIlraith, S.A.: Towards a unified view of AI planning and reactive synthesis. In: Proceedings of the 29th International Conference on Automated Planning and Scheduling (ICAPS 2019), pp. 58–67. AAAI Press (2019). https://doi.org/10.1609/icaps.v29i1.3460

16. Cimatti, A., Pistore, M., Roveri, M., Traverso, P.: Weak, strong, and strong cyclic planning via symbolic model checking. Artif. Intell. **147**(1–2), 35–84 (2003). https://doi.org/10.1016/S0004-3702(02)00374-0

17. Darwiche, A.: SDD: A new canonical representation of propositional knowledge bases. In: Proceedings of the 22nd International Joint Conference on Artificial Intelligence, pp. 819–826. AAAI Press (2011). https://doi.org/10.5591/978-1-57735-516-8/IJCAI11-143

18. De Giacomo, G., Favorito, M.: Compositional approach to translate LTL_f/LDL_f into deterministic finite automata. In: Proceedings of the 31st International Conference on Automated Planning and Scheduling (ICAPS 2021), pp. 122–130 (2021). https://doi.org/10.1609/icaps.v31i1.15954

19. De Giacomo, G., Favorito, M.: Compositional approach to translate LTLf/LDLf into deterministic finite automata. In: Biundo, S., Do, M., Goldman, R., Katz, M., Yang, Q., Zhuo, H.H. (eds.) Proceedings of the 31'st International Conference on Automated Planning and Scheduling (ICAPS 2021), pp. 122–130. AAAI Press (2021) https://doi.org/10.1609/icaps.v31i1.15954

20. De Giacomo, G., Favorito, M., Li, J., Vardi, M.Y., Xiao, S., Zhu, S.: LTLf synthesis as AND-OR graph search: knowledge compilation at work. In: Raedt, L.D. (ed.) Proceedings of the 31st International Joint Conference on Artificial Intelligence (IJCAI 2022), pp. 2591–2598. International Joint Conferences on Artificial Intelligence Organization (2022). https://doi.org/10.24963/ijcai.2022/359

21. De Giacomo, G., Rubin, S.: Automata-theoretic foundations of fond planning for LTL_f/LDL_f goals. In: Proceedings of the 27th International Joint Conference on Artificial Intelligence (IJCAI 2018), pp. 4729–4735 (2018). https://doi.org/10.24963/ijcai.2018/657

22. De Giacomo, G., Vardi, M.Y.: Linear temporal logic and linear dynamic logic on finite traces. In: Proceedings of the 23rd International Joint Conference on Artificial Intelligence (IJCAI 2013), pp. 854–860. IJCAI 2013, AAAI Press (2013). https://doi.org/10.5555/2540128.2540252

23. De Giacomo, G., Vardi, M.Y.: Synthesis for LTL and LDL on finite traces. In: Proceedings of the 24th International Joint Conference on Artificial Intelligence (IJCAI 2015), pp. 1558–1564. AAAI Press (2015). https://doi.org/10.5555/2832415.2832466

24. Duret-Lutz, A.: Supporting material for "Engineering an LTLf Synthetizer Tool" (2025). https://doi.org/10.5281/zenodo.15752968

25. Duret-Lutz, A., et al.: From Spot 2.0 to Spot 2.10: What's new? In: Proceedings of the 34th International Conference on Computer Aided Verification (CAV 2022). LNCS, vol. 13372, pp. 174–187. Springer (2022). https://doi.org/10.1007/978-3-031-13188-2_9

26. Ehlers, R., Lafortune, S., Tripakis, S., Vardi, M.Y.: Supervisory control and reactive synthesis: a comparative introduction. Discrete Event Dynamic Systems **27**(2), 209–260 (2017). https://doi.org/10.1007/s10626-015-0223-0

27. Esparza, J., Křetínský, J., Sickert, S.: One theorem to rule them all: a unified translation of LTL into ω-automata. In: Dawar, A., Grädel, E. (eds.) Proceedings of the 33rd Annual ACM/IEEE Symposium on Logic in Computer Science (LICS 2018), pp. 384–393. ACM (2018). https://doi.org/10.1145/3209108.3209161

28. Favorito, M.: Forward LTLf synthesis: DPLL at work. In: Benedictis, R.D., et al. (eds.) Proceedings of the 30th Workshop on Experimental evaluation of algorithms for solving problems with combinatorial explosion (RCRA 2023). CEUR Workshop Proceedings, vol. 3585 (2023). https://ceur-ws.org/Vol-3585/paper7_RCRA4.pdf

29. Favorito, M., Zhu, S.: LydiaSyft: a compositional symbolic synthesis framework for LTLf specifications. In: Proceedings of the 31st International Conference on Tools and Algorithms for the Construction and Analysis of Systems (TACAS 2025). LNCS, vol. 15696, pp. 295–302. Springer (2025). https://doi.org/10.1007/978-3-031-90643-5_15

30. Finkbeiner, B.: Synthesis of reactive systems. In: Javier Esparza, Orna Grumberg, S.S. (ed.) Dependable Software Systems Engineering, NATO Science for Peace and Security Series — D: Information and Communication Security, vol. 45, pp. 72–98. IOS Press (2016). https://doi.org/10.3233/978-1-61499-627-9-72

31. Finkbeiner, B., Geier, G., Passing, N.: Specification decomposition for reactive synthesis. In: Dutle, A., Moscato, M.M., Titolo, L., Muñoz, C.A., Perez, I. (eds.) NFM 2021. LNCS, vol. 12673, pp. 113–130. Springer, Cham (2021). https://doi.org/10.1007/978-3-030-76384-8_8

32. Fujita, M., McGeer, P.C., Yang, J.C.: Multi-terminal binary decision diagrams: an efficient data structure for matrix representation. Formal Methods Syst. Des. **10**(2/3), 149–169 (1997). https://doi.org/10.1023/A:1008647823331

33. Gabbay, D., Pnueli, A., Shelah, S., Stavi, J.: On the temporal analysis of fairness. In: Proceedings of the 7th ACM SIGPLAN-SIGACT Symposium on Principles of programming languages (POPL 1980), pp. 163–173. Association for Computing Machinery (1980). https://doi.org/10.1145/567446.5674

34. Gerevini, A., Haslum, P., Long, D., Saetti, A., Dimopoulos, Y.: Deterministic planning in the fifth international planning competition: PDDL3 and experimental evaluation of the planners. Artif. Intell. **173**(5–6), 619–668 (2009). https://doi.org/10.1016/j.artint.2008.10.012

35. Grädel, E.: Finite model theory and descriptive complexity. In: Finite Model Theory and Its Applications, chap. 3, pp. 125–230. Texts in Theoretical Computer Science an EATCS Series, Springer Berlin Heidelberg, Heidelberg (2007). https://doi.org/10.1007/3-540-68804-8_3

36. Henriksen, J.G., t al.: Mona: monadic second-order logic in practice. In: Brinksma, E., Cleaveland, W.R., Larsen, K.G., Margaria, T., Steffen, B. (eds.) First International Workshop on Tools and Algorithms for the Construction and Analysis of Systems (TACAS 1995), pp. 89–110. Springer, Heidelberg (1995). https://doi.org/10.1007/3-540-60630-0_5

37. Jacobs, S., Perez, G.A., Schlehuber-Caissier, P.: The temporal logic synthesis format TLSF v1.2. arXiV (2023). https://doi.org/10.48550/arXiv.2303.03839

38. Jacobs, S., et al.: The reactive synthesis competition (SYNTCOMP): 2018–2021. arXiV (2022). https://doi.org/10.48550/ARXIV.2206.00251

39. Klarlund, N., Møller, A.: MONA version 1.4, user manual. Tech. rep., BRICS (2001). https://www.brics.dk/mona/mona14.pdf

40. Lind-Nielsen, J.: BuDDy: a binary decision diagram package. User's manual (1999). https://web.archive.org/web/20040402015529/, http://www.itu.dk/research/buddy/

41. Liu, X., Smolka, S.A.: Simple linear-time algorithms for minimal fixed points. In: Larsen, K.G., Skyum, S., Winskel, G. (eds.) Proceedings of the 25th International Colloquium on Automata, Languages and Programming (ICALP 1998), pp. 53–66. Springer Berlin Heidelberg (1998). https://doi.org/10.1007/BFb0055035

42. Long, D.: BDD library. source archive. https://www.cs.cmu.edu/~modelcheck/bdd.html

43. Minato, S.: Representation of Multi-Valued Functions, pp. 39–47. Springer US, Boston, MA (1996). https://doi.org/10.1007/978-1-4613-1303-8_4

44. Piterman, N., Pnueli, A., Sa'ar, Y.: Synthesis of reactive(1) designs. In: Proceedings of the 7th international conference on Verification, Model Checking, and Abstract Interpretation (VMCAI 2006). LNCS, vol. 3855, pp. 364–380. Springer (2006). https://doi.org/10.1007/11609773_24

45. Pnueli, A., Rosner, R.: On the synthesis of a reactive module. In: Proceedings of the 16th ACM SIGPLAN-SIGACT symposium on Principles of Programming Languages (POPL1989). Association for Computing Machinery (1989). https://doi.org/10.1145/75277.75293

46. Renkin, F., Schlehuber-Caissier, P., Duret-Lutz, A., Pommellet, A.: Effective reductions of mealy machines. In: Proceedings of the 42nd International Conference on Formal Techniques for Distributed Objects, Components, and Systems (FORTE 2022). LNCS, vol. 13273, pp. 170–187. Springer (2022) https://doi.org/10.1007/978-3-031-08679-3_8

47. Renkin, F., Schlehuber-Caissier, P., Duret-Lutz, A., Pommellet, A.: Dissecting ltlsynt. Formal Methods Syst. Des. (2023). https://doi.org/10.1007/s10703-022-00407-6

48. Sickert, S., Meyer, P.: Modernizing strix (2021). https://www7.in.tum.de/~sickert/publications/MeyerS21.pdf

49. Somenzi, F.: CUDD: CU decision diagram package release 3.0.0 (2015). https://web.archive.org/web/20171208230728/, http://vlsi.colorado.edu/~fabio/CUDD/cudd.pdf

50. Xiao, S., Li, J., Zhu, S., Shi, Y., Pu, G., Vardi, M.: On-the-fly synthesis for LTL over finite traces. In: Proceedings of the 35th AAAI Conference on Artificial Intelligence (AAAI'21, Technical Track 7), pp. 6530–6537 (2021). https://doi.org/10.1609/aaai.v35i7.16809

51. Xiao, S., et al.: Model-guided synthesis for LTL over finite traces. In: Proceedings of the 25th International Conference on Verification, Model Checking, and Abstract Interpretation. LNCS, vol. 14499, pp. 186–207. Springer (2024). https://doi.org/10.1007/978-3-031-50524-9_9

52. Zhu, S., De Giacomo, G.: Synthesis of maximally permissive strategies for LTLf specifications. In: Raedt, L.D. (ed.) Proceedings of the 31st International Joint Conference on Artificial Intelligence (IJCAI 2022), pp. 2783–2789. ijcai.org (2022). https://doi.org/10.24963/IJCAI.2022/386

53. Zhu, S., Tabajara, L.M., Li, J., Pu, G., Vardi, M.Y.: Symbolic LTLf synthesis. In: Proceedings of the 26th International Joint Conference on Artificial Intelligence (IJCAI 2017), pp. 1362–1369 (2017). https://doi.org/10.24963/ijcai.2017/189

Subsequence Matching and Analysis Problems for Automata with Translucent Letters

Szilárd Zsolt Fazekas[1]([⊠]), Béla Klein[2], Tore Koß[2], Florin Manea[2], Robert Mercaş[3], and Timo Specht[2]

[1] Akita University, Akita, Japan
szilard.fazekas@ie.akita-u.ac.jp
[2] University of Göttingen, Göttingen, Germany
{bela.klein,tore.koss,timo.specht}@uni-goettingen.de,
florin.manea@informatik.uni-goettingen.de
[3] Loughborough University, Loughborough, UK
R.G.Mercas@lboro.ac.uk

Abstract. In this paper, we settle a series of algorithmic problems related to the (sets of) subsequences occurring in the strings of a given formal language, represented by a deterministic finite automaton with translucent letters accepting it, which were left open in [Fazekas et al., ISAAC 2024]. Firstly, we show that one can decide in polynomial time whether all elements of a given language have a given string as subsequence. We continue by considering the problems of deciding for a given language $L \subset \Sigma^*$ and a given positive integer $k > 0$ (respectively, for all integers $k > 0$) whether there exists a k-universal string in L (i.e., a string which has all possible length-k strings over Σ as subsequences). For both problems we show that they are NP-hard, and in the case of the second problem we give an NP-upper bound and thus achieve NP-completeness. For the other problem, i.e. for the question whether there exists a k-universal word in L given a language $L \subset \Sigma^*$ and a positive integer $k > 0$, we present a PSPACE algorithm.

Keywords: Subsequence · Universality · Automata with translucent letters

1 Introduction

A string v is a subsequence of a string w, denoted $v \leq w$, if there exist (possibly empty) strings $x_1, \ldots, x_{\ell+1}$ and $v_1, \ldots, v_\ell$ such that $v = v_1 \cdots v_\ell$ and $w = x_1 v_1 \cdots x_\ell v_\ell x_{\ell+1}$. That is, v can be obtained from w by removing some of its letters.

Szilárd Zsolt Fazekas was supported by JSPS Kakenhi Grant No. 23K10976. *Florin Manea* was supported by the German Research Foundation (Deutsche Forschungsgemeinschaft, DFG) in the framework of the Heisenberg Programme project number 466789228. *Robert Mercaş* was partially supported by a Return Fellowship from the Alexander von Humboldt Foundation for a research stay at University of Göttingen, Germany.

G. Castiglione and S. Mantaci (Eds.): CIAA 2025, LNCS 15981, pp. 148–164, 2026.
https://doi.org/10.1007/978-3-032-02602-6_11

Following [1,2,5], the focus of this paper is the study of the subsequences of strings of a formal language, the main idea behind this series of works being to extend the fundamental problems related to matching subsequences in a string and to the analysis of the sets of subsequences of a single string to the case of sets of strings. Firstly, [1] and its predecessor [2] approached algorithmic matching and analysis problems related to the universality of regular languages. More precisely, a string over Σ is called k-universal if its set of subsequences includes all strings of length k over Σ, and the respective papers presented algorithmic results and hardness results for problems asking whether a given regular language contains a k-universal string (for some given k, or, respectively, for all $k > 0$), and if all the words of a given regular language are k-universal (for a given $k > 0$). As a direct predecessor of this paper, [5] proposed a more systematic approach to such problems and defined a series of five algorithmic matching and analysis problems (listed below) related to matching subsequences in the words of regular, context-free, and context-sensitive languages and the universality of such languages, and showed results related to their decidability and complexity. The reader interested in the motivation behind the study of these subsequence-centered matching and analysis problems for formal languages, and their connections to other areas, as well as the study of subsequences in general, is referred to [5,8] and the references therein.

The Approached Problems and an Overview of our Results: This work is a direct continuation of [5], where the following five problems were defined and, as mentioned above, thoroughly considered for regular, context-free and context-sensitive languages.

Problem 1 (∃-Subsequence). Given language L by a machine (grammar) M accepting (resp., generating) it and a string w, is there a string $v \in L$ such that $w \le v$?

Problem 2 (∀-Subsequence). Given language L by a machine (grammar) M accepting (resp., generating) it and a string w, do all strings $v \in L$ satisfy $w \le v$?

Problem 3 (∃-k-universal). Given language L by a machine (grammar) M accepting (resp., generating) it and integer k, is there a k-universal string in L?

Problem 4 (∀-k-universal). Given language L by a machine (grammar) M accepting (resp., generating) it and integer k, are all strings of L k-universal?

Problem 5 (∞-universal). Given language L by a machine (grammar) M accepting (resp., generating) it, are there k-universal strings in L, for all integers $k > 0$?

To give some intuition on our terminology, Problems 1 and 3 can be seen as *matching problems* (asking to find a string which contains a certain subsequence or set of subsequences), while the other three problems are *analysis problems* (asking to decide properties concerning multiple strings of the language).

In [5], the authors conclude that the complexity of the approached problems is, to a certain extent, similar when the input languages are given by finite automata and context-free grammars: Problems 1, 2, 4, and 5 can be solved in polynomial time, while Problem 3 is NP-hard and FPT with respect to the size of the input alphabet as parameter in the case of CFL, but NP-complete in the regular case. Moreover, all problems become undecidable for context-sensitive languages. As such, a well-motivated, natural question is to identify classes of languages (ideally, strictly included in the class of context-sensitive languages, and also generalizing the class of regular languages) which exhibit a different behaviour. To this end, they suggested considering the class of languages accepted by deterministic finite automata with translucent letters (a novel, non-classical model of automata, see, e.g., [10,12] and the references therein). As an initial result, in [5], it was shown that Problem 1 is NP-complete in this setting; the other problems were left open. Here, we make significant progress in the investigation of these problems, when the input language is given as a deterministic finite automaton with translucent letters. Firstly, we give a polynomial-time algorithm for Problem 2. Secondly, we show the NP-hardness for Problems 3 and 5, by giving a reduction from the Hamiltonian cycle problem. Thirdly, we present a PSPACE algorithm for Problem 3 and in the last section of this paper we focus on the proof that Problem 5 is contained in NP. In particular, the P-membership of Problem 2 and the NP-membership of Problem 5 are obtained by rather involved, novel combinatorial insights and arguments.

2   Preliminaries

Let $\mathbb{N} = \{1, 2, \ldots\}$ denote the natural numbers and set $\mathbb{N}_0 = \mathbb{N} \cup \{0\}$ as well as $[n] = \{1, \ldots, n\}$ and $[i, n] = \{i, i+1, \ldots, n\}$, for all $i, n \in \mathbb{N}_0$ with $i \leq n$.

An *alphabet* $\Sigma = \{1, 2, \ldots, \sigma\}$ is a finite set of symbols, called *letters*. A *word* or *string* w is a finite concatenation of letters from a given alphabet with the number of these letters giving its *length* $|w|$. The word with no letters is the *empty word* ε of length 0. The set of all finite words over the alphabet Σ, denoted by Σ^*, is the free monoid generated by Σ with concatenation as operation. A subset $L \subset \Sigma^*$ is called a *(formal) language*. Let Σ^n denote all words in Σ^* exactly of length $n \in \mathbb{N}_0$.

For $1 \leq i \leq j \leq |w|$, denote the ith letter of w by $w[i]$ and the *factor* of w starting at position i and ending at position j as $w[i, j] = w[i] \cdots w[j]$. If $i = 1$ the factor is also called a prefix, while if $j = |w|$ it is called a suffix of w. For each $a \in \Sigma$, let $|w|_a = |\{i \in [|w|] \mid w[i] = a\}|$ denote the number of occurrences of letter a in w.

Let alph(w) denote the set of all letters of Σ occurring in w. A string $u \in \Sigma^*$ is called a *subsequence* of w, denoted $u \leq w$, if there exist $w_1, \ldots, w_{n+1} \in \Sigma^*$ such that $w = w_1 u[1] w_2 u[2] \cdots w_n u[n] w_{n+1}$. For $k \in \mathbb{N}_0$, a string $w \in \Sigma^*$ is called k-*universal* (w.r.t. Σ) if every $u \in \Sigma^k$ is a subsequence of w. The *universality-index* $\iota(w)$ is the largest k such that w is k-universal.

Definition 1. *Let Σ be some finite alphabet and $A \subset \Sigma$. The projection onto A is the word morphism $\pi_A : \Sigma^* \to A^*$, defined by $\pi_A(a) := a$ if $a \in A$; and ε otherwise.*

Definition 2. *The arch factorization of $w \in \Sigma^*$ is given by $w = ar_1(w)\cdots$ $ar_{\iota(w)}(w)r(w)$ with $\iota(ar_i(w)) = 1$ and $ar_i(w)[|ar_i(w)|] \notin \mathrm{alph}(ar_i(w)[1,|ar_i (w)| - 1])$, for all $i \in [1, \iota(w)]$. Furthermore, $\mathrm{alph}(r(w)) \subsetneq \Sigma$ applies. The strings $ar_i(w)$ are called* arches *and $r(w)$ is called* the rest *of w.*

As an example, in the factorization $w = (bca)\cdot(accab)\cdot(cab)\cdot b$ of $w \in \{a,b,c\}^*$ it holds $\iota(w) = 3$, the parentheses mark the three arches and $r(w) = b$. For more details on the arch factorization and the universality index see [3,7].

A k-universal string of length $k\sigma$, over $\Sigma = \{a_1,\ldots,a_\sigma\}$, is called *short k-universal string.* For a short k-universal string w over Σ, there exist $\tau_1,\ldots,\tau_k$ permutations of $[\sigma]$ such that $w = (a_{\tau_1(1)}\cdots a_{\tau_1(\sigma)})\cdots(a_{\tau_k(1)}\cdots a_{\tau_k(\sigma)})$. The following holds.

Lemma 1. *Let $v \in \Sigma^*$. The string v is k-universal if and only if there exists $w \in \Sigma^*$ with $w \leq v$ and w is a short k-universal string.*

This work focuses on the class of languages accepted by a non-classical model of automata, namely, the deterministic finite automata with translucent letters (or, for short, translucent finite automaton – TFA). This model generalizes the classical deterministic finite automata, DFA, by allowing the processing of the input string in an order which is not necessarily the usual sequential left-to-right order, without the help of an explicit additional storage unit. These automata, first considered in [11] (see also the survey [12] for a discussion on their properties and motivations), are strictly more powerful than classical finite automata and are part of a larger class of automata-models that are allowed to jump symbols in their processing, e.g., see [9] or [4]. From our perspective, these automata and the class of languages they accept are interesting because, on the one hand, they seem to be a generalization of regular languages which is orthogonal to the classical generalization provided by context-free languages, and, on the other hand, the problems considered in [5] become harder for them, compared to the class of regular and context-free languages, while staying decidable (unlike the case of context-sensitive languages). We first define the TFA model, following the formalization from [6].

Definition 3. *A finite automaton with translucent letters (TFA) M is a tuple $M = (Q, \Sigma, q_0, F, \delta)$, just as in the case of DFA. The partial relation $\circlearrowleft$ on the set $Q \times \Sigma^*$ of configurations of M is defined as: $(p, xay) \circlearrowleft_M (q, xy)$ if $\delta(p,a) = q$, and $\delta(p,b)$ is not defined for any $b \in \mathrm{alph}(x)$, where $p,q \in Q$, $a,b \in \Sigma$, $x,y \in \Sigma^*$. The subscript M is omitted when it is understood from the context. The reflexive and transitive closure of $\circlearrowleft$ is $\circlearrowleft^*$ and the language accepted by M is defined as $L(M) = \{w \in \Sigma^* \mid (q_0, w) \circlearrowleft^* (f, \varepsilon) \text{ for some } f \in F\}$.*

Remark 1. For the rest of this paper we fix the notations $M = (Q, \Sigma, q_0, F, \delta), n = |Q|$ and $\sigma = |\Sigma|$. For all considered problems, the input is such a TFA.

In this model, letters a such that $\delta(p,a)$ is not defined are called *translucent*, for state p, hence the name of the model. The machine reads, and erases from the tape, the letters of the input one-by-one. Note that the definition requires that every letter of the input is read before the respective input can be accepted. This differs slightly from the original definition [11], which did not require reading all letters and used an unerasable endmarker on the tape. A TFA under our definition is trivially simulated by a machine using the original one, thus our hardness results also apply to that model. We adopted the definition from [10] as it is, in our view, simpler and equally illustrates the difficulty of the subsequence matching problems for nonsequential machine models, but with very few changes our algorithmic results stand also for the original TFA model.

We note that, in terms of execution, in each step a TFA reads (and consumes) the leftmost unconsumed symbol which allows a transition (i.e., that was not previously read, and there is a transition labeled with it from the current state). Therefore, for every individual letter, the order of the processing of its occurrences in the TFA is that in which they appear in a string. The non-deterministic version of this automata model accepts all rational trace languages, and all accepted languages have semi-linear Parikh images. Moreover, the class of languages accepted by this model is incomparable to the class of CFL, while still being CS. The class of languages accepted by the more restrictive TFA strictly includes the class REG and is still incomparable with CFL. The survey [12] overviews the extensive literature regarding variations of these types of machines.

3   Polynomial Time Algorithm for Problem 2

We first give an intuitive overview of the decision procedure, then present the pseudocode for the algorithm and finally a sketch of the proof of its correctness and time complexity. The question we need to decide for a given TFA M with n states and a string w is whether all words accepted by M contain w as a subsequence. Our approach is to look for a counterexample $v \in L(M)$ with $w \not\leq v$ up to a given length and if we cannot find any, the answer to the Problem 2 instance is YES. Following the definition of TFA, if a word v' is accepted along a walk on its state transition graph, then there is a $v \leq v'$ such that $|v| < n$ and $v \in L(M)$, obtained by removing the (not necessarily contiguous) letters of v' that are read in a loop along the accepting walk, to obtain v. If a counterexample v that does not contain w exists, then its length must be less than n.

For the analogous problem in the case of DFA, having a counterexample is equivalent to the existence of a simple path leading to a final state, whose label does not contain w as a subsequence. In the case of TFA, that is still a sufficient condition, but not necessary, as even if all paths from the initial state to final states contain w as a subsequence, a counterexample may exist. As a simple illustration consider the TFA with 3 states and transitions $\delta(q_0, a) = q_1$, $\delta(q_1, b) = q_2$ with q_0 initial and q_2 final, and let $w = ab$. Note that the TFA accepts the language $\{ab, ba\}$, so the answer to this Problem 2 instance is NO, but the only accepting path in the transition graph has label ab which contains w. In general, that can happen if some letter of the pattern w is read 'out-of-order' from the input, as the a was above in the input ba. Intuitively, checking for that requires remembering the 'jumps' (transitions from a state with some translucent letters) the TFA makes when matching letters of the pattern against the input. The difficulty is that remembering all such jumps (and the possible factors jumped over) would potentially lead to intractability, among other reasons, because the number of different paths of a given length in an automaton can be exponential in the length. From the viewpoint of an adversary looking for a counterexample $v \in L(M)$ with $w \not\leq v$ the following ideas reduce the search space enough that a polynomial time algorithm exists.

Store Only the Worst *Walk per State.* Fix a state q and a length bound ℓ. Now, order all simple walks from q_0 to q of length at most ℓ increasingly by the length of the longest prefix of w occurring as subsequence in their label and let $\pi_{\min}$ be the smallest such path. Extend every such walk by the same suffix. Because the suffix starts in the *same* state q, all extended walks read the *same* continuation; thus the extension of $\pi_{\min}$ still matches *no more* of w than any other. Therefore it suffices to remember only the "worst so far" matched prefix of w.

Record only Jumps Involving an Adjacent *pair of Pattern Letters* (a_i, a_{i+1}). Reading letters that do not occur in w or jumping over them can be disregarded as they do not affect w being a subsequence of the input (with the exception of discarding jumps in the next paragraph). To find a counterexample, the machine must eventually read $w = a_1 \ldots a_k$ out of order. Any long-range inversion implies an *adjacent* inversion: if the matched a_i in the input is to the right of a factor containing $a_{i+1}, \ldots, a_{i+j}$, then, in particular it is to the right of a_{i+1}. Hence it is enough to track whether the pair of consecutive pattern letters (a_i, a_{i+1}) have been matched by a factor $a_{i+1} \ldots a_i$. This can only happen if there is a transition reading a_i from a state p where a_{i+1} is translucent. We must remember p, because the next a_{i+1} read during the computation may have been to the left of the a_i, as above. Technically, we do not have to remember the state itself: it suffices to know the set of translucent letters in p, however, for a simpler presentation of the algorithm it is more convenient to store the state and look up its translucent letters as needed.

We Can Discard a Remembered Jump. The assumption "a_{i+1} was skipped" remains plausible until it is refuted by a transition $q \xrightarrow{c} r$ satisfying *both* (1.) c is not translucent in p, and (2.) the still-pending letter a_{i+1} is not translucent in q. Then c forms a left barrier: no unread a_{i+1} can hide left of c (and of the earlier a_i), so the adversary can discard p.

Algorithm 1: ALLSUBSEQ-TFA(M, w)

Input: Deterministic TFA $M = (Q, \Sigma, q_0, F, \delta)$, pattern $w = a_1 \dots a_k$
Output: YES iff every $v \in L(M)$ contains w as a subsequence

1 **if** $k \geq |Q|$ **then**
2 $\quad \lfloor$ **return** NO

3 **foreach** $q \in Q$ **do**
4 $\quad \lambda(q) \leftarrow \infty$
5 $\quad \mathcal{S}(q) \leftarrow \emptyset$

6 $\lambda(q_0) \leftarrow 0$ // none of the pattern was matched
7 $round \leftarrow 0$
8 **while** $round < |Q|$ **do**
9 $\quad$ **foreach** $p \in Q$ **do**
10 $\quad\quad$ **foreach** $c \in \Sigma$ with $\delta(p, c) = r$ **do**
11 $\quad\quad\quad \lfloor$ UPDATE(p, c, r)

12 $\quad round \leftarrow round + 1$;

13 **if** $\forall f \in F: \lambda(f) = k$ **then return YES**
14 **else return** NO

Algorithm 2: UPDATE(p, c, r)

1 $d \leftarrow \lambda(p)$
2 $\ell \leftarrow \lceil d \rceil + 1$ // index of upcoming pattern letter
3 **if** d *is integer* $< k$ **and** $c = a_\ell$ **then**
4 $\quad$ **if** $\delta(p, a_{\ell+1}) = \emptyset$ **then**
5 $\quad\quad \hat\lambda \leftarrow d + \frac{1}{2}$ // Case A: matched a_ℓ possibly jumping
6 $\quad\quad \lfloor \hat{\mathcal{S}} \leftarrow \{p\}$
7 $\quad$ **else**
8 $\quad\quad \hat\lambda \leftarrow d + 1$ // Case A': matched a_ℓ without jump
9 $\quad\quad \lfloor \hat{\mathcal{S}} \leftarrow \emptyset$

10 **else if** $d = m + \frac{1}{2}$ **and** $\delta(p, a_\ell) \neq \emptyset$ **then**
11 $\quad \hat{\mathcal{S}} \leftarrow \mathcal{S}(p)$ // Case B
12 $\quad$ **foreach** $q \in \mathcal{S}(p)$ **do**
13 $\quad\quad$ **if** $\delta(q, c) \neq \emptyset$ **then**
14 $\quad\quad\quad \lfloor \hat{\mathcal{S}} \leftarrow \hat{\mathcal{S}} \setminus \{q\}$

15 $\quad$ **if** $\hat{\mathcal{S}} = \emptyset$ **then**
16 $\quad\quad \lfloor \hat\lambda \leftarrow m + 1$ // forced to match $a_{\ell-1}$ on all walks to p

17 **else**
18 $\quad \hat\lambda \leftarrow d$ // Case C
19 $\quad$ **if** d *is half-integer* **then**
20 $\quad\quad \hat{\mathcal{S}} \leftarrow \mathcal{S}(p)$
21 $\quad$ **else**
22 $\quad\quad \lfloor \hat{\mathcal{S}} \leftarrow \emptyset$

23 **if** $\hat\lambda < \lambda(r)$ // Case D
24 **then**
25 $\quad \lambda(r) \leftarrow \hat\lambda$ // the new walk to r matches less of the pattern
26 $\quad \mathcal{S}(r) \leftarrow \hat{\mathcal{S}}$

27 **else if** $\hat\lambda = \lambda(r) = m + \frac{1}{2}$ **then**
28 $\quad \lfloor \mathcal{S}(r) \leftarrow \mathcal{S}(r) \cup \hat{\mathcal{S}}$ // jumping matches along multiple walks

Theorem 1. *Problem 2 can be decided in* $\mathcal{O}(n^3\sigma)$ *time by Algorithm 1.*

Proof (Proof Sketch). For a state q the value $\lceil \lambda(q) \rceil$ is the length of the longest prefix of w contained by *all* words read along any walk of length *round* from q_0 to q. The value is a half-integer when the state that matched the most recent letter from the pattern may have jumped over the subsequent pattern letter, and $\mathcal{S}(q)$ is the set of states where such jumps may have occurred on the walk to q. The dynamic programming technique for remembering only the 'worst walks' with the shortest matched pattern prefix is standard (Case D). The bookkeeping of states where pattern letters are matched by jumps (Case A) and the subsequent discarding of such states (Case B) is illustrated in Fig. 1. Case C corresponds to merely propagating values along transitions that do not involve matching current pattern letters. The algorithm terminates after round n (checking all walks of length $< n$). If $\lambda(f) = k$ for each final state f, then every accepted word of length $n - 1$ contains w and we answer YES, otherwise we answer NO.   □

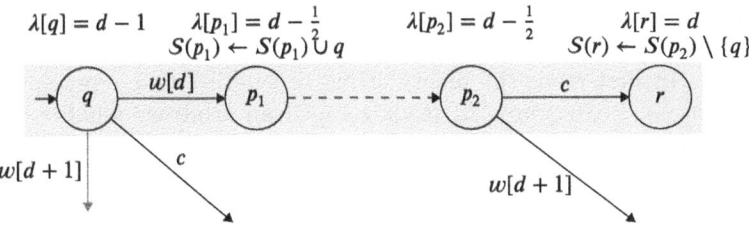

Fig. 1. Entering and exiting half-integer λ phase (red is undefined transition, gray rectangle illustrates considered path, $d \in [|w| - 1]$). When the computation reaches r through p_2, the c just read from the tape can not occur before the $w[d]$ read in q, since c is not transparent in q. As we did not take $w[d+1]$ transition from p_2, any remaining occurrences of $w[d+1]$ in the input are to the right of that c and hence to the right of the $w[d]$ occurrence read in q, so we must consider them as matches for the upcoming letter of w. (Color figure online)

4   NP-Hardness of Problems 3 and 5

Theorem 2. *Problem 1, Problem 3, and Problem 5 are NP-hard for TFA.*

Proof. To prove that the considered problems are NP-hard we give a reduction from the Hamiltonian cycle problem. This reduction is a modified version of the one used in [5], where the hardness was already shown for Problem 1. For the Hamiltonian cycle problem, we are given a simple undirected graph $G = (V, E)$, where $V = \{v_1, \ldots, v_m\}$ is the set of vertices, and E is the set of edges. If G has a cycle that visits each vertex exactly once, which is called a Hamiltonian cycle, the problem is answered positively; otherwise, a negative answer is given. This problem is known to be NP-complete. Therefore, we construct from G a TFA A_G that accepts certain words if and only if G admits a Hamiltonian cycle. We assume w.l.o.g. that we start at v_1.

Let $A_G = (Q, \Sigma, q_0, F, \delta)$, where $\Sigma = \{v_1, \ldots, v_m, x_1, \ldots, x_m, 0, 1\}$, Q is the set of states, q_0 is the initial state, F is the set of final states, and δ is the transition function. The automaton consists of several gadgets $M_1, \ldots, M_{m+1}$, where M_i decides on the $(i+1)$-th vertex on the cycle. Within M_i the decision depends on the reachable vertices from the i-th vertex. On the one hand, M_i has states $u_{i,j}$ and $v_{i,j}$, for $i, j \in [1, m]$, and on the other hand there are states that allow the transition from $u_{i,j}$ to $v_{i,j'}$ if and only if $(j, j') \in E$. The automaton transitions from $u_{i,j}$ to $v_{i,j'}$ by reading the sequence $b_{j'} = bin^{\ell}(j'-1)$, where $bin^{\ell}(j'-1)$ is the binary representation of length ℓ of the integer $j'-1$ over $\{0, 1\}$ with $\ell = \lceil log_2(m) \rceil$. Hence, the transition from $u_{i,j}$ to $v_{i,j'}$ means that $v_{j'}$ is the next vertex we visit. Reading a sequence of length ℓ, representing $j'-1$, such that $j' > m$ or $(j, j') \notin E$ leads to a sink state z_i, because in the case of $(j, j') \notin E$ we do not have an edge between the two vertices and, for $j' > m$, no vertex exists. Furthermore, M_i contains the states $w_{i,j}^x$, $w_{i,j}^0$, $w_{i,j}^1$ with the associated transitions $\delta(v_{i,j}, x_j) = w_{i,j}^x$, $\delta(v_{i,j}, v_j) = z_i$, $\delta(w_{i,j}^x, v_j) = w_{i,j}^0$, $\delta(w_{i,j}^0, 1) = z_i$, $\delta(w_{i,j}^0, 0) = w_{i,j}^1$, $\delta(w_{i,j}^1, 0) = z_i$, $\delta(w_{i,j}^1, 1) = v_{i,j}$, and $\delta(w_{i,j}^x, x_j) = u_{i+1,j}$. Hence, to transition from $u_{i,j}$ to $u_{i+1,j'}$ there needs to be an edge $(j, j') \in E$. If this is the case, A_G first reads the binary string of length ℓ that represents $j'-1$ to end up in $v_{i,j'}$. Afterward A_G reads $x_{j'}v_{j'}01$ an arbitrary number of times. Lastly, it reads $x_{j'}$ twice to reach $u_{i+1,j'}$. Thus, this cycle is used to pump up the universality which is important for Problem 3 and Problem 5.

The gadget M_{m+1} consists only of the states $u_{m+1,1}, \ldots, u_{n+1,n}$, and within M_{m+1} there are no transitions. The single accepting state is $u_{m+1,1}$, since we need to end up at v_1 to close the cycle. Thus, A_G accepts the language below (where ⊔ means shuffle)

$$L(A_G) = \{(x_{i_1}, v_{i_1})^{k_1} x_{i_1} ⊔ \cdots ⊔ (x_{i_m}, v_{i_m})^{k_m} x_{i_m} ⊔ b_{i_1}(01)^{k_1} \cdots b_{i_m}(01)^{k_m} \mid$$
$$k_1, \ldots, k_m \in \mathbb{N}, b_j = bin^{\ell}(j-1), \text{ for } j = i_1, \ldots, i_m, \text{ and } v_{i_1} \cdots v_{i_m}$$
$$\text{a walk on } G\}.$$

The numbers of states and transitions in each gadget are in $\mathcal{O}(m^2)$. The same holds for the number of transitions between gadgets. So, overall, the numbers of states and transitions of A_G are in $\mathcal{O}(m^3)$. The automaton accepts a supersequence of $v_1 \ldots v_m$ if and only if G admits a Hamiltonian cycle. Additionally, for every $k > 0$, we get that A_G accepts a k-universal string if and only if G admits a Hamiltonian cycle. □

Theorem 3. *Problem 3 is in PSPACE.*

Proof (Proof Sketch). The algorithm is non-deterministic and runs in $\mathcal{O}(\sigma \log k + \log n)$ space. What we need to store are the following:

1. **Counters** left(a) **for every letter** $a \in \Sigma$. Initially left$(a) = k$. Intuitively left(a) is the number of arches in which we still have to read a (needs $\mathcal{O}(\sigma \log k)$ bits).
2. **Current state** $q \in Q$ (needs $\mathcal{O}(\log n)$ bits), initially $q = q_0$.

3. **Stagnation counter** $S \in [0, n]$ recording how many steps ago *any* left$(\cdot)$ value was decremented (needs $\mathcal{O}(\log n)$ bits).

In each step, the machine non-deterministically picks a letter a and a transition $\delta(q, a) = q'$ such that, for every letter b with left$(b) >$ left(a), the letter b must be *translucent* in state q. Note that the condition is satisfied by any letter with maximal left$(\cdot)$ counter among the non-translucent ones in q. Once a is picked with its corresponding transition, we non-deterministically choose between two options:

(A) Use a in a *later* arch. If left$(a) > 0$, decrease it by 1 and reset $S \leftarrow 0$.
(B) Use a in the *current* arch. Leave left$[a]$ unchanged and set $S \leftarrow S + 1$.

Finally we update the current state $q \leftarrow q'$, and pick the next letter according as before, unless one of the following halting conditions apply:

- **Accept**, if $q \in F$ and left$(a) = 0$, for every $a \in \Sigma$.
- **Reject**, if $S > n$ (*no counter was decremented for more than n consecutive steps*).

Correctness. The algorithm simulates accepting computations of (not necessarily all) strings that have some factorization into k factors, each containing the whole alphabet. If the algorithm accepts, then one can construct an at least k-universal string from the accepting path, letter-by-letter: whenever an a is chosen in the algorithm, we insert a into the first available position in arch $k - $ left$(a) + 1$.

For the converse, we need to argue that the rejection conditions and stopping the stagnation counter at 0 do not exclude *all* k-universal strings. If none of the left(a) values was decremented in the last n steps, it means that we reached the same state p twice during that period and each of the letters read between those two visits were from an arch from which we already read such letters previously, so they can be removed without changing the universality of the string. The string with the fewest arches $\geq k$ may actually have some $\ell > k$ arches. However, note that for such a string, whenever left$(b) > 0$, for some b, and we read c such that left$(c) = 0$, we may 'move' that c to the k-th arch, without affecting the acceptance of the string or reducing its universality below k, modeled in the algorithm by stopping the counter S at 0.   □

5   NP Inclusion of Problem 5

In the case of DFA, infinite universality reduces to checking whether there exists an accepting computation that includes a 1-universal cycle (i.e., containing all letters of the alphabet). In the case of TFA that is sufficient but not necessary, as illustrated by the example in Fig. 2. Using translucent transitions, it is possible to read a given letter from every arch before processing any of the other letters. If a word with universality index larger than n is accepted, then either we have a 1-universal cycle, or we have several cycles that 'add up' to 1-universality. This property (defined later as UCSC) can be intuitively understood in terms of the component graph of the TFA, that is the graph whose nodes are the strongly connected components of the transition graph. Say, there is a word $w \in L(M)$, for some TFA M, such that $\iota(w) > n$, and there is no 1-universal cycle in M. The accepting computation for w corresponds to a simple path in the component graph, visiting components that consume letters from only a 'few' arches and others that consume from 'many' arches. Let the latter be $Q_1, \ldots, Q_m$. Consider the last component on this path, Q_m. Since Q_m is not 1-universal, there is at least one letter $a \in \Sigma$ that does not occur in Q_m, so earlier components must consume all (or most, as the paths connecting the components can also consume some) occurrences of a. For the sake of simplicity, assume there is one such component, Q_i with $i < m$. Since we assumed that Q_m consumes letters from multiple arches, those letters must still be on the tape when reaching Q_m; but that is only possible if they were to the right of the previously read letters or, crucially, if they have been jumped over by earlier transitions. In our simplified scenario, Q_i reads all the a's from the arches, so it must read the a in the last arch; but that means jumping over all letters to be read later by Q_m, from previous arches. This implies Q_i having some state in which a is not translucent, but all the letters to be read in Q_m are. Analogously, for each of the letters we must have some component reading it from arbitrarily many arches, thereby having to jump over letters read in later components. The proof formalizes this idea: having such a sequence of components that can read letters while jumping over the alphabet of later components is both sufficient and necessary to accept words of arbitrarily high universality index.

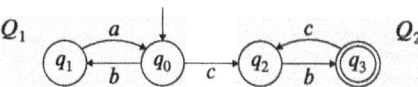

Q_1 and Q_2 are the cycles referred to after Definition 6 for this particular example.

Fig. 2. TFA with unbounded universality (inf-ESU): for any k, the k-universal word $b^k c (abc)^k$ is accepted, even though there is no 1-universal cycle in the TFA.

In this chapter, we denote the subclass of TFA which accept words of arbitrary universality by inf-ESU. Theorem 2 established the NP-hardness of inf-ESU, thus, for NP-completeness, it remains to show that the problem is contained in NP. We introduce the class UCSC, of TFA which *universally constructs subsequences on cycles*, we show that UCSC is in NP, and finish by showing UCSC = inf-ESU. To start we make the observation that the automaton $M \in$ inf-ESU if and only if every word $w \in \Sigma^*$ is a subsequence of a word accepted by M.

Lemma 2. *Let Σ be an alphabet and $L \subset \Sigma^*$ be some language over Σ. Then, every word $w \in \Sigma^*$ is a subsequences of a word in L if and only if for every integer $k \in \mathbb{N}$ there is a word $w_k \in L$ with universality index larger or equal to k.*

Before we give a formal definition of UCSC we introduce the following notations.

Definition 4. Let $\Sigma = \bigsqcup_{i=1}^{m} \Sigma_i$ be a disjoint decomposition (partition) of Σ, for some $m \in \mathbb{N}$. For $i \in [1, m]$, we set $\sigma_i = |\Sigma_i|$ and define:
$$\Sigma_i^< = \bigcup_{j=1}^{i-1} \Sigma_j, \quad \Sigma_i^\leq = \bigcup_{j=1}^{i} \Sigma_j, \quad \Sigma_i^> = \bigcup_{j=i+1}^{m} \Sigma_j, \quad \Sigma_i^\geq = \bigcup_{j=i}^{m} \Sigma_j.$$

Definition 5. Let M be a TFA, and consider the states $p, p' \in Q$ and word $v \in \Sigma^*$, such that there is the computation $(p, v) \vdash_M^* (p', \varepsilon)$. Let $A, B \subset \Sigma$.

1. We say that this computation reads A independently of B, if, during the computation, whenever some letter of A is read, then all letters of B are translucent. Formally:
   Let $(q, u\alpha u') \vdash_M (q', uu')$ be a single step of this computation for some states $q, q' \in Q$, words $u, u' \in \Sigma^*$ and a letter $\alpha \in \Sigma$, i.e., $(p, v) \vdash_M^* (q, u\alpha u') \vdash_M (q', uu') \vdash_M^* (p', \varepsilon)$. Then, $\alpha \in A$ implies $\delta(q, a) = \emptyset$, for all $a \in B$.
2. We say this computation reads A independently, if it reads A independently of $\Sigma \setminus A$.

To decide whether some TFA M is in inf-ESU, we examine for which states there are paths that contain some subsets independently from other subsets. The following lemma states that we only need to consider paths of a certain length.

Lemma 3. Let M be a TFA. Let $w \in \Sigma^*$, $p, q \in Q$ and $A, B, C \subset \Sigma$, such that $A \subset \text{alph}(w)$ and there is the computation $(p, w) \vdash_M^* (p', \varepsilon)$ that reads B independently of C. Then there is a subsequence $v \leq w$ of length $|v| \leq n \cdot |A| + (n-1)$, such that

1. $(p, v) \vdash_M^* (p', \varepsilon)$ reads B independently of C, and   2. $A \subset \text{alph}(v)$.

This brings us to the central definition of this section that combines all the observations about automata in inf-ESU, we made so far.

Definition 6. Let M be a TFA. We say that M universally constructs subsequences on cycles if, for some $m \leq |\Sigma| + 1$, there are a disjoint decomposition $\Sigma = \bigsqcup_{i=1}^{m} \Sigma_i$ with $|\Sigma_i| = \sigma_i$, some states $p_1, ..., p_{m-1} \in Q$ and a final state $p_m \in F$ such that the following conditions hold for $i \in [1, m]$:

1. There is a word $v_i \in \Sigma^*$ of length $|v_i| \leq n\sigma_i + (n-1)$, such that $\Sigma_i \subset \text{alph}(v_i)$ and we have the computation $(p_i, v_i) \vdash_M^* (p_i, \varepsilon)$, that reads $\Sigma_i^<$ and $\Sigma_i^\leq$ independently.
2. There is a word $w_i \in \Sigma^*$ of length $|w_i| \leq n-1$, such that there is the computation $(p_{i-1}, w_i) \vdash_M^* (p_i, \varepsilon)$ that reads $\Sigma_i^<$ independently.

We call the class of all TFA that universally construct subsequences on cycles UCSC.

Intuitively, by Definition 6 we can split a given TFA into cycles $Q_1, \ldots, Q_m$ such that Q_i consumes a subset Σ_i of Σ and $\bigcup_{j=1}^{m} \Sigma_j = \Sigma$. Furthermore, there is a subset $A_i \subset \Sigma_i$ such that A_i is read independently. Moreover, on the path between Q_{i-1} and Q_i whenever we get to a letter from $A_1 \sqcup \ldots \sqcup A_{i-1}$, all letters not in this set are translucent. Our goal is to show that this definition is in fact equivalent to TFA in inf-ESU. We first prove that we can decide whether some TFA M universally constructs cycles in NP-time.

Proposition 1. *For a TFA M, the decision problem $M \in$ UCSC is in NP.*

Proof. If $M = (Q, \Sigma, p_0, F, \delta) \in$ UCSC, then, for an $m \leq |\Sigma| + 1$, states $p_1, \ldots, p_m \in Q$, partition $\Sigma_1, \ldots, \Sigma_m$ of Σ, and words $v_1, \ldots, v_m \in \Sigma^{\leq n\sigma_i + n - 1}$ and $w_1, \ldots, w_m \in \Sigma^{\leq n-1}$, we can check in polynomial time that the conditions of Definition 6 are satisfied. $\square$

Next, consider a partition $\{\Sigma_1, \Sigma_2\}$ of Σ with $w_1 \in \Sigma_1^*, w_2 \in \Sigma_2^*$ and a TFA M, such that there is a computation $(p, w_1 w_2) \vdash_M^* (p', \varepsilon)$ that reads Σ_1 independently. However, since the computation reads Σ_1 independently of Σ_2, all letters contained in Σ_2 are translucent if a letter from Σ_1 is read. Hence, we can rearrange the letters of $w_1 w_2$ obtaining the same computation, e.g. $(p, w_2 w_1) \vdash_M^* (p', \varepsilon)$. The next lemma formalizes this.

Lemma 4. *Let M be a TFA and $\bigsqcup_{j=1}^{m} \Sigma_j = \Sigma$ be a disjoint decomposition of Σ, for some $m \in \mathbb{N}$. Furthermore, consider the states $p, p' \in Q$ and a word $w \in \Sigma^*$, such that there is the computation $(p, w) \vdash_M^* (p', \varepsilon)$ that reads $\Sigma_i^{\leq}$ independently, for every $i \in [1, m]$, and let $w_i := \pi_{\Sigma_i}(w)$ and $u_i \in \left(\Sigma_i^{\geq}\right)^*$, for every $i \in [1, m]$. Then, there also exists the computation $(p, w_m u_m \cdots w_1 u_1) \vdash_M^* (p', u_m \cdots u_1)$.*

Finally, we consider a TFA $M \in$ UCSC and prove that $M \in$ inf-ESU by constructing, for arbitrary $k \in \mathbb{N}$, a word $s \in L(M)$ with universality index k. We obtain s by using the words described in Definition 6 in combination with the results from Lemma 4.

Proposition 2. *Let M be a TFA that universally constructs subsequences on m cycles, for some $m \in \mathbb{N}$. Then, for every $k \in \mathbb{N}$, there is a k-universal word $s_k \in L(M)$.*

Now, we show that any TFA in inf-ESU is in UCSC, as well. Let $M \in$ inf-ESU be a TFA over the finite alphabet $\Sigma = \{a_1, \ldots, a_\sigma\}$. As M is in inf-ESU, for every $d \in \mathbb{N}$, there is a supersequence v_d of the word $w_d = (a_1 \cdots a_\sigma)^d$ that is accepted by M. If, during the computation of v_d by M, some letters are read a large number of times, while others are not read, this suggests the existence of a cycle that contains the letters read. Before we delve deeper into this consideration, we introduce a notion to specify in how many arches a letter was read.

Definition 7. *Let M be a TFA and $v \in L(M)$ be a word with the arch fac-torization $v = \mathrm{arch}_1(v) \cdots \mathrm{arch}_{\iota(v)}(v)\, \mathrm{r}(v)$. For some $t \in \mathbb{N}$, let v_t be the word* *obtained by performing t single computation steps with the automaton M on the input v, that is $v \vdash_M^t v_t$. For each $i \in [1, \iota(v)]$, let $\mathrm{arch}_{i,t}(v)$ be the subsequence of letters left from $\mathrm{arch}_i(v)$, and let $\mathrm{r}_t(v)$ be the subsequence of letters left from $\mathrm{r}(v)$, respectively, after computing t single computation steps on input v by M. That is $v_t = \mathrm{arch}_{1,t}(v) \cdots \mathrm{arch}_{\iota(v),t}(v) \, \mathrm{r}_t(v)$. For some letter $\alpha \in \Sigma$ we define the number of arches in which α was read by M on input v after t single computation steps as*

$$\mathrm{read}_M(v, \alpha, t) := |\{i \in [1, \iota(v)] \mid |\mathrm{arch}_i(v)|_\alpha \neq |\mathrm{arch}_{i,t}(v)|_\alpha\}|.$$

During the computation of a word by some TFA, any letter read in a single computation step is always the first letter visible in the current state (any earlier occurrence of this letter in the word remaining on the tape is visible and, therefore, consumed beforehand).

Remark 2. For some TFA M, $v \in L(M)$, $\alpha \in \Sigma$ and some $t \in \mathbb{N}$, let $\mathrm{read}_M(v, \alpha, t) = i$. Then, for every $j < i$, it holds that every occurrence of α in $\mathrm{arch}_j(v)$ was read after t single computation steps of M.

To infer some structure of TFA in inf-ESU we use a special form of pumping. We look at the computation of some word $v \in \Sigma^*$ by some automaton M. If between the t_1-th and the t_2-th single computation steps each letter of a subset $A \subset \Sigma$ is read in a large number of arches and at no point in between these computation steps any letter of A is read in significantly more arches than another letter of A, then there is a cycle in M in which all letters of A are read. If further, there is a subset $B \subset \Sigma$, so that in between those steps every letter of B is read in fewer arches than every letter of A, then we conclude that there is a cycle reading A independently of B. Next lemma presents this result.

Lemma 5. *Let M be a TFA, $t_1, t_2 \in \mathbb{N}$, $p_1, p_2 \in Q$ and $v, v_{t_1}, v_{t_2} \in \Sigma^*$, such that there is the computation $(q_0, v) \vdash_M^{t_1} (p_1, v_{t_1}) \vdash_M^{t_2 - t_1} (p_2, v_{t_2})$. Let $\Sigma_1 \sqcup \Sigma_2 \sqcup \Sigma_3 = \Sigma$ be a disjoint decomposition of Σ. Let $v_t \leq v$ be the word obtained by doing t single computation steps with M on input v, that is $(q_0, v) \vdash_M^t (q, v_t)$, for some $q \in Q$.*

Consider also that there are $y \in \mathbb{N}$ and $\beta \in \Sigma_2$, such that, for every $t \in \mathbb{N}$ with $t_1 \leq t \leq t_2$ and every $\alpha \in \Sigma_1, \beta' \in \Sigma_2, \gamma \in \Sigma_3$ it holds that

1. $\mathrm{read}_M(v, \gamma, t) < \mathrm{read}_M(v, \beta, t) < \mathrm{read}_M(v, \alpha, t)$,
2. $|\mathrm{read}_M(v, \beta, t) - \mathrm{read}_M(v, \beta', t)| < y$, *and*
3. $\mathrm{read}_M(v, \beta, t_2) - \mathrm{read}_M(v, \beta, t_1) = 2yn$.

Then, there is a state $q \in Q$, and words $u, w_1, w_2 \in \Sigma^*$ with lengths $|u| \leq n|\Sigma_2| + (n-1)$, and $|w_1|, |w_2| \leq n - 1$, such that $\Sigma_2 \subset \mathrm{alph}(u)$ and there are computations $(p_1, w_1) \vdash_M^* (q, \varepsilon)$, $(q, u) \vdash_M^* (q, \varepsilon)$ and $(q, w_2) \vdash_M^* (p_2, \varepsilon)$ that read Σ_1 and $\Sigma_1 \cup \Sigma_2$ independently.

Let M be some TFA in inf-ESU, i.e., we can pick some word accepted by M with arbitrarily large universality index. To prove that M is in UCSC, we look at the computation of words that have a particular universality index. If at some point during the computation, for some letters, the number of arches in which they were read differs by a certain amount, we can conclude that there is a cycle reading some of these letters.

Proposition 3. *Let $M = (Q, \Sigma, q_0, F, \delta)$ be a TFA with an alphabet of size $|\Sigma| = \sigma$ and $|Q| = n$ states, such that for every $k \in \mathbb{N}$ there is a k-universal word $s_k \in L(M)$. Then M universally constructs subsequences on m cycles for some $m \leq \sigma + 1$.*

Due to the length of the full proof, we only give a brief sketch here. Intuitively, to prove that a TFA $M \in$ inf-ESU is also in UCSC, we analyse the computation of words with large universality index $(d \in \Omega(n^{\sigma^2}))$ which are guaranteed to exist because $M \in$ inf-ESU. We specify a partition $\{\Sigma_1, \ldots, \Sigma_m\}$ of Σ aiming to ensure that the conditions of Definition 6 and Lemma 5 can be satisfied. During the computation of the given word we monitor, for each letter separately, the number of arches in which they are consumed entirely. If we can partition Σ into two subsets A, B such that this number for any letter $a \in A$ exceeds those numbers for all $b \in B$ by a certain threshold, then there is a cycle reading the letters of A independently from those of B. We let $\Sigma_1 = A$ and partition B similarly as we previously did for Σ.

To be more precise, assume we already have defined $\Sigma_1, \ldots, \Sigma_{i-1}$ and computation steps $t_1 < \ldots < t_{i-1}$, such that after t_1 steps all letters in Σ_1 have been consumed in approximately the same number of arches, and so on, similarly, for t_2 up to t_{i-1}. To identify Σ_i let $\overline{\Sigma_i} = \Sigma \setminus \Sigma_i^<$, i.e., $\overline{\Sigma_i}$ contains all letters that are not covered by $\Sigma_1, \ldots, \Sigma_{i-1}$. Next, we set an upper bound $t_{i,max}$ on the number of computation steps such that after $t_{i,max}$ steps all letters of $\overline{\Sigma_i}$ have been consumed in a specific amount of arches. To obtain Σ_i, we search for a subset $\Sigma_i' \subset \overline{\Sigma_i}$ such that all letters in Σ_i' are consumed in approximately the same amount of arches after t steps, where $t_{i-1} < t \leq t_{i,max}$, i.e., they lay in the same cycle. If further the number of arches they have been consumed in exceeds those number for the letters contained in $\overline{\Sigma_i} \setminus \Sigma_i'$ by some threshold then the letters contained in $\overline{\Sigma_i} \setminus \Sigma_i'$ possibly lay on the path to the cycle but are not contained in it.

If, on the one hand, no proper subset of $\overline{\Sigma_i}$ satisfies these conditions then we can end the partitioning of Σ by choosing $\Sigma_i = \overline{\Sigma_i}$. On the other hand, if there are several candidates for Σ_i', we choose one with minimal positive cardinality to become Σ_i.

Because we construct per iteration an alphabet which has at least one element, the size of $\overline{\Sigma_i}$ decreases after every iteration. Hence, this process terminates after a finite amount of iterations. Furthermore, the partition constructed above satisfies Lemma 5, which guarantees the existence of words that satisfy the properties of Definition 6.

6   Conclusion

The main open research direction from [5] was to identify relevant classes of languages for which the complexity of the five problems approached in this paper differs significantly from the complexity bounds achieved for regular and context-free languages, on the one hand, and their undecidability for the class of context-sensitive languages. While [5] suggested the class of translucent finite automata, the respective paper only characterized the complexity (NP-complete) of Problem 1; this problem was in P for regular and context-free languages. We now characterize the complexity of the remaining problems. More precisely, we showed that Problem 3 and Problem 5 are NP-hard. Further, we gave a proof, which required an involved analysis, for the NP-membership of Problem 5, whereas the NP-membership of Problem 3 stays open, since we only gave a PSPACE algorithm solving this particular problem; Problem 3 is also NP-hard for regular and context-free languages, while Problem 5 was in P in those cases. Interestingly, for context-free languages it is also still open if Problem 3 is contained in NP since in [5] only a PSPACE algorithm was given to solve this problem. Finally, we gave a polynomial-time algorithm solving Problem 2, just as for context-free languages. Problem 4 is the only one whose complexity is still completely open. In that case, we just mention a simple argument to show that it is Co-NP: if there exists a non-k-universal word in the language accepted by the TFA $M = (Q, \Sigma, q_0, F, \delta)$, then, by a pumping argument, the shortest such word has length at most $|Q|$; we can guess such a word, and then check whether, indeed, it is not-k-universal in polynomial time. However, we conjecture that Problem 4 can be, in fact, solved in polynomial time.

References

1. Adamson, D., Fleischmann, P., Huch, A., Koß, T., Manea, F.: k-universality of regular languages revisited. In: Chau, V., Dürr, C., Li, M., Lu, P. (eds.) Proceedings of 6th International Joint Conference on Theoretical Computer Science – 19th Frontier of Algorithmic Wisdom (IJTCS FAW 2025). Lecture Notes in Computer Science, vol. 15828, pp. 16–32. Springer, Singapore (2025). https://doi.org/10.1007/978-981-96-8312-3_2
2. Adamson, D., Fleischmann, P., Huch, A., Koß, T., Manea, F., Nowotka, D.: k-universality of regular languages. In: Proceedings of 34th International Symposium on Algorithms and Computation (ISAAC 2023). LIPIcs, vol. 283, pp. 4:1–4:21 (2023). https://doi.org/10.4230/LIPIcs.ISAAC.2023.4. https://arxiv.org/abs/2311.10658
3. Barker, L., Fleischmann, P., Harwardt, K., Manea, F., Nowotka, D.: Scattered factor-universality of words. In: Jonoska, N., Savchuk, D. (eds.) DLT 2020. LNCS, vol. 12086, pp. 14–28. Springer, Cham (2020). https://doi.org/10.1007/978-3-030-48516-0_2
4. Chigahara, H., Fazekas, S.Z., Yamamura, A.: One-way jumping finite automata. Int. J. Found. Comput. Sci. **27**(3), 391–405 (2016). https://doi.org/10.1142/S0129054116400165
5. Fazekas, S.Z., Koß, T., Manea, F., Mercaş, R., Specht, T.: Subsequence matching and analysis problems for formal languages. In: Proceedings of 35th International Symposium on Algorithms and Computation, (ISAAC 2024). LIPIcs, vol. 322, pp. 28:1–28:23 (2024). https://doi.org/10.4230/LIPICS.ISAAC.2024.28
6. Fazekas, S.Z., Mitrana, V., Paun, A., Paun, M.: Jump complexity of deterministic finite automata with translucent letters. In: Anutariya, C., Bonsangue, M.M. (eds.) Proceedings of 21st International Colloquium on Theoretical Aspects of Computing, (ICTAC 2024). Lecture Notes in Computer Science, vol. 15373, pp. 62–77. Springer, Heidelberg (2024). https://doi.org/10.1007/978-3-031-77019-7_4

7. Hébrard, J.J.: An algorithm for distinguishing efficiently bit-strings by their subsequences. Theor. Comput. Sci. **82**(1), 35–49 (1991). https://doi.org/10.1016/0304-3975(91)90170-7

8. Kosche, M., Koß, T., Manea, F., Siemer, S.: Combinatorial algorithms for subsequence matching: a survey. In: Proceedings of 12th International Workshop on Non-Classical Models of Automata and Applications, (NCMA 2022). EPTCS, vol. 367, pp. 11–27 (2022). https://doi.org/10.4204/EPTCS.367.2

9. Meduna, A., Zemek, P.: Jumping finite automata. Int. J. Found. Comput. Sci. **23**(7), 1555–1578 (2012). https://doi.org/10.1142/S0129054112500244

10. Mitrana, V., Păun, A., Păun, M., Sánchez-Couso, J.: Jump complexity of finite automata with translucent letters. Theor. Comput. Sci. **992**, 114450 (2024). https://doi.org/10.1016/j.tcs.2024.114450

11. Nagy, B., Otto, F.: Finite-state acceptors with translucent letters. In: Proceedings of AI Methods for Interdisciplinary Research in Language and Biology (BILC 2011), pp. 3–13 (2011). https://doi.org/10.5220/0003272500030013

12. Otto, F.: A survey on automata with translucent letters. In: Proceedings of 27th International Conference on Implementation and Application of Automata, (CIAA 2023), vol. 14151, pp. 21–50 (2023). https://doi.org/10.1007/978-3-031-40247-0_2

Shape Preserving Tree Transducers

Paul Gallot$^{(\boxtimes)}$ and Sebastian Maneth

Department of Mathematics and Informatics, University of Bremen,
Bremen, Germany
gallopaul@hotmail.fr

Abstract. It is shown that shape preservation is decidable for top-down tree transducers, bottom-up tree transducers, and for compositions of total deterministic macro tree transducers. Moreover, if a transducer is shape preserving, then it can be brought into a particular normal form, where every input node creates exactly one output node.

1 Introduction

Tree transducers are a well established formalism for describing translations on finite ordered ranked trees. They were invented in the 1970s in the context of compilers and mathematical linguistics. An important type of problem that has been of continuing interest are so called "definability problems": given two classes X and Y of translations we want to know whether or not it is decidable for any $\tau \in X$, whether τ is definable in the class Y, i.e., whether or not $\tau \in Y$. If such decidability holds then we say that "Y is decidable within X". Knowing that $\tau \in Y$ may be beneficial for several reasons. For instance elements from Y may be more efficient to implement than elements from X; or, Y may enjoy better closure properties than X.

Examples of decidable definability problems on strings are: the class of regular languages is decidable within the class of deterministic context-free languages [25] and the class of one-way string transductions is decidable within the class of functional two-way transductions [2,16]. For tree transducers: the class of macro tree translations [14] of linear size increase is decidable within the class of macro tree translations [11] (in fact, even within the composition closure of the latter [9]); this result has recently been generalized to linear height increase and to linear size-to-height increase [18,19]. It was shown that linear deterministic top-down tree translations are decidable within top-down tree translations [21].

In this paper we study *shape preservation*: a tree translation is shape preserving if the shape of every output tree coincides with the shape of the corresponding input tree. Thus, the output and input trees have the same nodes and edges, but may only differ in the labels of the nodes. It was shown by Fülöp and Gazdag [17] that the class of shape preserving top-down tree translations is decidable within the class of top-down tree translations [22,23]. They also show that this class can be realized by "relabling TOPs" (see below). Their proofs go through a series of technically involved constructions and lemmas.

In this paper we give an alternative proof of the decidability of shape preservation for top-down tree transducers (TOPs). It is based on the facts that

G. Castiglione and S. Mantaci (Eds.): CIAA 2025, LNCS 15981, pp. 165–179, 2026.
https://doi.org/10.1007/978-3-032-02602-6_12

(i) functionality of TOPs is decidable and that (ii) equivalence of functional TOPs is decidable. Both were proven originally by Ésik [15]. The idea is extremely simple: for a given TOP, we change the labels of all output nodes to a fixed label. The resulting TOP must be functional, if the given one is shape preserving. Moreover, it must be equivalent to a fixed shape preserving transducer that only depends on the domain of the given transducer. This "proof scheme" can readily be applied to other classes: e.g., we can show that shape preservation for bottom-up tree transducers is decidable. In fact, we can even show that shape preservation for (compositions of) total deterministic macro tree transducers is decidable.

Next, we want to show that every shape preserving TOP can be realized by a relabeling TOP, i.e., one where every rule is of the form $q(f(x_1, \ldots x_k)) \to f'(q_1(x_1), \ldots, q_k(x_k))$. This was already shown by Fülöp and Gazdag, however only for *linear TOPs*. It can be tempting to assume that every shape preserving TOP is linear, but it is not true:

$$
\begin{array}{ll}
q_0(a(x_1)) \to a(f(q_1(x_1), q_2(x_1))) & q_1(f(x_1, x_2)) \to q_0(x_1) \\
q_0(e) \quad\ \to e & q_2(f(x_1, x_2)) \to q_0(x_2)
\end{array}
$$

This TOP is shape preserving (it realizes the identity on trees with paths of odd length and consisting of alternating a- and f-nodes and a final e-node). Its first rule is *not* linear: the input variable x_1 appears more than once in the right-hand side. The TOP is also *not* nondeleting (not every x_i from a left-hand side appears in the right-hand side). Lemma 3.1 of [17] states that every shape preserving TOP is nondeleting. However, this lemma is true only for linear TOPs and is invalidated by the example above for non-linear ones. To fix this omission, we prove that for every shape preserving TOP, an equivalent linear TOP can be constructed. Together with the results of [17] this establishes that every shape preserving TOP is effectively equivalent to a relabeling TOP.

Let us now consider shape preservation for total deterministic macro tree transducers (MTTs). The property is decidable, because linear size increase of (compositions of) MTTs is decidable, and in the affirmative case we are in a class of translations that has decidable equivalence [12]. A question that remains: is there a normal form for MTTs that is akin to relabeling TOPs? Consider the following example of an MTT. It takes as input trees of the form $\$(t)$ where t is a monadic tree over unary labels a, b, and c (and the nullary label e).

$$
\begin{array}{lll}
q_0(\$(x_1)) & \to \$(q_a(x_1, q_b(x_1, q_c(x_1, e)))) & \\
q_\delta(\gamma(x_1), y) & \to \delta(q_\delta(x_1, y)) & \text{for } \delta, \gamma \in \{a, b, c\} \text{ with } \delta = \gamma \\
q_\delta(\gamma(x_1), y) & \to q_\delta(x_1, y) & \text{for } \delta, \gamma \in \{a, b, c\} \text{ with } \delta \neq \gamma \\
q(e, y) & \to y & \text{for } q \in \{q_0, q_a, q_b, q_c\}
\end{array}
$$

For instance, the input tree $\$(c(a(b(a(a(a(b(e)))))))$ is translated to the output tree $\$(a(a(a(b(b(c(e)))))))$. Our corresponding normal form of "relabeling MTT" requires that each input node produces exactly one output node; but this output node may be anywhere in the rule, including a "parameter position".

2   Preliminaries

The set $\{0, 1, \dots\}$ of natural numbers is denoted by $\mathbb{N}$. For $k \in \mathbb{N}$ we denote by $[k]$ the set $\{1, \dots, k\}$; thus $[0] = \emptyset$. We fix the set $X = \{x_1, x_2, \dots\}$ of variables and the set $Y = \{y_1, y_2, \dots\}$ of parameters and assume these sets to be disjoint from all other alphabets and sets. For $k \geq 1$ let $X_k = \{x_1, \dots, x_k\}$ (and similarly for Y). A ranked alphabet (set) consists of an alphabet (set) Σ together with a mapping $\mathrm{rank}_\Sigma : \Sigma \to \mathbb{N}$ that assigns to each symbol $\sigma \in \Sigma$ a natural number called its "rank". By $\Sigma^{(k)}$ we denote the symbols of Σ that have rank k.

The set T_Σ of (finite, ranked, ordered) trees over Σ is the smallest set of strings S such that if $\sigma \in \Sigma^{(k)}$, $k \geq 0$, and $s_1, \dots, s_k \in S$, then also $\sigma(s_1, \dots, s_k) \in S$. We write σ instead of $\sigma()$. Let A be a set that is disjoint from Σ. Then the set $T_\Sigma(A)$ of trees over Σ indexed by A is defined as $T_{\Sigma'}$ where $\Sigma' = \Sigma \cup A$ and $\mathrm{rank}_{\Sigma'}(a) = 0$ for $a \in A$ and $\mathrm{rank}_{\Sigma'}(\sigma) = \mathrm{rank}_\Sigma(\sigma)$ for $\sigma \in \Sigma$. For a tree $s = \sigma(s_1, \dots, s_k)$ with $\sigma \in \Sigma^{(k)}$, $k \geq 0$, and $s_1, \dots, s_k \in T_\Sigma$, we define the set $V(s) \subseteq \mathbb{N}^*$ of nodes of s as $\{\varepsilon\} \cup \{iu \mid i \in [k], u \in V(s_i)\}$; thus, nodes are strings over positive integers, where ε denotes the root node of s, and for a node u, ui denotes the i-th child of u. For $u \in V(s)$ we denote by $s[u]$ the label of u in s and by s/u the subtree rooted at u. We denote the prefix order on paths by $\leq$. Formally, let $s = \sigma(s_1, \dots, s_k)$ and define $s[\epsilon] = \sigma$, $s[iu] = s_i[u]$, $s/\varepsilon = s$, and $s/iu = s_i/u$ for $\sigma \in \Sigma^{(k)}$, $k \geq 0$, $s_1, \dots, s_k \in T_\Sigma$, $i \geq 1$ and $u \in V(s_i)$ such that $iu \in V(s)$. For trees s, t and $u \in V(s)$ we denote by $s[u \leftarrow t]$ the tree obtained from s by replacing its subtree rooted at u by the tree t. Similarly, for trees $t_1, \dots t_k$ we denote by $s[x_i \leftarrow t_i]_{i \leq k}$ the tree obtained from s by replacing each occurrence of x_i by t_i.

For each $n \in \mathbb{N}$ we fix the shape alphabet $\Psi_n = \{\#_0, \#_1, \dots, \#_n\}$ with $\mathrm{rank}(\#_i) = i$ for all $i \leq n$ and, for each alphabet Σ of maximal arity n, the shape function $\mathrm{sh}_\Sigma : T_\Sigma \to T_{\Psi_n}$ on Σ inductively defined by, for all $k \geq 0$, $\sigma \in \Sigma^{(k)}$ and trees $t_1, \dots, t_k \in T_\Sigma$, $\mathrm{sh}_\Sigma(\sigma(t_1, \dots, t_k)) = \#_k(\mathrm{sh}_\Sigma(t_1), \dots, \mathrm{sh}_\Sigma(t_k))$. We note SH the set of all shape functions.

The domain of a relation f is noted $\mathrm{dom}(f)$. The composition of relations is noted ; and is defined so that $(x, z) \in f \,;\, g$ when $\exists y \in \mathrm{dom}(g), (x, y) \in f$ and $(y, z) \in g$. A tree relation $f \subseteq T_\Sigma \times T_\Delta$ over alphabets Σ and Δ is *shape preserving* if $f \,;\, \mathrm{sh}_\Delta \subseteq \mathrm{sh}_\Sigma$. **Convention:** All inclusions of transducer classes mentioned in this paper are meant effective.

A *top-down tree transducer* is a system $M = (Q, \Sigma, \Delta, q_0, R)$, where Q is a finite set of *states*, Σ and Δ are the *input and output ranked alphabets*, $q_0 \in Q$ is the *initial state*, and R is a finite set of *rules* of the form: $q(\sigma(x_1, \dots, x_k)) \to \mathrm{rhs}$, $\sigma \in \Sigma^{(k)}$, $k \geq 0$, $q \in Q$, and $\mathrm{rhs} \in T_\Delta(\langle Q, X_k \rangle)$, with $\langle Q, X_k \rangle = \{q'(x) \mid q' \in Q, x \in X_k\}$. For every state $q \in Q$ and tree $t \in T_\Sigma(X)$, $q(t)$ is called a *state call* of q on t. The set of state calls of M is written $\langle Q, T_\Sigma(X) \rangle$. For a rule r of the form $q(\sigma(x_1, \dots, x_k)) \to \mathrm{rhs}$ in R and tree $t \in T_\Delta(\langle Q, T_\Sigma(X) \rangle)$ such that $t/u = q(\sigma(t_1, \dots, t_k))$ for some $u \in V(t)$ and trees $t_1, \dots, t_k \in T_\Sigma(X)$, we write $t \xrightarrow{r, u} t[u \leftarrow \mathrm{rhs}[x_i \leftarrow t_i]_{i \leq k}]$.

A run of a state $q \in Q$ on a tree $t \in T_\Sigma(X)$ is of the form $q(t) \xrightarrow{r_1, u_1} t_1 \ldots \xrightarrow{r_n, u_n} t_n$. We also write $t \xrightarrow{q} t_n$ or $t \to^* t$ and call t_n the *output* of the run. If $t_n \in T_\Delta$ then the run is called *valid*. Let $M_q = \{(t, t') \mid$ there is a valid run of q on t with output $t'\}$. The *translation realized by M* is M_{q_0}. For each valid run r on t with output t' and for each $u \in V(t)$, there exists two runs $t[u \leftarrow x_1] \to^* t_1$ and $t_1[x_1 \leftarrow t/u] \to^* t'$ with $t_1 \in T_\Delta(\langle Q, \{x_1\}\rangle)$ The first such run is called *provisional run* and t_1 the *provisional output* of ξ at path u. The sequence of states appearing in t_1 (in pre-order) is called *state sequence* of r at u. Provisional runs only contain one variable x_1, we usually write it x to avoid confusion with variables $x_1, \ldots, x_k$ used in the rules of M.

Example 1. Consider the TOP from the Introduction and number its rules from left to right and then from top to bottom. Then

$$q_0(a(f(e,e))) \xrightarrow{r_1, \varepsilon} a(f(q_1(f(e,e)), q_2(f(e,e)))) \xrightarrow{r_2, 11} a(f(q_0(e), q_2(f(e,e))))$$
$$\xrightarrow{r_4, 12} a(f(q_0(e), q_0(e))) \xrightarrow{r_3, 1} a(f(e, q_0(e))) \xrightarrow{r_3, 2} a(f(e, e)).$$

Let ID denote the class of all total identity functions on T_Σ, for any ranked alphabet Σ

Lemma 1. *All mappings in ID and in SH can be realized by total deterministic nondeleting and linear top-down tree transducers.*

Proof. Let Σ be a ranked alphabet and let n be the maximal rank of symbols in Σ. For $i \in \{1, 2\}$ define $M_i = (\{q\}, \Sigma, \Delta_i, q, R_i)$ where $\Delta_1 = \Sigma$ and $\Delta_2 = \Psi_n$. For every $\sigma \in \Sigma^{(k)}$ with $k \geq 0$ let $q(\sigma(x_1, \ldots, x_k)) \to \sigma(q(x_1), \ldots, q(x_k))$ be in R_1 and let $q(\sigma(x_1, \ldots, x_k)) \to \#_k(q(x_1), \ldots, q(x_k))$ be in R_2. Then M_1 realizes the total identity on T_Σ and M_2 realizes sh_Σ.   □

3   Decidability of Shape Preservation

Let $\mathcal{C}$ be a class of relations on trees, i.e., for any $c \in \mathcal{C}$, $c \subseteq T_\Sigma \times T_\Delta$ for some ranked alphabets Σ and Δ. For $c \in \mathcal{C}$ let dom(c) denote the domain of c and let DOM(c) denote this domain, seen as a partial identity function, i.e., DOM$(c) = \{(s, s) \mid s \in \text{dom}(c)\}$. We denote by DOM$(\mathcal{C})$ the class of domains of $\mathcal{C}$, seen as partial identity functions, DOM$(\mathcal{C}) = \bigcup_{c \in \mathcal{C}} \text{DOM}(c)$.

Theorem 1. *Let $\mathcal{C}$ be a class of relations on trees. Shape preservation is decidable for every element of $\mathcal{C}$ if:*

1. *$ID \subseteq \mathcal{C}$,*
2. *$\mathcal{C}\,;SH \subseteq \mathcal{C}$,*
3. *$DOM(\mathcal{C})\,;\mathcal{C} \subseteq \mathcal{C}$,*
4. *functionality is decidable for any $c \in \mathcal{C}$, and*
5. *equivalence is decidable for all functions in $\mathcal{C}$.*

Proof. Let $c \in \mathcal{C}$ and let Σ, Δ be ranked alphabets (of minimal size) such that $c \subseteq T_\Sigma \times T_\Delta$. Define $d = c \,;\, \mathsf{sh}_\Delta$. By Property (2), $d \in \mathcal{C}$. By Property (4) it is decidable whether or not d is a function. If d is not a function, then c is not shape preserving. If d is a function, then let $e = \mathrm{DOM}(c) \,;\, \mathsf{sh}_\Sigma$. It follows from Properties (1) and (2) that $\mathsf{sh}_\Sigma \in \mathcal{C}$ and hence $e \in \mathcal{C}$ by Property (3). Clearly, c is shape preserving if and only if d is equivalent to e; the latter is decidable by Property (5). □

Let FTA denote all partial identity functions on regular tree languages. Let us now consider the class $\mathcal{C} = \mathrm{FTA} \,;\, \mathrm{TOP}$. By Lemma 1, $\mathrm{ID} \subseteq \mathrm{TOP}$ and therefore $\mathrm{ID} \subseteq \mathcal{C}$. Hence, Property (1) is satisfied. Since every element of SH can be realized by a linear and nondeleting top-down tree transducer, it follows from Theorem 1 of [1] that Property (2) is satisfied. Property (3) follows from the fact that FTA is closed under composition (Lemma 3.4 of [6]). Property (4) follows from the proof of Theorem 8 of [15], where it is shown that functionality of $\mathcal{C}$ is decidable. By the (unique) Theorem of [8] the restriction of TOP to functions is included in the class of deterministic top-down tree transducers with regular look-ahead. Equivalence of the latter class is decidable (see [13]), so Property 5 is satisfied.

Corollary 1. *Shape preservation is decidable for top-down tree translations (restricted to arbitary regular input tree languages).*

Note that for two tree relations r_1, r_2 such that r_1 is shape preserving it holds that $r_1 \,;\, r_2$ is shape preserving if and only if r_2 is shape preserving on the range of r_1. Top-down tree transducers *with regular look-ahead* are included in the composition of deterministic bottom-up finite state relabelings and TOP by Theorem 2.6 of [7]. Such relabelings are shape preserving and their ranges are effectively regular. Thus, decidability of shape preservation can be reduced from top-down tree transducers with regular look-ahead to $\mathcal{C} = \mathrm{FTA} \,;\, \mathrm{TOP}$, which is solved by Corollary 1. Similarly, by Theorem 3.15 of [6], *bottom-up tree translations* are included in compositions of bottom-up finite state relabelings and tree homomorphisms; since tree homomorphisms are particular TOPs, decidability of shape preservation again reduces to that of $\mathcal{C}$.

Corollary 2. *Shape preservation is decidable for (1) top-down tree translations with regular look-ahead and for (2) bottom-up tree translations.*

Let MTT denote the class of total deterministic macro tree transducers (full definition in Sect. 5) and let LMTT denote the restriction of MTT to linear size increase (LSI). Since every shape preserving translation is of LSI, we first check the LSI property. This is possible even for compositions of MTTs and in the affirmative case we obtain one equivalent LMTT, according to Theorems 44 and 43 of [9]. That Theorem 44 states that one "dTTsu" transducer can be constructed. This is a "single-use tree-walking transducer (with regular look-around)", which is the same as "single use restricted attributed tree transducer with look-ahead"; according to Theorem 7.1 of [10] they are the same as (deterministic) MSO definable tree transductions which are the same as LMTTs by Theorem 7.1 of [11].

We consider the class $\mathcal{C} = \text{FTA}\,;\text{LMTT}$. These are exactly the partial deterministic MSO tree translations, because LMTT coincides with the total deterministic MSO tree translations (see above) and the domains of partial MSO tree translations are exactly the regular tree languages (because an explicit closed "MSO domain formula" defines the domain and MSO definable tree languages are exactly the regular tree languages [4, 24]). Since every total deterministic TOP is also an MTT, and mappings in ID are LSI, it follows from Lemma 1 that Property (1) of Theorem 1 holds. Property (2) follows from the fact that (deterministic) MSO definable tree translations are closed under composition by Proposition 3.2 of [3]. Property (3) holds because FTA is closed under composition (see above). Property (4) is void, because $\mathcal{C}$ contains functions only. Equivalence of $\mathcal{C}$ is decidable [12] thus giving us Property (5).

Corollary 3. *Shape preservation is decidable for compositions of total deterministic macro tree transducers.*

4   Normalization of *non-Linear* shape preserving TOPs

The paper [17] provides a normalization algorithm for linear TOPs, i.e. showing how to transform a shape preserving linear TOP into an equivalent nondeterministic top-down relabeling. We only provide here a way to transform a non-linear shape preserving TOP into an equivalent linear shape preserving TOP. Our construction relies on determining which states of a TOP can only produce a bounded number of outputs, computing these outputs using a form of look-ahead automaton, and proving that other state calls on the same input subtree can be merged into one single state call.

Throughout this section we will use as running example the non-linear shape preserving TOP M_0 with set of states $Q_0 = \{q_0, q_1, q_2, q_3, q_\ell\}$, both input and output alphabet $\Sigma = \{f, g, h, a, b\}$, initial state q_0, and the following rules:

$$q_0(h(x)) \rightarrow q_1(x) \qquad\qquad q_1(h(x)) \rightarrow h(h(f(q_2(x), q_3(x))))$$
$$q_2(f(x_1, x_2)) \rightarrow f(q_2(x_1), q_\ell(x_2)) \qquad\qquad q_3(f(x_1, x_2)) \rightarrow q_3(x_1)$$
$$q_2(g(x_1, x_2)) \rightarrow q_\ell(x_2) \qquad\qquad q_3(g(x_1, x_2)) \rightarrow q_\ell(x_1)$$
$$q_\ell(a) \rightarrow a \qquad\qquad q_\ell(b) \rightarrow b$$

Here is an example of a run of M_0 on an input tree:

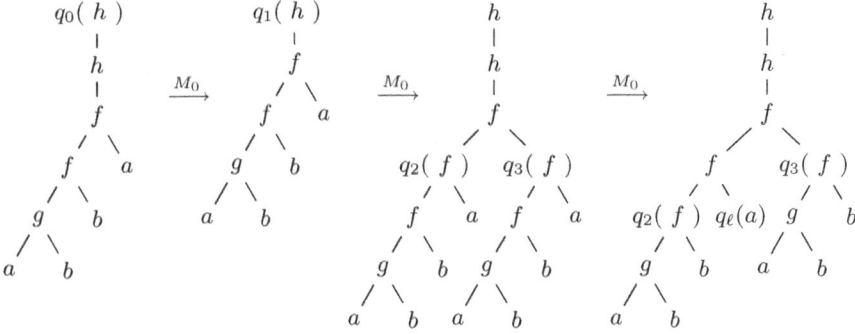

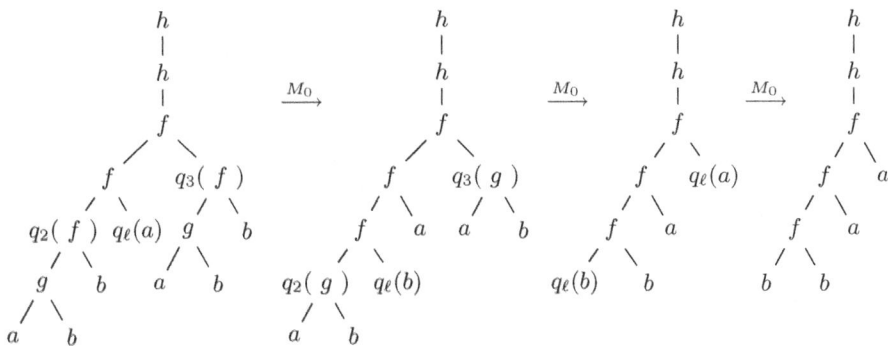

Effectively, the TOP M_0 checks that its input if of the right form (two h nodes at the root, a line of f nodes ending with a g node, and leafs a and b everywhere else) and outputs a tree of the same shape, replacing the g node with an f, and permuting leafs (moving each leaf to the next leaf spot to its left).

We assume given a non-linear shape preserving TOP M on tree alphabets (Σ, Δ), and we assume all states and rules which do not appear in a valid run of M have been removed.

As hinted at in our TOP example M_0, there is a one-to-one correspondence between leaves in the input and leaves in the output:

Lemma 2. *If $t \xrightarrow{M} t'$ with $t \in T_\Sigma$ and $t' \in T_\Delta$ then each input leaf is processed by exactly one state call, which itself produces an output tree containing exactly one output leaf.*

Proof. If a leaf is not processed by any state of M, then it can be replaced with a larger tree without changing the output, which contradicts the shape preserving property. Each state call to a leaf must produce at least one leaf in the output, so each input leaf produces at least one output leaf. The shape preserving implies that the number of leaves in the input is the same as in the output, so each input leaf produces exactly one output leaf.   □

Lemma 2 has a number of interesting implications. For instance, it implies that any rule on a non-leaf input node cannot produce any leaf; if it did, then the lemma implies that the output tree has more leaves than the input tree – a contradiction. It also implies that no state can be called twice on the same input node in a valid run. Indeed, if the same state appeared twice in a state sequence, then both state calls could have the same run on their input. In that run, at least one output leaf is produced when processing an input leaf u. This would imply that the input leaf u is processed twice, thus contradicting the lemma.

Lemma 3. *For each state sequence* $q_1, \ldots, q_n$ *of* M *and* $1 \leq i < j \leq n$: $q_i \neq q_j$.

This allows us to talk about *state sets* of M instead of state sequences of M. We denote by S_Q the set of *state sets* of M, i.e. $S_Q = \{\{q_1, \ldots, q_n\} \subseteq Q \mid q_1, \ldots, q_n$ is a state sequence of $M\}$. Note that the set S_Q is computable from M. For each $S \in S_Q$ we define $I(S) = \bigcap_{q \in S} \mathrm{dom}(M_q)$ and $\mathrm{fin}(S)$ as the set of states $q \in S$ such that $M_q(I(S))$ is a finite set of trees. For each $q \in Q$ the domain $\mathrm{dom}(M_q)$ is regular and so is $I(S)$. For each $S \in S_Q$, by deciding the finiteness of ranges of TOPs [5] we can compute $\mathrm{fin}(S)$ and $M_q(I(S))$ for all $q \in \mathrm{fin}(S)$. Finally, for every $S \in S_Q$, every $q \in \mathrm{fin}(S)$ and every $t \in M_q(I(S))$, the tree language $M_q^{-1}(\{t\})$ is regular and we can compute a nondeterministic top-down tree automaton for this language.

For example on TOP M_0 we get: $S_Q = \{\{q_0\}, \{q_1\}, \{q_2, q_3\}, \{q_\ell\}\}$, with $\mathrm{fin}(\{q_0\}) = \emptyset$, $\mathrm{fin}(\{q_1\}) = \emptyset$, $\mathrm{fin}(\{q_2, q_3\}) = \{q_3\}$ and $\mathrm{fin}(\{q_\ell\}) = \{q_\ell\}$. We have $M_{0q_3}(I(\{q_2, q_3\})) = \{a, b\}$ and $M_{0q_\ell}(I(\{q_\ell\})) = \{a, b\}$.

For every provisional run $t[u \leftarrow x] \rightarrow t'[u_i \leftarrow q_i(x)]_{i \leq n}$ and for every $i \leq n$, we say that u is an *origin* of the state call $q_i(x)$ at path u_i, and we say that this state call has *shape distance* n, if n is the distance in the tree t between the origin u and the node at path u_i.

We can then show that the shape distance is bounded except for state calls of states in $\mathrm{fin}(S)$. We can see that the exception of states in $\mathrm{fin}(S)$ is necessary on the example M_0:

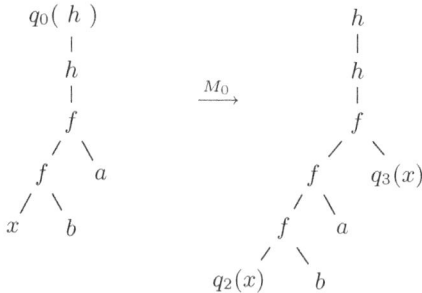

Here we can see that the state call $q_3(x)$ is far from its origin $u = 1111$, and indeed we can make state calls $q_3(x)$ arbitrarily far from its origin by lengthening the string of f nodes in the input. On the other hand, $q_2(x)$ is always one step below its origin.

Formally, we prove a bound on the shape distance for state calls of states not in $\mathrm{fin}(S)$, and we show that the path u' to such a state call in the output is either a prefix of its origin u, or u is a prefix of u'.

Lemma 4. *For every input tree t, run r of M on t and its provisional output $t[u \leftarrow x] \xrightarrow{M} t'[u_i \leftarrow q_i(x)]_{i \leq n}$ at a path $u \in V(t)$, noting S the state set of run r at path u, for every i such that $q_i \notin \mathrm{fin}(S)$ we have u and u_i are comparable wrt. the prefix ordering and the corresponding shape distance is $\leq \max(\mathsf{rhs}) \cdot 2^{|Q|}$ where $\max(\mathsf{rhs})$ is the maximum height of the right-hand side of a rule of M.*

With this lemma we can merge state calls on copies of input subtrees:

Theorem 2. *For every non-linear shape preserving TOP M, we can compute a shape preserving linear TOP M' equivalent to M.*

We show here how the construction works on an example and give intuition for it. The idea of this construction is to use nondeterminism to guess the outputs of state calls whose set of possible outputs is finite (states $q \in \mathrm{fin}(S)$ where S is the state set of the current input node). The reverse images of possible outputs of such states are regular, so we can check them using a nondeterministic top-down automaton, which we can simulate with our TOP since it has to visit every input node (according to Lemma 2). We then use Lemma 4 to merge remaining state calls on copied input subtrees. Indeed Lemma 4 implies that for each input subtree t, there is a subtree t' in the output which contains all state calls $q(t)$ on input t with $q \in S \setminus \mathrm{fin}(S)$, and there is a bound on the depth of $q(t)$ in t'. So all such state calls can be merged into one state call which outputs t'.

The construction on example TOP M_0: Note that the output of state q_3 is always either a or b. So, for state set $S = \{q_2, q_3\}$, $\mathrm{fin}(S) = \{q_3\}$ and we can check the output of q_3 by checking a regular property on the input. Similarly, for $S' = \{q_\ell\}$, we have $\mathrm{fin}(S_\ell) = \{q_\ell\}$.

We now build the linear TOP M' equivalent to M_0. States of M' are of the form (ϕ, v, t), with the state call $(\phi, v, t)(t_0)$ meaning:

- ϕ is a regular property checked on t_0, it allows us to produce the output of state calls of M_0 on t_0 whose output is bounded (states in $\mathrm{fin}(S)$),
- t is an output subtree containing all state calls of M_0 on t_0, except for states in $\mathrm{fin}(S)$,
- v is a path such that, if $t_1[u \leftarrow x] \xrightarrow{M'} t_1'[u' \leftarrow (\phi, v, t)(x)]$, then $u = u'v$.

To put it simply, for a state call at path u' with origin at path u, if u is a prefix of u' then $v = \varepsilon$ and t gives the subtree at path u in the provisional output; but if u' is a prefix of u then $u = u'v$ and $t = q(x)$ where $q(x)$ is the only state call with origin u. Here are some of the rules for M':

$$(\emptyset, \varepsilon, q_0(x))(h(x_1)) \qquad\qquad \to (\emptyset, 1, q_1(x))(x_1)$$
$$(\emptyset, 1, q_1(x))(h(x_1)) \qquad\qquad \to h(h((q_3(x) = a, \varepsilon, f(q_2(x), a))(x_1)$$
$$(\emptyset, 1, q_1(x))(h(x_1)) \qquad\qquad \to h(h((q_3(x) = b, \varepsilon, f(q_2(x), b))(x_1)$$
$$(q_3(x) = a, \varepsilon, f(q_2(x), a))(f(x_1, x_2)) \to f((q_3(x) = a, \varepsilon, f(q_2(x), b))(x_1),$$
$$(q_l(x) = b, \varepsilon, a)(x_2))$$

Theorem 3. *For every shape preserving TOP M, we can construct an equivalent (nondeterministic) top-down relabeling equivalent to M.*

Proof. We can construct a linear TOP M' from M by applying Theorem 2. According to Theorem 3.30 from [17], which works on linear shape preserving TOPs, there exists a top-down relabeling equivalent to M'.   □

The construction in [17] uses properties of shape preserving TOPs similar to the ones we use here. The two constructions are similar enough that we believe they could be merged into a single-step transformation transforming an arbitrary shape preserving TOP directly into a relabeling TOP.

5   Normalization of Shape Preserving MTTs

In this section we show how to normalize shape preserving total deterministic MTTs. As opposed to TOPs, one cannot transform any shape preserving MTT into an equivalent relabeling. In fact, as can be seen from the example in the Introduction, we cannot even associate a linear MTTs with every shape preserving one, i.e., copying of input variables is an essential power of a shape preserving MTT that *cannot* be avoided! Instead, we find a class of MTTs that we call *one-to-one* which characterizes shape preserving MTTs: for a one-to-one MTT, each input node creates exactly one output node. A total deterministic *macro tree transducer* (MTT) M is a tuple $(Q, P, \Sigma, \Delta, q_0, R)$, where Q is a ranked alphabet of *states*, Σ and Δ are ranked alphabet of *input* and *output symbols*, $q_0 \in Q^{(0)}$ is the *initial state*, and R is the *set of rules*, where for each $q \in Q^{(m)}$, $m \geq 0$, $\sigma \in \Sigma^{(k)}$ and $k \geq 0$ there is exactly one rule of the form $\langle q, \sigma(x_1, \ldots, x_k)\rangle(y_1, \ldots, y_m) \to t$ with $t \in T_{\Delta \cup \langle Q, X_k \rangle}(Y_m)$. The right-hand side t of such a rule is denoted by $\mathrm{rhs}_M(q, \sigma)$ We use a notation that is slightly different from the one used in the Introduction: instead of, e.g., $q_2(x_2, q_{\mathrm{id}}(x_1))$ we write $\langle q_2, x_2\rangle(\langle q_{\mathrm{id}}, x_1\rangle)$. Thus, we use angular brackets $\langle \ldots \rangle$ to indicate a state call on an input subtree, and use round brackets (after the angular brackets), to indicate the parameter arguments of the particular state call. Formally, if Γ is a ranked alphabet and A a set, then $\langle \Gamma, A \rangle$ is a ranked alphabet of symbols $\langle \gamma, a \rangle$ of rank k for all $\gamma \in \Gamma^{(k)}$, $k \geq 0$, and $a \in A$.

The semantics of a MTT M (as above) is defined as follows. We define the derivation relation $\Rightarrow_M$ as follows. For two trees $\xi_1, \xi_2 \in T_{\Delta \cup \langle Q, T_\Sigma \rangle}(Y)$, $\xi_1 \Rightarrow_M \xi_2$ if there exists a node u in ξ_1 with $\xi_1/u = \langle q, s\rangle(t_1, \ldots, t_m)$, $q \in Q^{(m)}$, $m \geq 0$, $s = \sigma(s_1, \ldots, s_k)$, $\sigma \in \Sigma^{(k)}$, $k \geq 0$, $s_1, \ldots, s_k \in T_\Sigma$, $t_1, \ldots, t_m \in T_{\Delta \cup \langle Q, T_\Sigma \rangle}(Y)$, and $\xi_2 = \xi_1[u \leftarrow \xi]$ where ξ equals

$$\zeta[\![\langle q', x_i\rangle \leftarrow \langle q', s_i\rangle \mid q' \in Q, i \in [k]]\!][y_j \leftarrow t_j \mid j \in [m]]$$

and $\zeta = \mathrm{rhs}_M(q,\sigma)$. The double angular brackets in the displayed formular merely means that each occurrence of x_i is replaced by s_i. Since M is total deterministic (i.e., for every state q, input symbol $\sigma \in \Sigma^{(k)}$ and $k \geq 0$, M contains exactly one corresponding rule) there is for every ξ_1 a unique tree $\xi' \in T_\Delta(Y)$ such that $\xi_1 \Rightarrow_M^* \xi'$. For every $q \in Q^{(m)}$, $m \geq 0$ and $s \in T_\Sigma$ we define the q-translation of s, denoted by $M_q(s)$, as the unique tree t in $T_\Delta(Y_m)$ such that $\langle q, s\rangle(y_1, \ldots, y_m) \Rightarrow_M^* t$. We denote the translation realized by M also by M, i.e., $M = M_{q_0}$ and for every $s \in T_\Sigma$, $M(s) = M_{q_0}(s)$ is the unique tree $t \in T_\Delta$ such that $\langle q_0, s\rangle \Rightarrow_M^* t$.

Let $s \in T_\Sigma$ and $u \in V(s)$. Consider the tree $M(s[u \leftarrow x])$. Since there are no rules for the input symbol x, this tree contains state calls in the form of occurrences of symbols of the form $\langle Q, \{x\}\rangle$. The state sequence of M on s at path u is the sequence of states $q_1, q_2, \ldots$ such that $\langle q_1, x\rangle, \langle q_2, x\rangle \ldots$ are all elements of $\langle Q, \{x\}\rangle$ that appear in $M(s[u \leftarrow x])$ in pre-order. We define the origin function $f : V(M(s)) \rightarrow V(s)$ associating with each output node the input node which produced it via a rule application; see [20] for more details and a precise definition. Intuitively, if we consider the output nodes that are inserted via rule applications into the tree $M(s[u \leftarrow x])$ when we build $M(s[u \leftarrow \sigma(x_1, \ldots, x_k)])$ for some $\sigma \in \Sigma^{(k)}$.

We illustrate this on an example MTT M_1 with the following rules:

$$\langle q_0, a(x)\rangle \rightarrow \langle q_a, x\rangle(\langle q_b, x\rangle) \qquad \langle q_a, a(x)\rangle(y) \rightarrow \langle q_a, x\rangle(a(y))$$
$$\langle q_0, b(x)\rangle \rightarrow \langle q_m, x\rangle(b(e)) \qquad \langle q_a, b(x)\rangle(y) \rightarrow \langle q_a, x\rangle(y)$$
$$\langle q_0, e\rangle \rightarrow e \qquad \langle q_a, e\rangle(y) \rightarrow y$$
$$\langle q_m, a(x)\rangle(y) \rightarrow \langle q_m, x\rangle(a(y)) \qquad \langle q_b, a(x)\rangle \rightarrow \langle q_b, x\rangle$$
$$\langle q_m, b(x)\rangle(y) \rightarrow \langle q_m, x\rangle(b(y)) \qquad \langle q_b, b(x)\rangle \rightarrow b(\langle q_b, x\rangle)$$
$$\langle q_m, e\rangle(y) \rightarrow y \qquad \langle q_b, e\rangle \rightarrow a(e)$$

The MTT M_1 translates monadic input trees of the form $b(w(e))$ into trees $w^r(b(e))$ where w is an arbitary sequence of a- and b-nodes and w^r denotes the reverse of the sequence. On input trees of the form $a(w(e))$ it outputs $a^m b^n(a(e))$ where m is the number of a-nodes in s and n is the number of b-nodes in w.

One-to-One MTT. An MTT M is one-to-one if, for every input tree $t \in T_\Sigma$, corresponding output tree t' and origin function $f : V(t') \rightarrow V(t)$: the origin function f is a bijection.

We can transform any MTT into a composition of a nondeterministic top-down relabeling with a one-to-one MTT. The construction relies on adjusting the rate of production of the output. We define a notion of delay in output production. This delay is bounded and we can track it using a top-down relabeling. We can then use a top-down relabeling to compute pieces of output to produce earlier in order to adjust the output production rate of the transducer.

For the remainder of this section, we assume given a shape preserving MTT M over the alphabets (Σ, Δ). We extend the definition of the size of trees to trees in $T_\Sigma(X \cup Y)$ such that $|t|$ is the number of nodes in t with labels in Σ. With this definition, substitution of variables or parameters preserves size.

We define the *delay* of M on $t \in T_\Sigma$ at path $u \in V(t)$ as the difference between the size of output produced from t/u and the size of t/u. We write it $\Delta(t, u)$, so $\Delta(t, u) = (\sum_{i=1}^n |q_i(t/u)|) - |t/u|$ where $q_1, \ldots, q_n$ is the state sequence at path u.

On the example M_1, the state sequences are $s_1 = q_0$, $s_2 = q_m$ and $s_3 = q_a, q_b$. Notice that each state sequence always has the same delay: s_1 has delay 0, s_2 has delay -1 and s_3 has delay 1. We can prove this in general. The delay is determined by the state sequence and the input subtree t/u, so if two subtrees t/u and t'/u' of two different input trees have the same state sequence but two different delays, then we could substitute t/u for t'/u' and get an output tree of a different size compared to the size of the input tree. So the state sequence always determines the delay, and since state sequences are computable by a relabeling, we can also compute the delay using a relabeling.

Lemma 5. *One can construct a nondeterministic top-down relabeling R associating, with each node at path $u \in V(t)$ in an input tree $t \in T_\Sigma$, the state sequence $q_1, \ldots, q_n$ and the delay $\Delta(t, u)$ at that node.*

For our construction, for state sequences with positive delay, we will need to produce output earlier than when it is produced by M. To do this we change the relabeling so that it predicts parts of the output of states, which we can then output earlier compared to M.

On the example MTT M_1, the bound on the delay is $b = 1$, so we only need to predict the node at the root of the output of each state. For example, the root node of $q_b(t)$ for any input tree t is b if t contains at least one b, but it is a if t contains no b. Each of these cases corresponds to t being in a regular tree language, which can be computed by a relabeling.

For the construction in general, we may need to predict larger parts of those outputs, but for every predicted node in the output of a state q, the reverse image is a regular tree language, which we can compute with a relabeling. Moreover, an important first step of the full construction is to make the given MTT *nondeleting in the parameters* and *nonerasing in the parameters*. This means that every parameter that appears in the left-hand side of a rule must also appear in the right-hand side of a rule, and, that there is no rule with a right-hand side of the form y_1. These two properties guarantee that each state call of the MTT will produce at least one output symbol; in fact, for every state q and every input tree s, it guarantees that $M_q(s)$ is of the form $\delta(t)$ for some $\delta \in \Delta$ and tree t.

We are now able to normalize M into a composition of a relabeling with a one-to-one MTT:

Theorem 4. *Given a shape preserving MTT M, one can effectively compute a relabeling R and a one-to-one MTT M' so that $R \, ; M'$ is equivalent to M.*

Note that being a one-to-one MTT is equivalent to having $\Delta(t, u) = 0$ for all $t \in T_\Sigma$ and $u \in V(t)$. We show the construction on our example M_1.

First, we compute the relabeling R_1 giving, for each input subtree, the state sequence, the delay and the root nodes of the outputs of states q_0, q_m, q_a and

q_b. Each input node with symbol a is relabeled with symbol (a, r) where r gives the state sequence, the delay and the root nodes of outputs of states.

Second, we decrease the delay of state sequences where it is positive. State sequence q_a, q_b has delay $+1$. We decrease the delay through q_b by replacing calls to $\langle q_b, x \rangle$ with the correct root node (given by the relabeling) and a new helper state which computes what appears below this root node. As we saw earlier, there are two possible root nodes for the output of q_b: a and b.

We replace the rule $\langle q_0, a(x) \rangle \rightarrow \langle q_a, x \rangle (\langle q_b, x \rangle)$ with the rules of the form:

- $\langle q_0, (a, r)(x) \rangle \rightarrow \langle q_a, x \rangle (a(\langle (q_b, 1), x \rangle)))$ for each r giving a as root node of $q_b(t)$,
- $\langle q_0, (a, r)(x) \rangle \rightarrow \langle q_a, x \rangle (b(\langle (q_b, 1), x \rangle)))$ for each r giving b as root node of $q_b(t)$.

The rules $\langle q_b, a(x) \rangle \rightarrow \langle q_b, x \rangle$ and $\langle q_b, b(x) \rangle \rightarrow b(\langle q_b, x \rangle)$ are replaced with:

- $\langle q_b, (a, r)(x) \rangle \rightarrow a(\langle (q_b, 1), x \rangle)$ and $\langle q_b, (b, r)(x) \rangle \rightarrow b(a(\langle (q_b, 1), x \rangle)))$ for r giving a as root node of $q_b(t)$,
- $\langle q_b, (a, r)(x) \rangle \rightarrow b(\langle (q_b, 1), x \rangle)$ and $\langle q_b, (b, r)(x) \rangle \rightarrow b(b(\langle (q_b, 1), x \rangle)))$ for r giving b as root node of $q_b(t)$.

The newly created helper state $(q_b, 1)$ computes the subtree at path 1 (first child of the root) of the output tree computed by q_b. To reflect this, we add rules for $(q_b, 1)$ computed from the rules for q_b:

$\langle q_b, (a, r)(x) \rangle \rightarrow a(\langle (q_b, 1), x \rangle)$ becomes: $\langle (q_b, 1), (a, r)(x) \rangle \rightarrow \langle (q_b, 1), x \rangle$
$\langle q_b, (b, r)(x) \rangle \rightarrow b(a(\langle (q_b, 1), x \rangle)))$ becomes: $\langle (q_b, 1), (b, r)(x) \rangle \rightarrow a(\langle (q_b, 1), x \rangle)$
$\langle q_b, e \rangle \rightarrow a(e)$ becomes: $\langle (q_b, 1), e \rangle \rightarrow e$

The rules above are for r giving a as root node of $q_b(t)$ (we need to add similar rules for r giving b as root node of $q_b(t)$). The state sequence q_a, q_b has been effectively replaced with the sequence $q_a, (q_b, 1)$ which has delay 0. Third, we increase the delay of state sequences where it is negative. State sequence q_m has delay -1. For each rule with $\langle q_m, x \rangle$ and an output symbol in the right-hand side, we transform the output symbol into a new helper state. For example, $\langle q_0, b(x) \rangle \rightarrow \langle q_m, x \rangle (b(e))$ becomes $\langle q_0, (b, r)(x) \rangle \rightarrow \langle q_m, x \rangle (b(\langle (1, e), x \rangle)))$, where $(1, e)$ is a helper state computing e. We now have the state sequence $q_m, (1, e)$ with delay 0. We then add rules for $(1, e)$ so that the state sequence $q_m, (1, e)$ keeps delay 0 when q_m calls itself. We add the rules $\langle (1, e), (a, r)(x) \rangle \rightarrow \langle (1, e), x \rangle$; $\langle (1, e), (b, r)(x) \rangle \rightarrow \langle (1, e), x \rangle$ and $\langle (1, e), (e, r) \rangle \rightarrow e$ for all label r. We have now built a one-to-one MTT which, precomposed with the relabeling R_1, is equivalent to M_1.

6   Conclusion and Open Problems

It was shown that shape preservation can be decided for various classes of tree transducers: nondeterministic top-down tree transducers (TOPs) with look-ahead, nondeterministic bottom-up tree transducers, and compositions of total deterministic macro tree transducers (MTT). The proofs are very short and concise, because the rely on previously known results. For the nondeterministic classes shape preservation is reduced to deciding functionality of such transducers and to deciding equivalence of functional such transducers (both of which are known from the literature). For MTTs (or compositions of MTTs) we first decide linear size increase and then decide equivalence of the resulting class.

It was shown how to construct for a shape preserving TOP an equivalent relabeling TOP. We showed how to go from non-linear to linear shape preserving TOPs and then rely on the result of [17] to go to a relabeling TOP. In the future it would be nice to have a self-contained procedure for the full normalization and analyze this procedure further: can it be shown that if the given shape preserving TOP is deterministic, then we can construct an equivalent deterministic relabeling? Can we normalize bottom-up transducers using this very same procedure? For MTTs a number of open problems remain, most prominently the question whether our results can be generalized to nondeterministic MTTs.

References

1. Baker, B.S.: Composition of top-down and bottom-up tree transductions. Inf. Control **41**(2), 186–213 (1979)
2. Baschenis, F., Gauwin, O., Muscholl, A., Puppis, G.: One-way definability of two-way word transducers. Log. Methods Comput. Sci. **14**(4) (2018)
3. Courcelle, B.: Monadic second-order definable graph transductions: a survey. Theor. Comput. Sci. **126**(1), 53–75 (1994)
4. Doner, J.: Tree acceptors and some of their applications. J. Comput. Syst. Sci. **4**(5), 406–451 (1970)
5. Drewes, F., Engelfriet, J.: Decidability of the finiteness of ranges of tree transductions. Inf. Comput. **145**(1), 1–50 (1998)
6. Engelfriet, J.: Bottom-up and top-down tree transformations - a comparison. Math. Syst. Theory **9**(3), 198–231 (1975)
7. Engelfriet, J.: Top-down tree transducers with regular look-ahead. Math. Syst. Theory **10**, 289–303 (1977)
8. Engelfriet, J.: On tree transducers for partial functions. Inf. Process. Lett. **7**(4), 170–172 (1978)
9. Engelfriet, J., Inaba, K., Maneth, S.: Linear-bounded composition of tree-walking tree transducers: linear size increase and complexity. Acta Informatica **58**(1–2), 95–152 (2021)
10. Engelfriet, J., Maneth, S.: Macro tree transducers, attribute grammars, and MSO definable tree translations. Inf. Comput. **154**(1), 34–91 (1999)
11. Engelfriet, J., Maneth, S.: Macro tree translations of linear size increase are MSO definable. SIAM J. Comput. **32**(4), 950–1006 (2003)
12. Engelfriet, J., Maneth, S.: The equivalence problem for deterministic MSO tree transducers is decidable. Inf. Process. Lett. **100**(5), 206–212 (2006)
13. Engelfriet, J., Maneth, S., Seidl, H.: Deciding equivalence of top-down XML transformations in polynomial time. J. Comput. Syst. Sci. **75**(5), 271–286 (2009)
14. Engelfriet, J., Vogler, H.: Macro tree transducers. J. Comput. Syst. Sci. **31**(1), 71–146 (1985)

15. Ésik, Z.: Decidability results concerning tree transducers I. Acta Cybern. **5**(1), 1–20 (1980)
16. Filiot, E., Gauwin, O., Reynier, P., Servais, F.: From two-way to one-way finite state transducers. In: 28th Annual ACM/IEEE Symposium on Logic in Computer Science, LICS 2013, New Orleans, LA, USA, 25–28 June 2013, pp. 468–477. IEEE Computer Society (2013)
17. Fülöp, Z., Gazdag, Z.: Shape preserving top-down tree transducers. Theor. Comput. Sci. **304**(1–3), 315–339 (2003)
18. Gallot, P., Lhote, N., Nguyên, L.T.D.: The structure of polynomial growth for tree automata/transducers and mso set queries (2025). https://arxiv.org/abs/2501.10270
19. Gallot, P., Maneth, S., Nakano, K., Peyrat, C.: Deciding linear height and linear size-to-height increase of macro tree transducers. In: Bringmann, K., Grohe, M., Puppis, G., Svensson, O. (eds.) 51st International Colloquium on Automata, Languages, and Programming, ICALP 2024, Tallinn, Estonia, 8–12 July 2024. LIPIcs, vol. 297, pp. 138:1–138:20. Schloss Dagstuhl - Leibniz-Zentrum für Informatik (2024)
20. Maneth, S., Seidl, H.: Deciding origin equivalence of weakly self-nesting macro tree transducers. Inf. Process. Lett. **180**, 106332 (2023)
21. Maneth, S., Seidl, H., Vu, M.: Definability results for top-down tree transducers. Int. J. Found. Comput. Sci. **34**(2&3), 253–287 (2023)
22. Rounds, W.C.: Mappings and grammars on trees. Math. Syst. Theory **4**(3), 257–287 (1970)
23. Thatcher, J.W.: Generalized sequential machine maps. J. Comput. Syst. Sci. **4**(4), 339–367 (1970)
24. Thatcher, J.W., Wright, J.B.: Generalized finite automata theory with an application to a decision problem of second-order logic. Math. Syst. Theory **2**(1), 57–81 (1968)
25. Valiant, L.G.: Regularity and related problems for deterministic pushdown automata. J. ACM **22**(1), 1–10 (1975)

Simulating Two-Way Nondeterministic Finite Automata Over Small Alphabets by One-Way Nondeterministic Automata

Viliam Geffert[1] and Alexander Okhotin[2]

[1] Department of Computer Science, P. J. Šafárik University, Košice, Slovakia
viliam.geffert@upjs.sk
[2] Department of Mathematics and Computer Science, St. Petersburg State University, 7/9 Universitetskaya Nab., Saint Petersburg 199034, Russia
alexander.okhotin@spbu.ru

Abstract. It is shown that an n-state two-way nondeterministic finite automaton (2NFA) over an alphabet Σ can be transformed to a one-way automaton (1NFA) with $|\Sigma| \cdot \binom{n}{1}_2 + 1$ states, where $\binom{n}{1}_2 \sim \sqrt{\frac{3}{4\pi n}} 3^n$ is a trinomial coefficient. A close lower bound of $\binom{n}{1}_2$ states is given for a four-symbol alphabet. To compare, for unrestricted alphabets, this transformation is known to require $\binom{2n}{n+1} \sim \frac{1}{\sqrt{\pi n}} 4^n$ states (Kapoutsis, "Removing bidirectionality from nondeterministic finite automata", MFCS 2005).

Keywords: Two-way finite automata · nondeterministic finite automata · state complexity

1 Introduction

Finite automata exist in different variants: deterministic and nondeterministic, one-way and two-way, etc. Most of them define exactly the regular languages, and automata of one type can be effectively transformed to automata to another time. These transformations, however, incur a blow-up in size, and much work has been done on reducing the size of resulting automata, as well as proving that a certain blow-up is necessary in the worst case.

For some of these transformations, the required number of states has been determined precisely: for instance, it is well-known that transformation of one-way nondeterministic finite automata (1NFA) to their deterministic counterparts (1DFA) requires exactly 2^n states in the worst case, where n is the number of states in a 1NFA. For some transformations, their complexity remains a long-standing open problem. Such is the complexity of transforming nondeterministic two-way finite automata (2NFA) to deterministic two-way automata (2DFA), for which the best lower bound is $\Omega(n^2)$, the best known upper bound is ca. $2^{O(n^2)}$, and in the case of a unary alphabet the upper bound is improved to $2^{O((\log n)^2)}$ [4]. The question of whether every 2NFA can be transformed to 2DFA with only a

© The Author(s), under exclusive license to Springer Nature Switzerland AG 2026
G. Castiglione and S. Mantaci (Eds.): CIAA 2025, LNCS 15981, pp. 180–192, 2026.
https://doi.org/10.1007/978-3-032-02602-6_13

polynomial blow-up is particularly important due to its relation to the L vs. NL and $L/poly$ vs. NL problems in the complexity theory, which was investigated by Kapoutsis [12], Geffert and Pighizzini [8], and Kapoutsis and Pighizzini [13].

What is understood much better about two-way finite automata, is the complexity of their transformation to one-way automata. The mere possibility of transforming 2DFA to a 1DFA recognizing the same language was established already by Rabin and Scott [20] and by Shepherdson [22], with Shepherdson's construction using at most $(n+1)^{n+1}$ states. Later, Kapoutsis [10,11] established the precise worst-case complexity of all four transformations from 2DFA and from 2NFA, to 1DFA and to 1NFA. Further results of this kind include bounds on the complexity of transforming 2DFA to sweeping two-way automata (Berman [1], Micali [16]) 2DFA to 1NFA recognizing the complement (Vardi [23]), two-way alternating automata to 1NFA (Geffert and Okhotin [6]), 2DFA and 2NFA to one-way unambiguous finite automata (Petrov and Okhotin [18,19]), etc. The complexity of such transformations in the case of a unary alphabet was studied by Chrobak [3], Mereghetti and Pighizzini [15], Geffert et al. [4], and Kunc and Okhotin [14].

For the transformation of n-state 2DFA to 1DFA, the precise number of states needed in the worst case, determined by Kapoutsis [11], is $n(n^n - (n-1)^n)$: this number of states is always sufficient, and for every n there is a 2DFA for which this number of states is necessary. However, the latter witness 2DFA is defined over an exponential-size alphabet. For alphabets of subexponential size, the required number of states is smaller: as proved by the authors [7], it is sufficient to use only $|\Sigma| \cdot \max_{k=1}^{n} k^{n-k+1} = n^{n - \frac{n \ln \ln n}{\ln n} + O(\frac{n}{\ln n})}$ states by having a 1DFA remember the last symbol read by the automaton, which allows remembering less information on the behaviour of the simulated 2DFA. It was also proved that transformation of 2DFA over a binary alphabet to 1DFA requires at least $\max_{k=1}^{n} k^{n-k+1}$ states, thus showing that the new construction is optimal up to a constant factor depending on the alphabet.

The same idea for improvement is also applicable to the 2DFA-to-1NFA transformation for small alphabets. For unrestricted alphabets, the precise succinctness tradeoff by Kapoutsis [10] is $\binom{2n}{n+1} \sim \frac{1}{\sqrt{\pi n}} 4^n$ states. For bounded alphabets it is sufficient to use $|\Sigma| \cdot \binom{n}{\lfloor n/2 \rfloor} = \Theta(\frac{1}{\sqrt{n}} 2^n)$ states by remembering the last symbol and a nondeterministically guessed set of states in which this symbol is visited by the 2DFA [5]. For this transformation, for 2DFA over the binary alphabet, at least $\binom{n-2}{\lfloor n/2 \rfloor}$ states are necessary in the worst case [5], which is again optimal up to a constant factor.

This paper investigates the transformation of 2NFA to 1NFA over small alphabets. For unrestricted alphabets, the precise tradeoff is the same as in the case of 2DFA, that is, $\binom{2n}{n+1} \sim \frac{1}{\sqrt{\pi n}} 4^n$ states [10]. However, the known improved construction for 2DFA over small alphabets is not applicable to the nondeterministic case: knowing the last symbol read and the states visited by the 2NFA is not enough to reconstruct the computation of the 2NFA on that symbol.

In this paper, a new transformation is proposed, which converts an n-state 2NFA over an alphabet Σ to a 1NFA with $|\Sigma| \cdot \binom{n}{1}_2$ states, where

$\binom{n}{1}_2 = \sum_{k=0}^{\lfloor \frac{n-1}{2} \rfloor} \frac{n!}{k!(k+1)!(n-2k-1)!}$ is the *trinomial coefficient*, which is asymptotically of the order $\binom{n}{1}_2 \sim \sqrt{\frac{3}{4\pi n}} 3^n$. For alphabets of subexponential size, this construction yields fewer states than the construction by Kapoutsis [10] (approximately 3^n vs. 4^n). At the same time, it is shown that at least $\binom{n}{1}_2$ states are necessary for this transformation, with witness languages defined over a four-symbol alphabet. Therefore, the bound is tight up to a constant factor depending on the alphabet, and hence a more complicated construction in comparison with the 2DFA-to-1NFA construction for small alphabets was in fact necessary.

2   Two-Way Automata

A *two-way nondeterministic finite automaton* (2NFA) is a quintuple $\mathcal{A} = (\Sigma, Q, Q_0, \delta, F)$, in which:

- Σ is a finite alphabet, which is extended with a left end-marker $\vdash \notin \Sigma$, and a right end-marker $\dashv \notin \Sigma$;
- Q is a finite set of states;
- $Q_0 \subseteq Q$ is the set of initial states;
- $\delta \colon Q \times (\Sigma \cup \{\vdash, \dashv\}) \to 2^{Q \times \{-1,+1\}}$ is the transition function, which lists possible transitions in a certain state while observing a certain tape symbol; $F \subseteq Q$ is the set of accepting states, effective on the right end-marker $\dashv$.

Given an input string $w \in \Sigma^*$, a 2NFA operates on a read-only tape containing this string enclosed within end-markers ($\vdash w \dashv$). A 2NFA begins its computation in any state $q_0 \in Q_0$ with the head observing the left end-marker ($\vdash$). While in a state q observing a symbol a, the automaton proceeds by choosing any pair from $\delta(q, a)$, and then entering a new state and moving its head by one position to the left or to the right. The automaton accepts if it ever comes to the right end-marker in any state $q \in F$. The set of strings, on which there is *at least one accepting computation*, forms the language recognized by the 2NFA, denoted by $L(\mathcal{A})$.

A 2NFA is *deterministic* (2DFA), if $|Q_0| = 1$ and $|\delta(q, a)| \leqslant 1$ for all $q \in Q$ and $a \in \Sigma \cup \{\vdash, \dashv\}$

The definition of a computation has a convenient interpretation in terms of directed graphs, which was proposed by Sakoda and Sipser [21]. For a 2NFA with a set of states Q, its computations on an input string $w = a_1 \ldots a_\ell$, delimited by end-markers $a_0 = \vdash$ and $a_{\ell+1} = \dashv$, are represented by paths in a graph (V, E), with $V = Q \times \{0, \ldots, \ell+1\}$ and $E = \{ \langle (q, i), (q', i + d) \rangle \mid (q', d) \in \delta(q, a_i) \}$. Every vertex (q, i) represents the automaton being in state q, with its head at position i in the input string, and every outgoing arc leads to a possible next configuration of the 2NFA. Then, the automaton accepts w if there is a path from any vertex $(q_0, 0)$, with $q_0 \in Q_0$, to any of the vertices $(q, \ell+1)$, with $q \in F$.

In this paper, 2NFA shall be simulated by *one-way nondeterministic finite automaton* (1NFA). A 1NFA is a quintuple $\mathcal{B} = (\Sigma, Q, Q_0, \delta, F)$, in which Σ is a finite alphabet; Q is a finite set of states; $Q_0 \subseteq Q$ is the set of initial states; $\delta \colon Q \times \Sigma \to 2^Q$ is a nondeterministic transition function that defines all possible next states; $F \subseteq Q$ is the set of accepting states. A computation of $\mathcal{B}$ on a string $w = a_1 \ldots a_\ell$ is any sequence of states $q_0, q_1, \ldots, q_\ell \in Q$ satisfying $q_0 \in Q_0$ and $q_i \in \delta(q_{i-1}, a_i)$ for all i. Such a string w is accepted if there exists a computation with $q_\ell \in F$.

3   The Known Simulation of 2NFA by 1NFA

Before presenting the proposed new simulation of 2NFA over small alphabets by 1NFA, the state-of-the-art transformation by Kapoutsis [10] ought to be recalled first. Generally speaking, it proceeds by guessing fragments of an accepting path of the original machine and linking them to form the entire path. This has to be done while reading the input string from left to right, and the simulating 1NFA needs to maintain some kind of data structure representing the connectivity of the graph of computations of the 2NFA restricted to the prefix of the tape seen by the 1NFA. Once an input string of length ℓ is read entirely, the 1NFA may have determined that there exists a desired path from the initial configuration $(q_0, 0)$ to an accepting configuration $(q, \ell+1)$, with $q \in F$. In this case, the 1NFA accepts.

For a 2NFA $\mathcal{A} = (\Sigma, Q, Q_0, \delta, F)$, the states of an equivalent 1NFA defined by Kapoutsis [10] are pairs (P, R), with $P, R \subseteq Q$ and $|P| + 1 = |R|$. When a 1NFA has read a prefix u of an input string uv, it is guessing an accepting computation of $\mathcal{A}$ on the whole string uv, and the current sets P and R contain the states of that computation reached on the boundary between u and v. More precisely, what the 1NFA remembers about u is that the computation graph of $\mathcal{A}$ on u contains $|R|$ paths that start in some specific configurations (namely, in one of the initial configurations, and in each of the configurations $(p, |u|)$, with $p \in P$, at the last symbol of u), and eventually arrive at the first symbol of the yet unread suffix v in pairwise distinct states from R (that is, in configurations $(r, |u| + 1)$, with $r \in R$).

Such a set of connections is illustrated in Fig. 1(left) in solid lines. Notably, the sets P and R refer to different positions in the input: P contains states visited at the last read symbol while going to the left, and R contains states to be visited at the next symbol to come. The 2NFA may have any other computation paths on u, but the simulating 1NFA has guessed this particular set.

Note that the simulating 1NFA does not remember which of the states in P are connected to which states in R inside u; the simulation works correctly for any permutation of these connections. Similarly, the 1NFA makes no guesses as to which states in R shall be connected to which states in P inside v; it shall determine it in due time. An important detail is that the number of states in R is always greater by one than the number of states in P, and that the verified computations on u form a bijection between P augmented with the initial configuration and R.

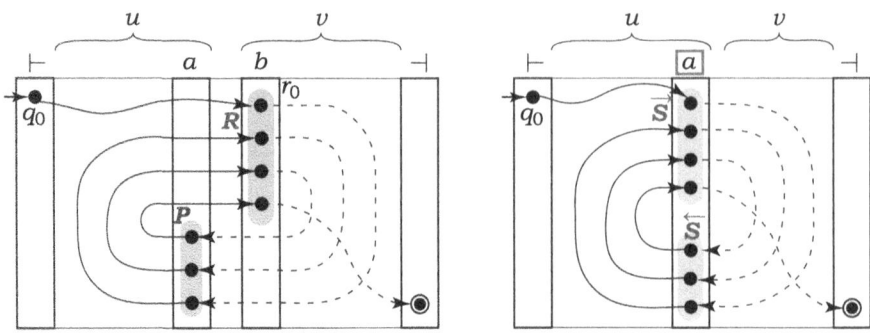

Fig. 1. Simulation of a 2NFA by a 1NFA: (left) state (P, R) in Kapoutsis's construction; (right) state $(\overrightarrow{S}, \overleftarrow{S}, a)$ in the proposed construction.

Lemma A (Kapoutsis [10]). *For every 2NFA $\mathcal{A} = (\Sigma, Q, Q_0, \delta, F)$, there exists a 1NFA $\mathcal{B} = (\Sigma, Q', Q'_0, \delta', F')$ recognizing the same language, which has the following set of states.*

$$Q' = \left\{ (P, R) \mid P, R \subseteq Q, \ |P| + 1 = |R| \right\}$$

The number of such pairs (P, R) is exactly $\binom{2n}{n+1}$. Having defined the 2NFA to 1NFA transformation described in Lemma A, Kapoutsis [10] also proved that his construction is optimal already for a transformation from a 2DFA to a 1NFA.

Theorem A (Kapoutsis [10]). *For every n-state 2NFA over an alphabet Σ, there exists a 1NFA with $\binom{2n}{n+1}$ states, which recognizes the same language. Conversely, for every n, there is such an alphabet Σ_n of size $\Theta(n^n)$, and such a language $L_n \subseteq \Sigma_n^*$ recognized by an n-state 2DFA, that every 1NFA recognizing L_n must have at least $\binom{2n}{n+1}$ states.*

An essential detail of the lower bound in Theorem A is the immense size of the alphabet, over which the witness languages are defined. For any smaller alphabets, Theorem A does not rule out improving the bound of $\binom{2n}{n+1}$ states.

It turns out that, under the assumption that the alphabet is smaller than exponential, the construction by Kapoutsis [10] can be improved upon, which shall be demonstrated in the next section.

4   Transformation for Small Alphabets

The proposed new simulation of a 2NFA by a 1NFA also guesses fragments of an accepting path of the 2NFA, and also requires this accepting path to be simple, in the sense that it never returns to any nodes it has visited earlier. The new simulations differs from Kapoutsis's simulation in several details. First of all, the 1NFA remembers states of a crossing sequence of the 2NFA *at a single symbol*, rather than on the boundary between two consecutive symbols; namely, it keeps the set of states of an accepting computation of the 2NFA at the last symbol the 1NFA has read. Secondly, for each state the 1NFA remembers, it additionally remembers the direction in which the 2NFA has crossed the last symbol in that state; this direction is unique, since each configuration is visited at most once, whereas states in which the 2NFA has turned back at that position are not remembered as all. And thirdly, the simulating 1NFA also remembers the last symbol itself, because it needs to know the 2NFA's transitions at that symbol in order to continue the simulation. The resulting combinatorial object representing the simulation of a 2NFA by a 1NFA, illustrated in Fig. 1(right), is a triple $(\overrightarrow{S}, \overleftarrow{S}, a)$, where $\overrightarrow{S}$ and $\overleftarrow{S}$ are disjoint subsets of states and a is the last symbol. The two subsets are disjoint, because, in each state, a possible accepting computation considered by the automaton may cross a either on the way from left to right, or on the way from right to left.

In the rest of this section, the following lemma is proved.

Lemma 1. *For every 2NFA* $\mathcal{A} = (\Sigma, Q, Q_0, \delta, F)$, *there exists a 1NFA* $\mathcal{B} = (\Sigma, Q', q_0', \delta', F')$ *that recognizes the same language, and uses the following set of states.*

$$Q' = \{ (\overrightarrow{S}, \overleftarrow{S}, a) \mid \overrightarrow{S}, \overleftarrow{S} \subseteq Q, \ \overrightarrow{S} \cap \overleftarrow{S} = \varnothing, \ |\overrightarrow{S}| = |\overleftarrow{S}| + 1, \ a \in \Sigma \} \cup \{q_0'\}$$

A single initial state q_0' is used to represent initial configurations for all initial states in Q_0, which could be systematically described by triples $(\{q_0\}, \varnothing, \vdash)$, with $q_0 \in Q_0$. Instead of adding all these triples to Q', it is sufficient to use one initial state.

The transitions of the 1NFA are defined by tracing the 2NFA's transition exchange between two consecutive symbols: the last symbol read, a, which is remembered in the state, and the symbol b being read by the transition. The condition to be verified, presented in the next definition, also involves two extra subsets of states, which the 1NFA will not remember: states S' in which the automaton comes to a from the right and leaves to the right, and states T' in which the automaton comes to b from the left and leaves to the left.

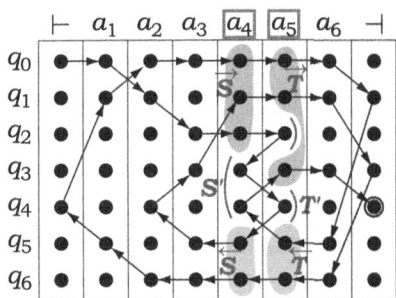

Fig. 2. $\tau(\overrightarrow{S}, \overleftarrow{S}, S', a_4, \overrightarrow{T}, \overleftarrow{T}, T', a_5)$: an example.

Definition 1. *For two consecutive symbols $a \in \Sigma \cup \{\vdash\}$ and $b \in \Sigma \cup \{\dashv\}$, let $\overleftarrow{S}, \overrightarrow{S}, S' \subseteq Q$ be three disjoint sets of states at a, and let $\overrightarrow{T}, \overleftarrow{T}, T' \subseteq Q$ be three disjoint sets of states at b. Assume that there exists a bijection $f \colon \overrightarrow{S} \cup S' \to \overrightarrow{T} \cup T'$ satisfying $(f(q), +1) \in \delta(q, a)$ for all $q \in \overrightarrow{S} \cup S'$, as well as another similarly defined bijection $g \colon \overleftarrow{T} \cup T' \to \overleftarrow{S} \cup S'$, with $(g(q), -1) \in \delta(q, b)$ for all $q \in \overleftarrow{T} \cup T'$. This condition is denoted by $\tau(\overrightarrow{S}, \overleftarrow{S}, S', a, \overrightarrow{T}, \overleftarrow{T}, T', b)$.*

An example for this condition is presented in Fig. 2; the corresponding transition of the 1NFA is from $(\overrightarrow{S}, \overleftarrow{S}, a_4)$ by a_5 to $(\overrightarrow{T}, \overleftarrow{T}, a_5)$.

Proof (of Lemma 1). The transition function of the new 1NFA is defined according to Definition 1.

$$\delta'((\overrightarrow{S}, \overleftarrow{S}, a), b) = \{ (\overrightarrow{T}, \overleftarrow{T}, b) \mid \exists S', T' \subseteq Q : \tau(\overrightarrow{S}, \overleftarrow{S}, S', a, \overrightarrow{T}, \overleftarrow{T}, T', b) \}$$

Transitions from the initial state q_0' require a separate definition that substitutes all possible initial states of the 2NFA.

$$\delta'(q_0', b) = \{ (\overrightarrow{T}, \overleftarrow{T}, b) \mid \exists q_0 \in Q_0, \ \exists S', T' \subseteq Q : \tau(\{q_0\}, \varnothing, S', \vdash, \overrightarrow{T}, \overleftarrow{T}, T', b) \}$$

The acceptance condition is defined by a virtual "transition" by the right endmarker ($\dashv$). In addition, the initial state is marked accepting if the original automaton accepts the empty string.

$$F' = \{(\overrightarrow{S}, \overleftarrow{S}, a) \mid \exists q_F \in F, \ \exists S', T' \subseteq Q : \tau(\overrightarrow{S}, \overleftarrow{S}, S', a, \{q_F\}, \varnothing, T', \dashv)\} \cup \underbrace{\{q_0'\}}_{\text{if } \varepsilon \in L(\mathcal{A})}$$

The correctness of the transformation is formally established as follows.

Claim 1.1. The 1NFA $\mathcal{B}$ constructed above can reach a state $(\overrightarrow{S}, \overleftarrow{S}, a)$ upon reading a string $u \in \Sigma^+$ if and only if the last symbol of u is a, and there exist a state $\widehat{q} \in \overrightarrow{S}$ and a bijection $\varphi \colon \overleftarrow{S} \to (\overrightarrow{S} \setminus \{\widehat{q}\})$, such that the 2NFA $\mathcal{A}$ has the following $|\operatorname{Im}\varphi| + 1$ disjoint computations on the tape $\vdash u$:

i. *a computation that begins in one of the 2NFA's initial configurations and ends at the rightmost symbol of u in the state $\widehat{q}$;*
ii. *for each state $q \in \overleftarrow{S}$, a computation that begins at the rightmost symbol of u in the state q and ends at the same symbol in the state $\varphi(q)$.*

The claim is proved by induction on the length of u. Once the claim is established, a separate argument is used to show that $\mathcal{B}$ accepts a string w if and only if $\mathcal{A}$ accepts w. Both proofs are omitted due to space constraints. □

In order to calculate the number of pairs $(\overrightarrow{S}, \overleftarrow{S})$ satisfying $\overrightarrow{S} \cap \overleftarrow{S} = \varnothing$ and $|\overrightarrow{S}| = |\overleftarrow{S}| + 1$, let $|\overleftarrow{S}| = k$ and $|\overrightarrow{S}| = k + 1$. For each k, the number of possible pairs is determined by the choice of $2k + 1$ states in $\overrightarrow{S} \cup \overleftarrow{S}$ from all n states, followed by another choice of k states in $\overleftarrow{S}$ out of these $2k + 1$ states. Altogether, this is expressed by the sum $\sum_{k=0}^{\lfloor \frac{n-1}{2} \rfloor} \binom{n}{2k+1} \binom{2k+1}{k}$, which is exactly the *trinomial coefficient* $\binom{n}{1}_2$.

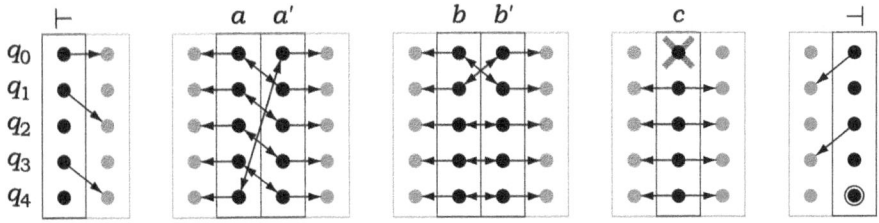

Fig. 3. Transitions of the 2NFA in Lemma 2, illustrated for $n = 5$.

Theorem 1. *For every n-state 2NFA over an alphabet Σ, there exists a 1NFA with $|\Sigma| \cdot \binom{n}{1}_2$ states that recognizes the same language.*

The growth rate of this trinomial coefficient (OEIS sequence A005717) is $\Theta(\frac{3^n}{\sqrt{n}})$. Therefore, as long as the alphabet grows slower than exponentially, the construction in Theorem 1 improves over the construction by Kapoutsis [10].

5   Lower Bound on the 2NFA to 1NFA Transformation

To see that the simulation of 2NFA by 1NFA described above is optimal up to a constant factor depending on the alphabet, consider the following lower bound.

Lemma 2. *For every $n \geqslant 1$, there exists an n-state 2NFA over the alphabet $\Sigma = \{a, a', b, b', c\}$, such that every 1NFA recognizing the same language must use at least $\sum_{k=0}^{\lfloor \frac{n-1}{2} \rfloor} \binom{n}{2k+1}\binom{2k+1}{k}$ states.*

Proof. This 2NFA $\mathcal{A} = (\Sigma, Q, q_0, \delta, F)$, with the set of states $Q = \{q_0, \dots, q_{n-1}\}$, is defined to use its nondeterminism in the following special way. On each string x constructed in this proof, the transition structure connects each state q_i at the first symbol of x with some state $f(q_i)$ at the last symbol of x, where f is a partial injective function. This connection is a *bi-directional chain of transitions* traversable in both ways. At every step, the only nondeterministic decision is whether to go forward or backward along the same pre-determined path. This property is ensured by using only strings built from the substrings aa', bb' and c, using the following transitions.

The transitions by the symbols a and a' are defined so that the string aa' implements a circular permutation of all states that is traversable in both directions (see Fig. 3).

$$\delta(q_i, a) = \{(q_i, -1), (q_{(i+1) \bmod n}, +1)\}, \qquad \text{for } i \in \{0, \dots, n-1\}$$
$$\delta(q_i, a') = \{(q_{(i-1) \bmod n}, -1), (q_i, +1)\}, \qquad \text{for } i \in \{0, \dots, n-1\}$$

Similarly, transitions by the symbols b and b' swap the first two states on the string bb', which is again traversable in both directions.

$$\delta(q_i, b) = \{(q_i, -1), (q_{1-i}, +1)\} \qquad \text{for } i \in \{0, 1\}$$
$$\delta(q_i, b) = \{(q_i, -1), (q_i, +1)\}, \qquad \text{for } i \in \{2, \ldots, n-1\}$$
$$\delta(q_i, b') = \{(q_{1-i}, -1), (q_i, +1)\} \qquad \text{for } i \in \{0, 1\}$$
$$\delta(q_i, b') = \{(q_i, -1), (q_i, +1)\}, \qquad \text{for } i \in \{2, \ldots, n-1\}$$

The last symbol c is used to cancel the path through the state q_0, while maintaining two-way traversals in all other states.

$$\delta(q_0, c) = \varnothing$$
$$\delta(q_i, c) = \{(q_i, -1), (q_i, +1)\}, \qquad \text{for } i \in \{1, \ldots, n-1\}$$

This bi-directional behaviour is interrupted at the end-markers, where deterministic transitions instruct the automaton to transfer from one bi-directional path to another—with no way back.

$$\delta(q_0, \vdash) = (q_0, +1),$$
$$\delta(q_i, \vdash) = (q_{i+1}, +1), \qquad \text{for odd } i \text{ less than } n-1$$
$$\delta(q_i, \dashv) = (q_{i+1}, -1), \qquad \text{for even } i \text{ less than } n-2$$

At the right end-marker ($\dashv$), the automaton accepts in the state q_{n-1}. The remaining transitions at the end-markers are left undefined.

The goal is to prove that every 1NFA that recognizes the same language as the 2NFA constructed above must have at least $\binom{n}{1}_2 = \sum_{k=0}^{\lfloor \frac{n-1}{2} \rfloor} \binom{n}{2k+1}\binom{2k+1}{k}$ states. This will be done using the fooling set method (Birget [2]; Glaister and Shallit [9]), according to which it is sufficient to present $\binom{n}{1}_2$ different pairs of strings (u_i, v_i), so that each concatenation $u_i v_i$ is in the language $L(\mathcal{A})$, whereas for each $i \neq j$ at least one of the mismatched concatenations $u_i v_j$ and $u_j v_i$ is not in $L(\mathcal{A})$. Once such a set of pairs is constructed, every 1NFA recognizing this language must have at least $\binom{n}{1}_2$ states.

In this proof, pairs of strings in the fooling set are indexed by crossing sequences, each represented by disjoint sets $\overrightarrow{S}$ and $\overleftarrow{S}$, as in the construction in Lemma 1. The plan is that for each pair $(u_{\overrightarrow{S}, \overleftarrow{S}}, v_{\overrightarrow{T}, \overleftarrow{T}})$, the accepting computation on their concatenation crosses the boundary between $u_{\overrightarrow{S}, \overleftarrow{S}}$ and $v_{\overrightarrow{T}, \overleftarrow{T}}$ from left to right in all states from $\overrightarrow{S}$, and from right to left in all states from $\overleftarrow{S}$.

The first step towards obtaining such a fooling set is constructing strings representing all partial injective functions on the set of states. Every string in $\{aa', bb', c\}^*$ naturally represents a partial injective function $f : \{q_0, \ldots, q_{n-1}\} \rightarrow \{q_0, \ldots, q_{n-1}\}$, where $f(p) = q$ if and only if the state p at the first symbol of the string is connected with q at the last symbol of the string by a bi-directional path. Besides these bi-directional connections, no other paths exist, except for broken bi-directional paths that start at the beginning or at the end of the string, and lead to a configuration with no transitions defined. An example of such a string is given in Fig. 4. It turns out that every partial injective function f can be implemented in this way.

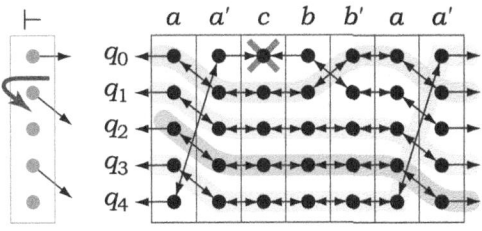

Fig. 4. A string $w_f \in \{aa', bb', c\}^*$ implementing a function f as in Claim 2.1, with $f(q_0) = q_1$, $f(q_1) = q_3$, $f(q_2) = q_4$, $f(q_3) = q_0$ and $f(q_4)$ undefined.

Claim 2.1. For every injective partial function $f\colon \{q_0, \ldots, q_{n-1}\} \rightarrow \{q_0, \ldots, q_{n-1}\}$, there is such a string $w_f \in \{aa', bb', c\}^+$ that:

i. *for every state q_i with $f(q_i)$ defined, there are computations beginning at the first symbol of w_f in the state q_i that exit the string to the right in the state $f(q_i)$, computations that exit to the left in the state q_i, and computations that loop; computations beginning at the last symbol of w_f in the state $f(q_i)$ have the same possible outcomes;*

ii. *for every state q_i with $f(q_i)$ undefined, there are computations beginning at the first symbol of w_f in the state q_i that exit the string to the left in the state q_i, computations that reject, and computations that loop;*

iii. *for every state q_j with $f^{-1}(q_j)$ undefined, computations beginning at the last symbol of w_f in the state q_j may only exit the string to the right in the state q_j, or reject, or loop.*

If f is a permutation, then it can be generated by composing cyclic permutations (aa') with swapping of the first two elements (bb'). An arbitrary injective partial function is then obtained from a permutation by making it undefined on some select input values. This is done by rotating any undesired paths to position 0 using several instances of the substring aa', then applying c, and finally rotating the paths back. This proves Claim 2.1.

Now the strings for the fooling set are constructed as follows. The first string in a pair corresponding to sets $\overrightarrow{S}$ and $\overleftarrow{S}$ implements a bi-directional path from the initial configuration to the first state in $\overrightarrow{S}$, as well as paths from every i-th element of $\overleftarrow{S}$ to the $(i+1)$-th element of $\overrightarrow{S}$.

Claim 2.2. For every two disjoint sets $\overrightarrow{S}, \overleftarrow{S} \subseteq \{q_0, \dots, q_{n-1}\}$ satisfying $|\overrightarrow{S}| = |\overleftarrow{S}| + 1$, let $\overrightarrow{S} = \{q_{i_0}, \dots, q_{i_m}\}$ and $\overleftarrow{S} = \{q_{j_0}, \dots, q_{j_{m-1}}\}$, where $0 \leqslant i_0 < \dots < i_m \leqslant n - 1$ and $0 \leqslant j_0 < \dots < j_{m-1} \leqslant n - 1$. Then there exists such a string $u_{\overrightarrow{S}, \overleftarrow{S}}$, that the 2NFA makes the following computations on the tape $\vdash u_{\overrightarrow{S}, \overleftarrow{S}}$.

   i. A computation beginning at the left end-marker ($\vdash$) in the state q_0 may exit the string to the right in the state q_{i_0} or loop.
   ii. For each $t \in \{0, \dots, m-1\}$, a computation beginning at the last symbol of $\vdash u_{\overrightarrow{S}, \overleftarrow{S}}$ in the state q_{j_t} may exit the string to the right in either of the states q_{j_t} and $q_{i_{t+1}}$, or reject, or loop.
   iii. A computation beginning at the last symbol of $\vdash u_{\overrightarrow{S}, \overleftarrow{S}}$ in any state not in $\overleftarrow{S}$ may exit the string to the right in the same state, or reject, or loop.

The string w_f illustrated in Fig. 4 can actually serve as $u_{\overrightarrow{S}, \overleftarrow{S}}$, for $\overrightarrow{S} = \{q_1, q_4\}$ and $\overleftarrow{S} = \{q_3\}$. Indeed, if the left end-marker $\vdash$ is put in front of this string, then the path from the initial configuration leads to q_1 at the right border of the string, whereas the path from q_3 at the right border goes all the way to the left, then turns using the transition $\delta(q_1, \vdash) = (q_2, +1)$ at the left end-marker, and then proceeds to q_4 at the right border.

Claim 2.2 is proved by applying Claim 2.1 to a function f that encodes the desired connectivity between states at the last symbol of the string.

The right counterpart of $u_{\overrightarrow{S}, \overleftarrow{S}}$ is defined symmetrically: this will be a string implementing a path from the maximal element of $\overleftarrow{S}$ to an accepting configuration, as well as paths from every i-th element of $\overleftarrow{S}$ (except the maximal one) to the i-th element of $\overrightarrow{S}$. The proof is again by using Claim 2.1.

Claim 2.3. Let $\overrightarrow{S} = \{q_{i_0}, \dots, q_{i_m}\}$ and $\overleftarrow{S} = \{q_{j_0}, \dots, q_{j_{m-1}}\}$ be two disjoint sets, as in Claim 2.2. Then there exists such a string $v_{\overrightarrow{S}, \overleftarrow{S}}$, that a computation beginning at the first symbol of $v_{\overrightarrow{S}, \overleftarrow{S}} \dashv$ can have any of the following outcomes.

   i. Beginning in each state q_{i_t}, with $0 \leqslant t \leqslant m-1$, the automaton may exit the string to the left in either of the states q_{i_t} and q_{j_t}, or reject, or loop.
   ii. Beginning in the state q_{i_m}, the automaton may exit the string to the left in the states q_{i_m}, or accept, or loop.
   iii. Beginning in any state not in $\overrightarrow{S}$, the automaton may exit the string to the left in the same state, or reject, or loop.

With the fooling set constructed, the first fact to prove is that all concatenations of matching left and right strings are accepted.

Claim 2.4. For every two disjoint sets $\overrightarrow{S}, \overleftarrow{S} \subseteq \{q_0, \dots, q_{n-1}\}$, with $|\overrightarrow{S}| = |\overleftarrow{S}| + 1$, the concatenation $u_{\overrightarrow{S}, \overleftarrow{S}} v_{\overrightarrow{S}, \overleftarrow{S}}$ is in $L(A)$.

Let $\overrightarrow{S} = \{q_{i_0}, \ldots, q_{i_m}\}$ with $0 \leqslant i_0 < \ldots < i_m \leqslant n - 1$, and $\overleftarrow{S} = \{q_{j_0}, \ldots, q_{j_{m-1}}\}$, with $0 \leqslant j_0 < \ldots < j_{m-1} \leqslant n - 1$. Then an accepting computation on $u_{\overrightarrow{S},\overleftarrow{S}} v_{\overrightarrow{S},\overleftarrow{S}}$ first reaches the boundary between the substrings in the state q_{i_0}, then gets to q_{j_0} by an excursion over $v_{\overrightarrow{S},\overleftarrow{S}} \dashv$, then to q_{i_1} by an excursion over $\vdash u_{\overrightarrow{S},\overleftarrow{S}}$, etc., until it finally reaches the boundary in the state q_{i_m}, and then proceeds to the right end-marker ($\dashv$), where it accepts in the state q_{n-1}.

The second property to be established is that at least one of the two mismatched concatenations is rejected.

Claim 2.5. *Let the four subsets* $\overrightarrow{S}, \overleftarrow{S}, \overrightarrow{T}, \overleftarrow{T} \subseteq \{q_0, \ldots, q_{n-1}\}$ *satisfy* $|\overrightarrow{S}| = |\overleftarrow{S}| + 1$ *and* $|\overrightarrow{T}| = |\overleftarrow{T}| + 1$, *and assume that* $(\overrightarrow{S}, \overleftarrow{S}) \neq (\overrightarrow{T}, \overleftarrow{T})$. *Then at least one of the two strings* $u_{\overrightarrow{S},\overleftarrow{S}} v_{\overrightarrow{T},\overleftarrow{T}}$ *and* $u_{\overrightarrow{T},\overleftarrow{T}} v_{\overrightarrow{S},\overleftarrow{S}}$ *is not in* $L(A)$.

By Claims 2.4 and 2.5, the set of pairs $(u_{\overrightarrow{S},\overleftarrow{S}}, v_{\overrightarrow{S},\overleftarrow{S}})$ is a fooling set. Therefore, every 1NFA recognizing this language needs to have at least as many states as there are pairs of disjoint subsets $\overrightarrow{S}, \overleftarrow{S} \subseteq \{q_0, \ldots, q_{n-1}\}$, with $|\overrightarrow{S}| = |\overleftarrow{S}| + 1$, which is the exact expression given in the statement of the lemma. $\square$

Theorem 2. *Let an alphabet* Σ *contain at least five symbols. Then, for every* $n \geqslant 2$, *the number of states in a 1NFA necessary to represent every language over* Σ *that is recognized by an n-state 2NFA, is at least* $\binom{n}{1}_2 = \sum_{k=0}^{\lfloor \frac{n-1}{2} \rfloor} \binom{n}{2k+1} \binom{2k+1}{k}$.

The size of the alphabet in the lower bound can be reduced to four by a slight adjustment of transitions of the 2NFA. This will be presented in the full version of this paper.

Thus, in conjunction with Theorem 1, the 2NFA-to-1NFA tradeoff for small alphabets has been determined up to a constant factor depending on the alphabet. For instance, in the case of a five-symbol alphabet, the upper bound is $5\binom{n}{1}_2$ and the lower bound is $\binom{n}{1}_2$. A problem suggested for future research is to investigate the dependence of the size of the resulting 1NFA on the size of the alphabet.

References

1. Berman, P.: A note on sweeping automata. In: de Bakker, J., van Leeuwen, J. (eds.) ICALP 1980. LNCS, vol. 85, pp. 91–97. Springer, Heidelberg (1980)
2. Birget, J.-C.: Intersection and union of regular languages and state complexity. Inf. Process. Lett. **43**, 185–190 (1992)
3. Chrobak, M.: Finite automata and unary languages. Theor. Comput. Sci. **47**, 149–158 (1986)
4. Geffert, V., Mereghetti, C., Pighizzini, G.: Converting two-way nondeterministic unary automata into simpler automata. Theor. Comput. Sci. **295**(1–3), 189–203 (2003)
5. Geffert, V., Okhotin, A.: One-way simulation of two-way finite automata over small alphabets. In: NCMA 2013, Umeå, Sweden, 13–14 August 2013

6. Geffert, V., Okhotin, A.: Transforming two-way alternating finite automata to one-way nondeterministic automata. In: Csuhaj-Varjú, E., Dietzfelbinger, M., Ésik, Z. (eds.) MFCS 2014. LNCS, vol. 8634, pp. 291–302. Springer, Heidelberg (2014)

7. Geffert, V., Okhotin, A.: Deterministic one-way simulation of two-way deterministic finite automata over small alphabets. Int. J. Found. Comput. Sci. (2025, to appear)

8. Geffert, V., Pighizzini, G.: Two-way unary automata versus logarithmic space. Inf. Comput. **209**(7), 1016–1025 (2011)

9. Glaister, I., Shallit, J.: A lower bound technique for the size of nondeterministic finite automata. Inf. Process. Lett. **59**, 75–77 (1996)

10. Kapoutsis, C.A.: Removing bidirectionality from nondeterministic finite automata. In: Jędrzejowicz, J., Szepietowski, A. (eds.) MFCS 2005. LNCS, vol. 3618, pp. 544–555. Springer, Heidelberg (2005)

11. Kapoutsis, C.A.: Algorithms and lower bounds in finite automata size complexity. Ph.D. thesis, Massachusetts Institute of Technology (2006)

12. Kapoutsis, C.A. : Two-way automata versus logarithmic space. Theory Comput. Syst. **55**(2), 421–447 (2014)

13. Kapoutsis, C.A., Pighizzini, G.: Two-way automata characterizations of L/poly versus NL. Theor. Comput. Syst. **56**(4), 662–685 (2015)

14. Kunc, M., Okhotin, A.: Describing periodicity in two-way deterministic finite automata using transformation semigroups. In: Mauri, G., Leporati, A. (eds.) DLT 2011. LNCS, vol. 6795, pp. 324–336. Springer, Heidelberg (2011)

15. Mereghetti, C., Pighizzini, G.: Optimal simulations between unary automata. SIAM J. Comput. **30**(6), 1976–1992 (2001)

16. Micali, S.: Two-way deterministic finite automata are exponentially more succinct than sweeping automata. Inf. Process. Lett. **12**(2), 103–105 (1981)

17. Moore, F.R.: On the bounds for state-set size in the proofs of equivalence between deterministic, nondeterministic, and two-way finite automata. IEEE Trans. Comput. **20**, 1211–1214 (1971)

18. Petrov, S., Okhotin, A.: On the transformation of two-way finite automata to unambiguous finite automata. Inf. Comput. **295A** (2023). Article 104956

19. Petrov, S., Okhotin, A.: On the transformation of two-way nondeterministic finite automata to unambiguous finite automata. In: Ko, S., Manea, F. (eds.) DLT 2025. LNCS, vol. 16036 (2025)

20. Rabin, M.O., Scott, D.: Finite automata and their decision problems. IBM J. Res. Dev. **3**, 114–125 (1959)

21. Sakoda, W.J., Sipser, M.: Nondeterminism and the size of two way finite automata. In: 10th ACM Symposium on Theory of Computing (STOC 1978), pp. 275–286 (1978)

22. Shepherdson, J.C.: The reduction of two-way automata to one-way automata. IBM J. Res. Dev. **3**, 198–200 (1959)

23. Vardi, M.: A note on the reduction of two-way automata to one-way automata. Inf. Process. Lett. **30**(5), 261–264 (1989)

A New Approach for Showing Termination of Parameterized Transition Systems

Roland Herrmann[1]([✉])[iD] and Philipp Rümmer[1,2][iD]

[1] University of Regensburg, Regensburg, Germany
roland.herrmann@ur.de
[2] Uppsala University, Uppsala, Sweden

Abstract. We investigate the termination problem of regular transition systems, i.e., of transition systems whose transition relations can be represented by finite-state automata. Leveraging this automata-theoretical perspective, we propose a new, efficient approach to termination analysis. Our method encodes termination arguments, called statements, as words in an auxiliary language, enabling the verification of whether configurations satisfy those termination arguments via an interpretation automaton. The choice of a fixed interpretation automaton enables the synthesis of new automata dedicated to searching for statements and verifying properties within any given regular transition system. The central contribution of this work is the construction of a specialized finite-state automaton that recognizes statements ensuring the system's termination according to predefined conditions. Depending on the fragment of regular transitions considered, we define two sets of sufficient conditions for termination and provide an automata-based method to verify them within a given regular transition system.

1 Introduction

We consider the termination problem of transition systems. The termination problem is an instance of a liveness property and is, in general, undecidable, as illustrated by the halting problem [15]. Consequently, we must either identify suitable fragments of transition systems that retain decidability, or rely on sound but incomplete procedures that deliver good results in practical applications. In this paper, we examine fragments of regular transition systems (RTS), which can be viewed as an instance of parameterized transition systems. The transition relation in an RTS is modeled by a length-preserving transducer, with the length of the initial configuration acting as an implicit parameter of the system.

The standard approach for proving termination involves over-approximating the transition relation with a well-founded relation. However, finding such a relation efficiently is often challenging. To tackle this problem, we leverage an automata-theoretical framework. Recently, a new perspective on analyzing RTS was introduced by Welzel-Mohr [17], which we adopt. It reasons about properties of the configurations of the system. Those properties are represented by words

G. Castiglione and S. Mantaci (Eds.): CIAA 2025, LNCS 15981, pp. 193–207, 2026.
https://doi.org/10.1007/978-3-032-02602-6_14

over an auxiliary alphabet and called statements. To verify whether a given configuration satisfies a statement, a fixed interpretation automaton is used. In this way, the entire verification process is described by automata. The synthesis of the relevant automata leads to a framework that transforms the process of invariant synthesis into automata queries, in particular checking non-emptiness.

In previous work, this approach was utilized only for analyzing safety properties. In order to apply the framework for termination analysis, we adapt the alphabet to reason about relations between configurations. In particular, then a statement can be viewed as a relation by considering the set of configurations that satisfy the statement according to the interpretation automaton. The search for well-founded relations is incorporated into the synthesized automaton. The challenge is no longer the automation of the process, but rather to find suitable preconditions that the synthesized automata should search for. We give two example automata that verify (1) that a regular transition system is lexicographically ordered, or (2) that it is letter-wise ordered, which both imply termination. The approach is sound for any given RTS and complete for the respective fragment. Our work can be seen as a recipe for transferring the approach to other areas.

Related Work. Our work lies within the broader field of **regular model checking,** which was first developed by Bouajjani et al. [7]. In general, questions in regular model checking can be formulated as second-order formulae over some automatic structure [12]. The most common focus is on **safety** properties, that is, whether error states can be reached from initial states. A typical approach involves searching for inductive invariants, e.g., via abstract regular model checking [6] or active automata learning using the L^* algorithm [5,8]. An alternative approach, utilized and extended in this paper, was recently introduced by Welzel-Mohr [17], determining inductive invariants with the help of interpretation automata.

In contrast, **liveness** properties, such as **termination**, have received less attention in regular model checking so far. Podelski et al. [13] introduce the notion of a transition invariant as a tool to prove termination of programs, which directly applies to transition systems as well. A combination with predicate abstraction for automation is described in [14]. A key result shown is that a program terminates if and only if a disjunctively well-founded transition invariant exists, which is a generalization of well-founded relations. However, in our work, we are only concerned with ordinary well-founded relations.

To compute transition invariants of **parameterized transition systems,** it is necessary to over-approximate the transitive closure of the transition relation, which already by itself is a hard problem. An approach that avoids this construction is described by Abdullah et al. [3], where instead backward calculation is used to determine which states necessarily lead to terminating states, leveraging existing methods for safety analysis. Closely related to well-founded relations is the concept of ranking functions. Fang et al. [10] use a heuristic called projection & generalize to construct a ranking function, which is intended to verify that the transition relation eventually makes progress towards termination by reducing the rank of the current state at each step. Lin et al. [11] use a CEGAR approach to show termination of parameterized transition systems modeling two-player reachability games, applying both Angluin's L^*-algorithm and SAT-solving.

Structure of the Paper. In Sect. 2, we give preliminaries on finite-state automata, regular transition systems, relations and fix some notation. We recall the approach from [17] in Sect. 3, which is initially used to prove safety properties. Our main contribution is to carry over the approach from [17] to liveness properties, more precisely to termination analysis. In Sect. 4, we consider two fragments of regular transition systems that we prove terminating: (1) lexicographically ordered systems and (2) letter-wise ordered systems. Both sections follow a common pattern of identifying preconditions that are sufficient to prove termination and then constructing an automaton that searches for witnesses for these conditions. We conclude in Sect. 5 with future work.

2   Preliminaries

Finite-State Automata. We assume basic familiarity with finite automata. A non-deterministic finite-state automaton (NFA) is a tuple $\mathcal{A} = (Q_\mathcal{A}, \Sigma_\mathcal{A}, s_\mathcal{A}, \delta_\mathcal{A}, F_\mathcal{A})$, where $Q_\mathcal{A}$ is the set of states, $\Sigma_\mathcal{A}$ the finite alphabet, $s_\mathcal{A} \in Q_\mathcal{A}$ the initial state, $\delta_\mathcal{A} \subseteq Q_\mathcal{A} \times \Sigma_\mathcal{A} \times Q_\mathcal{A}$ the transition relation, and $F_\mathcal{A} \subseteq Q_\mathcal{A}$ the set of accepting states. We will stick to this subscript notation for the corresponding components of an automaton $\mathcal{A}$. The language accepted by $\mathcal{A}$ is denoted by $\mathcal{L}(\mathcal{A})$. We assume that for all $p \in Q_\mathcal{A}, s \in \Sigma_\mathcal{A}$, there exists a $q \in Q_\mathcal{A}$ such that $(p, s, q) \in \delta_\mathcal{A}$, since otherwise we may simply introduce a fresh non-accepting "sink state". For the sake of presentation we will not explicitly depict the sink state in figures.

A Σ_1-Σ_2-transducer for two alphabets Σ_1, Σ_2 is an NFA with alphabet $\Sigma_1 \times \Sigma_2$. Let $\mathcal{T}$ be a Σ_1-Σ_2-transducer. Let $u = u_1 \cdots u_n \in \Sigma_1^*, v = v_1 \cdots v_n \in \Sigma_2^*$ be two words, then we write $u \otimes v$ for the word $(u_1, v_1) \ldots (u_n, v_n)$ in $(\Sigma_1 \times \Sigma_2)^*$ and $\binom{u_i}{v_i}$ for its letters instead of (u_i, v_i). In case $u \otimes v \in \mathcal{L}(\mathcal{T})$, we interpret u as the input and v as the output of $\mathcal{T}$. If $\Sigma_1 = \Sigma_2 = \Sigma$, then $\mathcal{T}$ defines a relation $R_\mathcal{T}$ on Σ^* by $(u, v) \in R_\mathcal{T} :\Leftrightarrow u \otimes v \in \mathcal{L}(\mathcal{T})$.[1] When writing $u \otimes v$, we we implicitly assume that the lengths of the words u, v is equal. Furthermore, our notion of transducer does not allow ε-transitions, neither in the input nor in the output, as opposed to other definitions from the literature.

Regular Transition Systems
Definition 1 (regular transition system [1,2]). *A regular transition system (RTS) is a tuple $(\Sigma, \mathcal{T})$, where Σ is a finite alphabet and $\mathcal{T}$ is a Σ-Σ transducer. We call $\mathcal{T}$ the transition relation, the letters $s \in \Sigma$ states, words $w \in \Sigma^*$ configurations and the i-th position of a configuration an agent. In other words, a configuration is a finite sequence of the current states of agents.*

---

[1] We write "$A :\Leftrightarrow B$" to denote that A is defined to be equivalent to B.

Remark 1 (initial configurations). We consider termination from any configuration. Therefore, we do not specify initial configurations. In this regard, our definitions deviate from other standard terminology also found in the literature, which usually emphasizes the origin from transition systems and then additionally demands the initial configurations and transition relation to be a regular expression.

Let $(\Sigma, \mathcal{T})$ be a RTS, $a = a_1 \ldots a_n \in \Sigma^*$, $b = b_1 \ldots b_n \in \Sigma^*$ two words. Then we write $a \to_{\mathcal{T}} b$, or simply $a \to b$ if $\mathcal{T}$ is clear from the context, if we have $\binom{a_1}{b_1} \ldots \binom{a_n}{b_n} \in \mathcal{L}(\mathcal{T})$ and call it a transition from a to b. The number of agents in a configuration is invariant under the transition relation, since transducers are defined to be length-preserving in our setting. One may view the number of agents as a parameter, making RTSs an instance of parametrized transition systems. Furthermore, since our alphabet is finite, there are at most $|\Sigma|^n < \infty$ reachable configurations from any initial configuration $w \in \Sigma^n$ for n agents. We call transition systems in which for any configuration the set of reachable configurations is finite, weakly-finite [9].

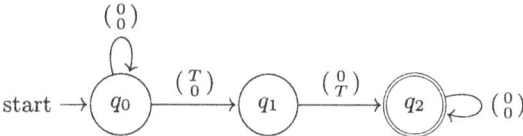

Fig. 1. Transition relation $\mathcal{T}$ for token passing.

Example 1 (Token passing). Let $\Sigma = \{0, T\}$. Consider the transition relation described by the Σ-Σ-transducer depicted in Fig. 1. To illustrate the transition system, we also give all transitions by hand, starting from a specific configuration $T000$ and the parameter value $n = 4$:

$$T000 \to 0T00 \to 00T0 \to 000T \tag{1}$$

The transition relation can equivalently be defined using the corresponding regular expression:

$$\binom{0}{0}^* \binom{T}{0} \binom{0}{T} \binom{0}{0}^*. \tag{2}$$

The RTS $(\Sigma, \mathcal{T})$ models a token passing protocol. The state T represents that at this position there is a token and 0 represents that there is none. The transition relation $\mathcal{T}$ models that a token is handed over from the i-th to the $(i + 1)$-th agent, starting at the first and ending in the last. The transition relation assumes that there exists only one token (otherwise there are no transitions). The token passing protocol serves as an easy running example of a terminating RTS.

Relations. We define some terms and notation on relations for later use.

Definition 2. *Let $R \subseteq S \times S$ be a relation. We call R*

- *reflexive if for all $x \in S$, we have $(x, x) \in R$.*
- *transitive if for all $x, y, z \in S$ with $(x, y), (y, z) \in R$, we have $(x, z) \in R$.*
- *irreflexive if for all $x \in S$, we have $(x, x) \notin R$.*
- *total if for all $x, y \in S$ with $x \neq y$, we have $(x, y) \in R$ or $(y, x) \in R$.*
- *partial if it is not total.*
- *preorder if it is reflexive and transitive.*

Remark 2. Let $\preceq$ be a preorder on a set S. Then

$$a \prec b :\Leftrightarrow a \preceq b \wedge \neg(b \preceq a). \tag{3}$$

defines an irreflexive, transitive relation and every irreflexive, transitive relation arises in this way. This motivates the following notation.

Notation 1. Let $R \subseteq S \times S$ be a relation. For $x, y \in S$, we write

- $x \geq_R y$ if $(x, y) \in R$.
- $x >_R y$ if $(x, y) \in R$ and $(y, x) \notin R$.
- $x =_R y$ if $(x, y), (y, x) \in R$.

We abbreviate $\geq_i, >_i, =_i$ for $\geq_{R_i}, >_{R_i}, =_{R_i}$ for a relation R_i with index i.

As seen in [13], well-founded relations play a crucial role for termination.

Definition 3 (well-founded [16]). *Let S be a set. A relation $R \subseteq S \times S$ is called well-founded if there exists no sequence $(s_n)_{n \in \mathbb{N}}$ such that $(s_n, s_{n+1}) \in R$ for all $n \in \mathbb{N}$.*

Lemma 1. *Let R be an irreflexive, transitive relation on a finite set S. Then R is well-founded.*

Proof. Assume for a contradiction that R is not well-founded, i.e., there exists a sequence $(w_i)_{i \in \mathbb{N}}$ such that $(w_i, w_{i+1}) \in R$ for all $i \in \mathbb{N}$. Since S is finite there exists $i, j \in \mathbb{N}$ such that $i \neq j$ and $w_i = w_j$. Without loss of generality we may assume $i < j$. Then transitivity of R along $w_i \geq_R w_{i+1} \geq_R \cdots \geq_R w_j$ implies $w_i \geq_R w_j = w_i$ which contradicts irreflexivity of R.

3   Properties of RTS as Statements

We revise the approach presented in [17] for reasoning about configurations in an RTS. Statements are formulated as words over an auxiliary alphabet and an interpretation automaton justifies whether a configuration satisfies a statement.

Fix a new (finite) alphabet Γ for statements. A statement is then a word $I \in \Gamma^*$. An interpretation automaton is a Σ-Γ-transducer $\mathcal{V}$. A statement $I \in \Gamma^n$

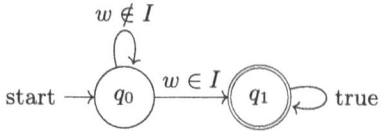

Fig. 2. The interpretation automaton $\mathcal{V}_{Trap}$.

holds in a configuration $w \in \Sigma^n$ if and only if $w \otimes I \in \mathcal{L}(\mathcal{V})$. For $\Gamma = 2^\Sigma$, an example of an interpretation automaton is depicted in Fig. 2 by $\mathcal{V}_{Trap}$.

The labels "$w \in I$" and "$w \notin I$" summarize the exponentially many transition labels $\{(w, I) \in \Sigma \times 2^\Sigma \mid w \in I\}$, respectively $\{(w, I) \in \Sigma \times 2^\Sigma \mid w \notin I\}$ and the transition "true" is always available. In natural language $\mathcal{V}_{Trap}$ checks whether at least one of the states of the configuration is included in the corresponding set (at same position) in the statement. We can define the set of configurations that satisfy I as $\{w \in \Sigma^* \mid w \otimes I \in \mathcal{L}(\mathcal{V})\}$. That way, I naturally describes an invariant candidate.

Example 2. We come back to our token passing example. Consider the word $I = \{T\}\{T\}\{T\}\{T\} \in (2^\Sigma)^*$. It can be easily verified that $\Sigma^4 \setminus \{0000\}$ is the set of all configurations that satisfy I in the interpretation $\mathcal{V}_{Trap}$. In natural language, I can be reformulated as "has at least one token".

The main result of this construction is now that we can automatically verify whether a statement holds (for certain words) and therefore also automatically search for statements that have certain properties. In the work of [17], this is done for inductive statements. In other words, the statement alphabet and the interpretation automaton yield a set of invariant candidates (one for each statement) and we need to search within this set for actual invariants with the desired properties, like being inductive. The challenge in this approach is no longer the automation of the search but rather finding appropriate alphabets for statements and interpretations. Ideally, one tries to avoid a choice of alphabet and interpretation that is specific for one single RTS.

4   Termination of RTS

In this section we show how the approach from [17] can be used to show termination of regular transition systems. The proofs of this section are straightforward and rather technical, hence they are deffered to the appendix.

Definition 4 (Termination of a RTS). *Let $(\Sigma, \mathcal{T})$ be an RTS. We say that $(\Sigma, \mathcal{T})$ terminates on $w_0 \in \Sigma^*$ if there exists no sequence $(w_i)_{i \in \mathbb{N}}$ such that $w_i \otimes w_{i+1} \in \mathcal{L}(\mathcal{T})$ for all $i \in \mathbb{N}_0$. We say that $(\Sigma, \mathcal{T})$ terminates if it terminates for all $w_0 \in \Sigma^*$.*

Theorem 1. *The termination problem of RTS is undecidable in general.*

Proof. Consider the halting problem whether a turing machine M halts on input x. This can be reduced to termination of a turing machine M' using at most N tape cells on inputs of at most N cells of memory for each $N \in \mathbb{N}$ uniformly. M' can be modelled by a length preserving transducer $\mathcal{T}$. Hence, M' halts for all $N \in \mathbb{N}$ and every input if and only if the RTS $(\Sigma, \mathcal{T})$ terminates where Σ represents both the tape content and the state of the turing machine M'.

As a first step, we exploit the fact that regular transition systems are weakly-finite. Therefore, it is sufficient for termination that there exists an irreflexive, transitive relation on the set of reachable configurations for each value of the parameter, over-approximating the transition relation. This relation is then immediately well-founded by Lemma 1. Formally, we have the following Theorem.

Theorem 2. *Let $(\Sigma, \mathcal{T})$ be an RTS, $n \in \mathbb{N}$. Let R be a well-founded relation that over-approximates the transition relation, i.e. $(\Sigma^n \times \Sigma^n) \cap R_{\mathcal{T}} \subseteq (\Sigma^n \times \Sigma^n) \cap R$. Then the RTS terminates on any input $w \in \Sigma^n$.*

In order to reason about relations, we consider $\Sigma \times \Sigma$ as underlying alphabet and $\Gamma = 2^{\Sigma \times \Sigma}$ as alphabet for statements. For a fixed interpretation automaton $\mathcal{V}$, a statement $I \in \Gamma^*$ then defines a relation on the configurations by

$$R_I = \{(w_1, w_2) \in \Sigma^* \times \Sigma^* \mid ((w_1 \otimes w_2) \otimes I) \in \mathcal{L}(\mathcal{V})\} \subseteq \Sigma^{|I|} \times \Sigma^{|I|}. \quad (4)$$

We give two frameworks to prove termination of fragments of regular transition systems in the following.

4.1 Lexicographically Ordered Systems

We first treat lexicographically ordered systems. Intuitively, this is how words in a dictionary are arranged.

Lemma 2 (lexicographical order). *Let $R_\Sigma \subseteq \Sigma \times \Sigma$ be a relation on Σ. Then R_Σ induces a relation R_{lex} on Σ^* by*

$$u_1 \ldots u_n R_{lex} v_1 \ldots v_m :\Leftrightarrow \exists i \in \{1, \ldots, n\}. \; (u_i R_\Sigma v_i \wedge \forall j < i. \; u_j = v_j)$$
$$\vee (n > m \wedge u_1 \ldots u_m = v_1 \ldots v_m). \quad (5)$$

If R_Σ has any of the properties in Definition 2, then so does R_{lex}. If R_Σ is a total, irreflexive, transitive relation, we call R_{lex} lexicographical order. We call an RTS $(\Sigma, \mathcal{T})$ lexicographically ordered if there exists a lexicographical order R_{lex} such that $R_{\mathcal{T}} \subseteq R_{lex}$.

Let $\Delta := \{(\begin{smallmatrix} x \\ x \end{smallmatrix}) \mid x \in \Sigma\}$. We define the interpretation automaton that captures the essential steps to recognize lexicographical orders as the automaton $\mathcal{V}_{lex}$ depicted in Fig. 3a over the alphabet $(\Sigma \times \Sigma) \times 2^{\Sigma \times \Sigma}$. In particular $I \cup \Delta, I \setminus \Delta \in 2^{\Sigma \times \Sigma}$.

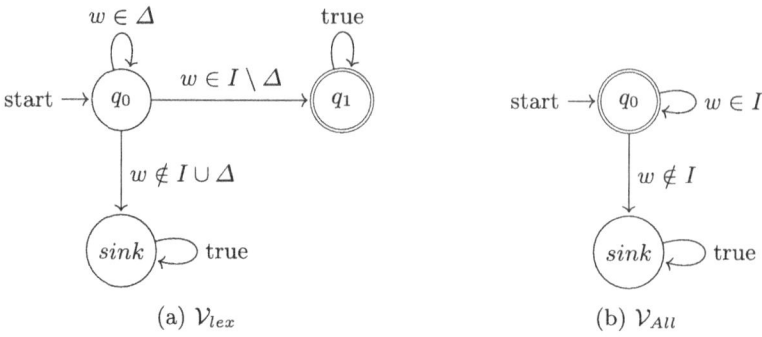

Fig. 3. Interpretation automata for lexicographical, respectively letterwise order.

The automaton $\mathcal{V}_{lex}$ demands for some $i \in \mathbb{N}$ the first $i - 1$ letters to be equal and then make a transition according to our input statement on the i-th position in order to reach the accepting state.

For a fixed statement $I \in (2^{\Sigma \times \Sigma})^*$, we consider the induced relation R_I from (4). By Lemma 1 and Theorem 2 it suffices to find for all $n \in \mathbb{N}$ an irreflexive, transitive relation that over-approximates the transition relation on Σ^n, i.e.,

(i) Irreflexive: $R_I \subseteq \{(x,y) \mid x \neq y\} \cap (\Sigma \times \Sigma)^n$.
(ii) Transitive: $\{(x,z) \mid \exists y.\{(x,y),(y,z)\} \subseteq R_I\} \subseteq R_I$.
(iii) over-approximates transition relation $R_T \cap (\Sigma \times \Sigma)^n \subseteq R_I$

Example 3. The token passing protocol is lexicographically ordered. We simply define $R_\Sigma = \{(T,0)\}$. The corresponding statement that recognises this lexicographical order is then $I_n = \{\binom{T}{0}\}^n$ for all $n \in \mathbb{N}$ and we have

$$R_{I_n} = \mathcal{L}\left(\left(\binom{0}{0} + \binom{T}{T}\right)^* \binom{T}{0} (\Sigma \times \Sigma)^*\right) \cap (\Sigma \times \Sigma)^n. \tag{6}$$

We now construct for a given RTS (Σ, T) an automaton $\mathcal{A}_{lex}$ that searches within the set of relations R_I for one that satisfies the conditions (i)-(iii). $\mathcal{A}_{lex}$ should have $2^{\Sigma \times \Sigma}$ as alphabet and accept a statement $I \in (2^{\Sigma \times \Sigma})^*$ if and only if R_I satisfies (i)-(iii). However, the conditions (i)-(iii) are universally quantified. In order to eliminate the universal quantifier on our three conditions, we first construct the complement of the desired automaton, that is, an automaton that accepts I if and only if there exist $x, y, z \in \Sigma^*$ such that one of (i)-(iii) does not hold, i.e.

$$(xR_Iy \wedge x = y) \vee (xR_Iy \wedge yR_Iz \wedge \neg xR_Iz) \vee (xTy \wedge \neg xR_Iy). \tag{7}$$

Any of the occurring predicates can be expressed by an automaton, e.g., xR_Iy evaluates to true if and only if $((x \otimes y) \otimes I) \in \mathcal{L}(\mathcal{V}_{lex})$, $\mathcal{V}_{lex}$ just ignores the z argument. Hence, we need three copies of $\mathcal{V}_{lex}$ for the predicates xR_Iy, yR_Iz, xR_Iz, each one ignoring the argument that does not occur in it. For the equality check, $x = y$, we take a two state automaton that accepts a pair (x,y) if and only if $x = y$. Lastly xR_Ty is expressed by ignoring the z input and have a run in the transducer $\mathcal{T}$. We can check which of these predicates hold simultaneously by going through the product of all these automata states in parallel, that is, we have

$$Q_{\mathcal{A}_{lex}^C} = \underbrace{Q_\mathcal{T}}_{xR_Ty} \times \underbrace{Q_=}_{x=y} \times \underbrace{Q_{\mathcal{V}_{lex}}}_{xR_Iy} \times \underbrace{Q_{\mathcal{V}_{lex}}}_{yR_Iz} \times \underbrace{Q_{\mathcal{V}_{lex}}}_{xR_Iz}. \tag{8}$$

The transition relation is then given according to the transition relations of the respective factors, i.e.,

$$((p_\mathcal{T}, p_=, p_1, p_2, p_3), (x_i, y_i, z_i, I_i), (q_\mathcal{T}, q_=, q_1, q_2, q_3)) \in \delta \tag{9}$$

if and only if all of the following conditions hold:

$$\begin{array}{ll}
(p_\mathcal{T}, (x_i, y_i), q_\mathcal{T}) \in \delta_\mathcal{T} & (xTy) \\
(p_=, (x_i, y_i), q_=) \in \delta_= & (x = y) \\
(p_1, (x_i, y_i, I_i), q_1) \in \delta_{\mathcal{V}_{lex}} & (xR_Iy) \\
(p_2, (y_i, z_i, I_i), q_2) \in \delta_{\mathcal{V}_{lex}} & (yR_Iz) \\
(p_3, (x_i, z_i, I_i), q_3) \in \delta_{\mathcal{V}_{lex}} & (xR_Iz)
\end{array}$$

We can now translate our formula one to one to the accepting states of $\mathcal{A}_{lex}^C$, where we use that the formula 7 is already in disjunctive normal form (each clause stands for the negation of one of the conditions (i)–(iii)):

$$\begin{array}{ll}
F_{\mathcal{A}_{lex}^C} = Q_\mathcal{T} \times F_= \times F_{\mathcal{V}_{lex}} \times Q_{\mathcal{V}_{lex}} \times Q_{\mathcal{V}_{lex}} & \neg(i) \\
\cup\, Q_\mathcal{T} \times Q_= \times F_{\mathcal{V}_{lex}} \times F_{\mathcal{V}_{lex}} \times (Q_{\mathcal{V}_{lex}} \setminus F_{\mathcal{V}_{lex}}) & \neg(ii) \\
\cup\, F_\mathcal{T} \times Q_= \times (Q_{\mathcal{V}_{lex}} \setminus F_{\mathcal{V}_{lex}}) \times Q_{\mathcal{V}_{lex}} \times Q_{\mathcal{V}_{lex}} & \neg(iii)
\end{array}$$

That way, our automaton accepts a tuple (x, y, z, I) if and only if at least one of the conditions (i)–(iii) fails. In order to get rid of the first three input tapes, we project onto the statement input tape, ignoring x_i, y_i, z_i in the transition relation δ. We can do this because x, y, z are existentially quantified in the negated statements of (i)–(iii). Let $\delta_{\mathcal{V}_{lex}^C}$ be the transition relation that we obtain from δ by projecting onto the statement tape. Finally, we can define

$$\mathcal{A}_{lex}^C = (Q_{\mathcal{A}_{lex}^C}, \Gamma, (s_\mathcal{T}, s_=, s_{\mathcal{V}_{lex}}, s_{\mathcal{V}_{lex}}, s_{\mathcal{V}_{lex}}), \delta_{\mathcal{A}_{lex}^C}, F_{\mathcal{A}_{lex}^C}). \tag{10}$$

$\mathcal{A}_{lex}^C$ accepts a statement $I \in \Gamma^*$ if and only if there exist $x, y, z \in \Sigma^*$ such that one of (i)-(iii) does not hold. We obtain the following theorem:

Theorem 3. Let $(\Sigma, \mathcal{T})$ be a RTS, $\mathcal{A}_{lex}^C$ the corresponding automaton according to our construction above. If $(\mathcal{A}_{lex}^C)^C = \mathcal{A}_{lex}$ accepts a word of every length, i.e. $\mathcal{L}(\mathcal{A}_{lex}) \cap \Gamma^n \neq \emptyset$ for all $n \in \mathbb{N}$, then $(\Sigma, \mathcal{T})$ terminates.

Proof. Suppose $\mathcal{A}_{lex}$ accepts a word of every length. Let $w_0 \in \Sigma^*$, $I \in \mathcal{L}(\mathcal{A}_{lex})$ with $|w_0| = |I|$. Assume for a contradiction that there exists $(w_i)_{i \in \mathbb{N}}$ with $w_i \otimes w_{i+1} \in \mathcal{L}(\mathcal{T})$. By construction R_I over-approximates the transition relation, hence, $(w_i, w_{i+1}) \in R_I$. Furthermore, R_I is irreflexive and transitive since $\mathcal{A}_{lex}$ accepts only statement with these properties. In particular, R_I is well-founded which is a contradiction to $(w_i, w_{i+1}) \in R_I$ for all $i \in \mathbb{N}$.

Remark 3. Let $\mathcal{A}$ be an NFA. In order to show that $\mathcal{A}$ accepts a word of every length we can apply the following construction. Construct the automaton $\mathcal{A}_*$ which is the same as $\mathcal{A}$ except that we replace the alphabet by the singleton $\Sigma_{\mathcal{A}_*} = \{*\}$ and accordingly $(p, *, q) \in \delta_{\mathcal{A}_*}$ if and only if there exists an $s \in \Sigma_{\mathcal{A}}$ such that $(p, s, q) \in \delta_{\mathcal{A}}$. Then $\mathcal{A}$ accepts a word of every length if and only if $\mathcal{A}_*$ is universal which is easy to check.

Corollary 1. *Let $(\Sigma, \mathcal{T})$ be a lexicographically ordered RTS. Then $\mathcal{L}(\mathcal{A}_{lex}) \cap \Gamma^n \neq \emptyset$ for all $n \in \mathbb{N}$.*

Proof. Let $>_\Sigma$ be the irreflexive, transitive relation on Σ that induces the lexicographical order on Σ^*. Then $(>_\Sigma)^n \in \mathcal{L}(\mathcal{A}_{lex})$ for all $n \in \mathbb{N}$.

Remark 4 (positionwise different orders). The interpretation automaton does even recognise more than only lexicographically ordered systems. We are allowed to consider different orders on every position. For example, we can take the system where even agents count down from some fixed $n \in \mathbb{N}$, and odd agents count up from zero to n, i.e. the transition relation is given by

$$\left(\left(\left(\begin{smallmatrix} 0 \\ 1 \end{smallmatrix} \right) + \cdots + \left(\begin{smallmatrix} n-1 \\ n \end{smallmatrix} \right) \right) \left(\left(\begin{smallmatrix} n \\ n-1 \end{smallmatrix} \right) + \cdots + \left(\begin{smallmatrix} 1 \\ 0 \end{smallmatrix} \right) \right) \right)^*. \tag{11}$$

Then $\mathcal{A}_{lex}$ would recognise all words with the following letters on odd, respectively even positions

$$I_{odd} = \{ \left(\begin{smallmatrix} x-1 \\ x \end{smallmatrix} \right) \mid x \in \{1, \ldots, n\} \} \qquad I_{even} = \{ \left(\begin{smallmatrix} x \\ x-1 \end{smallmatrix} \right) \mid x \in \{1, \ldots, n\} \} \tag{12}$$

In particular, the system is proved terminating by our automaton $\mathcal{A}_{lex}$.

4.2   Letterwise Ordered Systems

In lexicographically ordered systems, one only needs to wait for the next letter to strictly progress towards some terminating configuration. In this section we will add letterwise loops, i.e., the i-th agent of the system may execute a loop but the RTS still terminates, since the loop can only be executed finitely often. The following example explains the class of systems we want to tackle.

Example 4 (polite Mexican standoff). Let n agents be alive (state A) as an initial configuration of our RTS. Every agent is armed and wants to kill all remaining agents, such that the agent is the last one alive (such situations are called Mexican standoff). In order to bring the whole thing to a neat and tidy end, they decide to randomly pick two agents to have a transition into a shooting state

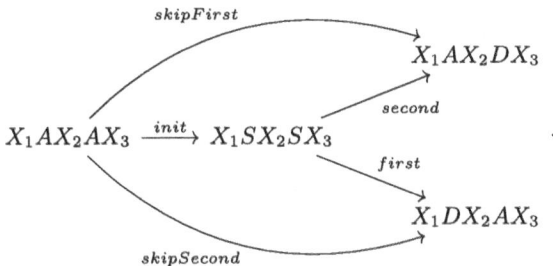

Fig. 4. transition graph of the polite Mexican standoff.

(S) and have a duel, while the others are waiting until the duel has finished. In the duel, randomly one of the two duelists dies (changes state to D) and the other stays alive (changes state back to A). Formally, let $\Sigma = \{A, S, D\}$ be the alphabet. The transition relation $\mathcal{T}$ is described in Fig. 4, where $X_i \in (A + D)^*$ for $i \in \{1, \ldots, 3\}$.

We labelled the transitions for later reference. It is now possible for the first agent who transitions into S to go back to A afterwards, as long as a later agent changes his state to D, contributing towards the terminating configuration D^*AD^*.

The polite Mexican standoff is not lexicographically ordered, since we do not know whether the first or the second agent of the duel dies. Formally, both choices $A > S$ and $S > A$ contradict either *second* or *init* in the induced lexicographical order. However, we can say that at some point we may progress from A to D and can never go back. Hence, to model just *first*, *second*, *skipFirst* and *skipSecond* of Fig. 4 we may set $A =_1 S >_1 D$ for a relation R_1 and observe that all agents of our configuration follow this order. This relation can be modelled by the set

$$\{(\tbinom{A}{A}), (\tbinom{S}{S}), (\tbinom{D}{D}), (\tbinom{A}{S}), (\tbinom{S}{A}), (\tbinom{A}{D}), (\tbinom{S}{D})\} \tag{13}$$

In order to verify that every agent follows this relation we define the interpretation $\mathcal{V}_{All}$, depicted in Fig. 3. A configuration $w_1 \cdots w_n \in \Sigma^n$ satisfies $I_1 \cdots I_n \in (2^\Sigma)^n$ in $\mathcal{V}_{All}$ if and only if $w_i \in I_i$ for all $i \in \{1, \ldots, n\}$. Consequently, the statements that describe the desired order relation on configuration is given by applying the Kleene-star to (13).

Remark 5. $\mathcal{V}_{All}$ does not subsume lexicographical orders. Consider counting down in binary. Then the statement $\{(\tbinom{0}{0}), (\tbinom{1}{1}), (\tbinom{1}{0})\}^n$ does not contain all transitions of counting down in binary in the interpretation $\mathcal{V}_{All}$, since it would demand the agents after the first change to follow this order, too.

In order to include the transition *init* from the polite Mexican standoff, we refine our relation by introducing another relation R_2 that specifies what happens if we are stuck in the case $x =_1 y$. Here, it suffices to demand $A >_2 S$. Since R_2 should only be asked if $=_1$ occurred, it is irrelevant how D is related (it could also be unrelated to A and S). Formally, the overall relation R_3 is defined by

$$x R_3 y :\Leftrightarrow x >_1 y \lor (x =_1 y \land x >_2 y). \tag{14}$$

Lemma 3. *Let* $R_1, R_2 \subset \Sigma \times \Sigma$ *be two preorders on a set* Σ. *Then* R_3 *defined by (14) is irreflexive and transitive.*

Theorem 2 and Lemma 3 result in the following Theorem.

Theorem 4. *Let* $(\Sigma, \mathcal{T})$ *be an RTS,* $n \in \mathbb{N}$. *Let* R_1, R_2 *be two preorders on* Σ^n *and let* R_3 *be defined as in (14), such that* R_3 *over-approximates the transition relation on* Σ^n, *i.e., we have* $(\Sigma \times \Sigma)^n \cap R_\mathcal{T} \subseteq R_3$. *Then the RTS terminates on any input* $w \in \Sigma^n$.

As in the case of lexicographical orders, we can construct an automaton $\mathcal{A}_{All}$ that searches for a relation that proves our system terminating. The construction works completely analogue to the previous section. The only difference is that now we need eight copies of $\mathcal{V}_{All}$ to model the predicates $R_i(x, y)$, $R_i(y, z)$, $R_i(x, z)$, $R_i(y, x)$ for each $i \in \{1, 2\}$. Formally we have

$$\mathcal{A}_{All}^C = (Q_\mathcal{T} \times Q_= \times (Q_{\mathcal{V}_{All}})^8, 2^{\Sigma \times \Sigma}, (s_\mathcal{T}, s_=, (s_{\mathcal{V}_{All}})^8), \delta_{\mathcal{A}_{All}^C}, F). \tag{15}$$

$\delta_{\mathcal{A}_{All}^C}$ is constructed analogously as for $\mathcal{A}_{lex}^C$, following the transition relations of each factor if there exists a triple $(x, y, z) \in \Sigma^3$ that permits a transition in each factor uniformly. The accepting states F correspond to the negated properties of being reflexive, respectively transitive for R_1 and R_2, and R_3 over-approximating the transition relation. For the sake of presentation, we give the accepting states for the negation of R_3 over-approximating the transition relation. The negation of being an over-approximation of R_3 looks as follows.

$$x R_\mathcal{T} y \land (\neg x R_1 y \lor y R_1 x) \land (\neg x R_1 y \lor \neg y R_1 x \lor \neg x R_2 y \lor y R_2 x). \tag{16}$$

Let W_1 be the set of literals in the second conjunct, W_2 the set of literals in the third conjunct of (16). Then the conjunctive normal form of (16) is given by

$$\bigvee_{w_1 \in W_1, w_2 \in W_2} x R_\mathcal{T} y \land w_1 \land w_2. \tag{17}$$

The corresponding accepting states for $\mathcal{A}_{All}^C$ are then given by the union of the accepting states corresponding to each disjunct of (17).

We may carry on this construction inductively. Let in the following R_i always be a preorder. Let $F_1 :\Leftrightarrow x R_1 y \land \neg y R_1 x$. We define

$$S_i \Leftrightarrow \left(\bigwedge_{k=1}^{i} x =_k y \right) \land x >_{i+1} y. \tag{18}$$

Then we can replace the relation from (14) in Theorem 4 by $F_{i+1} :\Leftrightarrow F_i \lor S_i$.

Lemma 4. *Let $n \in \mathbb{N}$. Let R_i be preorders on a set Σ for all $i \in \mathbb{N}$ with $i \leq n$. Let R_{n+1} be the relation defined by F_n. Then R_{n+1} is irreflexive and transitive.*

Theorem 5. *Let $(\Sigma, \mathcal{T})$ be an RTS. Let $\mathcal{A}_{All,i} = (\mathcal{A}_{All,i}^C)^C$ be the complement of the automaton constructed above with the modifications according to F_i. If there exists an $i \in \mathbb{N}$ such that $\mathcal{A}_{All,i}$ accepts a word of every length, then $(\Sigma, \mathcal{T})$ terminates.*

Proof. Let $i \in \mathbb{N}$. The automaton $\mathcal{A}_{All,i}$ accepts exactly those words whose induced relation is given by F_i and over-approximates the transition relation by construction. Since F_i is well-founded by Lemma 4, $(\Sigma, \mathcal{T})$ terminates if $\mathcal{A}_{All,i}$ accepts a word of every length.

5   Conclusion and Future Work

We provided the constructions of automata that can prove termination of lexicographically, respectively letterwise ordered systems, carrying over the approach of [17] from safety to liveness properties. The implementation of the construction is ongoing work. A natural continuation would be to tackle other fragments of regular transition systems that can not be proved terminating with the methods from Sect. 4, e.g. a "waving" token passing

$$T000 \rightarrow 0TTT \rightarrow 0T00 \rightarrow T0TT \rightarrow 00T0 \rightarrow TT0T \rightarrow 000T. \qquad (19)$$

On the other hand, one can loosen restrictions on the alphabet, allowing infinite alphabets. This can be done by using parameterized automata [4] to model the transition relation. An instance of such a transition system is Dijkstra's self-stabilizing protocol. However, these systems have two difficulties to overcome: (1) they are no longer weakly-finite and therefore need more preparation to formulate terminating conditions to search for and (2) complements of parameterized automata do not exist in general and even if they exist, it is hard to complement them but complementing is a crucial step to eliminate universal quantifiers involved in the conditions that are needed to prove termination.

Furthermore, the flexibility of the approach from [17] may allow to tackle completely different verification problems apart from safety and termination. To this end, the interpretation automaton and the corresponding alphabet for statements have to be varied. This seems to be the challenging part. Many choices may result in constructing empty or universal automata in the end.

Acknowledgement. This work was supported by the Swedish Research Council through grant 2021-06327, and by the Knut and Alice Wallenberg Foundation through project UPDATE. The authors thank the CIAA reviewers for insightful comments!

Disclosure of Interests. The authors have no competing interests to declare that are relevant to the content of this article.

References

1. Abdulla, P., Jonsson, B., Nilsson, M., Saksena, M.: A survey of regular model checking. In: Gardner, P., Yoshida, N. (eds.) CONCUR 2004 - Concurrency Theory. CONCUR 2004. LNCS, vol. 3170, pp. 35–48 (2004). https://doi.org/10.1007/978-3-540-28644-8_3
2. Abdulla, P.A.: Regular model checking. Int. J. Softw. Tools Technol. Transfer **14**(2), 109–118 (2012). https://doi.org/10.1007/s10009-011-0216-8
3. Abdulla, P.A., Jonsson, B., Rezine, A., Saksena, M.: Proving Liveness by Backwards Reachability. In: Hutchison, D., et al. (eds.) CONCUR 2006 – Concurrency Theory, vol. 4137, pp. 95–109. Springer, Heidelberg (2006). https://doi.org/10.1007/11817949_7
4. Alur, R., Henzinger, T.A., Vardi, M.Y.: Parametric real-time reasoning. In: Proceedings of the twenty-fifth annual ACM symposium on Theory of Computing, pp. 592–601. STOC 1993, Association for Computing Machinery, New York, NY, USA (1993). https://doi.org/10.1145/167088.167242
5. Angluin, D.: Learning regular sets from queries and counterexamples. Inf. Comput. **75**(2), 87–106 (1987). https://doi.org/10.1016/0890-5401(87)90052-6, https://www.sciencedirect.com/science/article/pii/0890540187900526
6. Bouajjani, A., Habermehl, P., Vojnar, T.: Abstract regular model checking. In: Alur, R., Peled, D.A. (eds.) Computer Aided Verification, 16th International Conference, CAV 2004, Boston, MA, USA, July 13-17, 2004, Proceedings. LNCS, vol. 3114, pp. 372–386. Springer, Heidelberg (2004). https://doi.org/10.1007/978-3-540-27813-9_29
7. Bouajjani, A., Jonsson, B., Nilsson, M., Touili, T.: Regular Model Checking. In: Emerson, E.A., Sistla, A.P. (eds.) Computer Aided Verification, pp. 403–418. Springer, Heidelberg (2000). https://doi.org/10.1007/10722167_31
8. Chen, Y.F., Hong, C.D., Lin, A.W., Rcmmer, P.: Learning to prove safety over parameterised concurrent systems. In: Proceedings of the 17th Conference on Formal Methods in Computer-Aided Design, pp. 76–83. FMCAD 2017, FMCAD Inc, Austin, Texas, October 2017
9. Esparza, J., Gaiser, A., Kiefer, S.: Proving termination of probabilistic programs using patterns. In: Madhusudan, P., Seshia, S.A. (eds.) Computer Aided Verification, pp. 123–138. Springer, Heidelberg (2012). https://doi.org/10.1007/978-3-642-31424-7_14
10. Fang, Y., Piterman, N., Pnueli, A., Zuck, L.: Liveness with Invisible Ranking. In: Steffen, B., Levi, G. (eds.) Verification, Model Checking, and Abstract Interpretation, pp. 223–238. Springer, Heidelberg (2004). https://doi.org/10.1007/978-3-540-24622-0_19
11. Lin, A.W., Rümmer, P.: Liveness of randomised parameterised systems under arbitrary schedulers. In: Chaudhuri, S., Farzan, A. (eds.) Computer Aided Verification, pp. 112–133. Springer International Publishing, Cham (2016). https://doi.org/10.1007/978-3-319-41540-6_7
12. Lin, A.W., Rümmer, P.: Regular model checking revisited. In: Olderog, E.R., Steffen, B., Yi, W. (eds.) Model Checking, Synthesis, and Learning: Essays Dedicated to Bengt Jonsson on The Occasion of His 60th Birthday, pp. 97–114. Springer International Publishing, Cham (2021). https://doi.org/10.1007/978-3-030-91384-7_6
13. Podelski, A., Rybalchenko, A.: Transition invariants. In: Proceedings of the 19th Annual IEEE Symposium on Logic in Computer Science, 2004. pp. 32–41. IEEE, Turku, Finland (2004). https://doi.org/10.1109/LICS.2004.1319598

14. Podelski, A., Rybalchenko, A.: Transition invariants and transition predicate abstraction for program termination. In: Abdulla, P.A., Leino, K.R.M. (eds.) Tools and Algorithms for the Construction and Analysis of Systems, pp. 3–10. Springer, Heidelberg (2011). https://doi.org/10.1007/978-3-642-19835-9_2
15. Sipser, M.: Introduction to the Theory of Computation, 2nd edn. Thomson Course Technology, Cambridge (2006)
16. Takeuti, G., Zaring, W.M.: Introduction to Axiomatic Set Theory. Springer-Verlag, New York (1982). http://archive.org/details/introductiontoax00take
17. Welzel-Mohr, C.: Inductive Statements for Regular Transition Systems. Ph.D. thesis, Technische Universität München (2024). https://mediatum.ub.tum.de/1721365

An Earley-Based Universal
Error-Correcting Parser

Maurice Herwig$^{(\boxtimes)}$, Norbert Hundeshagen, and Martin Lange

University of Kassel, Kassel, Germany
{maurice.herwig,hundeshagen,martin.lange}@uni-kassel.de

Abstract. Motivated by educational applications, we present an extension of the Earley parser in order to compute all possible corrections that take a given word into a context-free target language. Unlike classical error-correcting parsers, which compute a single correction or a set of minimal corrections, our approach targets the possibly infinite set of all corrections represented as a shared packed parse forest. It is achieved by extending the theory of syntactic corrections and normal-forms thereof. We introduce optimisations that eliminate redundant computations and thus, indicate a runtime improvement.

Keywords: corrections · error-correcting parsing · Earley parser · Shared Packed Parse Forests

1 Introduction

Error-correcting parsing is a well-established concept in the theory of formal languages, in particular context-free ones, with applications in interactive development environments (auto completion, syntax error recovery), natural language processing (parsing input with grammatical errors, speech recognition under noise), bioinformatics (DNA sequence alignment under mutations), text recognition (optical character recognition or handwriting), etc. It describes a relaxed version of the well-known parsing problem for a word w and a language described by, say, a context-free grammar (CFG) G. In error-correcting parsing, w is not necessarily required to belong to $L(G)$. In fact, it is generally assumed *not* to belong to $L(G)$. The required output of an error-correcting parser is then a *correction* ρ of w, i.e. a description of how to modify it using insertions, deletions and replacements of symbols, as well as a syntax tree for $\rho(w)$, i.e. the word that is obtained from w by applying the correction to it.

This description of a computation problem clearly contains some degree of freedom since any word can always be "corrected" into any other word simply by deleting and inserting enough symbols. This is why traditional error-correcting parsing typically restricts the valid solution space either explicitly through parametrisation ("replace at most k symbols") or implicitly by demanding that w is corrected to a word in $L(G)$ that is closest w.r.t. some metric on words like Levenshtein distance [9].

G. Castiglione and S. Mantaci (Eds.): CIAA 2025, LNCS 15981, pp. 208–222, 2026.
https://doi.org/10.1007/978-3-032-02602-6_15

The problem of error-correcting parsing in this form has already been studied in the 70's (cf. [7] pg. 521–544). Solutions generally make use of the following observation: given a CFG G over alphabet Σ, it can be modified to a CFG G^* that generates all words w' which can be obtained from a word $w \in L(G)$ by deleting or modifying each of its symbols independently and inserting arbitrary symbols at any position. Necessarily, we get $L(G^*) = \Sigma^*$, and from a parse tree for a word $w' \in L(G^*)$ we can obtain a parse tree for a word $w \in L(G)$ and a correction ρ such that $\rho(w) = w'$. An error-correcting parser can then be built by internalising the modification on the grammar and implementing the extraction of a correction. This has successfully been done for the most prominent parsing algorithms for context-free grammars like the CYK algorithm or the Earley parser, cf. [1,11].

Error-correcting parsing is a viable approach to parsing under uncertainty, i.e. when the input word is expected to contain errors, hence the name. This applies well to scenarios with data that may have been transmitted through lossy channels or has been obtained from faulty sources. Then it makes sense to let the parsing algorithm or perhaps the underlying metric essentially recover the correct word and therefore determine the fix. However, recent work [5] argued for applications in educational setups where this form of error-correcting parsing is too restrictive, because a correction can be seen as feedback information to improve a given solution. Providing feedback on learners' mistakes through immediate error-messages is known to be beneficial [2], and this principle should guide the construction of learning tools. This has led to the extended *universal* error-correcting parsing problem, asking not just for *one* correction but for *all* (reasonable) corrections.

Here we address the same problem, and we employ the theory of corrections developed in [5] that provides normal-forms for equivalent corrections. In order to yield a finite solution space, [5] introduced a notion of minimality that is motivated by such particular applications in learning scenarios. We show that such a restriction is not necessary. The generally infinite solution space to the universal error-correcting parsing problem for context-free languages can be finitely represented using shared packed parse forests (SPFF) and computed using a modification of the Earley parser.

The paper is organised as follows. In Sect. 2 we recall basics from the theory of parsing and corrections. In Sect. 3 we lay the formal foundations of the study of the universal error-correcting parsing problem. In Sect. 4 we present an algorithmic solution based on Earley's parser using SPPFs as a data structure to store a potentially infinite amount of corrections. In Sect. 5 we discuss possibilities to optimise this algorithm in practice and present empirical data from some benchmarks. In Sect. 6 we conclude with remarks on further work.

2   Preliminaries

We recall some preliminaries regarding formal languages, parsing and error correcting. An alphabet Σ is a finite, non-empty set of symbols. A word w over some Σ is a finite sequence of letters $w = a_0 a_1 \ldots a_{n-1}$, where $w(i) = a_i$ denotes the i-th letter, $w(i, j) = a_i \ldots a_{j-1}$ represents the subword from i up to j, and $|w| = n$ is the length of the word. As usual, ε denotes the empty word, and Σ^* and Σ^+ represent the sets of all words and all non-empty words over Σ, respectively.

Parsing. A language is a subset of Σ^* and belongs to the class of context-free languages (CFL) if it is generated by a context-free grammar (CFG), denoted as the 4-tuple (N, Σ, P, S), where N is a finite, non-empty set of nonterminals with $N \cap \Sigma = \emptyset$, P is a set of production rules $P \subseteq N \times (N \cup \Sigma)^*$, and $S \in N$ is the start symbol. We write $L(G)$ for the language of the CFG G. A one-step derivation $\Rightarrow_G$ is the application of a production rule $\alpha A \beta \Rightarrow_G \alpha \gamma \beta$, where $A \rightarrow \gamma \in P$. A derivation $\Rightarrow_G^*$ is the reflexive and transitive closure of $\Rightarrow_G$. A word w belongs to the language $L(G)$ iff there exists a derivation $S \Rightarrow_G^* w$, which can be represented by a syntax tree.

It is well-known that a word can have multiple syntax trees w.r.t. a single CFG; the concept is known by the name *ambiguity*. We are not primarily interested in detecting or resolving ambiguity but in error-correcting parsing. Here, a similar effect arises: for a given word there generally are multiple, in fact infinitely many ways to correct it to a word of a given CFL. A data structure that can be used to efficiently store a potentially infinite number of different syntax trees is that of a *Shared Packed Parse Forest* (SPPF) [13, 15]. It is based on two main principles:

- Nodes of different syntax trees that have the same subtree are *shared*.
- Nodes representing the same subword but different one-step derivations from the same nonterminal are *packed*.

Corrections. We use notation similar to [5] in order to represent the "classical" *edit-operations* (cf. [9]) of insertion, deletion and replacement of letters in words by a single operation $v/_i u$ for $u, v \in \Sigma^*$ and $i \in \mathbb{N}$.

Let $w = a_0 \ldots a_{n-1} \in \Sigma^*$. Then $v/_i u$ arises as the replacement of u at position i in w by v as follows:

$$(v/_i u)(w) := \begin{cases} a_0 \ldots a_{i-1} v a_{i+|u|} \ldots a_{n-1} & , \text{ if } i + |u| \leq n, u = a_i \ldots a_{i+|u|-1}, \\ \bot & , \text{ otherwise.} \end{cases}$$

Thus, $a/_i \varepsilon$ denotes the insertion of letter a at position i, $\varepsilon/_i a$ the deletion of letter a there, and $a/_i b$ the replacement of a by b at this position, where $a, b \in \Sigma$.

A *correction* ρ is a sequence $\rho = \langle \alpha_1, \ldots \alpha_m \rangle$ of edit-operations α_i with $1 \leq i \leq m$. The empty correction is denoted as $\langle \rangle$. The *application of a correction* is a mapping on $\Sigma^* \cup \{\bot\}$. Let $w \in \Sigma^*$, then

$$\langle \rangle(w) := w \quad \text{and} \quad \langle \rho, \alpha \rangle(w) := \rho(\alpha(w)).$$

Furthermore, $\rho(\bot) := \bot$ for any correction ρ.

Two corrections ρ, ρ' are called *equivalent*, written $\rho \equiv \rho'$, if $\rho(w) = \rho'(w)$ for all $w \in \Sigma^*$. It is easy to see that $\equiv$ is indeed an equivalence relation on the set of all corrections. With $dom(\rho) := \{u \mid \rho(u) \neq \bot\}$ and $rng(\rho) := \bigcup_{u \in dom(\rho)} \rho(u)$ we denote the *domain* and *range* of correction ρ.

3 Normalised Corrections

We are interested in the following computational problem ALLCORRECT of "computing all corrections":

> $given$: a (so-called) $target - language\ L \subseteq \Sigma^*$ and a word $w \in \Sigma^*$,
> $compute$: the set $\mathcal{A}_w^L := \{\rho \mid \rho(w) \in L\}$.

Two observations are rather obvious. First, $L = \emptyset$ if and only if $\mathcal{A}_w^L = \emptyset$. This follows from the fact that for every $w, v \in \Sigma^*$ there is ρ such that $v = \rho(w)$, for example $\rho = v/_0 w$. An immediate consequence of this and the undecidability of the emptiness problem for context-sensitive languages is the non-computability of ALLCORRECT for context-sensitive target languages, regardless of the representation of $\mathcal{A}_w^L$ [5]. We therefore restrict our attention to context-free target languages.

Second, $|\mathcal{A}_w^L| = \infty$ if and only if $L \neq \emptyset$. This means that any solution to ALLCORRECT needs to employ a way to finitely represent infinite sets of corrections. To this end, [5] developed a theory of rewriting and normalisation of corrections that captures equivalence in the sense that $\rho \equiv \sigma$ iff ρ and σ have the same (so-called strong) normal-forms. Hence, it suffices to limit $\mathcal{A}_w^L$ to normalised corrections only. It does not yield finiteness of the solution space, though, as already witnessed by the simple example of $w = a$ and $L = b^+$. Then $\rho_i = b^i/_0 a$ for $i \in \mathbb{N}$ are all mutually inequivalent corrections in strong normal form. This is why [5] introduces the notion of a *minimal* correction motivated by applications in educational settings and shows, using Higman's Lemma, that there are only finitely many minimal and strongly normalised corrections in $\mathcal{A}_w^L$ for any $w \in \Sigma^*$, $L \subseteq \Sigma^*$. Hence, their version of ALLCORRECT asks for the computation of all minimal corrections, unlike the version defined above.

We aim for a more general approach that does not restrict $\mathcal{A}_w^L$ artificially in order to make ALLCORRECT well-defined. We adopt the syntactic notion of normalisation of corrections and continue to restrict attention to normalised ones or, likewise, to equivalence classes of corrections under $\equiv$. Note that two equivalent corrections σ, ρ are merely syntactically different representations of the same operation on any word. We then deviate from [5] and show how to represent an infinite set of corrections using SPPFs as a data structure.

Definition 1. *Let $w = a_0 \dots a_{n-1}$ be a word over Σ. A correction ρ_w is called Word-ordered, if*

$$\rho_w := \langle u_0/_0\varepsilon,\ x_0/_0 a_0,\ u_1/_1\varepsilon,\ x_1/_1 a_1, \dots, x_{n-1}/_{n-1}a_{n-1},\ u_n/_n\varepsilon\rangle$$

with $u_i \in \Sigma^*$, and $x_i \in \Sigma \cup \{\varepsilon\}$ for $0 \leq i \leq n$.

A Word-ordered correction ρ_w is composed of insertions $U_i/_i\varepsilon$ at the beginning, end, and in between every two letters of w, as well as an edit-operation for every letter signalling whether it is kept $(a_i/_ia_i)$, replaced by a letter from the alphabet $(b/_ia_i)$, or deleted $(\varepsilon/_ia_i)$. Hence, ρ_w is of fixed length $2|w|+1$ and $dom(\rho_w) = \{w\}$.

The restriction to singleton domains is not harmful for the use of such Word-ordered corrections in a solution to ALLCORRECT, as we are only interested in corrections that are applied to the input word w.

Definition 1 essentially corresponds to the notion of weak normalised form [5]. There, a distinction is drawn between weak normalisation into a form that applies edit-operations in an ordered fashion, and the more restrictive strong normal form which is needed to obtain unique normalised representations of equivalence classes.

Example 1. Let $\Sigma = \{a, b\}$ and $w = bab$. There are multiple corrections ρ such that $\rho(w) = babab$, for instance

$$\rho_1 = \langle \varepsilon/_0\varepsilon, b/_0b, ab/_1\varepsilon, a/_1a, \varepsilon/_2\varepsilon, b/_2b, \varepsilon/_3\varepsilon \rangle$$
$$\rho_2 = \langle babab/_0\varepsilon, \varepsilon/_0b, \varepsilon/_1\varepsilon, \varepsilon/_1a, \varepsilon/_2\varepsilon, \varepsilon/_2b, \varepsilon/_3\varepsilon \rangle$$
$$\rho_3 = \langle a/_1\varepsilon, b/_0\varepsilon \rangle$$

Correction ρ_3 is not Word-ordered. It does not apply its operations from right-to-left, and it is not a complete description of what happens at every position in a word. Corrections ρ_1 and ρ_2 are both Word-ordered, but they build *babab* in two different ways. According to [5], only ρ_1 would be called strongly normalised.

As stated above, we aim for a general solution to the problem ALLCORRECT and therefore need to explain finite representability of the solution space.

To this end, we encode corrections as words over a certain alphabet. As the order and type of edit-operations in Word-ordered corrections is defined by the word to be corrected, we can omit index information and simply write $u^\uparrow$ for $u/_i\varepsilon$, $\bar{a}$ for $a/_ia$, $b^{/a}$ for $b/_ia$ with $b \neq a$, and $a^\downarrow$ for $\varepsilon/_ia$. For a string encoding we further identify each insert $u^\uparrow$ of a word $u = a_0 \ldots a_{k-1}$ as the sequence of single letter insertions $a_0^\uparrow \ldots a_{k-1}^\uparrow = u^\uparrow$, and omit empty insertions $\varepsilon^\uparrow$. It is not hard to see that every Word-ordered correction can be represented as a string over symbols of this form, and can also be recovered uniquely from such a string.

Example 2. The string representation of the Word-ordered correction ρ_1 from Ex. 1 above is $b^-, a^\uparrow, b^\uparrow, a^-, b^-$.

Finite representability of the solution space to ALLCORRECT for an arbitrary word w and context-free language L is then given by the following theorem which states that in this case, the set of all (representations of) corrections is context-free over the alphabet $\widehat{\Sigma} := \{x^\uparrow, x^-, x^\downarrow, y^{/x} \mid x, y \in \Sigma\}$. We write $C_w(L)$ to denote the set $\{\rho \mid \rho \text{ is Word-ordered and } \rho(w) \in L\}$. When the target language L is fixed and clear from context we also simply write C_w for this set.

Theorem 1. *Let $L \subseteq \Sigma^*$ be a context-free language and $w \in \Sigma^*$ a word. Then $C_w(L)$ is a context-free language over Σ.*

Proof. The class of context-free languages is closed under intersections with regular languages and substitutions with context-free languages. In the following we use these closure properties to derive C_w from L.

For that, we define a substitution $s : \Sigma \rightarrow \mathcal{P}(\widehat{\Sigma}^*)$ as $s(x) := \{x^\uparrow, x^-, x^\downarrow, y^{/x} \mid y \in \Sigma\}$. Naturally, we associate the extension onto words and languages of the latter substitution with s. Then, $s(L)$ is context-free and it is easy to see that $s(L)$ contains all Word-ordered corrections into the target language L, that is, $s(L) = \bigcup_{w \in \Sigma^*} C_w(L)$, simply because it contains all ways to delete or replace a symbol and pad it with additional symbols. Finally, for a given word $w = a_0 \ldots a_{n-1}$ let $L_w := \{u_0 x_0 u_1 \ldots x_{n-1} u_n \mid u_i \in \{y^\uparrow \mid y \in \Sigma\}^*, x_i \in \{a_i^-, a_i^\downarrow, y^{/a_i} \mid y \in \Sigma\}\}$. Obviously, L_w is regular and can be interpreted as the set of all Word-ordered corrections of w into Σ^*. Thus, $s(L) \cap L_w$ is context-free, and since each Word-ordered correction is a unique identifier for the word to be corrected and L_w contains all such possibilities, it immediately follows that $s(L) \cap L_w = C_w(L)$. $\square$

This opens up the possibility for an algorithmic solution to ALLCORRECT (for context-free target languages). In the following sections we explore such a solution based on an extension of Earley's parsing algorithm and the use of SPPFs to represent infinite sets of corrections.

4   Computing All Corrections

Theorem 1 is constructive by first building the grammar for the substitution $s(L)$ similar to [1], and then computing the intersection with the regular language

$$L_w := \{u_0 x_0 u_1 \ldots x_{n-1} u_n \mid u_i \in \{y^\uparrow \mid y \in \Sigma\}^*, x_i \in \{a_i^-, a_i^\downarrow, a_i^{/y} \mid y \in \Sigma\}\} \ .$$

Earley's parsing method can be interpreted as an implementation of Bar-Hillel et al.'s proof that the class of context-free languages is closed under intersection with regular languages [3], [7, pp. 425–442]. Instead of computing the substitution and intersection mentioned above, we internalise these steps into an implementation of a generalised Earley parser [14]. Furthermore, the latter implementation allows us to collect all corrections in form of an SPPF [13,15], a data structure used to represent all (infinitely) many parse trees for a given word and context-free language. Note that this is also the main difference to the approach in [1], where the substitution grammar is used directly and then only the correction with a minimal number of edit-operations is computed by using an Early parser.

The All-Corrections Earley Parser (ALLCEP *).* To improve the readability in the following description and optimisation of ALLCEP in Sect. 5, we use standardised notation. For a given word $w = a_0 \ldots a_{n-1}$ and a context-free grammar $G = (N, \Sigma, P, S)$, uppercase letters A, B represent nonterminal symbols from

N, lower case letters represent terminal symbols over Σ, and greek letters words over $(\Sigma \cup N)^*$. ALLCEP computes *Earley sets* Q_j for each position $0 \leq j \leq n$ in w. These sets contain *Earley items* of the form

$$[A \rightarrow \alpha \bullet \beta, i, v],$$

where $A \rightarrow \alpha\beta \in P$, $0 \leq i \leq j$ is the position in w where the parse of α started (hence the $\bullet$), and v is used to store certain information on the parse, which will later be used to output all possible parse trees as shown below. We simply use v, v', v'' to denote these information and only define them in detail when necessary. Furthermore, an item of the form $[A \rightarrow \alpha\bullet, i, v]$ is called *final*. Intuitively, each Earley set Q_j contains all possible partial derivations of a rule $A \rightarrow \alpha\beta$ in G for the subword $w(i, j)$.

The parsing steps are done as follows, similar to those done by the well-known Earley parser.

1. Add $[S \rightarrow \bullet\alpha, null]$ to Q_0 for all $S \rightarrow \alpha \in P$.
2. For each $0 \leq j \leq n$ apply the following operations to Q_j until nothing changes:
   - predictor: if $[A \rightarrow \alpha \bullet B\beta, i, v] \in Q_j$ add $[B \rightarrow \bullet\gamma, j, null]$ to Q_j for all $B \rightarrow \gamma \in P$.
   - completer: if a final Earley item $[A \rightarrow \alpha\bullet, i, v''] \in Q_j$ add $[B \rightarrow \beta A \bullet \gamma, k, v']$ to Q_j for all items $[B \rightarrow \beta \bullet A\gamma, k, v] \in Q_i$.
   - insertion: if $[A \rightarrow \alpha \bullet b\beta, i, v] \in Q_j$ add $[A \rightarrow \alpha b \bullet \beta, i, v']$ to Q_j.
3. For each $0 \leq j < n$ apply the following operations to Q_j once.
   - scanner: if $[A \rightarrow \alpha \bullet w(j)\beta, i, v] \in Q_j$ add $[A \rightarrow \alpha w(j) \bullet \beta, i, v']$ to Q_{j+1}.
   - replacement: if $w(j) \neq b$ and $[A \rightarrow \alpha\bullet b\beta, i, v] \in Q_j$ add $[A \rightarrow \alpha b\bullet\beta, i, v']$ to Q_{j+1}.
   - deletion: if $[A \rightarrow \alpha \bullet \beta, i, v] \in Q_j$ add $[A \rightarrow \alpha \bullet \beta, i, v']$ to Q_{j+1}.

The predictor-, completer-, and scanner-steps together form a generalised Earley parser, c.f. [14].

The scanner-step reads the next input symbol $w(j)$, if a respective rule is present in the current Early set, that is, such a symbol may be interpreted as $w(j)\hat{}$ in the correction to be computed. The newly introduced replacement-step then can be seen as a replacement of the j-th symbol $w(j)$ by b ($b^{/w(j)}$), if a rule exists in Q_j permitting that b is read. Accordingly, the deletion-step then simply moves Earley items from one Earley set to the next, indicating that the symbol $w(j)$ is deleted ($w(j)^{\downarrow}$). Hence, scanner, replacement, and deletion together realise every possible sequence of the respective edit-operations on the input word.

Finally, the insertion-step updates the Earley item Q_j by reading every terminal symbol b in every rule in Q_j of the form $[A \rightarrow \alpha \bullet b\beta, i, v]$ without scanning the input, that is, the new Earley item $[A \rightarrow \alpha b \bullet \beta, i, v']$ is inserted in the same Earley set Q_j. Thus, this step may be interpreted as a consecutive insert $b^{\uparrow}$ of every possible terminal symbol derivable in the current Earley set without scanning the input. The latter then realises the insertions of subwords derivable by G between input symbols.

Recall that for an original Earley parser on input $w = a_0 \ldots a_{n-1}$ and G the following equivalence holds: $S \Rightarrow_G^* a_0 \ldots a_{i-1} A\gamma \Rightarrow_G a_0 \ldots a_{i-1} \alpha\beta\gamma \Rightarrow_G^* a_0 \ldots a_{i-1} a_i \ldots a_{j-1}\beta\gamma$ if, and only if $[A \to \alpha \bullet \beta, i, v] \in Q_j$, e.g. in [1].

Now, let $s(G)$ be the context-free grammar derived from G by applying the substitution $s(x) := \{x^\uparrow, x^-, x^\downarrow, y^{/x} \mid y \in \Sigma\}$ given in Theorem 1, that is, $s(G)$ simply contains a new rule set of the form $\{A \to s(\alpha) \mid A \to \alpha \in P\}$. Thus, these new rules in $s(G)$ reflect that every terminal symbol from G is replaced by each edit-operation symbol. Based on the above observations on the newly introduced steps to the Earley parser and the correctness argument above we may immediately conclude the following statement for an Earley item Q_j computed by ALLCEP.

Proposition 1. *We have* $[A \to \alpha \bullet \beta, i, v] \in Q_j$ *iff*

$$S \Rightarrow_{s(G)}^* u_0 x_0 \ldots u_{i-1} x_{i-1} u_i' A\gamma \Rightarrow_{s(G)}^* u_0 x_0 \ldots u_{i-1} x_{i-1} u_i' \alpha\beta\gamma$$

$$\Rightarrow_{s(G)}^* u_0 x_0 \ldots u_{i-1} x_{i-1} u_i x_i \ldots u_{j-1} x_{j-1} u_j \beta\gamma$$

where $0 \leq i \leq j \leq n$, $u_i \in \{y^\uparrow \mid y \in \Sigma\}^*$, u_i' *is a prefix of* u_i, *and* $x_i \in \{a_i^-, a_i^\downarrow, a_i^{/y} \mid y \in \Sigma\}$.

It follows that ALLCEP realizes the intersection of $s(G)$ with $L_w := \{u_0 x_0 u_1 \ldots x_{n-1} u_n \mid u_i \in \{y^\uparrow \mid y \in \Sigma\}^*, x_i \in \{a_i^-, a_i^\downarrow, a_i^{/y} \mid y \in \Sigma\}\}$.

SPPF Computed by ALLCEP. It remains to elaborate how the additional context information v, v' and v'' present in the Earley items above are used to efficiently store all possible parse trees in form of an SPPF. As the construction of the SPPF is a slightly modified version of the original one, c.f. [14], we only recall the basic steps and the modifications needed to represent every possible correction of the input word w.

Here, an SPPF is a directed graph with three different types of nodes. *Symbol-* and *intermediate*-nodes are used to represent shared syntax trees of different derivations, and *packed*-nodes are used to represent parts of different derivations of the same subword.

For an Earley item $[A \to \alpha \bullet \beta, i, v] \in Q_j$, the value v refers to either the intermediate node $(A \to \alpha \bullet \beta, i, j)$, indicating that the subgraph below this node contains all syntax trees for the partial derivation of $w(i, j)$ by α, the symbol node (A, i, j) if the item is final, or *null*, in which case the item is not associated with any node in the SPPF. This last case occurs if no derivation for any subword can be obtained from the Earley item. We refer to the pair (i, j) of an intermediate- or symbol-node as the *extent* of that node. Note that each intermediate-node represents exactly one non-final Earley item, whereas a symbol-node can represent different final Earley items.

Figure 1 now illustrates the process of constructing the SPPF for the j-th Earley set Q_j, where an intermediate-node is represented by a square, a symbol-node by rounded corners, and a packed-node as an empty circle. Observe that the node v' is introduced by the different steps of AllCEP, provided that v and v'' are already present in the SPPF.

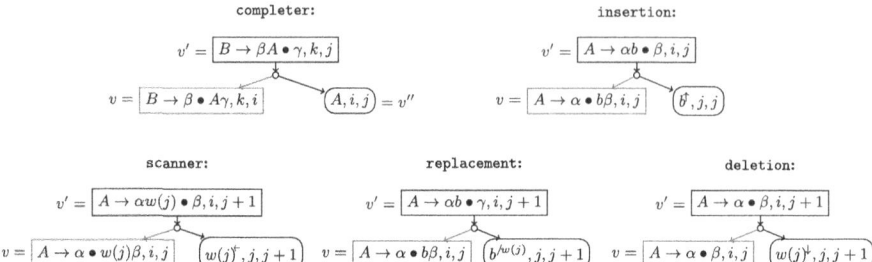

Fig. 1. A graphical representation of the construction of an SPPF by the AllCEP operations.

Figure 1 illustrates this construction process for all operations, provided that $v \neq null$. In the case that $v = null$ only the non-red nodes are created/linked by the respective operations. To allow the SPPF to represent sequences of consecutive `deletion`-steps, we extend the definition of intermediate-nodes $(A \rightarrow \alpha \bullet \gamma, j, i)$ to permit $\alpha = \varepsilon$, deviating from [13,15]. Note that during the parsing process, nodes already contained in the SPPF are reused.

Example 3. Let $G = (\{S\}, \{a, b\}, \{S \rightarrow aSb \mid ab\}, S)$ be the grammar for the target language and b the word to be corrected. Figure 2 shows the SPPF computed by AllCEP. Furthermore, the different edge-styles represent the different syntax trees present in the SPPF, where dotted edges and nodes define the Word-ordered correction $A^{/b}b^{\uparrow}$ and blue edges and nodes the corrections $a^{\downarrow}b^{\downarrow}a^{\uparrow m}b^{\uparrow m}$ for $m \in \mathbb{N}_{\geq 0}$.

As illustrated in the previous example, all syntax trees contained in an SPPF computed by AllCEP can be derived from the root symbol node $(S, 0, n)$ by choosing exactly one packed node for each symbol- and intermediate-node. Then, the respective syntax tree represents a Word-ordered correction by reading the leaves of the syntax tree from left to right.

Based on the latter observations, we may conclude the following.

Proposition 2. *For a given context-free target language $L(G)$ and a word $w \in \Sigma^*$, AllCEP computes $C_w(L(G)) = \{\rho \mid \rho \text{ is Word-ordered and } \rho(w) \in L(G)\}$ in time $\mathcal{O}(n^3)$.*

AllCEP constructs an SPPF with $\mathcal{O}(n^3)$ nodes and edges for any fixed CFG G. Note that on unambiguous grammars, a traditional Earley parser has a runtime of $\mathcal{O}(n^2)$ only. This is not the case for AllCEP as the Grammar $s(G)$ is highly ambiguous.

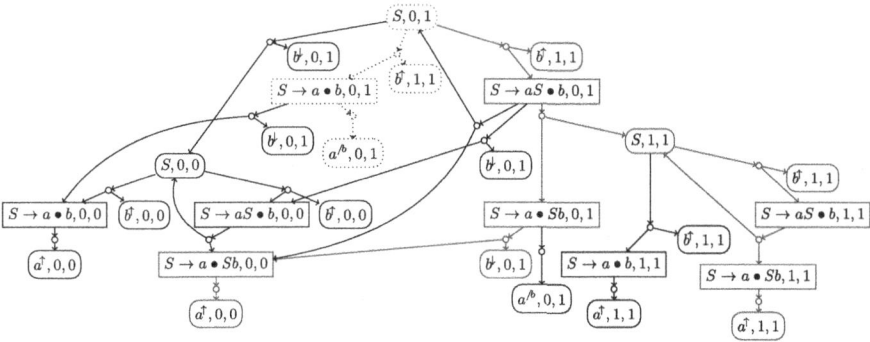

Fig. 2. An exemplary SPPF. Where the blue and dotted parts represent different syntax trees contained in SPPF. (Color figure online)

5 Optimisation and Empirical Evaluation

The modifications introduced in ALLCEP in Sect. 4 lead to seemingly redundant computations. The reason for this is given by `insertion`-steps that, regardless of the current input symbol, perform every possible derivation of the grammar in every computed Earley set.

We leverage this to optimise ALLCEP so that it computes certain Earley items only once and "copies" them to consecutive Earley sets. This then leads to a significant runtime speedup in practice.

Again for the following, let the input for ALLCEP be a word $w = a_0 \ldots a_{n-1}$ and a CFG $G = (N, \Sigma, P, S)$ such that every rule in P is reachable. Let $s(G)$ be the grammar derived by applying the substitution from Theorem 1 (cf. Section 4), and let Q_j $(0 \le j \le n)$ be the Earley sets computed by ALLCEP. Moreover, we use further notation implicitly defined by the context, as presented in Sect. 4.

Lemma 1. *If* $[A \to \alpha \bullet \beta, i, v] \in Q_j$, *then* $[A \to \alpha \bullet \beta, 0, v'] \in Q_0$ *for* $0 \le i \le j \le n$.

Proof. (*sketch*) Based on the observations made for ALLCEP in Sect. 4 it follows from $[A \to \alpha \bullet \beta, i, v] \in Q_j$, that there is a derivation

$$S \Rightarrow^*_{s(G)} u_0 x_0 \ldots u_{i-1} x_{i-1} u'_i A\gamma \Rightarrow^*_{s(G)} u_0 x_0 \ldots u_{i-1} x_{i-1} u'_i \alpha\beta\gamma$$

$$\Rightarrow^*_{s(G)} u_0 x_0 \ldots u_{i-1} x_{i-1} u_i x_i \ldots u_{j-1} x_{j-1} u_j \beta\gamma$$

for $0 \le i \le j \le n$, and all u_i, u'_i are insertions of the form $\{y^\uparrow \mid y \in \Sigma\}^*$ and all x_i are read- $a_i^\leftarrow$, replacement- $a_i^{/y}$, or deletion-symbols $a_i^\downarrow$.

In $s(G)$ every terminal symbol of G is replaced by either one of the above edit-operation-symbols (including insertions) in every rule, so there is also a derivation of the form

$$S \Rightarrow^*_{s(G)} u_0 a_0^\uparrow \ldots u_{i-1} a_{i-1}^\uparrow u'_i A\gamma \Rightarrow^*_{s(G)} u_0 a_0^\uparrow \ldots u_{i-1} a_{i-1}^\uparrow u'_i \alpha\beta\gamma$$

$$\Rightarrow^*_{s(G)} u_0 a_0^\uparrow \ldots u_{i-1} a_{i-1}^\uparrow u_i a_i^\uparrow \ldots u_{j-1} a_{j-1}^\uparrow \beta\gamma \Rightarrow^*_{s(G)} u'_0 \beta\gamma,$$

where all former input symbols are already inserted in the beginning of the derivation. Again, we conclude from the observations in Sect. 4 that by repeated applications of `predictor`-,`completer`-, and `insertion`-steps (cf. part 2 of ALLCEP) all possible derivations of $s(G)$ are already computed and thus, $[A \rightarrow \alpha \bullet \beta, 0, v'] \in Q_0$. Furthermore, v' is derived from v by updating the extent of the respective SPPF node to $(0,0)$.   □

Conversely, we need to consider different cases regarding whether the start symbol S is present in a rule's right-hand side, or not. For that purpose we define T as follows:

$$T := \begin{cases} \{[S \rightarrow \alpha \bullet \beta, i, v] \mid 0 \le i \le n, [S \rightarrow \alpha \bullet \beta, 0, v] \in Q_0\}, & \text{if } A \rightarrow \alpha S \beta \in P; \\ \emptyset, & \text{otherwise.} \end{cases}$$

Lemma 2. *If $[A \rightarrow \alpha \bullet \beta, 0, v] \in Q_0 \setminus T$, then $[A \rightarrow \alpha \bullet \beta, i, v'] \in Q_j$ for all $0 \le i \le j \le n$.*

Proof. (sketch) Observe that the `deletion`-step in ALLCEP moves every Early item in $[A \rightarrow \alpha \bullet \beta, 0, v] \in Q_0$ to Q_1. Hence, after j steps we have $[A \rightarrow \alpha \bullet \beta, 0, v''] \in Q_j$ holds.

It remains to show how the index 0 of item $[A \rightarrow \alpha \bullet \beta, 0, v''] \in Q_j$ can be adjusted to any $i \le j$. Observe that only the `predictor`-step introduces new items with indices updated to the current Earley set index. Thus, for $i < j$ we claim that, if $[A \rightarrow \alpha \bullet \beta, i, v_1] \in Q_j \setminus T$, then $[A \rightarrow \alpha \bullet \beta, i+1, v_2] \in Q_j$. This can be proven inductively by applying a similar argument as in Lemma 1. Roughly speaking, if there is an item $[A \rightarrow \alpha \bullet \beta, i, v_1] \in Q_j$, then there is a derivation of A starting at position i and reading, replacing or deleting $j - i$ symbols. Hence, there is the same derivation from A except for inserting the i-th symbol. It follows that also $[A \rightarrow \alpha \bullet \beta, i+1, v_2] \in Q_j$ and thus, $[A \rightarrow \alpha \bullet \beta, i, v'] \in Q_j$ for every $0 \le i \le j \le n$. Again, v' is derived from v by updating the extend of the respective SPPF node to (i,j).   □

Lemmas 1 and 2 state that all Earley items occuring during a computation of ALLCEP are already computed in Q_0. This is used in the following to optimise ALLCEP. To ease the description of the following algorithm, we introduce the sets X_i of Earley items defined for $0 \le i \le n - 1$, as follows:

$$X_0 := Q_0 \setminus T,$$
$$X_{i+1} := \{[A \rightarrow \alpha \bullet \beta, i+1, v'] \mid [A \rightarrow \alpha \bullet \beta, i, v] \in X_i\}.$$

The *optimised all-correction Earley parser*(OPTALLCEP) performs the following steps for w and G:

1. Perform part 1 and part 2 of ALLCEP to compute Q_0.
2. For each $1 \le j \le n$ determine Q_j as $Q_j = Q_0 \cup \bigcup_{i=1}^{j} X_i$.
3. For each $0 \le j \le n$ add the nodes of the Earley items to the SPPF as follows:
   - for each $[A \rightarrow \alpha \bullet \beta, i, v] \in Q_j$ add v to the SPPF if v is not yet included.

4. For each node v in the SPPF link them to all possible nodes as illustrated in Fig. 1.

Remember that ALLCEP computes the Earley sets by applying operations to already existing Earley items, thereby incrementally building up the SPPF through these operations. In contrast, OPTALLCEP does not use these operations. Instead, it relies on the previous observations (Lemma 1 and Lemma 2) to derive all Earley sets directly from Q_0 and then the SPPF is computed in a subsequent step. Since, in practice, adjusting indices is significantly faster than recomputing all items from scratch, we expect OPTALLCEP to perform much faster than ALLCEP. This is also confirmed by a benchmark test, presented below.

In the original Earley parser [6], the computation of each Earley set requires a runtime of $\mathcal{O}(n^2)$, due to the $\mathcal{O}(n^2)$ potential applications of the completer-step. However, under the condition that the size of the grammar is constant, only a constant number of completer step can be applied to items in Q_0. Hence, OPTALLCEP computes every Earley set in constant time, disregarding the time needed to update the index information when "copying" Q_0 to Q_j and thus, implying linear runtime. This is not very surprising, as OPTALLCEP essentially can be interpreted as a parser for Σ^*. Here, the complexity is hidden in the construction of the SPPF and especially in the computation of edges that still requires time $\mathcal{O}(n^3)$ due to its cubic size.

An implementation of ALLCEP and OPTALLCEP, written in Python3 and based on the lark-parser[1] is publicly available.[2] In Fig. 3, we compare the run-times of ALLCEP and OPTALLCEP on the word $+^m$ ($m \in \mathbb{N}$) and the grammar of arithmetic expressions $G_{aexpr} = (\{S, T, F\}, \{-, +, *, /, (,), n\}, P, S)$ with P defined as follows:

$$S \rightarrow S - T \mid S + T \mid T,$$
$$T \rightarrow T * F \mid T/F \mid F,$$
$$F \rightarrow (S) \mid n.$$

The tests were run on an Intel-Core I7-12700H prozessor with 16 GB RAM and CPU speed up to 2.3 GHz. We observe that OPTALLCEP achieves a significant runtime improvement compared to ALLCEP.

---

[1] https://github.com/lark-parser/lark.
[2] https://github.com/maurice-herwig/alcep.

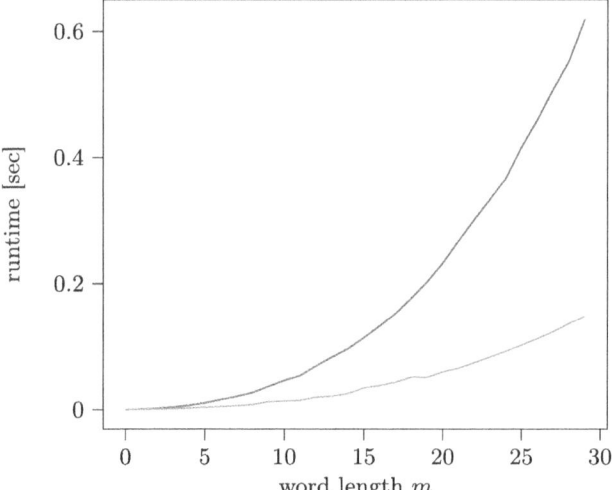

Fig. 3. Running times for ALLCEP (Blue) and OPTALLCEP (Orange) for the words $+^m$ and the language of all arithmetic expressions G_{aexpr} (Color figure online).

6   Conclusion

Based on theory of error-correcting parsing, we developed ALLCEP, an Early parser capable of computing all corrections of a given word into a context-free target language. Furthermore, we showed how ALLCEP can be optimised by avoiding certain redundant computations present in the calculation of corrections. From an algorithmic perspective we suggest parallelisation to be applied in order to compute the links of the item nodes (cf. step 4 of OPTALLCEP) independently.

The main motivation of this paper is the computation of corrections in an educational environment to serve as a basis for feedback generation in response to incorrect solutions, for example entered into a automatic assessment tool by a student. In such a setting one of the main open questions is: what correction should be used to produce the feedback information provided to the student? [5] showed that it is not necessarily the one with smallest edit distance that should be chosen. However, edit distance can also be used as a measure of correctness of a solution. It is used as such in didactic applications for finite automata construction [4,12] or programming [10].

Clearly, computing a data-structure that all corrections is only half the way. The next step demands an investigation into the possibilities of how to filter an SPPF (cf. [8]) to output only corrections useful in the specific context. For instance, certain domain-specific restrictions can be used to reduce the number of corrections in the SPPF, similar to [5]. Such restrictions then need to be formalised with respect to the question whether they are computable, feasible, and SPPF-preserving. From the perspective that corrections actually formalise feedback in natural language, one can aim for an approximated solution of the problem on how to filter the SPPF. Such an approximation might be achieved by using pre-trained large language models to filter the set of all-corrections for those most suitable in the specific context.

References

1. Aho, A.V., Peterson, T.G.: A minimum distance error correcting parser for context-free languages. SIAM J. Comput. **1**(4), 305–312 (1972). https://doi.org/10.1137/0201022
2. Anderson, J.R., Corbett, A.T., Koedinger, K.R., Pelletier, R.: Cognitive tutors: lessons learned. J. Learn. Sci. **4**(2), 167–207 (1995)
3. Bar-Hillel, Y., Perles, M., Shamir, E.: On formal properties of simple phrase structure grammars. Zeitschrift für Phonetik, Sprachwissenschaft und Kommunikationsforschung 14, 143–172 reprinted in Y. Bar-Hillel. (1964). Language and Information: Selected Essays on their Theory and Application, Addison-Wesley **1964**, 116–150 (1961)
4. Bruse, F., Herwig, M., Lange, M.: A similarity measure for formal languages based on convergent geometric series. In: Proceedings of 26th International Conference on Implementation and Application of Automata, CIAA 2022. LNCS, vol. 13266, pp. 80–92. Springer, Cham (2022). https://doi.org/10.1007/978-3-031-07469-1_6
5. Bruse, F., Lange, M.: Computing all minimal ways to reach a context-free language. In: Kovács, L., Sokolova, A. (eds.) Reachability Problems. RP 2024. LNCS, vol. 15050, pp. 38–53. Springer, Cham (2024). https://doi.org/10.1007/978-3-031-72621-7_4
6. Earley, J.: An efficient context-free parsing algorithm. Commun. ACM **13**, 94–102 (1970). https://doi.org/10.1145/362007.362035
7. Grune, D., Jacobs, C.J.H.: Parsing Techniques: A Practical Guide, 2nd edn. Springer, New York (2008). https://doi.org/10.1007/978-0-387-68954-8
8. Klint, P., Visser, E.: Using filters for the disambiguation of context-free grammars. In: Proceedings of ASMICS Workshop on Parsing Theory, pp. 1–20. Milan, Italy (1994)
9. Levenshtein, V.I.: Binary codes capable of correcting deletions, insertions and reversals. Soviet Phys. Doklady **10**(8), 707–710 (1966)
10. Price, T., Zhi, R., Barnes, T.: Evaluation of a data-driven feedback algorithm for open-ended programming. International Educational Data Mining Society (2017)
11. Rajasekaran, S., Nicolae, M.: An error correcting parser for context free grammars that takes less than cubic time. In: Proceedings 10th International Conference on Language and Automata, Theory and Applications, LATA 2016. LNCS, vol. 9618, pp. 533–546. Springer, Heidelberg (2016). https://doi.org/10.1007/978-3-319-30000-9_41
12. Rivers, K., Koedinger, K.R.: Data-driven hint generation in vast solution spaces: a self-improving python programming tutor. Int. J. Artif. Intell. Educ. **27**, 37–64 (2017). https://doi.org/10.1007/s40593-015-0070-z
13. van der Sanden, B.: Parse Forest Disambiguation. Master's thesis, Eindhove University of Technology (2014). https://pure.tue.nl/ws/portalfiles/portal/46998704/784691-1.pdf

14. Scott, E.: SPPF-style parsing from earley recognisers. Electron. Notes Theoret. Comput. Sci. **203**(2), 53–67 (2008). https://doi.org/10.1016/j.entcs.2008.03.044
15. Tomita, M.: Efficient Parsing for Natural Language: A Fast Algorithm for Practical Systems, vol. 8. Springer Science & Business Media, Heidelberg (1985). https://doi.org/10.1007/978-1-4757-1885-0

More on Language Families with a Decidable Pumping-Problem (Extended Abstract)

Markus Holzer[1](✉) and Christian Rauch[2]

[1] Institut für Informatik, Universität Giessen, Arndtstr. 2, 35392 Giessen, Germany
holzer@informatik.uni-giessen.de
[2] School of Electrical Engineering and Computer Science, University of Kassel, Kassel 34121, Germany
christian.rauch@uni-kassel.de

Abstract. The decidability of the PUMPING-PROBLEM—determining if a language given by an automaton or a grammar satisfies a pumping lemma w.r.t. a given pumping constant—exhibits a striking contrast: it is solvable for regular languages, even under context-free pumping constraints, but undecidable for (linear) context-free languages [H. GRUBER and M. HOLZER and C. RAUCH. The Pumping Lemma for Context-Free Languages is Undecidable. *DLT 2024*, pp. 141–155]. This work investigates which subfamilies of context-free languages retain decidability. We first prove that the PUMPING-PROBLEM is decidable for k-rated linear languages, a generalization of even linear languages that balances structural regularity with expressive power. As a byproduct of this research we develop an automaton model for k-rated linear languages. Further, we show decidability for well-matched visibly pushdown languages, a subfamily of the well-known visibly pushdown languages (VPL), which attracted a lot of attention, especially in verification, during the last decade.

1 Introduction

The PUMPING-PROBLEM was introduced in [7]. This is the problem of deciding for an automaton A (or a grammar G, resp.) and a value p, whether the language $L(A)$ ($L(G)$, resp.) satisfies a previously fixed pumping lemma w.r.t. the value p. Although every (linear) context-free language satisfies Bar-Hillel's pumping lemma [4], the PUMPING-PROBLEM is undecidable. In contrast to that, the PUMPING-PROBLEM for regular languages becomes decidable, not only for the regular pumping lemma of [9], but also for Bar-Hillel's pumping lemma. This is the starting point of our investigation.

To better understand the boundaries of decidability for the PUMPING-PROBLEM, we investigate it within two subfamilies of context-free languages: the k-rated linear languages and the well-matched visibly pushdown languages (wm-VPL):

G. Castiglione and S. Mantaci (Eds.): CIAA 2025, LNCS 15981, pp. 223–236, 2026.
https://doi.org/10.1007/978-3-032-02602-6_16

- The k-rated linear languages were introduced in [3] as a generalization of even linear languages [2]. A linear grammar is even linear if every production is of the form $A \to u$ or $A \to vBw$ with $|v| = |w|$. These languages lie strictly between regular and general linear context-free languages and admit an algebraic characterization *via* congruences, similar to regular languages [2]. In k-rated linear grammars, the ratio $|v|/|w|$ is fixed to a non-negative rational k for all productions involving nonterminals. For $k = 1$, we recover the even linear languages. While a general algebraic characterization of k-rated linear languages exists [3], it is non-constructive.
- The well-matched visibly pushdown languages form a subfamily of visibly pushdown languages (VPL), where push and pop operations are balanced. VPLs were formalized as visibly pushdown automata in [1]. They extend regular languages while preserving many of their closure properties and support nested structures, such as those in recursive programs or XML documents. Well-matched VPLs have recently been characterized algebraically using Ext-algebras [5,6].

For both k-rated linear languages and well-matched visibly pushdown languages, we prove that the PUMPING-PROBLEM—with respect to their language-specific variants—is decidable. Our approach relies on the algebraic characterization of each family: for k-rated linear languages, we build on the congruence-based characterization of even linear languages [2], enabling an automata-theoretic view. For well-matched VPLs, we use the Ext-algebra framework of [5,6], which builds on forest algebra techniques. These results extend the known frontier of decidability for the PUMPING-PROBLEM toward the full family of (deterministic) context-free languages—though the precise boundary remains open. Due to space constraints almost all proofs are omitted; they can be found in the full version of this paper.

2   Preliminaries

We assume the reader to be familiar with the basic notions on grammars and languages as contained in [9]. In particular, a *context-free grammar* (CFG) is a 4-tuple $G = (N, T, P, S)$, where N and T are disjoint alphabets of *nonterminals* and *terminals*, resp., $S \in N$ is the *axiom*, and P is a finite set of *productions* of the form $A \to \alpha$, where $A \in N$ and $\alpha \in (N \cup T)^*$. As usual, the transitive closure of the derivation relation $\Rightarrow_G$ is written as $\Rightarrow_G^*$. If there is no danger of confusion, we simply write $\Rightarrow$ ($\Rightarrow^*$, resp.) instead of $\Rightarrow_G$ ($\Rightarrow_G^*$, resp.). The *language generated* by G is defined as

$$L(G) = \{ w \in T^* \mid S \Rightarrow_G^* w \}.$$

A context-free language is a language generated by a context-free grammar. A context-free grammar is said to be *linear context-free* (LIN) if the productions are of the form $A \to \alpha$, where $A \in N$ and $\alpha \in T^*(N \cup \{\varepsilon\})T^*$—here ε refers to the *empty word*, and moreover, a context-free grammar is said to be *right-linear* or *regular* (REG) if the productions are of the form $A \rightarrow \alpha$, where $A \in N$ and $\alpha \in T^*(N \cup \{\varepsilon\})$. A linear context-free language (regular language, resp.) is a language generated by a linear context-free grammar (right-linear or regular grammar, resp.). The family of linear context-free languages properly contains the a family of regular languages and is properly contained in the family of context-free languages.

3   On k-Rated Linear Languages

As already mentioned in the introduction, the family of k-rated linear languages, for k in general, can be non-constructively characterized by a congruence in similar vein as regular languages. We first show that the algebraic characterization for even and k-rated linear languages can be obtained constructively. In addition, we also obtain a deterministic automaton model for k-rated linear languages. Then, we prove that the PUMPING-PROBLEM for k-rated linear language family is decidable w.r.t. pumping lemmata that are particularly designed for this language family.

A linear context-free grammar G is said to be *k-rated linear* (k-rLIN) if the productions are of the form $A \rightarrow u$ or $A \rightarrow vBw$, where $A, B \in N$ and $u, v, w \in T^*$ where the ratio of the lengths of v and w is a fixed non-negative rational number k, i.e., $k = |v|/|w|$. In case $k = 1$ we speak of an *even linear* (ELIN) grammar. A k-rated linear language is a language generated by a k-rated linear grammar—similarly this holds for even linear languages and grammars.

3.1   An Automaton Characterization of k-Rated Linear Languages

At first, we show that for k-rated linear grammar there is a normal form that is deterministic in its derivation, which helps us to develop a machine characterization of this language family. The normal form is an improved version of a normal form presented in [10], which in turn is based on a straight-forward generalization of a normal form of even linear grammars introduced in [2].

Theorem 1. *Let $G = (N, T, P, S)$ be a k-rated linear grammar with $k = g/h$ for g and h being coprime. Then one can effectively construct a k-rated linear grammar $G = (N', T, P', S')$ with $L(G) = L(G')$ such that*

1. *each production of G' of the form $A \rightarrow vBw$ with $v, w \in T^*$ and fulfills $|v| = g$ and $|w| = h$, while each of its productions of the form $A \rightarrow u$ with $u \in T^*$ satisfies $|u| < g + h$, and*
2. *for all $\alpha \in (N' \cup T)^*$ and all $A, B \in N' \setminus \{S'\}$ we have*
    (a) *if $(A \rightarrow \alpha) \in P'$ and $(B \rightarrow \alpha) \in P'$, then $A = B$, and*
    (b) *if $(S' \rightarrow \alpha) \in P'$, then $\alpha \in N'$.*

One essential tool for studying the PUMPING-PROBLEM for k-rated linear languages will be the following relation: given a language $L \subseteq \Sigma^*$ we define the relation C_L^k by setting $(w, w') \in C_L^k$ if for all $v, v' \in \Sigma^*$ with $|v|/|v'| = k$ or $v = v' = \lambda$ we have $vwv' \in L$ if and only if $vw'v' \in L$. The reader may confirm that C_L^k is an equivalence relation, i.e., it is reflexive, symmetric, and transitive. An equivalence relation $R \subseteq \Sigma^* \times \Sigma^*$ is said to be a k-quasi-congruence if for each $(w, w') \in R$ and $(v, v') \in \Sigma^g \times \Sigma^h$ with $k = g/h$, we have $(vwv', vw'v') \in R$.

By following the lines of [2], where the case $k = 1$ is studied, we show analogous results for C_L^k. First, we obtain the following lemma as a direct consequence of the fact that all tuples $(v, v') \in \Sigma^g \times \Sigma^h$ fulfill $|v|/|v'| = g/h$.

Lemma 2. *The equivalence relation C_L^k is a k-quasi-congruence.* $\square$

For utilizing the relation C_L^k for the PUMPING-PROBLEM it is essential that there are no words in L which are in relation to words in $\overline{L}$.

Lemma 3. *If the language L is a k-rated linear language then L is partitioned by equivalence classes of C_L^k, i.e., the words $w \in L$ cannot be in the same equivalence class as words $w' \in \overline{L}$.*

The following theorem marks the significance of the relation C_L^k.

Theorem 4. *Every k-quasi-congruence R that partitions L is a refinement of C_L^k.*

Next, we define the concept of k-TFAs for which we will see that they are closely related to k-rated linear languages. Recall, $k = g/h$, for g and h being coprime. A *two-sided finite automaton with k-window* (k-TFA[1]) is a tuple $A = (Q, \Sigma, \delta, Q_I, F, f)$, where Q is the finite set of *states*, Σ is the finite set of *input symbols*, $Q_I \subseteq Q$ is the set of *initial states* with $|Q_I| \leq |\Sigma^{<g+h}|$, $F \subseteq Q$ is the set of *accepting states*, the *transition function* δ maps $\Sigma^g \times Q \times \Sigma^h$ to Q, and f maps $\Sigma^{<g+h}$ to Q_I. The *language accepted* by the k-TFA A is defined as

$$L(A) = \{\, w \in \Sigma^* \mid \delta(w', f(\underline{w}), w'') \in F,$$

$$\text{where } w = w' \underline{w} w'' \text{ with } |\underline{w}| = |w| \bmod (g + h),$$

$$|w'| = g \cdot \lfloor \frac{|w|}{g+h} \rfloor \text{ and } |w''| = h \cdot \lfloor \frac{|w|}{g+h} \rfloor \,\},$$

where the transition function is recursively extended to $(\Sigma^g)^* \times Q \times (\Sigma^h)^* \to Q$ in the following way. For all $q \in Q$ and all $(w_1', w_1''), (w_2', w_2'') \in (\Sigma^g)^* \times (\Sigma^h)^*$ with $|w_1'| = g$ and $|w_1''| = h$ we have

- $\delta(\lambda, q, \lambda) = q$ and
- $\delta(w_2' w_1', q, w_1'' w_2'') = \delta(w_2', \delta(w_1', q, w_1''), w_2'')$.

---

[1] In [2] the case $g/h = k = 1$ is handled, where the according automata are called TFAs.

For $w \in \Sigma^*$ we will denote $\underline{w}$ as the subword of w which fulfills $w = w'\underline{w}w''$ where $|w'| = g \cdot \lfloor\frac{|w|}{g+h}\rfloor$, $|w''| = h \cdot \lfloor\frac{|w|}{g+h}\rfloor$, and $|\underline{w}| = |w| \bmod (g+h)$. Note that $\underline{w} \in \Sigma^{<g+h}$ for each $w \in \Sigma^*$. The k-TFA A works on w as follows; it starts in the state $f(\underline{w})$, then in each computation step it reduces the suffix of w' and the prefix of w'' by g and h letters, resp., while changing its state accordingly, and finally after $\lfloor\frac{|w|}{g+h}\rfloor$ steps A accepts if its state is accepting and rejects otherwise.

The next theorem plays a crucial role for the PUMPING-PROBLEM.

Theorem 5. *The language L is accepted by a k-TFA A if and only if there exists a k-quasi-congruence R of finite index such that L is partitioned by equivalence classes of R.*

Next, we prove a relation between k-quasi congruences and k-rated linear languages. The next theorem is due to [3], where it was proven in a non-constructive way. Here we give a constructive proof, since our decidability proof on the PUMPING-PROBLEM requires an effective construction.

Theorem 6. *Let L be a language over the alphabet Σ. There exists a k-quasi-congruence R of finite index on Σ^* such that the language L is partitioned by equivalence classes of R if and only if L is a k-rated linear language.*

Proof. The implication from left to right is seen as follows: Assume that $\Sigma^*/R = \{[w_0], [w_1], \ldots, [w_{n-1}]\}$. Define the grammar $G = (\Sigma^*/R \cup \{S\}, \Sigma, P, S)$, where S is a symbol not contained in Σ^*/R and P contains the productions[2]

$$
\begin{aligned}
[w_i] &\to v[w_j]v', &&\text{if } [vw_jv'] = [w_i], \\
[w_i] &\to \underline{w}, &&\text{if } \underline{w} \in [w_i], \\
S &\to [w_i], &&\text{if } [w_i] \subseteq L,
\end{aligned}
$$

for all $\underline{w} \in \Sigma^{<g+h}$, $(v, v') \in \Sigma^g \times \Sigma^h$, and $0 \le i, j \le n - 1$. We observe that G is a k-rated linear grammar with $L(G) = L$.

Conversely, we argue as follows: let $G = (N, \Sigma, P, S)$ be a k-rated linear grammar in normal form w.r.t. Theorem 1 with $L(G) = L$. Next we define a grammar G' by adding a new non-terminal and productions to G which have the new non-terminal as their left side such that these productions are never applied in any derivation starting from the axiom. These productions are needed to induce an equivalence relation R on all words regardless of whether the words are in the language $L(G)$ or not. Therefore, we add three types of productions: (i) those which end in an unreachable terminal word, (ii) those which have a right side which never appears in a derivation[3] of G, and (iii) those which reproduce the symbol S' by concatenating two arbitrary words (of length g

---

[2] The reader may easily confirm that the choice of the representatives of the equivalence classes has no influence on the argumentation of this proof.

[3] Since G is in normal form the property 2(a) implies that no derivation using one of these productions can produce a word which is in $L(G)$.

and h, resp.) to the left and right side of S'. Roughly speaking, each word which is not in $L(G)$ can be produced by a derivation starting in S'. Hence, we define the grammar[4] $G' = (N \cup \{S'\}, \Sigma, P', S)$, where S' is a new symbol not included in $N \cup \Sigma$, and P' will be defined as follows: production set P' is obtained by adding for all $(v, v') \in \Sigma^g \times \Sigma^h$, all $\underline{w} \in \Sigma^{<g+h}$, and all $B \in N$ the productions

$$\begin{array}{ll} S' \to \underline{w}, & \text{if } \nexists A \in N \text{ s.t. } (A \to \underline{w}) \in P, \\ S' \to vBv', & \text{if } \nexists A \in N \text{ s.t. } (A \to vBv') \in P, \\ S' \to vS'v', & \text{for all } (v, v') \in \Sigma^g \times \Sigma^h, \end{array}$$

to the set P. The reader may easily confirm that $L(G) = L(G')$ and that G' is also in normal form w.r.t. Theorem 1. Now we set

$$R = \{ (w, w') \mid \exists A \in (N \cup \{S'\}) \setminus \{S\} : A \Rightarrow_{G'}^* w \text{ and } A \Rightarrow_{G'}^* w' \}$$

to be a binary relation on Σ^*. The proof of the following claim is left to the interested reader.

Claim. The relation R is an equivalence relation.

Due to the finiteness of N we observe that R is an equivalence relation of finite index on Σ^* and $\Sigma^*/R = \{ R_A := \{ w \mid A \Rightarrow_{G'}^* w \} \mid A \in (N \cup \{S'\}) \setminus \{S\} \}$. Hence, the language $L(G)$ can be represented as the finite union $L(G) = L(G') = \bigcup_{\substack{A \in N \\ (S \to A) \in P'}} R_A$. It remains to show that R is a k-quasi-congruence. Therefore, we observe that for all $(v, v') \in \Sigma^g \times \Sigma^h$ and each non-terminal $B \in N'$ there exists exactly one non-terminal $A \in N'$ such that $(A \to vBv') \in P'$ which directly implies that $vR_Bv' \subseteq R_A$, i.e., we have that the relation R is a k-quasi-congruence. $\square$

As a direct consequence of the Theorems 5 and 6 we obtain the next result.

Corollary 7. *The family of k-rated linear languages is equal to the family of languages accepted by k-TFAs.*

As for DFAs it is possible to minimize k-TFAs.

Theorem 8. *Let $A = (Q, \Sigma, \delta, Q_I, F, f)$ be a k-TFA. Then there exists a unique (up to isomorphism) equivalent k-TFA $A' = (Q', \Sigma, \delta', Q'_I, F', f')$, i.e., $L(A') = L(A)$, and there is no k-TFA B with less than $|Q'|$ states that accepts $L(A)$.*

3.2 The Pumping-Problem for k-Rated Linear Languages

In this section we show the decidability of the PUMPING-PROBLEM for the languages in k-rLIN w.r.t. two pumping lemmata presented in [10]. The first one can be seen as the pumping lemma for linear context-free languages with additional length constraints, while the second one could be understood as the k-rated linear version of the Bar-Hillel pumping lemma for context-free languages.

---

[4] As already mentioned the non-terminal S' shall not be reached in any derivation. Therefore, S is the axiom of G', too.

Lemma 9. *Let L be a k-rated linear language with $k = g/h$ over Σ.*

1. *Then there is a constant p (depending on L) such that the following holds: If $z \in L$ and $|z| \geq p$, then there are words $u, v, w, x, y \in \Sigma^*$ such that $z = uvwxy$, $0 < |u|, |v| \leq p\frac{g}{g+h}$, $0 < |x|, |y| \leq p\frac{h}{g+h}$, $\frac{|u|}{|y|} = \frac{|v|}{|x|} = \frac{g}{h} = k$, and $uv^t wx^t y \in L$ for $t \geq 0$.*
2. *Then there is constant p (depending on L) such that the following holds: If $z \in L$ and $|z| \geq p$, then there are words $u, v, w, x, y \in \Sigma^*$ such that $z = uvwxy$, $0 < |v| \leq p\frac{g}{g+h}$, $0 < |x| \leq p\frac{h}{g+h}$, $0 < |w| \leq p$, $\frac{|u|}{|y|} = \frac{|v|}{|x|} = \frac{g}{h} = k$, and $uv^t wx^t y \in L$ for $t \geq 0$.*

Let $\mathrm{mpk}_{lin}(L)$ denote the smallest number p that satisfies the first of the above statements, while $\mathrm{mpk}_{con}(L)$ refers to the smallest p that fulfills the second one. We show that the pumping constants mpk_{lin} and mpk_{con} are incomparable.

Lemma 10. *Let $k = g/h$ be a rational number where g and h are coprime. Moreover, let p_ℓ and p_c be two natural numbers greater or equal to $3 \cdot (g + h)$ satisfying $p_\ell \bmod (g + h) = 0$. Then there is a k-rated linear language L such that $\mathrm{mpk}_{lin}(L) = p_\ell$ and $\mathrm{mpk}_{con}(L) = p_c$.*

Next we show decidability of the PUMPING-PROBLEM w.r.t. Lemma 9.1. To this end we prove some helpful statements beforehand. The following result is a direct consequence of the definition of k-TFAs and the fact that they have a finite state set.[5]

Lemma 11. *Let $A = (Q, \Sigma, \delta, Q_I, F, f)$ be a k-TFA. Then for each $(w', w'') \in \Sigma^g \times \Sigma^h$ and each state $q \in Q$ there exist natural numbers $\theta, \pi \leq |Q|$ such that for all integers $j \geq \theta$ we have $\delta(w'^j, q, w''^j) = \delta(w'^{j+\pi}, q, w''^{j+\pi})$.*

Next, we show that pumpability of a word is decidable.

Lemma 12. *Let A be a k-TFA, p be a natural number, and z be a word in the language $L(A)$. Then, it is decidable whether a given decomposition $z = uvwxy$ fulfills the conditions of Lemma 9.1 (Lemma 9.2, resp.) w.r.t. $L(A)$ and the value p.*

For a minimal DFA A, each state of A can be reached from any other state within $n - 1$ letters, where n is the number of states of A. Next we prove an analogous result for k-TFAs. Indeed, this requires that all states are reachable but this can be assumed for k-TFAs due to Theorem 8.

Lemma 13. *Let A be a minimal k-TFA and q a state of A. There is a word w of length at most $|Q| \cdot (g + h) - 1$ such that A is in state q after parsing w.*

Now we are able to prove the first main result of this section.

---

[5] This result illustrates the close connection of k-TFAs and DFAs. The most significant difference is the way these two types of automata parse words.

Theorem 14. *Let A be a k-TFA A and p a natural number. Then it is decidable whether for the language $L(A)$ the statement of Lemma 9.1 holds w.r.t. the value p.*

Next we turn our focus to Lemma 9.2 for which we will prove an analogous result. Therefore, we need the following helpful lemma which is a direct consequence of the definition of k-TFAs.

Lemma 15. *Let L be a k-rated linear language accepted by the k-TFA $A = (Q, \Sigma, \delta, Q_I, F, f)$, i.e., $L = L(A)$. Then the transition monoid of A is finite, that is, there exists a finite number of mappings from Q to Q such that each pair of words in $\Sigma^g \times \Sigma^h$ induces one of these mappings.* □

Indeed, there are at most $|Q|^{|Q|}$ mappings from Q to Q and if there is a pair of words in $(\Sigma^g \times \Sigma^h)^*$ which induces a particular mapping from Q to Q then there is a word in $(\Sigma^g \times \Sigma^h)^{|Q|^{|Q|}}$ inducing the same mapping.

Theorem 16. *Let A be a k-TFA and p a natural number. Then it is decidable whether for the language $L(A)$ the statement of Lemma 9.2 holds for the value p.*

Proof. First we assume that $A = (Q, \Sigma, \delta, Q_I, F, f)$. We will show that only a finite number of decompositions $uvwxy$ has to be tested to decide whether $\langle L(A), p \rangle$ is a positive instance of the PUMPING-PROBLEM or not. On the other side, we observe that each word $w = w'\underline{w}w''$ can be exchanged by the shortest word $\tilde{w} = \tilde{w}'\underline{\tilde{w}}\tilde{w}''$ in the equivalence class of w with $\underline{w}, \underline{\tilde{w}} \in \Sigma^{<g+h}$ because

$$\delta(uv^iw', f(\underline{w}), w''x^iy) = \delta(uv^i, \delta(w', f(\underline{w}), w''), x^iy)$$
$$= \delta(uv^i, \delta(\tilde{w}', f(\underline{\tilde{w}}), \tilde{w}''), x^iy) = \delta(uv^i\tilde{w}', f(\underline{\tilde{w}}), \tilde{w}''x^iy),$$

where the first and third equation follow from the definition of k-TFAs and the second equation is implied by w being in the equivalence class of $\tilde{w}$. Therefore, we have that $|w| \leq |Q| \cdot (g+h) - 1$. Additionally, we observe that u and y induce one of the mappings from Q to Q due to Lemma 15. Hence, we have

$$\delta(uv^i, w, x^iy) = \delta(u, \delta(v^i, w, x^i), y) = \delta(\tilde{u}, \delta(v^i, w, x^i), \tilde{y}) = \delta(\tilde{u}v^i, w, x^i\tilde{y})$$

for $(\tilde{u}, \tilde{y}) \in (\Sigma^g \times \Sigma^h)^{\leq |Q|^{|Q|}}$ being one pair of words, not necessarily unique, which induces the same mapping on Q as (u, y). Therefore, it is sufficient to check whether the

$$|\Sigma|^{g^{|Q|^{|Q|}}} \cdot |\Sigma|^{p\frac{g}{g+h}} \cdot (|Q| \cdot (g+h) - 1) \cdot |\Sigma|^{p\frac{h}{g+h}} \cdot |\Sigma|^{h^{|Q|^{|Q|}}}$$

decompositions $\tilde{u}vwx\tilde{y}$ fulfill the requirements of Lemma 9.2 to decide whether the input $\langle A, p \rangle$ is a positive instance of the PUMPING-PROBLEM or not.[6]   □

---

[6] We check whether for a given word at least one of its decomposition encodings fulfills the requirements of Lemma 9.2.

Due to the Theorems 5 and 6 the two previous theorems are also valid if the k-rated linear language is given by its grammar G. Then we use the procedure described in the proof of Theorem 6 to construct a k-quasi-congruence R of finite index such that $L(G)$ is partitioned by equivalence classes and afterwards we use the proof of Theorem 5 to construct a k-TFA A such that $L(A) = L(G)$.

4   On Well-Matched Visibly Pushdown Languages

First we introduce the concept of Ext-algebras and their connection to visibly pushdown languages (VPLs) and well-matched visibly pushdown languages (wm-VPLs) in particular, as presented mostly in [5]. Indeed, we will see that well-matched visibly pushdown languages can be characterized by finite Ext-algebras. This finiteness enables us to reduce the PUMPING-PROBLEM w.r.t. Lemma 20 and its linear context-free variant to the decidable problem of determining whether two (well-matched) visibly pushdown languages are equal. As a direct consequence we obtain that the PUMPING-PROBLEM for well-matched VPLs is also decidable w.r.t. regular pumping. We conclude this section by showing that the family of well-matched VPLs and the family of k-rated linear languages are incomparable.

We call $\Sigma = \Sigma_{call} \cup \Sigma_{int} \cup \Sigma_{ret}$ a *visibly pushdown alphabet* if the alphabets Σ_{call}, Σ_{int}, and Σ_{ret} are pairwise disjoint. For each visibly pushdown alphabet Σ we say that $w \in \Sigma^*$ is a *well-matched word* if one of the following conditions holds:

1. $w \in \{\lambda\} \cup \Sigma_{int}$,
2. $w = aw'b$, for $a \in \Sigma_{call}$, $b \in \Sigma_{ret}$, and w' being a well-matched word,
3. $w = w'w''$, for w' and w'' being non-empty well-matched words.

The *languages of well-matched words* over the visibly pushdown alphabet Σ is denoted by Σ^Δ. For each pair of words u, v in Σ^* we say that (u, v) is a context if $uv \in \Sigma^\Delta$. The set of all contexts is denoted by $\mathrm{Con}(\Sigma)$. The concatenation of two contexts $(u, v), (x, y) \in \mathrm{Con}(\Sigma)$ is defined by $(u, v) \circ (x, y) = (ux, yv)$. For a context $\sigma \in \mathrm{Con}(\Sigma)$ and a natural number k the k-fold composition is denoted by σ^k. For each context $(u, v) \in \mathrm{Con}(\Sigma)$ and each well-matched word $w \in \Sigma^\Delta$ we set $(u, v)w = uwv$. Let an equivalence[7] relation $\equiv^\Delta$ on $\mathrm{Con}(\Sigma)$ be given. Then $\equiv^\Delta$ is called a *congruence* on the set of contexts $\mathrm{Con}(\Sigma)$, if $\sigma \equiv^\Delta \tau$ implies $\xi \circ \sigma \circ \xi' \equiv^\Delta \xi \circ \tau \circ \xi'$, for all contexts $\xi, \xi', \sigma, \tau \in \mathrm{Con}(\Sigma)$. Recall, that $[\sigma]_{\equiv^\Delta}$ denotes the equivalence class of σ w.r.t. $\equiv^\Delta$. For any language of well-matched words $L \subseteq \Sigma^\Delta$ we write $\sigma \equiv_L^\Delta \tau$ if for each context $\xi \in \mathrm{Con}(\Sigma)$ and each $w \in \Sigma^\Delta$ the word $(\xi \circ \sigma)w$ is in L if and only if $(\xi \circ \tau)w$ is in L. The reader may confirm that $\equiv_L^\Delta$ is a congruence.[8] Next we define the model of deterministic visibly pushdown automata. A *deterministic visibly pushdown*

---

[7] The superscript Δ will be used to emphasize that the relation is defined for well-matched words or for contexts of well-matched words.

[8] The congruence $\equiv_L^\Delta$ introduced can be understood as a restricted form of the syntactic congruence $\equiv_L$.

automaton (DVPA) is a tuple $A = (Q, \Sigma, \Gamma, \delta, q_0, F, \bot)$, where Q is a finite set of states, Σ is a visibly pushdown alphabet, the input alphabet, Γ is a finite alphabet, the stack alphabet, $q_0 \in Q$ is the initial state, $F \subseteq Q$ is the set of final states, $\bot \in \Gamma$ is the bottom-of-stack symbol, and $\delta : Q \times \Sigma \times \Gamma \rightarrow Q \times \{\lambda\} \cup \Gamma \cup (\Gamma \setminus \{\bot\})\Gamma$ is the transition function such that for all $q \in Q, a \in \Sigma, \alpha \in \Gamma$: (i) if $a \in \Sigma_{call}$, then $\delta(q, a, \alpha) \in Q \times (\Gamma \setminus \{\bot\})\alpha$, (ii) if $a \in \Sigma_{ret}$, then $\delta(q, a, \alpha) \in Q \times \{\lambda\}$, and (iii) if $a \in \Sigma_{int}$, then $\delta(q, a, \alpha) \in Q \times \{\alpha\}$. The extended transition function $\hat{\delta} : Q \times \Sigma^* \times \Gamma^* \rightarrow Q \times \Gamma^*$ is defined inductively by

- $\hat{\delta}(q, \lambda, \beta) = (q, \beta)$, for all $q \in Q$ and $\beta \in \Gamma^*$,
- $\hat{\delta}(q, \hat{w}, \lambda) = (q, \lambda)$, for all $q \in Q$ and $w \in \Sigma^+$, and
- $\hat{\delta}(q, aw, \alpha\beta) = \hat{\delta}(p, w, \gamma\beta)$, where we have $\delta(q, a, \alpha) = (p, \gamma)$ for all $q \in Q$, $a \in \Sigma$, $w \in \Sigma^*$, $\alpha \in \Gamma$ and $\beta \in \Gamma^*$.

The language accepted by the DVPA A is the language

$$L(A) = \{\, w \in \Sigma^* \mid \hat{\delta}(q_0, w, \bot) \in F \times \{\bot\} \,\}.$$

Each language accepted by a DVPA A is called a *well-matched visibly pushdown language* (wm-VPL). Indeed, we observe that $L(A) \subseteq \Sigma^\Delta$.

4.1   An Algebraic Characterization of Well-Matched VPLs

First we introduce a few algebraic terms: let $(M, \cdot, 1_M)$ be a monoid. For each element $m \in M$ we denote the left-multiplication map $x \mapsto m \cdot x$ and the right-multiplication map $x \mapsto x \cdot m$ by left_m and right_m, resp. Building on this, an *Ext-algebra* $(R, O, \cdot, \circ)$ consists of two monoids $(R, \cdot, 1_R)$ and $(O, \circ, 1_O)$, where the latter one is a submonoid of $(R^R, \circ)$ which contains the maps left_r and right_r for each element $r \in R$. Let (R, O) and (S, P) be two Ext-algebras. Given two monoid morphisms $\phi : R \rightarrow S$ and $\psi : O \rightarrow P$, where for all $e \in O$ and all $r \in R$ we have $\psi(e)(\phi(r)) = \phi(e(r))$ and $\psi(\mathrm{left}_r) = \mathrm{left}_{\phi(r)}$ and $\psi(\mathrm{right}_r) = \mathrm{right}_{\phi(r)}$, then the tuple (ϕ, ψ) is an *Ext-algebra morphism* from (R, O) to (S, P), which will be denoted by $(\phi, \psi) : (R, O) \rightarrow (S, P)$. A morphism (ϕ, ψ) is called *surjective* (*bijective*, resp.) if both ϕ and ψ are surjective (bijective, resp.). Whenever there is no danger of confusion, we write morphism for an Ext-algebra morphism.

Let Σ be a fixed visibly pushdown alphabet. We define the function $ext_{u,v} : \Sigma^\Delta \rightarrow \Sigma^\Delta$ with $ext_{u,v}(x) = uxv$ for each context $(u, v) \in \mathrm{Con}(\Sigma)$. In addition, define $\mathcal{O}(\Sigma^\Delta) \subseteq (\Sigma^\Delta)^{\Sigma^\Delta}$ as the set $\{\, ext_{u,v} \mid (u, v) \in \mathrm{Con}(\Sigma) \,\}$. Observe that $\mathcal{O}(\Sigma^\Delta)$ is closed under composition. Hence, the monoid $(\mathcal{O}(\Sigma^\Delta), \circ)$ is a submonoid of $(\Sigma^\Delta)^{\Sigma^\Delta}$. For each well-matched word $w \in \Sigma^\Delta$ the mappings left_w and right_w are equal to $ext_{w,\lambda}$ and $ext_{\lambda,w}$, resp. Hence, for all $w \in \Sigma^\Delta$ the mappings left_w and right_w are contained in $\mathcal{O}(\Sigma^\Delta)$, which implies that $(\Sigma^\Delta, \mathcal{O}(\Sigma^\Delta), \cdot, \circ)$ is an Ext-algebra. We say that a language of well-matched words $L \subseteq \Sigma^\Delta$ is *recognized by an Ext-algebra* (R, O) if there exists a morphism $(\phi, \psi) : (\Sigma^\Delta, \mathcal{O}(\Sigma^\Delta)) \rightarrow (R, O)$ such that $L = \phi^{-1}(F)$ for some subset F of R. The following result shows the deep connection between Ext-algebras and well-matched visibly pushdown languages [6, Theorem 12].

Theorem 17. *A language $L \subseteq \Sigma^{\Delta}$ is a VPL if and only if it is recognized by a finite Ext-algebra.*

For deciding the PUMPING-PROBLEM for regular languages we made use of the syntactic monoid, see [8]. Hence, we derive an analogous structure for visibly pushdown languages. For each Ext-algebra (R, O) we say that $\sim$ is an *equivalence relation on* (R, O) if it is an equivalence relation on R. In addition $\sim$ is said to be a congruence on (R, O) if for all $e \in O$ and all $x, y \in R$ we have that $x \sim y$ implies that $e(x) \sim e(y)$. Then we denote by $(R, O)/\sim$ the pair $(R/\sim, O')$, where

$$O' = \{\, e' \in (R/\sim)^{R/\sim} \mid \exists e \in O \, \forall x \in R : e'([x]_{\sim}) = [e(x)]_{\sim} \,\}.$$

We define the syntactic congruence $\sim_L$ of a language $L \subseteq \Sigma^{\Delta}$ as the congruence on $(\Sigma^{\Delta}, \mathcal{O}(\Sigma^{\Delta}))$ which is defined by $u \sim_L v$ for each pair of well-matched words $u, v \in \Sigma^{\Delta}$ whenever $xuy \in L$ if and only if $xvy \in L$ for each context $(x, y) \in \mathrm{Con}(\Sigma)$. Then we define the syntactic Ext-algebra of L to be $(R_L, O_L) = (\Sigma^{\Delta}, \mathcal{O}(\Sigma^{\Delta}))/\sim_L$ and the syntactic morphism of L to be the morphism (ϕ_L, ψ_L) where we have $\phi_L : R_L \to R_L/\sim$ and $\phi_L(r) = [r]_{\sim}$ and moreover $\psi_L : O_L \to O'_L$ and $\psi_L(e)([r]_{\sim}) = [e(r)]_{\sim}$, for all $e \in O$ and $r \in R$. We call (ϕ_L, ψ_L) the morphism associated to $\sim_L$. In [6, Lemma 3.12] it is shown that (ϕ_L, ψ_L) is a surjective morphism from (R, O) to $(R, O)/\sim$. Observe that L is recognized by the syntactic Ext-algebra (R_L, O_L) *via* the syntactic morphism (ϕ_L, ψ_L) because for each pair of well-matched words $u, v \in \Sigma^{\Delta}$ we know that if $u \sim_L v$ holds, the word u is in L if and only if v is in L, which in turn implies that $L = \phi_L^{-1}(\phi_L(L))$. Indeed, it is sufficient to study the syntactic Ext-algebra (R_L, O_L) with the syntactic morphism (ϕ_L, ψ_L) for any well-matched visibly pushdown language L, as the next lemma shows, which is due to [5, Proposition 6.1].

Lemma 18. *Let A be a DVPA and $L := L(A)$. Then one can compute the syntactic Ext-algebra (R_L, O_L) of L, its syntactic morphism (ϕ_L, ψ_L) and $\phi_L(L)$.*

We will reduce the PUMPING-PROBLEM for wm-VPLs to the problem of deciding whether specific visibly pushdown languages are equal. Therefore, we need the following statement.

Lemma 19. *Let $L \subseteq \Sigma^{\Delta}$ be a well-matched visibly pushdown language, a well-matched word $w \in \Sigma^{\Delta}$, and $e \in O_L$. Then the language $L_{e,w} := \{\, vwx \mid vx \in \Sigma^{\Delta} : ext_{v,x} \in \psi_L^{-1}(e) \,\}$ is a well-matched visibly pushdown language.*

4.2 The PUMPING-PROBLEM for Well-Matched VPLs

We show the decidability of the PUMPING-PROBLEM for well-matched visibly pushdown languages w.r.t. a pumping lemma that can be found in [9, page 125, Lemma 6.1] and is a variant of the well-known Bar-Hillel pumping lemma [4, page 154, Theorem 4.1]. Observe that the original Bar-Hillel pumping lemma uses two pumping constants, one for the length of the word and the other for the subword that can be pumped.

Lemma 20. *Let L be a context-free language over Σ. Then, there is a constant p (depending on L) such that the following holds: If $z \in L$ and $|z| \geq p$, then there are words $u, v, w, x, y \in \Sigma^*$ such that $z = uvwxy$, $|vx| \geq 1$, $|vwx| \leq p$, and $uv^t wx^t y \in L$ for $t \geq 0$—it is then said that v and x can be (simultaneously) pumped in z.*

The pumping lemma for linear-context free languages reads like the pumping lemma for context-free languages, but with one exception: instead of $|vwx| \leq p$, now the condition $|uvxy| \leq p$ is required—see [9, page 143, Exercise 6.11]. Further, if the decomposition is changed to $z = uvw$, $|v| \geq 1$, $|uv| \leq p$, and $uv^t w \in L$ for $t \geq 0$ we end up with a regular pumping lemma, which can be found in [9, page 56, Lemma 6.1].

Before we come to our main result of this subsection we need some more notions on Ext-algebras. For all $e \in O_L$ we denote a triple (v, w, x) as e-*pumpable* for $v, w, x \in \Sigma^\Delta$ if $(e \circ ext_{v,x}^i)(\phi_L(w)) \in \phi_L(L)$, for all $i \geq 0$. Now we are ready for the next theorem.

Theorem 21. *Let A be a DVPA and p a natural number. It is decidable whether for the language $L = L(A)$ the statement of Lemma 20 holds for the value p.*

Proof. First, we observe that any word z in a well-matched visibly pushdown language can only be decomposed into $z = uvwxy$ such that z can be pumping w.r.t. Lemma 20 by its subwords v and x if $(u, y), (v, x) \in \text{Con}(\Sigma)$ and $w \in \Sigma^\Delta$. Next, we see that the number of all triples (v, w, x) that satisfy the requirement $|vwx| \leq k$ is bounded from above[9] by $(|\Sigma| + 2)^{k+2}$. For each of these triples the set $\{(e \circ ext_{v,x}^i)(\phi_L(w))\}$ is finite because $ext_{v,x}^i$ is an element in the finite monoid O_L for each $i \geq 0$. These elements are easily seen to be computable. Hence, one can effectively compute the set P_e of all e-pumpable triples (v, w, x) with $|vwx| \leq k$. Since $L_{\leq k}$ is a finite and therefore regular language, we can also calculate it effectively. Observe that the statement of Lemma 20 holds for the language L w.r.t. the value p if and only if $L = L_{\leq k} \cup \tilde{L}$, where

$$\tilde{L} := \bigcup_{e \in O_L} \bigcup_{(v,w,x) \in P_e} \{ uvwxy \mid uy \in \Sigma^\Delta : ext_{u,y} \in \psi_L^{-1}(e) \}.$$

Next, we show that the language $\tilde{L}$ is also a well-matched visibly pushdown language. Since O_L and P_e are finite sets the language $\tilde{L}$ is the union of finitely many languages $L_{e,v,w,x} := \{ uvwxy \mid uy \in \Sigma^\Delta : ext_{u,y} \in \psi_L^{-1}(e) \}$. As a consequence of Lemma 19, we obtain that $L_{e,u,v,w}$ itself is a wm-VPL which directly implies that $L_{\leq k} \cup \tilde{L}$ is a wm-VPL, too. Hence, the PUMPING-PROBLEM w.r.t. L and the value k is equivalent to the problem of deciding whether L and $L_{\leq k} \cup \tilde{L}$ are equal. Indeed, the later problem is decidable as the EQUIVALENCE-PROBLEM

---

[9] This rough estimation can be obtained by introducing two additional symbols $\$, \#$ to Σ such that one marks the end of the subwords v and w, whereby the other one denotes a missing letter in the concatenation vwx because their length can be shorter than k.

for visibly pushdown languages with the same visibly pushdown alphabet is decidable, as proven in [1]. In conclusion, one can effectively decide the PUMPING-PROBLEM.   □

Indeed, it is also possible to decide the PUMPING-PROBLEM for well-matched visibly pushdown languages w.r.t. linear context-free pumping.

Theorem 22. *Let A be a DVPA and p a natural number. It is decidable whether for the language $L = L(A)$ the statement of the linear context-free variant of Lemma 20 holds for the value p. The statement remains true even if the regular pumping lemma is considered instead.*

Although well-matched visibly pushdown languages and k-rated linear languages contain the family of regular languages both families are incomparable.

Lemma 23. *The language family k-rLIN and the family of wm-VPLs are incomparable, i.e., there is an infinite family of well-matched visibly pushdown languages which are no k-rated languages and vice versa.*

Acknowledgements. The authors thank Benedek Nagy for highlighting the relevance of k-rated linear languages, and Stefan Göller for valuable discussions on the algebraic aspects of VPLs, in relation to the PUMPING-PROBLEM.

References

1. Alur, R., Madhusudan, P.: Visibly pushdown languages. In: Proceedings of the 36th Annual ACM Symposium on Theory of Computing, pp. 202–211. ACM, Chicago, IL, USA, June 2004
2. Amar, V., Putzolu, G.: On a family of linear grammars. Inform. Control **7**(3), 283–291 (1964). https://doi.org/10.1016/S0019-9958(64)90294-3
3. Amar, V., Putzolu, G.: Generalizations of regular events. Inform. Control **8**(1), 56–63 (1965). https://doi.org/10.1016/S0019-9958(65)90275-5
4. Bar-Hillel, Y., Perles, M., Shamir, E.: On formal properties of simple phrase structure grammars. Zeitschrift für Phonetik, Sprachwissenschaft und Kommunikationsforschung **14**, 143–177 (1961). https://doi.org/10.1524/stuf.1961.14.14.143
5. Göller, S., Grosshans, N.: The AC^0-complexity of visibly pushdown languages. arXiv:2302.13116v3 (2023). https://doi.org/10.48550/arXiv.2302.13116
6. Göller, S., Grosshans, N.: The AC^0-complexity of visibly pushdown languages. In: Beyersdorff, O., Kanté, M.M., Kupferman, O., Lokshtanov, D. (eds.) Proceedings of the 41st International Symposium on Theoretical Aspects of Compter Science. LIPIcs, vol. 289, pp. 38:1–38:18. Schloss Dagstuhl–Leibniz-Zentrum für Informatik, Dagstuhl, Germany, Clermont-Ferrand, France (2024). https://doi.org/10.4230/LIPIcs.STACS.2024.38
7. Gruber, H., Holzer, M., Rauch, C.: The pumping lemma for regular languages is hard. In: Nagy, B. (ed.) Proceedings of the 27th International Conference on Implementation and Application of Automata, LNCS, vol. 14151, pp. 128–140. Springer, Famagusta, Cyprus (2023). https://doi.org/10.1007/978-3-031-40247-0_9

8. Gruber, H., Holzer, M., Rauch, C.: The pumping lemma for context-free languages is undecidable. In: Day, J.D., Manea, F. (eds.) Proceedings of the 28th International Conference on Developments in Language Theory. LNCS, vol. 14791, pp. 141–155. Springer, Göttingen, Germany (2024). https://doi.org/10.1007/978-3-031-66159-4_11

9. Hopcroft, J.E., Ullman, J.D.: Introduction to Automata Theory. Languages and Computation. Addison-Wesley, Boston (1979)

10. Horváth, G., Nagy, B.: Pumping lemmas for linear and nonlinear context-free languages. Acta U. Sapien. Inform. **2**(2), 194–209 (2010)

Self-verifying Predicates in Büchi Arithmetic

Mazen Khodier[1] ⓘ, Luke Schaeffer[2] ⓘ, and Jeffrey Shallit[1](✉) ⓘ

[1] School of Computer Science, University of Waterloo, 200 University Ave. W., Waterloo, ON N2L 3G1, Canada
{mkhodier,shallit}@uwaterloo.ca
[2] Institute for Quantum Computing, University of Waterloo, 200 University Ave. W., Waterloo, ON N2L 3G1, Canada
lschaeffer@uwaterloo.ca

Abstract. We discuss a technique, based on Angluin's algorithm, for automatically generating finite automata for various kinds of useful first-order logic formulas in Büchi arithmetic. Construction in this way can be faster and use much less space than more direct methods. We discuss the theory and we present some empirical data for the free software `Walnut`.

1 Introduction

Let $k \geq 2$ be a fixed integer, and let $V_k(n) = \sup\{k^i \ : \ k^i \mid n\}$. The first-order logical theory $\langle \mathbb{N}, +, <, 0, 1, V_k \rangle$, is sometimes called *Büchi arithmetic* [6]. This theory, an extension of Presburger arithmetic, is powerful enough to express finite automata, and is algorithmically decidable [5]. Notice that addition and subtraction of integer variables are permitted in this theory, but not multiplication or division (although one can express multiplication and integer division by fixed integer constants).

In particular, this theory is quite useful in combinatorics on words, since it can be used to decide many claims about automatic sequences, provided the underlying numeration system is addable [17]. Here by an addable numeration system we mean one for which a finite automaton can compute the addition relation $x + y = z$. We say a sequence $(a(n))_{n \geq 0}$ is *automatic* if it takes its values in a finite set, and there is a finite automaton that, on input n expressed in the numeration system $\mathcal{N}$, reaches a state with output $a(n)$; see [1].

Example 1. It is decidable, given an automatic sequence $\mathbf{a} = (a(n))_{n \geq 0}$ defined on an addable numeration system, whether $\mathbf{a}$ is *squarefree*; that is, whether it contains no two consecutive identical nonempty blocks.

This follows because the property of not having a square is first-order expressible, as follows:

$$\neg \exists i, n \ (n \geq 1) \wedge \forall t \ (t < n) \implies a(i + t) = a(i + n + t).$$

Research of Mazen Khodier and Jeffrey Shallit is supported by NSERC Grant number 2024–03725.

G. Castiglione and S. Mantaci (Eds.): CIAA 2025, LNCS 15981, pp. 237–251, 2026.
https://doi.org/10.1007/978-3-032-02602-6_17

More precisely, the fundamental theorem of Büchi arithmetic is the following:

Theorem 1. *There exists an algorithm that takes a first-order logical formula φ about an automatic sequence in an addable numeration system, phrased in terms of addition and indexing of integer variables, and computes an automaton that accepts exactly those values of the free (unbound) variables that make φ true. If there are no free variables, the computed automaton is a single state that accepts everything (TRUE) or rejects everything (FALSE).*

However, the worst-case running time of the decision procedure is truly formidable; it is of the form

$$2^{2^{\cdot^{\cdot^{\cdot^{2^{p(N)}}}}}},$$

where the number of 2's in the exponent is equal to the number of quantifier alternations in the formula, p is a polynomial, and N is the size of the logical formula. This is because the algorithm depends on repeated applications of the subset construction from automata theory [10], each one of which can potentially result in an exponential blowup in the number of states.

Integers in Büchi arithmetic are represented by words x over a finite alphabet Σ^*. We also need to represent j-tuples of integers $(n_1, \ldots, n_j)$; this is done by using the alphabet Σ^j, and then n_i is the integer whose representation is the projection of the i'th coordinate of x. This may require padding of the shorter inputs with leading zeros. If $x \in (\Sigma^j)^*$, we let $[x]$ denote the tuple of integers represented by x. The canonical representation of an integer $n \in \mathbb{N}$ (that is, the one without leading zeros) is written (n). This is generalized to j-tuples by writing $(n_1, n_2, \ldots, n_j)$. Throughout the paper, we assume representations of integers are read with the most significant digit first.

The paper is organized as follows. In Sect. 2 we discuss the Walnut program and what it does. In Sect. 3 we review Angluin's algorithm. The next sections cover how we can use the self-verifying property of predicates to create an automaton for equality of factors (Sect. 4), equality of reversals of factors (Sect. 5), periods (Sect. 6), recognizing the addition relation (Sect. 7), and summation (Sect. 8). Section 9 discusses algorithm engineering issues and reports some computational results for our very preliminary implementation.

2   Walnut

An automaton-based decision procedure for Büchi arithmetic, as well as certain related theories based on other kinds of representations, has been implemented in the free software Walnut [12]. It has been used in over a hundred research papers and books to confirm old results, correct mistakes in the literature, and prove entirely new results [17].

The decision procedure for Büchi arithmetic implemented in Walnut allows representing integers in a variety of addable numeration systems, including base k for integers $k \geq 2$, Fibonacci (Zeckendorf) numeration system, Tribonacci numeration system, and so forth. The user can also define their own numeration system.

Given a first-order logical formula φ, `Walnut` produces a DFA (deterministic finite automaton) accepting the values of the free variables that make φ true. Furthermore, we are guaranteed that the automata that `Walnut` computes are minimal. `Walnut` can also produce DFAO's (deterministic finite automata with output), where the output is a specified function of the last state reached.

In some cases, `Walnut` returns no response to a query because its limit on space is exceeded (roughly, that automata cannot have more than 2^{32} transitions.) This can be the case even if the final result would be small, because intermediate results can be extremely large. This can occur even with a single quantifier and a very simple formula.

As an example of the kinds of space consumption that can occur, consider the following first-order logical formula for equality of factors of an infinite word $\mathbf{X}$:

$$\mathrm{EqFac}(i,j,n) := \forall t \ (t < n) \implies \mathbf{X}[i+t] = \mathbf{X}[j+t].$$

It is true precisely when the length-n factor beginning at position i is the same as that beginning at position j. (In this predicate, as in all others we discuss in this paper, the domain of integer variables is assumed to be $\mathbb{N} = \{0, 1, 2, \ldots\}$, the natural numbers.) Since `Walnut` implements Büchi arithmetic, it can form an automaton evaluating this logical formula for automatic sequences $\mathbf{X}$.[1] For example, for the Thue-Morse sequence, the resulting automaton takes the inputs i, j, n in parallel, has 14 states, and the largest intermediate automaton formed during the computation has 408 states. To construct it, one can use the `Walnut` command

```
def tm_eqfac "At (t<n) => T[i+t]=T[j+t]":
```

However, when we carry out the same construction for the so-called Tribonacci word [3, 21] $\mathbf{tr} = 0102010010201010201002 \cdots$, using the `Walnut` command

```
def trib_eqfac "?msd_trib At (t<n) => TR[i+t]=TR[j+t]":
```

the largest intermediate automaton has 323,831,403 states (!), while the final result has only 26 states.[2]

Therefore it is desirable to have alternative methods to compute some of these basic and useful automata. In this paper we develop a new technique for this, of both theoretical and practical interest. It uses the observation that some of the most useful predicates are "self-verifying" in a certain sense. In some cases, we can prove that our new algorithm runs in time polynomial in the size of the

---

[1] The need for Büchi arithmetic, as opposed to the simpler Presburger arithmetic, arises because we need to be able to express the i'th term of an automatic sequence $\mathbf{X}[i]$.

[2] For this particular example, some reformulations of the logical predicate result in much smaller intermediate automata.

automaton computing the original sequence **X** and the size of the final result. In particular, the idea is based on Angluin's algorithm, discussed in the next section.

The idea of using self-verifying predicates to verify automata was already discussed in a number of papers, such as [14,18]. The novelty of this paper is that no "guessing" of the automaton to be verified is needed; the automaton is, so to speak, "built from the ground up" in a completely deterministic fashion.

3   Angluin's Algorithm

Dana Angluin developed an algorithm [2], sometimes called the "L^* algorithm", that allows a *learner* to infer a finite automaton for a regular language $L \subseteq \Sigma^*$ from examples and counterexamples. In her algorithm, the learner interacts with a *teacher* in two different ways.

The learner can present a word $w \in \Sigma^*$ to the teacher and ask if $w \in L$; this is called a *membership query*. The learner can also present an hypothesized automaton A for the language L and ask if $L(A) = L$. If the teacher answers positively, the learner has now found an automaton for L and the algorithm terminates. If, on the other hand, the teacher answers negatively, they provide a counterexample to the learner; that is, a member of the symmetric difference $(L \setminus L(A)) \bigcup (L(A) \setminus L)$. We call all of this a *hypothesis query*. If the minimal automaton for L has N states, then the total number of membership queries and hypothesis queries, and their size, is bounded by a polynomial in N. More precisely, her result is the following:

Theorem 2 ([2]). *There exists an algorithm L^* that learns a regular language L by performing membership and hypothesis queries. If all negative hypothesis queries return counterexamples of length at most m and the regular language L has a minimal DFA of n states, then the total running time is polynomial in m and n.*

In Sect. 9.2 we will need a few more details about how Angluin's algorithm works to infer $L \subseteq \Sigma^*$. It maintains a finite prefix-closed set of strings S, a finite suffix-closed set of strings E, and a map T from $(S \bigcup S\Sigma)E$ to $\{0,1\}$, where the intent is that 1 means the string is in L and 0 otherwise. Prefix-closed means every prefix of S is in S, and similarly for the suffix-closed property of E. Thinking of T as a table with rows labeled by $S \bigcup S\Sigma$ and columns labeled by E, the table T is said to be closed if for each $t \in S\Sigma$ there is an $s \in S$ such that the row labeled s equals the row labeled t. It is called consistent if whenever two rows are equal, labeled s_1 and s_2, say, then the row labeled s_1a equals that labeled by s_2a, for all $a \in \Sigma$.

Normally the teacher and the learner are separate entities. However, for certain kinds of languages, corresponding to certain logical formulas about automatic sequences, we can use Theorem 1 to construct an algorithm that plays the role of *both* learner and teacher, as we will see in the next section. Currently

we have no general theory characterizing exactly for which kinds of logical formulas our technique works. Nevertheless, the general idea is widely applicable
and adaptable to a number of situations, as we intend to show in the following
sections.

4   Equality of Factors

One of the most useful of all predicates for understanding the properties of an
infinite word $\mathbf{X}$ is EqFac, the *equality of factors* predicate we saw in Sect. 1;
that is, given integers i, j, n, determine whether the length-n factors beginning
at positions i and j are the same. If we want to understand aspects of $\mathbf{X}$ such
as factor complexity (aka subword complexity), the number of distinct blocks of
length n appearing in $\mathbf{X}$, then finding an automaton for this EqFac is a critical
building block.

However, because of the appearance of the $\forall$ quantifier, the best estimate we
can find for the complexity of finding the automaton using Theorem 1, without more detailed analysis, is an exponential upper bound on the size of the
resulting automaton. We can potentially see exponential blowup in intermediate
automata, even if the final result is small.

In this section, by combining Angluin's algorithm with a self-verifying predicate, we prove the following result:

Theorem 3. *If* $\mathbf{X}$ *is an automatic sequence, we can construct a minimal
automaton A for the* EqFac *predicate in time polynomial in the size of the
automaton for* $\mathbf{X}$ *and the size of* A.

The language L in Angluin's algorithm is

$$L = \{x \in \mathcal{V}_3 \ : \ [x] = (i, j, n) \text{ and } \mathrm{EqFac}(i, j, n)\},$$

where $\mathcal{V}_j$ is the set of all valid representations of integer j-tuples in the underlying
numeration system. For example, if we are representing integers in base 2, then
$\mathcal{V}_3 = (\{0, 1\}^3)^*$. In particular, we note that a standard convention is that if an
input is accepted, then so are all inputs that start with an arbitrary number of
leading zeros. We enforce this by insisting that from the initial state, a transition
on the input $[0, 0, \ldots, 0]$ always leads back to the initial state.

4.1   Membership Queries

We now explain how the membership queries for L required by Angluin's algorithm can be carried out in polynomial time for EqFac. Of course, by "polynomial
time" we mean polynomial time in the *number of bits* of the integers involved.

Given x, the first thing we check is that we have used only legal representations in the underlying numeration system. For example, if we are working in
the Fibonacci numeration system, we have to check that the projection of x into

the binary strings representing three integers i_0, j_0, n_0 have no two consecutive 1's. If any of them do have two consecutive 1's, then $x \notin L$.

Now we know that the input is a legal representation of a triple of integers (i_0, j_0, n_0). We want to determine if $\mathbf{X}[i_0..i_0 + n_0 - 1] = \mathbf{X}[j_0..j_0 + n_0 - 1]$. A small difficulty is that given an automaton for $\mathbf{X}$ that maps n to $\mathbf{X}[n]$, it is not clear how many states are needed for $(i, n) \rightarrow \mathbf{X}[i + n]$. This is because the obvious logical formulation of this is $\exists t \ t = i + n \wedge \mathbf{X}[t]$, and the $\exists$ quantifier could create an NFA, which is then determinized. Thus, at least theoretically, this could introduce exponential blowup. We can avoid this using De Morgan's law to rewrite a $\forall$ query to a $\exists$ formula. Furthermore, we avoid computing the DFA for $\exists x \ \varphi(x)$; instead we compute the automaton for $\varphi(x)$ and use breadth-first search (BFS) to determine whether any string is accepted, and if so, what a shortest such string is.

Thus, the first step is create the automaton $E(i, j, n, t, u, v)$ for the expression

$$(t < n) \wedge (u = i + t) \wedge (v = j + t) \wedge (\mathbf{X}[u] \neq \mathbf{X}[v]).$$

The automata for the first three conjuncts each have a constant number of states, and so do the automata for the intersection of the corresponding languages. The last conjunct can be expressed as an automaton using $O(N^2)$ states, where N is the number of states in the automaton for $\mathbf{X}$. Constructing the automaton for $E(i, j, n, t, u, v)$ needs to be done only once, at the very beginning of the algorithm.

Now we want to evaluate $\mathrm{EqFac}(i_0, j_0, n_0)$ for some specific (i_0, j_0, n_0). To do this, we create automata accepting $0^*(i_0)$, $0^*(j_0)$, and $0^*(n_0)$. These individually require only $O(\log \max(i_0, j_0, n_0))$ states. With additional intersections we create an automaton E' accepting only those 6-tuples (i, j, n, t, u, v) for which i matches i_0, j matches j_0, and n matches n_0. We now perform breadth first search in E', to see if E' accepts anything; that is, if an accepting state is reachable from the start state. This can be done in linear time in the size of E', which is polynomial in $\log \max(i_0, j_0, n_0)$. If E' accepts anything, then $\mathrm{EqFac}(i_0, j_0, n_0)$ is false; otherwise it is true.

Thus we can perform membership queries in polynomial time.

4.2 Hypothesis Queries

Now we explain how to implement hypothesis queries efficiently. Given an automaton A, we want to check whether $L(A) = L$. The crucial point is that EqFac is *self-verifying*; that is, provided certain logical formulas about A hold, they provide us with a proof by mathematical induction that the automaton A is correct. In the case of EqFac, an easy induction on n shows that a putative automaton A correctly decides EqFac if and only if all of the following assertions hold:

1. A accepts no illegal representations for i, j, n (representations that are not permissible in the given numeration system);

2. the initial state of A transits to itself on input $[0, 0, 0]$;
3. $\forall i, j\ A[i, j, 0]$;
4. $\forall i, j, n\ A[i, j, n+1] \iff (A[i, j, n] \wedge \mathbf{X}[i+n] = \mathbf{X}[j+n])$.

Conditions 3 and 4 form the basis of a proof by induction on n that A is correct. Although it is perhaps not immediately obvious, we can check all of these conditions in polynomial time; in the case of conditions 3 and 4, by reformulating them with De Morgan's laws.

Condition 3 can be restated as

$$\neg \exists i, j, t\ (t = 0) \wedge (\neg A[i, j, t]).$$

This can be checked by complementing A, intersecting with the automaton accepting $(\Sigma^*, \Sigma^*, 0^*)$, and using BFS to check if there is a path to an accepting state. If there is none, then condition 3 holds; otherwise it fails and a counterexample $(i_0, j_0, 0)$ is given by the label of the path to any accepting state.

For Condition 4, we first rewrite it as

$$\forall i, j, n, t, u, v\ (t = n + 1 \wedge u = i + n \wedge v = j + n) \implies$$
$$(A[i, j, t] \iff (A[i, j, n] \wedge \mathbf{X}[u] = \mathbf{X}[v]))$$

and then use de Morgan's law to replace the $\forall$ with the conjunction of the following three statements:

$\neg \exists i, j, n, t\ (t = n + 1) \wedge (\neg A[i, j, n]) \wedge A[i, j, t]$
$\neg \exists i, j, n, t, u, v\ (t = n + 1) \wedge (u = i + n) \wedge (v = j + n) \wedge (\neg A[i, j, t]) \wedge A[i, j, n]$
$\wedge\ \mathbf{X}[u] = \mathbf{X}[v]$
$\neg \exists i, j, n, t, u, v\ (t = n + 1) \wedge (u = i + n) \wedge (v = j + n) \wedge \mathbf{X}[u] \neq \mathbf{X}[v] \wedge A[i, j, t].$

In all three cases we can test the conditions by forming intersections of languages created by the direct product of automata, followed by BFS to check if an input is accepted. If no input is accepted, then A is the correct automaton. Otherwise, A accepts some word that provides the needed counterexample (i_0, j_0, n_0). Actually, we do not know, a priori, whether it is $A[i_0, j_0, n_0]$ or $A[i_0, j_0, n_0 + 1]$ that is wrong, so some additional membership queries may be necessary to determine which value is incorrect.

We can now put everything together to prove Theorem 3.

Proof. As the analysis of Angluin's algorithm shows, if A has t states then we do at most polynomially many membership queries, and the construction of our automata for the membership queries shows that the size of the queries is also polynomially bounded. Furthermore, at most polynomially many hypotheses about A are evaluated, so the total cost is polynomial time.

Similarly, we can check formulas like

$$\forall t\ (t < n) \implies \mathbf{X}[i+t] = \mathbf{Y}[j+t]$$

for two automatic sequences $\mathbf{X}$ and $\mathbf{Y}$ defined over the same numeration system.

In what follows, we omit explicit mention of conditions 1 and 2 above, although we always need them to check that a hypothesized automaton A is correct.

5 Equality of Reversals of Factors

Another self-verifying predicate involves checking equality of a reversal of a factor. Let w^R denote the reversal of the word w so that, for example, $(\texttt{drawer})^R = \texttt{reward}$. Another useful predicate, EqRevFac, asserts that $\mathbf{X}[i..i + n - 1] = \mathbf{X}[j..j + n - 1]^R$. As a special case, we can use this predicate to test whether a factor is a palindrome.

We can use exactly the same technique as in the previous section, based on the fact that EqRevFac is self-verifying just the way EqFac is. More precisely, if A is a claimed automaton for EqRevFac, it is correct if and only if

1. $\forall i, j \ A[i, j, 0]$;
2. $\forall i, j, n \ A[i, j, n + 1] \iff (A[i + 1, j, n] \wedge \mathbf{X}[i] = \mathbf{X}[j + n])$.

We leave the details to the reader.

6 Periods of Factors

Let $x = x[1..n]$ be a finite word. We say that p is a *period* of x if $x[i] = x[i + p]$ for all i such that $1 \le i \le n - p$. From the definition we see that 0 is trivially a period of every word. Furthermore, if x is of length n, then every integer $\ge n$ is also trivially a period of x. Often these edge cases are not considered valid periods, but in order to show that period is self-verifying, we need to regard them as periods.

Consider the predicate $\mathrm{Per}(i, n, p)$ that asserts that p is a period of $\mathbf{X}[i..i + n - 1]$. This is a very basic assertion that can be used to determine whether a given factor is an overlap (of the form $axaxa$, where a is a single letter and x is possibly empty), a square (of the form xx with x nonempty), a cube (of the form xxx with x nonempty), and so forth, and it is very desirable to be able to find the corresponding automaton efficiently. We can write a first-order logic formula for this as follows:

$$\forall t \ (t + p < n) \implies \mathbf{X}[i + t] = \mathbf{X}[i + t + p],$$

and use Theorem 1 to find an automaton computing it. However, this may result in a large intermediate automaton, even when the final result is small.

Of course, we can also express $\mathrm{Per}(i, n, p)$ using the EqFac predicate, but it is possible that the automaton for Per is smaller and hence easier to compute directly by our method.

Suppose A is an automaton that is claimed to compute the predicate $\mathrm{Per}(i, n, p)$ that asserts that p is a period of $\mathbf{X}[i..i + n - 1]$. Then A is correct if and only if the following conditions hold.

1. $\forall i, p\ A[i, 0, p]$;
2. $\forall i, n, p\ A[i, n+1, p] \iff ((p \geq n+1) \lor (p \leq n \land A[i, n, p] \land \mathbf{X}[i+n] = \mathbf{X}[i+n-p]))$.

These can be translated into $\exists$ predicates just as we did for the case of EqFac. Notice that the subtraction can be handled by introducing a new variable t and imposing the condition $t + p = i + n$.

7   Creating an Adder for a Numeration System

Using `Walnut` requires choice of a numeration system. For some numeration systems, such as base k for $k \geq 2$, the numeration system is built in.

One of the crucial requirements for `Walnut`'s algorithm to succeed is that the numeration system be *addable*; that is, there is a finite automaton recognizing the addition relation $x + y = z$.

For more exotic numeration systems, such as those based on a Pisot number [4,8,9], constructing the adder can be a bit of an onerous task. However, Angluin's algorithm allows it to be constructed "from the ground up" when it exists, because the adder itself is self-verifying. This was already pointed out in [13, Remark 2.1], and we simply reprise this below.

The first step is that we need an incrementer; that is, an automaton Incr that accepts a pair of inputs (n, x) in parallel if and only if $x = n + 1$. This can be generated by combining the automaton recognizing the legal representations in the numeration system with an automaton that compares two inputs lexicographically, as discussed in [15, p. 37].

Membership queries are more or less trivial and do not require `Walnut`.

Now suppose we have an automaton A that we hypothesize computes the addition relation. We claim A is correct if and only if both of the following conditions hold (in addition to the conditions on the validity of the representations and leading zeros).

1. $\forall y, z\ A[0, y, z] \iff y = z$;
2. $\forall x, y, z, t, u\ (\mathrm{Incr}(x, t) \land \mathrm{Incr}(z, u)) \implies (A[x, y, z] \iff A[t, y, u])$.

These two conditions provide the basis and the induction step, respectively, for an induction proof on x that A is correct.

These two conditions can be rearranged, as we did above, to consist of existential claims that can be verified efficiently with BFS. If they fail, short counterexamples can be computed with BFS, too.

For other recent discussions of "automatically" obtaining an adder for a numeration system, see [11, §4.1] and [7].

8   Summation of Synchronized Sequences

We say a sequence (or function) $(b(n))_{n \geq 0}$ from $\mathbb{N}$ to $\mathbb{N}$ is *synchronized* if there exists an automaton B that accepts the representation of those pairs $(n, x) \in$

$\mathbb{N} \times \mathbb{N}$ for which $x = b(n)$. In this case we call B synchronized also. See [16] for more information about these sequences. Notice that it is possible that n is represented in one numeration system, while x is represented in some other numeration system. For an example of this, see [18], where n is represented in base 4 while x is represented in base 3.

Often we would like to do partial summation on such a sequence, defining $c(n) = \sum_{0 \le i < n} b(i)$. Unfortunately there are examples where a is synchronized but the partial sum sequence b is not. As an example, consider the synchronized sequence $\mathbf{p_2} = 11010001 \cdots$, which is 1 if $n + 1$ is a power of 2 and 0 otherwise. Then the partial sum sequence of $\mathbf{p_2}$ is given by $\lfloor \log_2 n \rfloor + 1$ for $n \ge 1$. However, this kind of growth rate is impossible for a synchronized sequence [16].

Nevertheless, in some cases the partial sum sequence is synchronized, and if it is, then we would like to construct the automaton for it. Luckily, the assertion that c is the partial sum sequence for b is self-verifying. In addition to the usual checks involving legal representations and leading zeros, in order to verify that an automaton C for c is correct, we only need to check that C satisfies the following:

(a) $\forall x \; C[0, x] \iff x = 0$;
(b) $\forall n, t, u, y, z \; (u = n + 1 \wedge z = t + y \wedge B[n, t]) \implies (C[n, y] \iff C[u, z])$.

Condition (a) provides the basis, and condition (b) the induction step for a proof by induction on n that $C[n, x]$ holds if and only if $c(n) = x$. These two conditions can be turned into existential claims that can be verified with BFS, as before.

There is also the issue of how to compute membership queries. Here we are presented with n and z and we want to determine whether $c(n) = z$. This can be done as follows: from the synchronized automaton B for b, first determine a linear representation for $b(n)$. Recall that a linear representation consists of a row vector v, a column vector w, and a matrix-valued morphism γ such that $b(n) = v \cdot \gamma(y) \cdot w$ for all representations y of n (including those with leading zeros). This linear representation can be trivially computed in linear time in the size of B, as explained in [17, §9.8].

Once we have the linear representation for b, we can compute a linear representation for c using a fairly simple transformation that only increases the size of the linear representation by a constant factor [20]. And from the linear representation we can compute $c(n)$ and check to see whether it equals x, in polynomial time in $\log n$.

Thus once again we can use Angluin's algorithm to compute the automaton C for the partial sum sequence c, if it exists. Unfortunately, we have no general method to know a priori whether C exists. If it does not, Angluin's algorithm will run forever. If it does halt, however, we are guaranteed that the computed automaton C is correct and that the running time is bounded by a polynomial in the size of B and the size of a minimal automaton for C.

In some cases Walnut can handle those sequences b that take negative integer values. For example, this is true for k-automatic sequences, which we can accept using an automaton defined over base $-k$, and for Fibonacci-automatic sequences, which we can accept using an automaton in the negaFibonacci system. In some cases, then we can compute automata for partial sums $\sum_{0 \leq i < n} b(i)$ for *integer* sequences, not just sequences of natural numbers. For more details about negative numbers in `Walnut`, see [19].

One of the most useful examples of this technique is the symbol-counting predicate $\text{Count}_a(n, x)$ which is true if and only if $x = |\mathbf{X}[0..n-1]|_a$, the number of occurrences of the symbol a in the length-n prefix of $\mathbf{X}$. If Angluin's algorithm produces an automaton for $\text{Count}_a(n, x)$, then we can also compute $|\mathbf{X}[i..i+j-1]|_a$ via subtraction. This is essential in understanding the abelian properties of an automatic sequence, such as the presence or absence of abelian powers. (Recall that a factor x is an abelian k'th power if $x = x_1 x_2 \cdots x_k$, with each x_i a permutation of x_1.)

We implemented Angluin's algorithm to successfully deduce the automaton for the partial sums of the Thue-Morse word, Fibonacci word, and Tribonacci word; see Table 2 for a summary of the computation. We also successfully deduced the rarefied sum automaton from [18], which accepts n represented in base 4 and $\sum_{0 \leq i < n} (-1)^{t(3i)}$ represented in base 3, where $t(i)$ is the i'th bit of the Thue-Morse sequence.

9   Algorithm Engineering Issues

9.1   Membership Queries

In practice (as opposed to theory), there are various algorithmic engineering techniques to speed up the approach we have outlined in this paper. For example, for the equality of factors predicate, membership queries involve testing whether $\mathbf{X}[i..i+n-1] = \mathbf{X}[j..j+n-1]$ for some specific triple $(i, j, n) = (i_0, j_0, n_0)$. There are three ways to evaluate this, as follows:

First, we can evaluate this predicate directly from the automaton for $\mathbf{X}$. In general, this uses $\Theta(n)$ operations on integers of magnitude $\Theta(\max(i, j, n))$, which is clearly not polynomial time in $\log n$. Nevertheless, in practice this is likely to be extremely fast, so if n_0 is small (say $n_0 < 10^7$) then we can evaluate the query more efficiently in this way.

Second, we can evaluate this predicate by substituting the specific values $(i, j, n) = (i_0, j_0, n_0)$ in the predicate and evaluating the resulting predicate in `Walnut`. This could end up taking even more time, asymptotically, since it involves creating intermediate automata computing the functions $t \to \mathbf{X}[i_0 + t]$ and $t \to \mathbf{X}[j_0 + t]$, which, when combined by $\mathbf{X}[i_0 + t] = \mathbf{X}[j_0 + t]$, could theoretically result in an automaton of size $\Theta(\max(i_0, j_0)^2)$. There is also another practical limitation here: in its current implementation, `Walnut` cannot process queries involving integers larger than $2^{31} - 1$. So it seems that this approach is unlikely to be competitive in practice with the first approach, except perhaps if i_0, j_0 are small (say $< 10^3$) and n_0 is large (say $10^7 < n_0 < 2^{31}$).

The third approach, previously outlined in Sect. 4.2, uses `Walnut` to build the "equality of factors" automaton without quantifiers once, and then intersects it with small automata that accept only the specific values of the tuples we wish to test. This can be done in polynomial time.

Similarly, computing partial sums $\sum_{0\leq i<n} b(i)$ of a synchronized sequence $(b(i))$ can be done either directly in $O(n)$ time, or via linear representations with $\log n$ multiplications of a vector times an $r \times r$ matrix, where r is the rank of the linear representation. This costs $\Theta(r^2 \log n)$. For small n the direct method will beat the linear representation, so some software engineering is needed to make sure membership queries are as efficient as possible.

Another way to compute the linear representation for the partial summation is as follows. If $b(n,x)$ is the synchronized automaton for b, then the `Walnut` command

```
def sumb "Ex $b(n,x) & i<n & j<x":
```

will directly compute the linear representation (as a `Maple` file) for $c(n) = \sum_{0\leq i<n} b(i)$. Unlike the method of the previous paragraph, because of the presence of the $\exists$ quantifier, it is conceivable that this could result in exponential blowup (in the size of the automaton for b).

9.2 Optimizing Performance

A primary bottleneck in our implementation of Angluin's algorithm is the consistency test. Aside from the fact that every table entry needs to be queried once to be filled, the check for consistency requires querying strings that are not even in the table. In the worst case, to check for consistency we must perform $\Theta(|S| \cdot \Sigma \cdot |E|)$ membership queries. However, many membership queries could actually be looking at the same string but split at different points into S and E. Thus, we cached query results into a hash table. This simple optimization technique reduced the total runtime of the algorithm on the equality of factors predicate for the Thue-Morse sequence from 1672.2 s to 124.6 s. Other runtime statistics, after using a hash table, are shown for different words in Table 1. The execution times in the table are measured in seconds.

9.3 Walnut Considerations

Several engineering details related to Walnut, and outside the core algorithm, can affect performance. First, reusing a single `Walnut` process throughout the run avoids repeated startup overhead. Second, we found `Walnut`'s built-in BFS in the `test` command to be inefficient. So, instead we used Python implemented a simple BFS for finding shortest counterexamples. Also, as mentioned in Sect. 4.2, to ensure each membership test remains polynomial-time, we rewrote key predicates in order to avoid `Walnut` running into exponential behavior.

We also faced a problem when trying to reject invalid representations. For example, we need to use `?msd_fib` to let `Walnut` know that we are currently using the Fibonacci numeration system. However, in the case of the Fibonacci word if we write all of the counterexample predicates using `?msd_fib`, we won't

Table 1. Performance of the EqFac predicate on four automatic sequences.

EqFac				
Metric	Thue-Morse	Baum-Sweet	Fibonacci	Tribonacci
Substitution method time	89.8	4402	323.1	2046.4
Intersection method time	439.5	20230.3	296.1	2687.1
# of unique queries	1672	75243	1032	4816
# of incorrect hypotheses	7	43	6	11
Longest counterexample	4	8	3	7
Longest queried string	8	15	6	11
Final # of states	15	130	12	27
Final $|S|$	26	210	16	40
Final $|E|$	9	51	9	17

Table 2. Performance of the partial and rarefied sum predicates on different automatic sequences.

Partial sums				Rarefied sums [18]
Metric	Thue-Morse	Fibonacci	Tribonacci	Thue-Morse
Total time (in seconds)	1.4	32.2	3243.4	6156.2
# of unique queries	132	146	12932	3548
# of incorrect hypotheses	3	3	23	9
Longest counterexample	3	4	11	4
Longest queried string	6	7	18	9
Final # of states	7	7	89	17
Final $|S|$	8	11	133	29
Final $|E|$	5	4	32	11

actually be able to reject the invalid representations. This is due to the fact that `Walnut` automatically does not accept these invalid representations, so they will never be suggested as counterexamples as they are not part of the test. To work around that, we added one predicate at the beginning for rejecting invalid representations and used `?msd_2` instead. This technique is needed no matter what numeration system we are dealing with. The reason is that `Walnut` is able to consider all binary sequences when using `?msd_2`, so we are able to isolate the invalid ones in this separate predicate.

Finally, a crucial consideration is the ordering of the counterexample predicates. Since the hypothesis tests are essentially doing an induction, we must ensure that the already accepted strings are in fact correct. Otherwise, we will be enforcing the algorithm to accept even more incorrect strings instead of rejecting them. For instance, the last hypothesis query mentioned in 4.2, was not written as a single test when programmed. Instead, each conjunction was used as a separate test and their orders were reversed. By using the third conjunction first, we ensure that we reject incorrect strings before building upon them as in the second conjunction.

Acknowledgments. We are grateful to three referees for their helpful comments.

References

1. Allouche, J.P., Shallit, J.: Automatic Sequences: Theory, Applications. Cambridge University Press, Generalizations (2003)
2. Angluin, D.: Learning regular sets from examples and counterexamples. Info. Comput. **75**, 87–106 (1987). https://doi.org/10.1016/0890-5401(87)90052-6
3. Barcucci, E., Bélanger, L., Brlek, S.: On Tribonacci sequences. Fibonacci Quart. **42**, 314–319 (2004). https://doi.org/10.1080/00150517.2004.12428402
4. Bruyère, V., Hansel, G.: Bertrand numeration systems and recognizability. Theoret. Comput. Sci. **181**, 17–43 (1997). https://doi.org/10.1016/S0304-3975(96)00260-5
5. Bruyère, V., Hansel, G., Michaux, C., Villemaire, R.: Logic and p-recognizable sets of integers. Bull. Belgian Math. Soc. **1**, 191–238 (1994). https://doi.org/10.36045/bbms/1103408547, corrigendum: vol. 1 (1994), 577
6. Büchi, J.R.: Weak second-order arithmetic and finite automata. Zeitschrift für mathematische Logik und Grundlagen der Mathematik **6**, 66–92 (1960), reprinted in Mac Lane, S., and Siefkes, D., (eds.), The Collected Works of J. Richard Büchi, Springer-Verlag, pp. 398–424 (1990)
7. Carton, O., Couvreur, J.M., Delacourt, M., Ollinger, N.: Linear recurrence sequence automata and the addition of abstract numeration systems. ArXiv preprint arXiv:2406.09868 [cs.FL] (2025), https://arxiv.org/abs/2406.09868
8. Frougny, C.: Representations of numbers and finite automata. Math. Syst. Theory **25**, 37–60 (1992). https://doi.org/10.1007/BF01368783
9. Frougny, C., Solomyak, B.: On representation of integers in linear numeration systems. In: Pollicott, M., Schmidt, K. (eds.) Ergodic Theory of $\mathbb{Z}^d$ Actions (Warwick, 1993–1994), London Mathematical Society Lecture Note Series, vol. 228, pp. 345–368. Cambridge University Press (1996)
10. Hopcroft, J.E., Ullman, J.D.: Introduction to Automata Theory, Languages, and Computation. Addison-Wesley (1979)
11. Mignoty, B., Renard, A., Rigo, M., Whiteland, M.A.: Automatic proofs in combinatorial game theory. Orbi reprint (2024). https://orbi.uliege.be/handle/2268/323845
12. Mousavi, H.: Automatic theorem proving in `Walnut`. ArXiv preprint arXiv:1603.06017 [cs.FL] (2016), http://arxiv.org/abs/1603.06017
13. Mousavi, H., Schaeffer, L., Shallit, J.: Decision algorithms for Fibonacci-automatic words, I: basic results. RAIRO Theoret. Inform. Appl. **50**, 39–66 (2016). https://doi.org/10.1051/ita/2016010
14. Rampersad, N., Shallit, J.: Rudin-Shapiro sums via automata theory and logic. Theor. Comput. Sci. **69**(9) (2025). https://doi.org/10.1007/s00224-025-10214-1
15. Schaeffer, L.: Deciding Properties of Automatic Sequences. Master's thesis, University of Waterloo (2013). available at https://cs.uwaterloo.ca/~shallit/thesisLukeSept4.pdf

16. Shallit, J.: Synchronized Sequences. In: Lecroq, T., Puzynina, S. (eds.) WORDS 2021. LNCS, vol. 12847, pp. 1–19. Springer, Cham (2021). https://doi.org/10.1007/978-3-030-85088-3_1

17. Shallit, J.: The Logical Approach To Automatic Sequences: Exploring Combinatorics on Words with Walnut, London Mathematical Society Lecture Note Series, vol. 482. Cambridge University Press (2023)

18. Shallit, J.: Rarefied Thue-Morse sums via automata theory and logic. J. Number Theory **257**, 98–111 (2024). https://doi.org/10.1016/j.jnt.2023.10.015

19. Shallit, J., Shan, S.L., Yang, K.H.: Automatic sequences in negative bases and proofs of some conjectures of Shevelev. RAIRO Theoret. Inform. Appl. **57**, Paper 4 (2023). https://doi.org/10.1051/ita/2022011

20. Shallit, J., Stipulanti, M.: Algorithms for linear representations of k-regular sequences (2025). manuscript in preparation

21. Tan, B., Wen, Z.Y.: Some properties of the Tribonacci sequence. Europ. J. Combin. **28**, 1703–1719 (2007). https://doi.org/10.1016/j.ejc.2006.10.014

State-Freezing Pushdown Automata

Martin Kutrib⬤, Andreas Malcher$^{(\boxtimes)}$⬤, and Priscilla Raucci⬤

Institut für Informatik, Universität Giessen, Arndtstr. 2, 35392 Giessen, Germany
{kutrib,andreas.malcher}@informatik.uni-giessen.de,
priscilla.raucci@uni-giessen.de

Abstract. Deterministic pushdown automata as well as deterministic one-counter automata with the restriction that their underlying state graph is quasi-acyclic are investigated. Here, quasi-acyclicity means that the graph is allowed to have self-loops, but removing the self-loops leads to an acyclic graph. Such automata are called state-freezing, since a state left can never be entered again. It turns out that the state-freezing property is a strong property, since the considered automata with the state-freezing property are less powerful than in the general case. This result holds in the real-time case, i.e., λ-moves are not allowed, as well as in the case of allowed λ-moves. We study the closure properties of the corresponding language families and can show that all families are anti-AFLs, whereas all families are closed under complementation. Finally, we look at decidability questions and obtain the undecidability of the inclusion problem already in the case of state-freezing real-time deterministic pushdown automata. However, the status of the inclusion problem in the case of state-freezing deterministic one-counter automata remains open.

Keywords: Deterministic pushdown automata · Deterministic counter automata · Quasi-acyclic state graphs · State-Freezing · Computational capacity · Closure Properties · Decidability problems

1 Introduction

In the field of automata theory, the study of variations of pushdown automata (PDA) is a broad and prolific area of research. Of particular interest are those variations that impose restrictions on different components of the model with the goal to obtain better results for the restricted model in terms of, for example, faster parsing algorithms, better positive closure properties, or undecidable questions becoming decidable in the restricted model. Some of such restricted models studied in the past are, for instance, *deterministic* pushdown automata (DPDA), deterministic *one-counter* automata [15] restricting the pushdown memory of the device, or *input-driven* pushdown automata [13] and *reversible* pushdown automata [11] constraining the behavior of the finite-state control. Each of these changes, structural or behavioral, affects the computational power, the closure properties, and the decision problems of the corresponding devices.

ⓒ The Author(s), under exclusive license to Springer Nature Switzerland AG 2026
G. Castiglione and S. Mantaci (Eds.): CIAA 2025, LNCS 15981, pp. 252–266, 2026.
https://doi.org/10.1007/978-3-032-02602-6_18

In this paper, we introduce a new variant of pushdown automata where the internal hardware of the devices is not changed, but some limitations are imposed on the transition function in such a way that the underlying state-graph is acyclic with the only exception of self-loops. As a result, whenever such devices leave a state, this state can be never re-entered in the same computation. Hence, the state is somehow blocked or frozen and we speak of *state-freezing* PDAs. Acyclic graphs with additional possible self-loops are also called *quasi-acyclic* in the literature. In the field of cellular automata, quasi-acyclic or freezing cellular automata have been introduced in [2,4], where the state of a cell can only increase according to some order on states which is equivalent to a quasi-acyclic state-graph of each cell. Moreover, such cellular automata turned out to be equivalent to cellular automata with *finite communication* introduced in [10,16]. In such cellular automata each two neighboring cells can exchange information only a finite number of times. Another reference for deterministic quasi-acyclic finite automata is the paper [1] on quantum finite automata, where such automata are called automata with *finite variation*.

Here, we start to investigate the property of being state-freezing for pushdown automata. First, we note that it is a well-known fact that any context-free language can be accepted by a nondeterministic pushdown automaton with a single state (see, for example, [8]). Hence, every context-free language can be accepted by a state-freezing PDA. This is no longer true for deterministic context-free languages, since a family of languages L_n is presented in [6] such that every deterministic pushdown automaton accepting L_n needs exactly n states. In this paper, we will therefore investigate deterministic state-freezing pushdown automata (stfDPDA). In addition, we will study also real-time deterministic state-freezing pushdown automata (stfDPDA$_{\lambda\text{-free}}$) which are stfDPDA where no λ-moves are allowed. While every nondeterministic pushdown automaton can be converted into an equivalent real-time pushdown automaton, this result is no longer true in the deterministic case and we will show that this holds in the state-freezing case as well. If the pushdown alphabet of a PDA is reduced to one, the pushdown store is reduced to a counter and we obtain one-counter automata. Thus, we extend our investigations on the state-freezing property also to state-freezing deterministic counter automata under real-time conditions (stfDCA$_{\lambda\text{-free}}$) and with allowed λ-moves (stfDCA), and we obtain similar results as in the case of an unrestricted pushdown. We also investigate the usually studied closure properties for the four introduced state-freezing language families together with the decidability questions for the corresponding automata.

The paper is organized as follows. After giving some necessary definitions and illustrating examples in the next section, we study in Sect. 3 the computational capacity of the four models introduced. It turns out that all four models are less powerful with the state-freezing property than without. Moreover, the real-time models are less powerful than the general models even with the state-freezing property. We also show that any stfDCA can be simulated by an stfDPDA$_{\lambda\text{-free}}$, that is, λ-moves in a one-counter automaton can be compensated by a pushdown store under state-freezing conditions. Finally, for regular languages we obtain the interesting result that stfDPDA$_{\lambda\text{-free}}$ can accept all regular languages, whereas this is not possible for any stfDCA even if they are not restricted to work in real time.

In Sect. 4, we analyze the closure (and non-closure) properties of the language families introduced and we obtain for all families that they are antiAFLs, i.e., they are not closed under union, concatenation, iteration, intersection with regular languages, homomorphism, and inverse homomorphism. On the other hand, all families are closed under complementation and almost all families are closed under marked concatenation. Finally, we address in Sect. 5 the decidability status of the inclusion problem for state-freezing DPDAs and we can show that it is already undecidable for stfDPDA$_{\lambda\text{-free}}$, but it remains open for stfDCA.

2   Preliminaries

We denote by Σ^* the set of all words on the finite alphabet Σ, including the empty word λ, and let $\Sigma^+ = \Sigma^* \setminus \{\lambda\}$. For any word $w \in \Sigma^*$, we let $|w|$ denote its length, w^R its reversal, and $|w|_a$ the number of occurrences of the symbol $a \in \Sigma$ in w. We use $\subseteq$ for inclusions, and $\subset$ for proper inclusion. Given a set S, we denote by 2^S its power set, by $|S|$ its cardinality, and by S_x the set $S \cup \{x\}$, for a given $x \notin S$. A language on Σ is any subset $L \subseteq \Sigma^*$. The complement of L is the language $\overline{L} = \Sigma^* \setminus L$, its reversal is $L^R = \{ w^R \mid w \in L \}$. Two language families $\mathscr{L}_1$ and $\mathscr{L}_2$ are said to be incomparable whenever $\mathscr{L}_1$ is not a subset of $\mathscr{L}_2$ and vice versa.

A *deterministic pushdown automaton* (DPDA) is defined as a system $M = \langle Q, \Sigma, \Gamma, \delta, q_0, \triangleleft, \bot, F \rangle$, where Q is a finite set of states, Σ is the finite input alphabet, Γ is a finite pushdown alphabet, $q_0 \in Q$ is the initial state, $\triangleleft \notin \Sigma$ is the right *endmarker*, $\bot \notin \Gamma$ is the *bottom-of-pushdown symbol*, which initially appears on the pushdown store, $F \subseteq Q$ is the set of accepting states, and δ is a partial mapping from $Q \times (\Sigma \cup \{\lambda, \triangleleft\}) \times \Gamma_\bot$ to $Q \times \Gamma_\bot^*$ called the transition function. There must never be a choice of using an input symbol or of using λ input. So, it is required that for all q in Q and Z in $\Gamma_\bot$: if $\delta(q, \lambda, Z)$ is defined, then $\delta(q, a, Z)$ is undefined for all a in $\Sigma_\triangleleft$. In addition, to simplify matters, we require that along computations the bottom-of-pushdown symbol appears exactly once at the bottom of the pushdown store, formally: if $\delta(q, a, Z) = (p, \beta)$ then either $Z \neq \bot$ and $\beta \in \Gamma^*$, or $Z = \bot$ and $\beta = \beta' \bot$ for $\beta' \in \Gamma^*$.

A *configuration* of a DPDA is a triple $(q, w\triangleleft, \gamma)$, where q is the current state, $w\triangleleft$ is the unread part of the input, and $\gamma \in \Gamma^* \bot$ denotes the current content of the pushdown store, the leftmost symbol of γ being the top symbol. On input w the initial configuration is defined to be $(q_0, w\triangleleft, \bot)$. To define the *successor configuration* of a configuration we differentiate two cases. First, for $q \in Q$, $a \in \Sigma_\lambda$, $w \in \Sigma^*$, $\gamma \in \Gamma_\bot^*$, and $Z \in \Gamma_\bot$, let $(q, aw\triangleleft, Z\gamma)$ be a configuration. Then its successor configuration is $(p, w\triangleleft, \beta\gamma)$, where $\delta(q, a, Z) = (p, \beta)$. We write $(q, aw\triangleleft, Z\gamma) \vdash (p, w\triangleleft, \beta\gamma)$ in this case. Note that this case includes possible λ-steps also on the endmarker. Second, the successor configuration of $(q, \triangleleft, Z\gamma)$ is $(p, \lambda, \beta\gamma)$, where $\delta(q, \triangleleft, Z) = (p, \beta)$, and the computation halts. The reflexive transitive closure of $\vdash$ is denoted by $\vdash^*$. The *language accepted* by M is

$$L(M) = \{\, w \in \Sigma^* \mid (q_0, w\triangleleft, \bot) \vdash^* (q, \lambda, \gamma), \text{ for some } q \in F \text{ and } \gamma \in \Gamma_\bot^* \,\}.$$

In general, the family of all languages that are accepted by some device X is denoted by $\mathscr{L}(X)$. The language family $\mathscr{L}(\mathsf{DPDA})$ is exactly the family of deterministic context-free languages (DCFL). It is well known that deterministic pushdown automata that are not allowed to perform λ-steps are weaker than DPDAs that may move on λ input [6]. Such automata are called *real-time* DPDAs and the corresponding language family is denoted by $\mathsf{DCFL}_{\lambda\text{-free}}$.

We will also consider *deterministic counter automata* (DCA), which are defined as DPDAs whose pushdown alphabet is restricted to be a singleton, and their real-time variant. The corresponding language families are denoted by DCL and $\mathsf{DCL}_{\lambda\text{-free}}$, respectively.

In this paper, we will investigate *state-freezing* pushdown automata which are defined as follows. We consider the state-graph G_M of a given DPDA $M = \langle Q, \Sigma, \Gamma, \delta, q_0, \triangleleft, \bot, F \rangle$ whose vertices are the state set Q. Moreover, there is an edge between p and q labeled with (a, Z), if there is a transition $\delta(p, a, Z) = (q, \beta)$ in M for $a \in \Sigma_\lambda \cup \{\triangleleft\}$, $Z \in \Gamma_\bot$, and $\beta \in \Gamma_\bot^*$. We say that a DPDA M is *state-freezing* if its underlying state-graph G_M is quasi-acyclic, which means that the only loops occurring are self-loops. It is an obvious consequence for state-freezing pushdown automata that whenever such an automaton leaves some state, then this state must never be entered again. In the following, we will denote the state-freezing variant of DPDAs and DCAs as well as their real-time counterparts and the corresponding language families with the prefix stf.

The following examples clarify the behavior of stfDPDAs.

Example 1. The language $L_1 = \{\, wcw^R \mid w \in \{a, b\}^* \,\}$ is accepted by a real-time stfDPDA. We can in fact easily describe a DPDA accepting L_1, whose computation is divided in two phases: first, w is read and its sequence of characters is stored in the pushdown. Then, the scanning of c marks the beginning of the second phase of the computation, where the machine checks whether the second part of the string matches the content of the pushdown. The string is accepted only if the pushdown is empty. To accomplish this, it will be necessary to have three states (one for reading w, one for checking the match of w^R and one for the acceptance) and it is straightforward to see that, in any point of the computation, the machine can either loop within the current state or move up to the next one, without possibility to re-enter in some of the previous states. Also, there are no λ-transitions, this proving that the described machine is a real-time stfDPDA.   ∎

Example 2. The language $L_2 = \{\, a^m b^n c^k d^k \mid m \geq n \geq 1 \text{ and } k \geq 0 \,\}$ is accepted by an stfDCA. Here, the basic idea is to use the counter to keep track of the parameters m, n and k. First, the counter is incremented for a's, then it is decremented for b's. When the next input symbol is the right endmarker, the input is accepted. Otherwise, when the first c is read, the counter is decremented to zero using λ-moves. Finally, the counter is incremented for the remaining c's and subsequently decremented for d's. The input is accepted reading the right endmarker and having the counter to zero. For what regards states, we will need six of them, one for each block of the input strings, plus one for the λ-loop, plus one checking the final acceptance of the input. According to the behavior of the automaton in any point of the computation, the machine can either loop within the current state or move up to the next one, and there is no way to return to a previous state.   ∎

3   Relationships

In this section, we clarify the relationship between the language families introduced and their general, not necessarily state-freezing counterpart. We use language $L_1 = \{\, wcw^R \mid w \in \{a,b\}^* \,\}$ from Example 1 to show that even in the real-time and state-freezing case a pushdown store is more powerful than a counter even if real-time and state-freezing is not required.

Lemma 3. *Language L_1 belongs to $\mathscr{L}(\mathsf{stfDPDA}_{\lambda\text{-free}})$, but L_1 is not accepted by any DCA.*

Proof. It is shown in Example 1 that L_1 belongs to $\mathscr{L}(\mathsf{stfDPDA}_{\lambda\text{-free}})$.

It remains to be argued towards the non-acceptance of L_1 by one-counter automata. It has been proved in [3] that any one-way counter automaton having k counters and accepting L_1 needs as computing time at least $2^{n/2k}$. But the special case $k = 1$ is not treated separately. In [5] it is already observed without obvious proof that L_1 cannot be accepted by any DCA. For the sake of completeness, here we present a proof.

If a given DCA M with state set Q performs more than $2|Q|$ consecutive λ-steps without decreasing its counter, it runs into an infinite loop. So, in any accepting computation, the maximal value of the counter is bounded from above by $c \cdot n$, for a constant c and input length n. Therefore, after having processed $w \in \{a,b\}^x$, M can be in one out of $|Q| \cdot c \cdot x$ different configurations. However, for x large enough, there are 2^x many different words in $\{a,b\}^x$. Thus, there exist two words $v, w \in \{a,b\}^x$ that drive M into the same configuration. Therefore, since wcw^R is accepted, the input wcv^R is accepted as well, a contradiction. □

Next, language L_2 from Example 2 shows that real-time DCA are less powerful than DCA in the general as well as in the state-freezing case.

Lemma 4. *Language L_2 belongs to the family $\mathscr{L}(\mathsf{stfDCA})$, but it is not accepted by any $\mathsf{DCA}_{\lambda\text{-free}}$.*

Proof. Language L_2 is accepted by an stfDCA due to Example 2. Now, we assume in contrast to the assertion that L_2 is accepted by some $\mathsf{DCA}_{\lambda\text{-free}}$ M. We consider an accepting computation on input $a^n bc^k d^k$ where n and k are large enough. In addition, n is very large in comparison with k. On computing the prefix $a^n b$, we get the configuration $(q_0, a^n bc^k d^k \triangleleft, \bot) \vdash^* (q, c^k d^k \triangleleft, Z^m \bot)$, since we may safely assume that the counter is increased while reading a's. Hence, the length m of the counter is in the order of n and, thus, very large in comparison with k. On computing the factor c^k, M may first enter a sequence of $t_1 \geq 0$ states, in which the counter is incremented or decremented, say its value is altered by $\ell_1 \geq 0$ or $\ell_1 \leq 0$, but eventually enters a loop of length $t_2 \geq 1$ in which the counter is incremented or decremented, say its value is altered by $\ell_2 \geq 0$ or $\ell_2 \leq 0$. Thus, we have

$$(q, c^k d^k \triangleleft, Z^m \perp) \vdash^* (p, c^{k-t_1-t_2} d^k \triangleleft, Z^{m+\ell_1+\ell_2} \perp) \vdash^* (q_f, \lambda, Z^i \perp)$$

with $q_f \in F$ and $i > 0$. But then we have also

$$(q, c^{k+t_2} d^k \triangleleft, Z^m \perp) \vdash^* (p, c^{k-t_1} d^k \triangleleft, Z^{m+\ell_1+\ell_2} \perp)$$
$$\vdash^* (p, c^{k-t_1-t_2} d^k \triangleleft, Z^{m+\ell_1+2\ell_2} \perp) \vdash^* (q_f, \lambda, Z^j \perp)$$

with $q_f \in F$ and $j > 0$. Hence, also $a^n b c^{k+t_2} d^k$ is accepted which is a contradiction. $\square$

The following result is similar to Lemma 4 and shows that real-time DPDA are less powerful than DPDA in the general as well as in the state-freezing case.

Lemma 5. *Language* $L_3 = \{ a^m b^n c^k a^m \mid m \geq 1 \text{ and } n \geq k \geq 1 \}$ *belongs to* $\mathscr{L}(\text{stfDPDA})$, *but* L_3 *is not accepted by any* $\text{DPDA}_{\lambda\text{-free}}$.

Proof. It is shown in [9] that L_3 is not accepted by any $\text{DPDA}_{\lambda\text{-free}}$. $\square$

The next lemma shows that all four variants of state-freezing automata are less powerful than their general, not necessarily state-freezing counterpart.

Lemma 6. *Language* $L_4 = \{ a^n b^n c \mid n \geq 1 \}^*$ *belongs to* $\mathscr{L}(\text{DCA}_{\lambda\text{-free}})$, *but it is not accepted by any* stfDPDA.

In the following lemma we show in a constructive way how a state-freezing DCA with possible λ-moves can be simulated by an equivalent state-freezing DPDA under real-time conditions. The result of Lemma 3 then gives that the inclusion shown in Lemma 7 is proper.

Lemma 7. *The family* $\mathscr{L}(\text{stfDCA})$ *is properly included in* $\mathscr{L}(\text{stfDPDA}_{\lambda\text{-free}})$.

Proof. We sketch the construction and note that the basic task is to handle λ-moves in an stfDCA M in an equivalent $\text{stfDPDA}_{\lambda\text{-free}}$ M' without λ-moves. First, we discuss the case when the given stfDCA has an infinite loop of λ-moves. Such an infinite loop can be increasing or decreasing. In the former case, the loop can never be left. Hence, we modify the transition function of the given stfDCA such that the transition remains undefined instead of entering an infinite increasing λ-loop. In the latter case, the λ-moves decrement the counter until it is decreased to zero. Now, the idea for an equivalent simulating $\text{stfDPDA}_{\lambda\text{-free}}$ is to push a new stack symbol which acts as bottom-of-counter symbol and to continue the computation with an "empty" counter instead of entering an infinite decreasing λ-loop.

It remains for us to handle the case of a constant number of consecutive λ-moves occuring in computations. Let m be the maximal number of consecutive λ-moves in M. Hence, the counter can maximally be decreased or increased by m in consecutive λ-moves. In addition, all states occurring in these sequences are pairwise different. The basic idea for the simulation by an $\mathsf{stfDPDA}_{\lambda\text{-free}}$ M' is to consider the next input symbol after consecutive λ-moves and to simulate at most $m + 1$ moves in M by one move in M' and to consider groups of at most $m + 1$ counter symbols of M by one pushdown symbol in M'.

To be a bit more formal, let us denote the pushdown symbols in M by $+$ and $\perp$ and in M' by $(1), (2), \ldots, (m + 1)$ together with $\perp$. For $a \in \Sigma$ and $Z \in \{+, \perp\}$, let p and q, respectively, be the start state and the end state, respectively, of a computation in M consisting of consecutive λ-moves followed by an a-move in which the counter is changed by n with $-(m+1) \leq n \leq m+1$.

If $n \geq 0$, we then define $\delta'(p, a, (\ell)) = (q, (\ell + n))$ if $\ell + n \leq m + 1$ and $\delta'(p, a, (\ell)) = (q, (\ell + n - m - 1)(m + 1))$ if $\ell + n > m + 1$. In this case, the definition of δ' does not introduce loops in the state-graph of M'. Hence, M' is still state-freezing.

If $n < 0$, we define $\delta'(p, a, (\ell)) = (q, (\ell + n))$ if $\ell + n > 0$, $\delta'(p, a, (\ell)) = (q, \lambda)$ if $\ell + n = 0$, and $\delta'(p, a, (\ell)) = ([q, \ell + n], \lambda)$ if $\ell + n < 0$. In the latter case, we introduced a new type of states where we store the number to be decreased from the counter in the second component of the state. In the following, M' works on these states taking into account the second component and charging this number, if possible, with the number seen on the pushdown in the next computation steps. It is clear that this number is only increased until it reaches the value 0. This behavior of M' does at most introduce self-loops in the state-graph of M'. Hence, M' is still state-freezing.

However, there might be a problem if, for example, another sequence of consecutive λ-moves has to be simulated next, before the second component introduced for simulating the first sequence of λ-moves has been counted to 0. But this is in fact no problem: if the next sequence of consecutive λ-moves plus input symbol increases the counter, the second component can be charged with the increasing of the counter. If the next sequence of consecutive λ-moves plus input symbol decreases the counter by, say $k \leq m + 1$, and the current second component of the state is $-m \leq k' \leq -1$, the topmost pushdown symbol is popped and the new second component of the state is removed, if $k' + m + 1 - k \geq 0$, or is set to $k' + m + 1 - k$ otherwise. Observe that $-m \leq k' + m + 1 - k \leq -1$ in the latter case. Since all states occurring in consecutive λ-moves are pairwise different, the resulting automaton is still state-freezing. This behavior works fine as long as the pushdown store does not get empty. In this case, the input may be rejected if the simulated counter machine tries to decrease the empty counter or the sequence of consecutive λ-moves has to be left since the simulated counter has been decreased to zero. This problem can finally be overcome by marking the first symbol pushed by M' after the empty pushdown store suitably. If

M' sees such a marked symbol as topmost pushdown symbol, M' knows that the pushdown store below is then empty and M' can then decide to reject the input or to continue the simulation accordingly. Again, this behavior does not introduce self-loops in the state-graph of M'. Altogether, we have that M' is a state-freezing real-time DPDA. □

The next lemma clarifies the relation to the regular languages.

Lemma 8.   *1. $REG \subset \mathscr{L}(stfDPDA_{\lambda\text{-}free})$.*
2. The language families $\mathscr{L}(stfDCA_{\lambda\text{-}free})$ and REG are incomparable.
3. The language families $\mathscr{L}(stfDCA)$ and REG are incomparable.

Finally, we have the following incomparability results. All inclusions and incomparabilities obtained in this section are summarized in Fig. 1.

Lemma 9.

1. *The language families $\mathscr{L}(stfDCA)$ and $\mathscr{L}(DCA_{\lambda\text{-}free})$ are incomparable.*
2. *The language families $\mathscr{L}(stfDPDA)$ and $\mathscr{L}(DPDA_{\lambda\text{-}free})$ are incomparable.*
3. *The language families $\mathscr{L}(stfDPDA)$ and $\mathscr{L}(DCA_{\lambda\text{-}free})$ are incomparable.*

Proof. By Lemma 4 we know that $L_2 = \{ a^m b^n c^k d^k \mid m \geq n \geq 1 \text{ and } k \geq 0 \}$ belongs to $\mathscr{L}(stfDCA)$, but is not accepted by any $DCA_{\lambda\text{-}free}$. By Lemma 5 we know that language $L_3 = \{ a^m b^n c^k a^m \mid m \geq 1 \text{ and } n \geq k \geq 1 \}$ belongs to $\mathscr{L}(stfDPDA)$, but is not accepted by any $DPDA_{\lambda\text{-}free}$. Hence, it is not accepted by any $DCA_{\lambda\text{-}free}$ as well. On the other hand, according to Lemma 6 language $L_4 = \{ a^n b^n c \mid n \geq 1 \}^*$ belongs to $\mathscr{L}(DCA_{\lambda\text{-}free}) \subset \mathscr{L}(DPDA_{\lambda\text{-}free})$, but not to $\mathscr{L}(stfDPDA)$ and, hence, not to $\mathscr{L}(stfDCA)$. This gives all incomparabilities claimed. □

4   Closure Properties

In this section, we investigate the closure properties of the state-freezing language families introduced. We start with the Boolean operations and obtain a positive result for complementation for all four families introduced. On the other hand, all families are not even closed under union or intersection with regular languages.

Lemma 10. *Let $X \in \{stfDCA, stfDCA_{\lambda\text{-}free}, stfDPDA, stfDPDA_{\lambda\text{-}free}\}$. Each language family $\mathscr{L}(X)$ is closed under complementation, but neither closed under union, nor under intersection, nor under union or intersection with regular languages.*

Proof. We first consider the closure under complementation for stfDPDA. The basic idea is to interchange accepting and non-accepting states. However, three problems may occur. First, the given stfDPDA may not read its input completely by either entering a configuration in which no next move is defined (1) or by

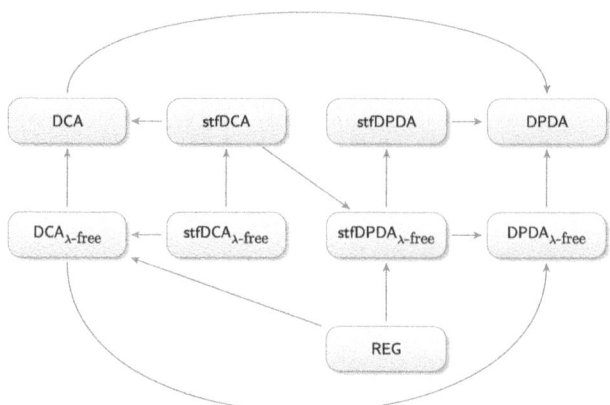

Fig. 1. Relationships between the language families discussed in this paper. An arrow $A \rightarrow B$ between families indicates a strict inclusion of $\mathscr{L}(A)$ in $\mathscr{L}(B)$. Families not connected by a path are incomparable. The proper inclusions $\mathsf{REG} \subset \mathscr{L}(\mathsf{DCA}_{\lambda\text{-free}})$, $\mathscr{L}(\mathsf{DCA}_{\lambda\text{-free}}) \subset \mathscr{L}(\mathsf{DPDA}_{\lambda\text{-free}})$, and $\mathscr{L}(\mathsf{DCA}) \subset \mathscr{L}(\mathsf{DPDA})$ are folklore.

entering an infinite λ-loop (2). Second, the given stfDPDA may perform λ-steps leading from an accepting state to a rejecting state and back (3). Since an input is only accepted after reading the endmarker and the computation is halting thereafter, we observe that problem (3) does not occur. To overcome problems (1) and (2) we consider the construction given in Lemma 10.3 in [8] which converts a given DPDA into an equivalent one that reads its input entirely. An inspection of the proof gives that the resulting DPDA is an stfDPDA if the given DPDA is an stfDPDA as well. Thus, for a given stfDPDA M we can effectively construct an stfDPDA accepting $\overline{L(M)}$ by interchanging the states obtained after reading the endmarker.

For $\mathsf{stfDCA}_{\lambda\text{-free}}$ and $\mathsf{stfDPDA}_{\lambda\text{-free}}$ problems (2) and (3) cannot occur. It is therefore sufficient for the transition functions to enter a state in which the remaining input is read, whenever an undefined transition occurs. This can clearly be realized with a state-freezing automaton and we obtain closure under complementation for the families $\mathscr{L}(\mathsf{stfDCA}_{\lambda\text{-free}})$ and $\mathscr{L}(\mathsf{stfDPDA}_{\lambda\text{-free}})$. Finally, we consider stfDCA and observe that problem (3) does not occur, whereas problem (1) can be solved in a way we have just discussed. It remains for us to consider infinite λ-loops. By inspection of the transition function we can find out whether an infinite λ-loop increases or decreases the counter. In the former case, we enter a state reading the remaining input as for problem (1), since an increasing infinite λ-loop can never be left. In the latter case, a decreasing infinite λ-loop eventually reaches an empty counter. If this transition is defined, no problem occurs. If the transition is undefined, the problem is of type (1) and has already been solved. Thus, the family $\mathscr{L}(\mathsf{stfDCA})$ is closed under complementation as well.

We are showing the non-closure under intersection. To this end, we consider the languages $\{\, a^n b^n c^m \mid n, m \geq 0 \,\}$ and $\{\, a^n b^m c^m \mid n, m \geq 0 \,\}$ which are both accepted by stfDCA$_{\lambda\text{-free}}$ in a straightforward way. Since the intersection of both languages is not even context free, we obtain that all language families are not closed under intersection. Since all language families are closed under complementation, we obtain the non-closure under union for all language families by De Morgan's law.

For the non-closure under intersection with regular languages, we consider the language $L = \{\, a^n b^n \mid n \geq 0 \,\} \sqcup \{c\}^* \in \mathscr{L}(\text{stfDPDA}_{\lambda\text{-free}})$ (cf. Lemma 6) and the regular language $R = \{\, a^n (b^k \sqcup c^m) \mid n, k \geq 0, m \equiv 0 \pmod 2 \,\}$. We have to show that $L \cap R = \{\, a^n (b^n \sqcup c^m) \mid n \geq 0, m \equiv 0 \pmod 2 \,\} \notin \mathscr{L}(\text{stfDPDA})$.

We argue similarly as in the proof of Lemma 6. Assume in contrast to the assertion that $L \cap R$ is accepted by some stfDPDA $M = \langle Q, \Sigma, \Gamma, \delta, q_0, \lhd, \bot, F \rangle$. During the computation of M on input prefixes a^+ no combination of state and content of the pushdown store may appear twice and, thus, the height of the pushdown increases arbitrarily. So, eventually M runs into a one-state loop while processing a's in which the sequence of symbols pushed becomes cyclic, say $Z_1, Z_2, \ldots, Z_k$, for some $k \geq 1$.

While M processes the input suffixes b^+, again, no combination of state and content of the pushdown store may appear twice and in any accepting computation the pushdown store has to be decreased until some symbol pushed before the Z_i-sequence appears. So, M runs into a one-state loop that decreases the height of the pushdown store while processing the b's.

Now we embed c's into the suffixes. Again, since M is state-freezing it will eventually run through a one-state loop, say p, while processing b's *and* c's.

Let M now be in state p with some Z_i on top of the pushdown. First, we consider the sub-computation on an input factor of the form b^+ until the next Z_i below the current Z_i appears at the top of the pushdown. Let M read the factor b^{ℓ_1}, for $\ell_1 \geq 1$, in this sub-computation. Second, we consider the sub-computation on an input factor of the form cb^* until the next Z_i below the current Z_i appears at the top of the pushdown. Let M read the factor cb^{ℓ_2}, for $\ell_2 \geq 0$, in this sub-computation. If $\ell_1 = \ell_2$ then we obtain a contradiction, since the factor b^{ℓ_1} can be replaced by cb^{ℓ_1} without changing the result of the computation. However, the c switches the parity of the number of c's in the suffix, a contradiction.

Next, we consider inputs of the form $a^n b^m (c b^{\ell_2} b^{\ell_1 \cdot \ell_2})^*$ that do belong to L, where n and m are large enough. If $\ell_1 > \ell_2$ then we replace all factors cb^{ℓ_2} by b^{ℓ_1} and conclude that the pushdown height at the end of the accepting computation is not decreased until some symbol pushed before the Z_i-sequence appears, a contradiction. Conversely, if $\ell_1 < \ell_2$ then we replace all factors $b^{\ell_1 \cdot \ell_2}$ by $(cb^{\ell_2})^{\ell_1}$ and conclude again that the pushdown height at the end of the accepting computation is not decreased until some symbol pushed before the Z_i-sequence appears, a contradiction as well.

So, we derive that all language families are not closed under intersection with regular languages. Again, their closure under complementation gives non-closure under union with regular languages. □

In contrast to the results obtained in Lemma 10 we obtain for all families the closure under union and intersection with regular languages under the condition that the given regular language can be described by a state-freezing DFA.

Lemma 11. *Let* $X \in \{\text{stfDCA}, \text{stfDCA}_{\lambda\text{-free}}, \text{stfDPDA}, \text{stfDPDA}_{\lambda\text{-free}}\}$. *Each language family* $\mathscr{L}(X)$ *is closed under union and intersection with regular languages if the regular language is accepted by a state-freezing* DFA.

Proof. Let $M = \langle Q, \Sigma, \Gamma, \delta, q_0, \triangleleft, \bot, F \rangle$ be an automaton of type X with $X \in \{\text{stfDCA}, \text{stfDCA}_{\lambda\text{-free}}, \text{stfDPDA}, \text{stfDPDA}_{\lambda\text{-free}}\}$ and $A = \langle Q', \Sigma \cup \{\triangleleft\}, \delta', q_0', F' \rangle$ be a state-freezing DFA, i.e., a DFA whose state-graph is quasi-acyclic. We tacitly assume that A handles the right endmarker suitably. Now, we apply the classical Cartesian product construction and define

$$M_A = \langle Q \times Q', \Sigma, \Gamma, \delta_A, (q_0, q_0'), \triangleleft, \bot, F \times F' \rangle$$

with $\delta_A((p,q), a, Z) = ((p', q'), \beta)$ if $\delta(p, a, Z) = (p', \beta)$ and $\delta'(q, a) = q'$ for $p, p' \in Q$, $q, q' \in Q'$, $a \in \Sigma_\lambda \cup \{\triangleleft\}$, $Z \in \Gamma_\bot$, and $\beta \in \Gamma_\bot^*$. It is clear that M_A is of type X if M is of type X. It remains to be argued that M_A is state-freezing. Let G be the state-graph of M_A. Since the state-graphs G_M and G_A are quasi-acyclic, G can only enter a loop when both components enter a loop. Since the latter can only enter self-loops, the loop entered in G is a self-loop as well. Therefore, G is quasi-acyclic and, hence, M_A is state-freezing. □

We continue by considering the operations concatenation, iteration, reversal, and length-preserving homomorphism and obtain that all families are not closed under these operations. This coincides with the results known for deterministic context-free languages.

Lemma 12. *Let* $X \in \{\text{stfDCA}, \text{stfDCA}_{\lambda\text{-free}}, \text{stfDPDA}, \text{stfDPDA}_{\lambda\text{-free}}\}$. *Each language family* $\mathscr{L}(X)$ *is not closed under concatenation, iteration, reversal, and length-preserving homomorphism.*

We now look at the operation marked concatenation and yield different results. We say that a language family $\mathscr{L}$ is closed under marked concatenation if for all $L_1, L_2 \in \mathscr{L}$ and a new symbol $\bullet$ holds $L_1 \bullet L_2 \in \mathscr{L}$. We obtain a positive result for (real-time) state-freezing DPDA and for (state-freezing) DCA as long as λ-moves are allowed, whereas we have a negative result for real-time (state-freezing) DCA.

Lemma 13. *The language families* $\mathscr{L}(\text{stfDCA})$, $\mathscr{L}(\text{stfDPDA})$, *and* $\mathscr{L}(\text{stfDPDA}_{\lambda\text{-free}})$ *are closed under marked concatenation. The language family* $\mathscr{L}(\text{stfDCA}_{\lambda\text{-free}})$ *is not closed under marked concatenation.*

Proof. The construction to show closure under marked concatenation for automata of type $X \in \{\text{stfDCA}, \text{stfDPDA}\}$ is straightforward and will be sketched only. The basic idea is first to simulate the first automaton and to enter a certain state q after reading the marking symbol #. In state q the automaton is emptying the counter respectively pushdown store using λ-moves. After reaching the $\perp$-symbol, the simulation of the second automaton is started.

The construction for $\text{stfDPDA}_{\lambda\text{-free}}$ is different. The basic idea here is first to simulate the first automaton and to push a copy $\perp'$ of the $\perp$-symbol on the pushdown store after reading the marking symbol #. Then, the simulation of the second automaton is started where the $\perp'$-symbol is interpreted as the original $\perp$-symbol.

For the non-closure under marked concatenation for $\mathscr{L}(\text{stfDCA}_{\lambda\text{-free}})$ we will consider the languages $\{\, a^m b^n \mid m \geq n \geq 1 \,\}$ and $\{\, c^k d^k \mid k \geq 0 \,\}$, where each of them is accepted by an $\text{stfDCA}_{\lambda\text{-free}}$. However, their marked concatenation gives the language $\{\, a^m b^n \# c^k d^k \mid m \geq n \geq 1 \text{ and } k \geq 0 \,\}$ which is basically language L_2 from Example 2. It can be shown almost literally as in the proof of Lemma 4 that the language is not accepted by any $\text{DCA}_{\lambda\text{-free}}$.   □

For the operation of marked iteration we yield non-closure for all four language families.

Lemma 14. *Let* $X \in \{\text{stfDCA}, \text{stfDCA}_{\lambda\text{-free}}, \text{stfDPDA}, \text{stfDPDA}_{\lambda\text{-free}}\}$. *Each language family* $\mathscr{L}(X)$ *is not closed under marked iteration.*

Finally, we look at the operation inverse homomorphism and obtain also here the non-closure under this operation for all four language families. This shows in particular that all four language families are anti-AFLs. All closure properties are summarized in Table 1.

Table 1. A summary of closure properties of the language families discussed in this paper.

Language Family	—	∪	∩	$\cap_{\text{reg}}$	·	*	$\cdot_m$	$*_m$	$h_{\text{len.pres.}}$	h^{-1}	R
$\mathscr{L}(\text{DPDA})$	✓	✗	✗	✓	✗	✗	✓	✓	✗	✓	✗
$\mathscr{L}(\text{DPDA}_{\lambda\text{-free}})$	✓	✗	✗	✓	✗	✗	✓	✓	✗	✓	✗
$\mathscr{L}(\text{stfDPDA})$	✓	✗	✗	✗	✗	✗	✓	✗	✗	✗	✗
$\mathscr{L}(\text{stfDPDA}_{\lambda\text{-free}})$	✓	✗	✗	✗	✗	✗	✓	✗	✗	✗	✗
$\mathscr{L}(\text{DCA})$	✓	✗	✗	✓	✗	✗	✓	✓	✗	✓	✗
$\mathscr{L}(\text{DCA}_{\lambda\text{-free}})$	✓	✗	✗	✓	✗	✗	✗	✗	✗	✓	✗
$\mathscr{L}(\text{stfDCA})$	✓	✗	✗	✗	✗	✗	✓	✗	✗	✗	✗
$\mathscr{L}(\text{stfDCA}_{\lambda\text{-free}})$	✓	✗	✗	✗	✗	✗	✗	✗	✗	✗	✗

Lemma 15. *Let* $X \in \{\text{stfDCA}, \text{stfDCA}_{\lambda\text{-free}}, \text{stfDPDA}, \text{stfDPDA}_{\lambda\text{-free}}\}$. *Each language family* $\mathscr{L}(X)$ *is not closed under inverse homomorphism.*

Proof. We consider the real-time deterministic linear context-free language $L = \{\, a^n b^n \mid n \geq 0 \,\}$ as witness for the non-closure. It follows from Lemma 6 that L is accepted by some $\mathsf{stfDCA}_{\lambda\text{-free}}$. Let $h \colon \{a, b, d\}^* \to \{a, b\}^*$ be the homomorphism defined as $h(a) = a$, $h(b) = b$, and $h(d) = bb$.

It remains to be shown that $h^{-1}(L)$ is not accepted by any $\mathsf{stfDPDA}$.

We argue similarly as in the proof of Lemma 10. Assume in contrast to the assertion that L is accepted by some $\mathsf{stfDPDA}$ $M = \langle Q, \Sigma, \Gamma, \delta, q_0, \vartriangleleft, \bot, F \rangle$. During the computation of M on input prefixes a^+ no combination of state and content of the pushdown store may appear twice and, thus, the height of the pushdown increases arbitrarily. So, eventually M runs into a one-state loop while processing a's in which the sequence of symbols pushed becomes cyclic, say $Z_1, Z_2, \ldots, Z_k$, for some $k \geq 1$. While M processes the input suffixes b^+, again, no combination of state and content of the pushdown store may appear twice and in any accepting computation the pushdown store has to be decreased until some symbol pushed before the Z_i-sequence appears. So, M runs into a one-state loop that decreases the height of the pushdown store while processing the b's.

Now we replace some factors bb by a d. Again, since M is state-freezing it will eventually run through a one-state loop, say p, while processing b's *and* d's. Let M now be in state p with some Z_i on top of the pushdown. We consider the sub-computation on an input factor of the form b^+ until the next Z_i below the current Z_i appears at the top of the pushdown. Let M read the factor b^{ℓ_1}, for $\ell_1 \geq 1$, in this sub-computation. Then, we consider the sub-computation on an input factor of the form db^* until the next Z_i below the current Z_i appears at the top of the pushdown. Let M read the factor db^{ℓ_2}, for $\ell_2 \geq 0$, in this sub-computation. If $\ell_1 = \ell_2 + 1$ then we obtain a contradiction, since the factor b^{ℓ_1} can be replaced by db^{ℓ_2} without changing the result of the computation, a contradiction.

Next, we consider inputs of the form $a^n b^m (db^{\ell_2} b^{\ell_1 \cdot (\ell_2 + 2)})^*$ that do belong to L, where n and m are large enough. If $\ell_1 > \ell_2 + 1$ then we replace all factors db^{ℓ_2} by $b^{\ell_2 + 2}$ and conclude that the pushdown height at the end of the accepting computation is not decreased until some symbol pushed before the Z_i-sequence appears, a contradiction. Conversely, if $\ell_1 < \ell_2 + 1$ then we replace all factors $b^{\ell_1 \cdot (\ell_2 + 2)}$ by $(db^{\ell_2})^{\ell_1}$ and conclude again that the pushdown height at the end of the accepting computation is not decreased until some symbol pushed before the Z_i-sequence appears, a contradiction as well. $\qquad\square$

5   Decidability Questions

In this section, we discuss which questions are decidable or undecidable for the four types of state-freezing automata introduced. Since the questions of emptiness, finiteness, universality, equivalence, and regularity are decidable for DPDA, they are decidable for the subclasses stfDCA, $\mathsf{stfDCA}_{\lambda\text{-free}}$, $\mathsf{stfDPDA}$, and $\mathsf{stfDPDA}_{\lambda\text{-free}}$ as well. We show that inclusion is undecidable for $\mathsf{stfDPDA}_{\lambda\text{-free}}$ which implies that the question remains undecidable for $\mathsf{stfDPDA}$ as well.

We use a reduction from Post's correspondence problem (PCP) which is known to be undecidable (see, for example, [14]). Let Σ be an alphabet and an instance of the PCP be given by two lists $\alpha = u_1, u_2, \ldots, u_k$ and $\beta = v_1, v_2, \ldots, v_k$ of words from Σ^+. Furthermore, let $A = \{a_1, a_2, \ldots, a_k\}$ be an alphabet with k symbols, and $\Sigma \cap A = \emptyset$. Now, consider two languages L_α and L_β.

$$L_\alpha = \{a_{i_1} a_{i_2} \cdots a_{i_m} \# u_{i_m}^R u_{i_{m-1}}^R \cdots u_{i_1}^R \mid m \geq 1, 1 \leq i_j \leq k, 1 \leq j \leq m\}$$
$$L_\beta = \{a_{i_1} a_{i_2} \cdots a_{i_m} \# v_{i_m}^R v_{i_{m-1}}^R \cdots v_{i_1}^R \mid m \geq 1, 1 \leq i_j \leq k, 1 \leq j \leq m\}$$

Lemma 16. *The languages L_α and L_β are accepted by stfDPDA$_{\lambda\text{-free}}$.*

Proof. The construction of stfDPDA$_{\lambda\text{-free}}$ is straightforward. When reading some symbol a_i from A, the corresponding string u_i and v_i, respectively, is pushed on the pushdown store. After reading the marker #, the content of the pushdown store is matched against the remaining input. The constructed DPDA$_{\lambda\text{-free}}$ are clearly state-freezing. □

Now, Lemma 16 can be used to show the undecidability of the inclusion problem for stfDPDA$_{\lambda\text{-free}}$.

Theorem 17. *Let M_1 and M_2 be two stfDPDA$_{\lambda\text{-free}}$. Then it is undecidable whether $L(M_1) \subseteq L(M_2)$.*

Proof. First, we show that it is undecidable to test whether $L(M_1) \cap L(M_2) = \emptyset$. Given an instance of the PCP we can construct due to Lemma 16 two stfDPDA$_{\lambda\text{-free}}$ whose intersection is empty if and only if the PCP has no solution. If we could decide the emptiness of the intersection, we could decide whether or not a PCP has a solution.

Obviously, $L(M_1) \subseteq L(M_2)$ if and only if $L(M_1) \cap \overline{L(M_2)} = \emptyset$. By Lemma 10 we know that $\mathscr{L}(\text{stfDPDA}_{\lambda\text{-free}})$ is closed under complementation. Let us assume that the inclusion $L(M_1) \subseteq L(M_2)$ is decidable. Then, we could decide the emptiness of intersection as well which is a contradiction. □

It is shown in [7,12] that inclusion is undecidable (and even not semidecidable) for DCA$_{\lambda\text{-free}}$. The idea of the proof seems not to be transferable to stfDCA$_{\lambda\text{-free}}$. Hence, it is currently an open problem whether or not the inclusion problem is decidable for stfDCA$_{\lambda\text{-free}}$.

References

1. Bertoni, A., Mereghetti, C., Palano, B.: Trace monoids with idempotent generators and measure-only quantum automata. Nat. Comput. **9**(2), 383–395 (2010). https://doi.org/10.1007/S11047-009-9154-8
2. Carton, O., Guillon, B., Reiter, F.: Counter machines and distributed automata: A story about exchanging space and time. In: Baetens, J.M., Kutrib, M. (eds.) Cellular Automata and Discrete Complex Systems (AUTOMATA 2018). LNCS, vol. 10875, pp. 13–28. Springer (2018)

3. Fischer, P.C., Meyer, A.R., Rosenberg, A.L.: Counter machines and counter languages. Math. Syst. Theory **2**(3), 265–283 (1968). https://doi.org/10.1007/BF01694011

4. Goles, E., Ollinger, N., Theyssier, G.: Introducing freezing cellular automata. In: Cellular Automata and Discrete Complex Systems, 21st International Workshop (AUTOMATA 2015). TUCS Lecture Notes, vol. 24, pp. 65–73 (2015)

5. Greibach, S.A.: A note on the recognition of one counter languages. RAIRO Inform. Théor. **9**, 5–12 (1975). https://doi.org/10.1051/ITA/197509R200051

6. Harrison, M.A.: Introduction to Formal Language Theory. Addison-Wesley, Reading (1978)

7. Hartmanis, J., Hopcroft, J.E.: What makes some language theory problems undecidable. J. Comput. Syst. Sci. **4**, 368–376 (1970). https://doi.org/10.1016/S0022-0000(70)80018-6

8. Hopcroft, J.E., Ullman, J.D.: Introduction to Automata Theory, Languages, and Computation. Addison-Wesley, Reading, Massachusetts (1979)

9. Igarashi, Y.: A pumping lemma for real-time deterministic context-free languages. Theor. Comput. Sci. **36**, 89–97 (1985). https://doi.org/10.1016/0304-3975(85)90032-5

10. Kutrib, M., Malcher, A.: Cellular automata with sparse communication. Theor. Comput. Sci. **411**(38–39), 3516–3526 (2010)

11. Kutrib, M., Malcher, A.: Reversible pushdown automata. J. Comput. Syst. Sci. **78**(6), 1814–1827 (2012)

12. Kutrib, M., Malcher, A., Wendlandt, M.: Input-driven pushdown, counter, and stack automata. Fund. Inform. **155**, 59–88 (2017). https://doi.org/10.3233/FI-2017-1576

13. Mehlhorn, K.: Pebbling mountain ranges and its application to DCFL-recognition. In: International Colloquium on Automata, Languages, and Programming. pp. 422–435. Springer (1980)

14. Salomaa, A.: Formal Languages. Academic Press, New York (1973)

15. Valiant, L.G., Paterson, M.S.: Deterministic one-counter automata. Springer (1973)

16. Vollmar, R.: On cellular automata with a finite number of state changes. Computing **3**, 181–191 (1981)

From Regular Expressions to Deterministic Finite Automata: $2^{\frac{n}{2}+\sqrt{n}(\log n)^{\Theta(1)}}$ States Are Necessary and Sufficient

Olga Martynova$^{(\boxtimes)}$ (ID) and Alexander Okhotin$^{(\boxtimes)}$ (ID)

Department of Mathematics and Computer Science, St. Petersburg State University,
7/9 Universitetskaya nab., Saint Petersburg 199034, Russia
st062453@student.spbu.ru, alexander.okhotin@spbu.ru

Abstract. It is proved that every regular expression of alphabetic width n, that is, with n occurrences of symbols of the alphabet, can be transformed into a deterministic finite automaton (DFA) with at most $2^{\frac{n}{2}+(\frac{\log_2 e}{2\sqrt{2}}+o(1))\sqrt{n \ln n}}$ states recognizing the same language (the best upper bound up to date is 2^n). At the same time, it is also shown that this bound is close to optimal, namely, that there exist regular expressions of alphabetic width n over a two-symbol alphabet, such that every DFA for the same language has at least $2^{\frac{n}{2}+(\sqrt{2}+o(1))\sqrt{\frac{n}{\ln n}}}$ states (the previously known lower bound is $\frac{5}{4}2^{\frac{n}{2}}$). The same bounds are obtained for an intermediate problem of determinizing nondetermistic finite automata (NFA) with each state having all incoming transitions by the same symbol.

1 Introduction

Transforming a regular expression of a given size to DFA with as few states as possible is a theoretically and practically important problem. This problem depends on the definition of size of a regular expression. Several possible measures of size were considered in the literature: the total number of characters in the string representation of a regular expression; the number of characters excluding brackets (the reverse Polish length), and the number of occurences of symbols of the alphabet, not counting brackets or operators (the alphabetic width). Ellul, Krawetz, Shallit and Wang [5] called the alphabetic width "the most useful in practice".

Using the alphabetic width as the measure of size, the problem is formally stated as follows: how many states are necessary in a DFA to recognize a language defined by a regular expression of alphabetic width n? In their survey of regular expressions, Ellul et al. [5] present this problem as Open Problem 1. In this paper, the alphabetic width of a regular expression will be called simply its *size*.

This work was supported by the Ministry of Science and Higher Education of the Russian Federation, agreement 075-15-2025-343.

G. Castiglione and S. Mantaci (Eds.): CIAA 2025, LNCS 15981, pp. 267–280, 2026.
https://doi.org/10.1007/978-3-032-02602-6_19

The only known upper bound on the number of states in a DFA sufficient for representing every regular expression of size n is $2^n + 1$ states. This bound was first proved by Glushkov [7, Thm. 16], and later it was independently obtained by Leiss [15,16], who has at the same time established a lower bound of the order $\Omega(2^{\frac{n}{3}})$. Leung [17, §3] proved a lower bound of $2^{\frac{n}{2}}$ states, which was improved to $\frac{5}{4} \cdot 2^{\frac{n}{2}}$ by Ellul et al. [5, Thm. 11]. These state-of-the-art bounds are mentioned in a survey by Gruber and Holzer [11, Thm. 11].

Clearly, the known lower bound $\frac{5}{4} \cdot 2^{\frac{n}{2}}$ and upper bound $2^n + 1$ are quite far apart! In this paper, a substantially more efficient transformation from a regular expression of size n to a DFA will be proved, which will use only $2^{\frac{n}{2} + (\frac{\log_2 e}{2\sqrt{2}} + o(1))\sqrt{n \ln n}}$ states. At the same time, the lower bound will also be improved to $2^{\frac{n}{2} + (\sqrt{2} + o(1))\sqrt{\frac{n}{\ln n}}}$ states. Thus it will be proved that the necessary and sufficient number of states is of the order $2^{\frac{n}{2} + \sqrt{n}(\log n)^{\Theta(1)}}$.

Besides the bounds on the size of DFA described above, there are a lot of other results on different transformations of regular expressions to automata. The most well-known is the classical Thompson's [22] construction, which converts a regular expression to an NFA with ε-transitions (ε-NFA); it is a convenient first step in a transformation to a DFA. Ilie and Yu [14] defined an alternative transformation of a regular expression to an ε-NFA that uses fewer transitions than Thompson's; assuming that the size of an automaton is the sum of the number of states and the number of transitions, they obtained a lower bound of $4n - 1$ and an upper bound of $9n - \frac{1}{2}$. Later, Gruber and Gulan [8] established a precise bound of $\frac{22}{5}n + 1$.

The complexity of direct transformation of regular expressions to NFA, bypassing the ε-NFA stage and aiming to minimize the number of transitions, has also been studied. Hromkovič et al. [13] reduced the number of transitions from the obvious $O(n^2)$ to $O(n(\log n)^2)$, and also presented an example of a regular expression, for which every automaton needs to have at least $\Omega(n \log n)$ transitions. Hagenah and Muscholl [12] developed a faster algorithm for doing this transformation. Lifshits [18] improved the lower bound on the number of transitions to $\Omega\left(\frac{n(\log n)^2}{\log \log n}\right)$. Finally, Schnitger [21] proved the lower bound of $\Omega(n(\log n)^2)$ transitions, which asymptotically matches the upper bound by Hromkovič et al. [13].

The case of a unary alphabet was investigated separately. Ellul et al. [5, Thm. 14] proved that every unary regular expression of size n can be transformed to a DFA with at most $g(n) + (n-1)^2 + 2$ states, where $g(n)$ is Landau's function, which is asymptotically of the order $e^{(1+o(1))\sqrt{n \ln n}}$. Also they proved a lower bound of $g(n)$ states.

Something is known on the complexity of transforming automata back to regular expressions. For the classical transformation of McNaughton and Yamada [20] Ellul et al. [5, Thm. 17] proved that an NFA with n states is transformed to a regular expression of size at most $|\Sigma| \cdot n4^n$. Ehrenfeucht and Zeiger [4] showed that a DFA with n states may require a regular expression of size at least 2^{n-1}; this lower bound was proved for an alphabet that grows quadratically in n. Gruber and Holzer [9], obtained a lower bound of $2^{\Theta(n)}$ using a binary alphabet. Also Gruber and Holzer [10] improved the upper bound in the case of a fixed alphabet; in particular, by their method, every DFA with n states over a two-symbol alphabet is transformed to a regular expression of size $O(1.742^n)$; later Edwards and Farr [3] improved this bound to $O(1.682^n)$. In the case of a unary alphabet, Martinez [19] proved an upper bound of $O(n^2)$, whereas Gawrychowski [6] improved it to $O(\frac{n^2}{(\log n)^2})$.

In this paper, the proposed improved upper bound on the transformation from a regular expression to a DFA is proved in two stages. At the first stage, described in Sect. 3, a regular expression is transformed to an intermediate model: an *NFA that remembers the last symbol it has read*. Next, this NFA is transformed to a DFA by the subset construction, and in Sect. 4, the number of reachable subsets is estimated. This number gives the upper bound. Finally, in Sect. 5, an example of a regular expression is constructed, so that its transformation to a DFA requires many states; the same language is also defined by an NFA that remembers the last symbol, so that the lower bound also applies to the transformation of NFA of this kind to DFA.

2   Regular Expressions and Finite Automata

We consider standard regular expressions over an alphabet Σ: every symbol $a \in \Sigma$ is a regular expression, $\varnothing$ is a regular expression, and for every regular expressions α and β, their concatenation $\alpha\beta$, their union $\alpha \mid \beta$ and the star α^* are regular expressions, with brackets used to specify the precedence of operations. The default precedence is first star, then concatenation, then union. As usual, the empty string ε can be defined by the regular expression $\varnothing^*$.

Definition 1. *The alphabetic width of a regular expression α, denoted by $|\mathrm{alph}(\alpha)|$, is the number of occurrences of all symbols of the alphabet in α, counted with multiplicity.*

In this paper, alphabetic width is called simply *size*, as no other succinctness measures are discussed.

Finite automata, deterministic and nondeterministic, are also defined in the standard way.

Definition 2. *A nondeterministic finite automaton (NFA) is a quintuple $A = (\Sigma, Q, Q_0, \delta, F)$, where*

- Σ *is a finite input alphabet,*
- Q *is a finite set of states,*
- $Q_0 \subseteq Q$ *is the set of initial states,*
- $\delta \colon Q \times \Sigma \to 2^Q$ *is the transition function,*
- $F \subseteq Q$ *is the set of accepting states.*

A computation of A on a string $w = a_1 \ldots a_m$ is a sequence of states $p_0, p_1, \ldots, p_m$, with $p_0 \in Q_0$ and $p_{i+1} \in \delta(p_i, a_{i+1})$ for all i. A computation is accepting if $p_m \in F$. The language defined by the automaton is the set of all strings with at least one accepting computation, and it is denoted by $L(A)$.

A deterministic finite automaton (DFA) is an NFA with $|Q_0| = 1$ and $|\delta(q,a)| = 1$ for all $q \in Q$ and $a \in \Sigma$.

Every NFA $A = (\Sigma, Q, Q_0, \delta, F)$ can be transformed to a DFA by the *subset construction*: the resulting DFA has the set of states 2^Q, with initial state $Q_0 \in 2^Q$, transition function $\delta'(S,a) = \bigcup_{q \in S} \delta(q,a)$, and accepting states $F' = \{\, S \subseteq Q \mid S \cap F \neq \varnothing \,\}$.

3   An Intermediate Model

The classical transformation from regular expressions to DFA first transforms them to NFA, and then to DFA by the subset construction. It turns out that if the first step (the transformation to NFA) is made carefully, then one can obtain an NFA of the following special form.

Definition 3. *An NFA $A = (\Sigma, Q, q_0, \delta, F)$ is said to* remember the last symbol, *if its set of states splits into disjoint subsets Q_a, for all $a \in \Sigma$, so that every state from Q_a can be entered only by the symbol a, that is, $\delta(q,a) \subseteq Q_a$ for all $q \in Q$ and $a \in \Sigma$.*

Lemma 1. *For every regular expression of size n, there is an NFA with $n+1$ states that remembers the last symbol and recognizes the same language.*

Actually, the NFA constructed by Leiss [15,16] does remember the last symbol, even though this property is not considered in the cited papers. Hence, Lemma 1 can be proved by examining his construction.

4   Upper Bound

In order to prove an upper bound on the size of a DFA recognizing the language of a regular expression of a given size, such a bound will first be proved for determinization of NFA that remember the last symbol. Then, due to Lemma 1, the bound for the case of regular expressions will follow immediately.

The notion of an NFA that remembers the last symbol is beneficial, because when the subset construction is applied to such an automaton, a lot of subsets turn out to be unreachable. Except for the initial subset, any other reachable subset consists of states that are enterable by the same symbol. It will be proved that the overall number of subsets is substantially smaller than the number of states necessary for determinizing NFA of the general form.

The plan for proving an upper bound is to calculate the number of reachable subsets by two different methods, and to take the minimum value in each case. The first calculation method gives a good upper bound in the case when the original NFA has no "dominating symbol", by which more than half of the states would be reachable (actually, a little bit more than a half).

Lemma 2 (The first method for subset calculation). *Let* $A = (\Sigma, Q, Q_0, \delta, F)$ *be an n-state NFA that remembers the last symbol. For each symbol $a \in \Sigma$, let $Q_a \subseteq Q$ be the subset of states that can be entered by transitions by a. Let n_1 be the maximal number of states enterable by the same symbol: $n_1 = \max_{a \in \Sigma} |Q_a|$. Then there exists a DFA with at most $\max(2^{\frac{n}{2}+1}, 2^{n_1+1})$ states that recognizes the same language.*

Proof. The standard subset construction is used, and a bound on the number of reachable subsets is derived. Let $\Sigma = \{a_1, \ldots, a_k\}$, and let the symbols be ordered by the number of states enterable by transitions by these symbols: $|Q_{a_1}| \geqslant \ldots \geqslant |Q_{a_k}| \geqslant 0$. Let $n_i = |Q_{a_i}|$. The proof is given separately for n_1 at most $\frac{n}{2}$ and for greater values of n_1.

- Let $n_1 \leqslant \frac{n}{2}$. The initial subset is always reachable: this is one subset. Any other reachable subset is reached by some symbol a_i, with $1 \leqslant i \leqslant k$. By reading this symbol, NFA A can get only to states from Q_{a_i}, and accordingly, this subset may contain only states from Q_{a_i}. Hence, there are at most $2^{n_i} - 1$ nonempty subsets reachable after reading a_i. Also, the empty subset may be reachable: it is one subset for all symbols. Therefore, the number of reachable subsets is bounded by $1 + 2^{n_1} + \ldots + 2^{n_k} - k + 1$. Since the largest of the exponents, n_1, is at most $\frac{n}{2}$, it can be proved that the sum does not exceed $2^{\frac{n}{2}+1}$ in this case, as the sets Q_{a_i} are disjoint.
- Assume that the symbol a_1 occurs more often than the rest of the symbols combined: $n_1 > \frac{n}{2}$. The subset construction gives the following reachable subsets: first, the set of initial states Q_0 (one subset), secondly, some subsets of Q_{a_1}, and thirdly, some subsets of $Q_{a_2} \cup \ldots \cup Q_{a_k}$. The number of subsets reachable by a_1 is at most 2^{n_1}, whereas the number of all other subsets is bounded by $2^{n_2 + \ldots + n_k} + 1 = 2^{n - n_1} + 1 < 2^{\lfloor \frac{n}{2} \rfloor} + 1$. Overall, there are at most $2^{n_1} + 2^{\lfloor \frac{n}{2} \rfloor} + 1 \leqslant 2^{n_1+1}$ subsets. $\square$

If much more than $\frac{n}{2}$ states are entered by some symbol $a \in \Sigma$, then the subset construction can possibly yield many subsets reachable by this symbol. In this case, the first method for subset calculation gives at least $2^{|Q_a|}$ subsets. However, the number of such subsets can be bounded by a different method, which uses the known results on determinization of unary NFA.

Unary NFA can be determinized by first transforming them to *Chrobak normal form*, and then applying the subset construction, which, for automata in this form, gives a predictable set of reachable subsets.

Definition 4 (Chrobak [2]). *A unary NFA is said to be in Chrobak normal form if it is of the form $A = (\{a\}, Q, q_0, \delta, F)$, where $Q = \{q_0, \ldots, q_{\ell-1}\} \cup \bigcup_{i=1}^{k} \{r_{i,0}, \ldots, r_{i,p_i-1}\}$, for some $\ell \geqslant 1$, $k \geqslant 1$ and $p_1, \ldots, p_k \geqslant 1$, and its transition function is as follows.*

$$\delta(q_i, a) = q_{i+1} \qquad\qquad (0 \leqslant i < \ell - 1)$$
$$\delta(q_{\ell-1}, a) = \{r_{1,0}, \ldots, r_{k,0}\}$$
$$\delta(r_{i,j}, a) = r_{i,j+1 \bmod p_i} \qquad (1 \leqslant i \leqslant k,\ 0 \leqslant j < p_i)$$

The accepting states can be arbitrary.

The subset construction applied to an NFA in Chrobak normal form produces a DFA with a tail of length ℓ and one cycle of length $\mathrm{lcm}(p_1, \ldots, p_k)$. The greatest possible length of this cycle for a given number of states in an NFA is expressed through *Landau's function*.

$$g(n) = \max\{\, \mathrm{lcm}(p_1, \ldots, p_k) \mid k \geqslant 1,\ p_1 + \ldots + p_k \leqslant n \,\} = e^{(1+o(1))\sqrt{n \ln n}}$$

In this paper, for an NFA over an arbitrary alphabet, the task is to estimate the number of subsets reachable by a string of the form wa^t, where $t \geqslant 1$ and w does not end with a, and a is the symbol with $|Q_a| > \frac{n}{2}$. The idea is that the number of subsets reachable by w will be bounded, and starting from each of these subsets, while reading a^t, the automaton works like a unary automaton. Once any symbol other than a is read, the transitions from the current subset will lead back to the non-unary part, that is, to some subset of $Q \setminus Q_a$. Hence, all that is needed is to estimate the number of different subsets reached in the unary automaton starting from any fixed subset.

If one simply transforms an NFA to Chrobak normal form, and then determinizes it, the result will only be a DFA recognizing the same language, which may have fewer states than the number of reachable subsets in the original NFA. In particular, such a DFA would not know the exact subset reached by the original NFA at the current moment, and hence would not be able to move on to the non-unary part upon reading a symbol other than a.

In order to estimate the number of reachable subsets, a small extension of Chrobak's [2, Lemma 4.3] construction will be proved. It will transform an NFA to a DFA in such a way that the states of the DFA are subsets *of the original NFA*.

Lemma 3. *Let A be a unary NFA without accepting states, and let C be a DFA made from A by the subset construction. Then C has at most $g(n)+O(n^2)$ states reachable from the initial state.*

Proof. The proof essentially uses the details of Chrobak's [2] construction. For every state q of the automaton $A = (\Sigma, Q, q_0, \delta)$, consider the automaton $A_q = (\Sigma, Q, q_0, \delta, \{q\})$, in which q is the unique accepting state. The automaton A_q is transformed into Chrobak normal form [2, Lemma 4.3], resulting in an NFA B_q that recognizes the same language, and has a tail of length at most $O(n^2)$ and cycles of length $p_1, \ldots, p_k$, where k is the number of strongly connected components in the automaton A, and for each i, the number p_i is the greatest common divisor of the lengths of all cycles in the i-th strongly connected component. It is important that the number of cycles and their lengths in each automaton B_q depend only on the automaton A, and do not depend on the choice of the accepting state q. One can assume that tails of all automata B_q are of the same length $\ell = O(n^2)$, by rolling back all cycles whenever the original tail was shorter. Then the automata B_q differ from each other only in their accepting states.

Next, for each q, let C_q be a DFA obtained from B_q by the subset construction. For different q, these automata also differ only in their accepting states, and the size of each of them is $g(n) + \ell$. Let R be the set of states of each automaton C_q.

It is claimed that for every string a^m, the set of states of the original NFA A reached by the string a^m contains exactly those states q, for which the corresponding DFA C_q accepts the string a^m. Indeed, a state q can be reached by the string a^m in the automaton A if and only if the automaton A_q accepts the string a^m. And this is equivalent to the acceptance of this string by automata B_q and C_q, because all of them recognize the same language.

Let C be the DFA obtained from the NFA A by the subset construction. Now it will be proved that C has at most $g(n) + \ell$ reachable states. Let $S \subseteq Q$ be a reachable subset in A, and let a^m be the string by which it is reached from the initial subset. Then, as proved above, $S = \{\, q \mid a^m \in L(C_q) \,\}$. Next, let $r \in R$ be the common state in all automata C_q, in which these automata come after reading a^m (this is the same state, because all C_q differ only in their accepting states). Then, $S = \{\, q \mid r \text{ is accepting in } C_q \,\}$. Therefore, the number of distinct subsets S does not exceed the number of states $r \in R$—and there are exactly $g(n) + \ell$ such states. $\qquad\qquad\square$

Now this construction will be used to estimate the number of reachable subsets in a given NFA over a non-unary alphabet, which remembers the last symbol. And if $|Q_a|$ is large enough, then the new method will be more efficient than the first one.

Assume that the NFA is either in its initial subset, or in a subset $S \subseteq Q_b$ enterable by some symbol $b \neq a$, and assume that it makes a transition by a. At this moment, it can potentially get into any subset of the unary part. However, the number of such transitions is bounded by the number of subsets entered not by a—and there are few of them by the assumption. Thus, we have to deal with a bounded number of starting subsets of the set Q_a, and for each of them one can work with a separate unary automaton that starts from the given subset of Q_a, and then reads symbols a, moving through different subsets of Q_a.

Lemma 4 (The second method for subset calculation). *Let $A = (\Sigma, Q, Q_0, \delta, F)$ be an NFA with n states that remembers the last symbol. For each symbol $a \in \Sigma$, let $Q_a \subseteq Q$ be all states enterable by a. Denote the maximal number of states enterable by the same symbol by $n_1 = \max_{a \in \Sigma} |Q_a|$. Then there is a DFA with at most $2^{n-n_1} \cdot (g(n_1) + O(n_1^2))$ states recognizing the same language.*

Proof. Let $\Sigma = \{a_1, \dots, a_k\}$, and let $|Q_{a_1}| \geqslant \dots \geqslant |Q_{a_k}| \geqslant 0$. Then $n_1 = |Q_{a_1}|$.

The subset construction gives the following reachable subsets: the set of initial states Q_0, some subsets of Q_{a_1}, and some subsets of $Q_{a_2} \cup \dots \cup Q_{a_k}$. At most 2^{n_1} subsets are entered by a_1, whereas the number of other subsets is bounded by $2^{n-n_1} + 1$.

The task is to improve the trivial 2^{n_1} bound on the number of subsets of Q_{a_1} in the NFA A that are reachable by some string. Every string, by which the NFA gets to states in Q_{a_1}, is of the form wa_1^t, with $t \geqslant 1$, and with the string w either empty or ending with a symbol different from a_1. Let P_w be the set of states of the NFA reachable after reading w. If w is empty, then $P_w = Q_0$, and if w ends with some symbol a_i, with $i = 2, \ldots, k$, then $P_w \subseteq Q_{a_i}$. These are all starting subsets P_w before the unary part of the string, and the number of such subsets is at most $2^{n-n_1} + 1$. It remains to estimate how many different subsets are obtained from each starting subset by reading suffixes from a_1^*. This is done by a unary NFA with n_1 states, which is a part of the original NFA. Then, Lemma 3 is applicable to this unary NFA, and it asserts that the subset construction, with the given starting subset as initial, gives at most $g(n_1) + O(n_1^2)$ reachable subsets.

Therefore, the overall number of reachable subsets in the constructed DFA does not exceed $(2^{n-n_1} + 1)(g(n_1) + O(n_1^2))$.   □

The minimum of the bounds given by the two methods is an upper bound on the size of a DFA. If n_1 is small, then the first method is better, and if it is large, then the second one. The desired minimum is estimated in the following theorem.

Theorem 1. *Let A be an NFA that remembers the last symbol, and let n be the number of states in it. Then there exists a DFA with at most $2^{\frac{n}{2} + \frac{\log_2 e}{2\sqrt{2}}\sqrt{n \ln n}(1+o(1))}$ states recognizing the same language.*

The theorem is proved by considering all possible values of n_1, and estimating the number of states given by the best of the two methods for this n_1. The worst-case n_1 is found to be of the order $\frac{n}{2} + (1+o(1))\frac{\log_2 e}{2\sqrt{2}}\sqrt{n \ln n}$, that is, for such n_1, both methods give asymptotically the same number of states, which is the upper bound stated in the theorem. The proof is omitted due to space constraints.

Corollary 1. *Let α be a regular expression of size n over an alphabet Σ. Then there is a DFA with at most $2^{\frac{n}{2} + \frac{\log_2 e}{2\sqrt{2}}\sqrt{n \ln n}(1+o(1))}$ states that recognizes the same language.*

5   Lower Bound

In this section, a lower bound on the number of states in a DFA necessary to recognize languages that are defined by regular expressions of a given size is obtained. Also, the same lower bound will be shown for the number of states necessary to determinize NFA that remember the last symbol.

Theorem 2. *For every $n \geqslant 2$ there exists a language defined by a regular expression of size n, and at the same time recognized by an NFA that remembers the last symbol, with $n-1$ states, such that every DFA recognizing this language has at least $2^{\frac{n}{2} + (1+o(1))\sqrt{2}\sqrt{\frac{n}{\ln n}}}$ states.*

The general form of regular expressions used in the proof of the theorem is as follows.

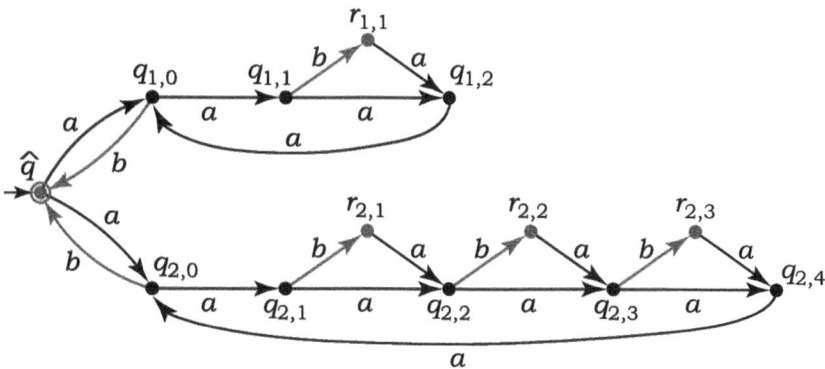

Fig. 1. NFA $A_{3,5}$.

Definition 5. *Let* $\pi_1, \ldots, \pi_k \geqslant 3$ *be relatively prime numbers, with* $\pi_1 = 3$. *For this set of numbers, a regular expression* $\alpha_{\pi_1,\ldots,\pi_k}$ *of the following form is defined:*

$$\alpha_{\pi_1,\ldots,\pi_k} = \left(a(\beta_{\pi_1} \mid \ldots \mid \beta_{\pi_k})b\right)^*, \qquad \text{where } \beta_\pi = (a((b \mid \varepsilon)a)^{\pi-2}a)^*.$$

The size of every subexpression β_π is $2\pi - 2$. Thus, the size of $\alpha_{\pi_1,\ldots,\pi_k}$ is $2 + \sum_{i=1}^{k}(2\pi_i - 2) = (2\sum_{i=1}^{k}\pi_i) - 2k + 2$.

It will be shown that, with a proper choice of numbers $\pi_1, \ldots, \pi_k$ satisfying the inequality $(2\sum_{i=1}^{k}\pi_i) - 2k + 2 \leqslant n$, the regular expression $\alpha_{\pi_1,\ldots,\pi_k}$ will be the desired example for the theorem. The numbers $\pi_1, \ldots, \pi_k$ will be chosen later.

To prove that a DFA for this language needs a lot of states, it would be easier to consider an NFA recognizing the same language. The suggested NFA slightly differs from the automaton given by the formal transformation from Lemma 1; in particular, it has two states less.

Definition 6. *An NFA* $A_{\pi_1,\ldots,\pi_k} = (\{a,b\}, Q, \widehat{q}, \delta, F)$ *has the set of states* $Q = \{\widehat{q}\} \cup \bigcup_{i=1}^{k} Q_i \cup \bigcup_{i=1}^{k} R_i$, *where* $Q_i = \{q_{i,0}, \ldots, q_{i,\pi_i-1}\}$ *and* $R_i = \{r_{i,1}, \ldots, r_{i,\pi_i-2}\}$. *It is defined to remember the last symbol, with* $Q_a = \bigcup_i Q_i$ *and* $Q_b = \{\widehat{q}\} \cup \bigcup_i R_i$.

The transitions are illustrated in Fig. 1 and defined as follows. The only transition in the initial state is

$$\delta(\widehat{q}, a) = \{q_{1,0}, \ldots, q_{k,0}\}$$

and this is the only nondeterministic transition in the automaton. Each subset Q_i *has a cycle by* a *of length* π_i, *and is called the* i-*th cycle:*

$$\delta(q_{i,j}, a) = q_{i,j+1 \bmod \pi_i}, \qquad \qquad \text{for all } j.$$

*For each state in Q_i, except the first and the last ones, there is a transition by b
to a separate state from R_i:*

$$\delta(q_{i,j}, b) = r_{i,j}, \qquad\qquad \text{with } 1 \leqslant j \leqslant \pi_i - 2.$$

From every such state the automaton moves by a to the next state in Q_i:

$$\delta(r_{i,j}, a) = q_{i,j+1}, \qquad\qquad \text{with } 1 \leqslant j \leqslant \pi_i - 2.$$

Finally, from the first state in each Q_i there is a transition by b to the initial
state:

$$\delta(q_{i,0}, b) = \widehat{q}.$$

*The only accepting state is the initial state: $F = \{\widehat{q}\}$. In each set $Q_i \cup R_i$ there
are $2\pi_i - 2$ states in total. Hence, the overall number of states in the automaton
is $2(\sum_{i=1}^{k} \pi_i) - 2k + 1$. The automaton $A_{3,5}$ is shown in Fig. 1.*

Lemma 5. *Let $\pi_1, \ldots, \pi_k \geqslant 3$ be any relatively prime integers, with $\pi_1 = 3$.
Then the automaton $A_{\pi_1,\ldots,\pi_k}$ and the regular expression $\alpha_{\pi_1,\ldots,\pi_k}$ define the
same language.*

The proof of Lemma 5 is by establishing a correspondence between cycles in
the automaton and subexpressions β_{π_i}.

Lemma 6. *Let $k \geqslant 2$, and let $\pi_1, \ldots, \pi_k \geqslant 3$ be any relatively prime integers,
where $\pi_1 = 3$. Then the minimal DFA that recognizes the same language as the
NFA $A_{\pi_1,\ldots,\pi_k}$, has at least $\prod_{i=1}^{k}(2^{\pi_i} - 2)$ states.*

Proof. The subset construction is applied to $A_{\pi_1,\ldots,\pi_k}$.

The proof uses subsets of the form $P_1 \cup \ldots \cup P_k$, where $P_i \subseteq Q_i$, in which,
from each Q_i, at least one state is taken and at least one state is not taken:
$\varnothing \subsetneq P_i \subsetneq Q_i$. There are $2^{\pi_i} - 2$ possible choices for each P_i, and hence there are
$\prod_{i=1}^{k}(2^{\pi_i} - 2)$ subsets of this form.

To subsets of such a form, one can conveniently apply an operation of *shift by
a vector of residues* $(d_1, \ldots, d_k)$, where $d_i \in \{0, \ldots, \pi_i - 1\}$: this means reading
a substring a^{ℓ}, where $\ell \equiv d_i \pmod{\pi_i}$ for all i. Since the cycle lengths $\pi_1, \ldots, \pi_k$
are relatively prime, such a number ℓ exists by the Chinese remainder theorem.
This operation transforms the subset $P_1 \cup \ldots \cup P_k$ into the subset $P_1' \cup \ldots \cup P_k'$,
where $P_i' = \{ q_{i,j+d_i \bmod \pi_i} \mid q_{i,j} \in P_i \}$ for all i.

The overall plan of the proof is standard: first, the reachability of all subsets
of the chosen form is established, and then a separating string is found for every
pair of distinct subsets of this form. Once this is done, this will imply that every
DFA recognizing this language must have at least as many states as there are
subsets of this form.

Claim. Each subset of the form $P_1 \cup \ldots \cup P_k$, with $\varnothing \subsetneq P_i \subsetneq Q_i$, is reachable
from the initial subset $\{\widehat{q}\}$ by some string.

Induction on the number of states in a subset.

Base case: $|P_1| \in \{1, 2\}$ and $|P_2| = \ldots = |P_k| = 1$. If $|P_1| = 1$, then the subset $\{q_{1,0}\} \cup \{q_{2,0}\} \cup \ldots \cup \{q_{k,0}\}$ is reached by a single transition by the symbol a, and then the desired subset is reached by applying a shift by the vector of residues $(d_1, \ldots, d_k)$, where $P_i = \{q_{i,d_i}\}$ for all i. If $|P_1| = 2$, then the subset $\{q_{1,0}, q_{1,2}\} \cup \{q_{2,0}\} \cup \ldots \cup \{q_{k,0}\}$ is reached by first moving by a, then applying a shift by the vector $(1, 0, \ldots, 0)$, and then reading ba. The rest of the subsets in the second part of the base case are reached from this one by applying another shift.

Induction step. Let $P_1 \cup \ldots \cup P_k$ be any subset satisfying the condition and not handled in the base case. Then $|P_m| > 1$ for some $m \geqslant 2$. Let this m be fixed. In each subset P_i, a state is chosen as follows. In the first cycle, the chosen state is the unique state $q_{1,s}$, such that $q_{1,s} \in P_1$ and $q_{1,s-1 \bmod 3} \notin P_1$. In each of the subsequent cycles Q_i, with $i \in \{2, \ldots, k\}$, any state q_{i,t_i} in P_i can be chosen, for which the next state is not in P_i, that is, $q_{i,t_i+1 \bmod \pi_i} \notin P_i$. Then the set $(P'_1, \ldots, P'_k)$, where $P'_m = P_m \setminus \{q_{m,t_m}\}$ and $P'_i = P_i$ for all $i \neq m$, satisfies the condition and contains fewer states than the desired subset. Therefore, it is reachable by the induction hypothesis.

It remains to reach the subset $(P_1, \ldots, P_k)$ starting from the subset $(P'_1, \ldots, P'_k)$. To this end, the subset $(P'_1, \ldots, P'_k)$ is first shifted by the vector of residues $(-s, -t_2 - 1, \ldots, -t_k - 1)$, resulting in the subset $\{q_{1,j-s \bmod \pi_1} \mid q_{i,j} \in P_1\} \cup \{q_{m,j-t_m-1 \bmod \pi_m} \mid q_{m,j} \in P_m \setminus \{q_{m,t_m}\}\} \cup \bigcup_{i \geqslant 2, i \neq m}^{k} \{q_{i,j-t_i-1 \bmod \pi_i} \mid q_{i,j} \in P_i\}$. In particular, the first cycle will contain the state $q_{1,0}$, but not the state $q_{1,2}$; in the m-th cycle there will be neither $q_{m,0}$ nor q_{m,π_m-1}; and in the rest of the cycles, for all $i \notin \{0, m\}$, there will be no $q_{i,0}$, but q_{i,π_i-1} will be present.

Then the string ba is read. It affects the first cycle as follows: if it contained only the state $q_{1,0}$, then this state will move to $\widehat{q}$ by b and return to $q_{1,0}$ by a. And if there were both states $q_{1,0}$ and $q_{1,1}$, then they move to $q_{1,0}$ and to $q_{1,2}$, respectively; that is, the cycle moves forward by two positions, and the resulting states are $\{q_{1,j-s+2 \bmod 3} \mid q_{i,j} \in P_1\}$.

In the m-th cycle, after reading ba, the following happens. All states $q_{m,1}, \ldots, q_{m,\pi_m-2}$ move by one position forward. After reading the first symbol b, the state $\widehat{q}$ emerges from the first cycle, and after reading a it gets copied to $q_{m,0}$. As the result, the states in the m-th cycle will be $\{q_{m,j-t_m \bmod \pi_m} \mid q_{m,j} \in P_m \setminus \{q_{m,t_m}\}\} \cup \{q_{m,0}\} = \{q_{m,j-t_m \bmod \pi_m} \mid q_{m,j} \in P_m\}$.

In every i-th cycle, with $i \notin \{1, m\}$, by reading ba, the states $q_{i,1}, \ldots, q_{i,\pi_i-2}$ are also shifted forward by one position, while the state q_{i,π_i-1} disappears after b but gets restored into $q_{i,0}$ upon reading a from $\widehat{q}$. Thus, the entire i-cycle is shifted forward by one position, and it will contain the states $\{q_{i,j-t_i \bmod \pi_i} \mid q_{i,j} \in P_i\}$.

In order to reach the desired subset $P_1 \cup \ldots \cup P_k$, it is enough to shift the subset obtained after reading ba either by the vector $(s, t_1, \ldots, t_k)$, if $|P_1| = 1$, or by $(s - 2, t_1, \ldots, t_k)$, if $|P_1| = 2$. This proves the induction step and the entire claim on the reachability of subsets.

It is left to prove that every two subsets can be separated.

Claim. For every two distinct subsets of the form $P_1 \cup \ldots \cup P_k$, with $\varnothing \subsetneq P_i \subsetneq Q_i$, there is a string that is accepted from exactly one of these subsets.

Let $P_1 \cup \ldots \cup P_k$ and $P'_1 \cup \ldots \cup P'_k$ be two such subsets, and let $q_{i,j}$ be the state on which they differ. Assume, without loss of generality, that $q_{i,j} \in P_i \setminus P'_i$. For each $t \neq i$, let j_t be such a residue modulo π_i, that $q_{t,j_t} \notin P'_i$. By the Chinese remainder theorem, there is such a length $\ell \geqslant 0$, that after reading a^ℓ the state $q_{i,j}$ moves to $q_{i,0}$, while every state q_{t,j_t} moves to $q_{t,0}$. Then the desired separating string is $a^\ell b$. After reading this string from the subset $P_1 \cup \ldots \cup P_k$, the automaton will accept, since the state $q_{i,j}$ moves to $q_{i,0}$ by a^ℓ, and then by b it moves to the accepting state $\widehat{q}$. But the automaton does not accept this string starting from the subset $P'_1 \cup \ldots \cup P'_k$, because after reading a^ℓ, in each t-th cycle, for $1 \leqslant t \leqslant k$, there will be no state $q_{t,0}$, and hence, none of the states can move to $\widehat{q}$ after reading b.

With the above two claims established, the lemma is proved. $\square$

In the lower bound expression in Lemma 6, the inconvenient summands -2 in all factors can be eliminated at the cost of a constant factor.

Lemma 7. *Let $\pi_1, \ldots, \pi_k \geqslant 3$ be any distinct integers. Then $\prod_{i=1}^{k} (2^{\pi_i} - 2) \geqslant \frac{1}{2} \prod_{i=1}^{k} 2^{\pi_i}$.*

The proof of Lemma 7 is omitted due to space constraints.

It remains to choose the optimal cycle lengths to obtain the desired lower bound for each n.

Let $n \geqslant 5$, and assume that k cycles of length $\pi_1, \ldots, \pi_k$ are used. Then the size of the regular expression is $S = 2 \sum_{i=1}^{k} \pi_i - 2k + 2$, and it should be not greater than n. Then $\sum_{i=1}^{k} \pi_i = \frac{S}{2} + k - 1$, and lower bound obtained can be represented as

$$\frac{1}{2} \cdot \prod_{i=1}^{k} 2^{\pi_i} = \frac{1}{2} \cdot 2^{\sum_{i=1}^{k} \pi_i} = \frac{1}{2} \cdot 2^{\frac{S}{2} + k - 1} = 2^{\frac{S}{2} + k - 2}.$$

If the size S of the regular expression is equal or almost equal to n, then the resulting lower bound is of the order $\Omega(2^{\frac{n}{2} + k})$, and hence in this case one should use as many cycles as possible, that is, maximize k.

Simply maximizing the number of cycles is easy: it is sufficient to use as many small odd primes as possible for cycle lengths. Let p_i be the i-th odd prime, that is, $p_1 = 3$, $p_2 = 5$, ..., and let $k = k(n)$ be the greatest such number that $S = (2 \sum_{i=1}^{k} p_i) - 2k + 2 \leqslant n$. If the cycles of length $p_1, \ldots, p_k$ are taken, then, since the gap between S and n can be large—of the order up to $\Theta(\sqrt{n \ln n})$—the exponent in the resulting lower bound may potentially be even less than $\frac{n}{2}$. The admissible gap in order to obtain the desired lower bound is $o\left(\sqrt{\frac{n}{\ln n}}\right)$. To ensure that S is close to n, it is sufficient to take the first $k - 1$ cycles to be $\pi_1 = p_1$, ..., $\pi_{k-1} = p_{k-1}$, and then choose the length of the last cycle as the largest prime p that fits into n, that is, $(2 \sum_{i=1}^{k-1} p_i) + 2p - 2k + 2 \leqslant n$. With such cycle lengths, the lower bound of $2^{\frac{n}{2} + (1 + o(1))\sqrt{2}\sqrt{\frac{n}{\ln n}}}$ states claimed in Theorem 2 is proved using the classical results on the asymptotics of primes and their sums, as well as a recent result by Baker et al. [1] on the distance between primes. The proof again has to be omitted to conserve space.

6   Conclusion

It has been proved that recognizing languages defined by regular expressions of alphabetic width n by a DFA requires at least $2^{\frac{n}{2}+(\sqrt{2}+o(1))\sqrt{\frac{n}{\ln n}}}$ states and at most $2^{\frac{n}{2}+(\frac{\log_2 e}{2\sqrt{2}}+o(1))\sqrt{n \ln n}}$ states. Therefore, the first term in the exponent, $\frac{n}{2}$ is known precisely, and the second term is known up to the power of the logarithm: $\sqrt{n}(\log n)^{\Theta(1)}$. Determining the exact power of this logarithm is an interesting open problem.

References

1. Baker, R.C., Harman, G., Pintz, J.: The difference between consecutive primes, II. Proc. London Math. Soc. **83**(3), 532–562 (2001). https://doi.org/10.1112/plms/83.3.532
2. Chrobak, M.: Finite automata and unary languages. Theoret. Comput. Sci. **47**, 149–158 (1986). https://doi.org/10.1016/0304-3975(86)90142-8
3. Edwards, K., Farr, G.: Improved upper bounds for planarization and series-parallelization of degree-bounded graphs. Electron. J. Comb. **19**(2), 2378 (2012). https://doi.org/10.37236/2378
4. Ehrenfeucht, A., Zeiger, P.: Complexity measures for regular expressions. J. Comput. Syst. Sci. **12**(2), 134–146 (1976). https://doi.org/10.1016/S0022-0000(76)80034-7
5. Ellul, K., Krawetz, B., Shallit, J., Wang, M.: Regular expressions: new results and open problems. J. Autom. Lang. Comb. **10**(4), 407–437 (2005). https://doi.org/10.25596/jalc-2005-407
6. Gawrychowski, P.: Chrobak normal form revisited, with applications. In: Bouchou-Markhoff, B., Caron, P., Champarnaud, J.-M., Maurel, D. (eds.) CIAA 2011. LNCS, vol. 6807, pp. 142–153. Springer, Heidelberg (2011). https://doi.org/10.1007/978-3-642-22256-6_14
7. Glushkov, V.M.: The abstract theory of automata. Russ. Math. Surv. **16**(5), 1–53 (1961). https://doi.org/10.1070/RM1961v016n05ABEH004112
8. Gruber, H., Gulan, S.: Simplifying regular expressions. In: Dediu, A.-H., Fernau, H., Martín-Vide, C. (eds.) LATA 2010. LNCS, vol. 6031, pp. 285–296. Springer, Heidelberg (2010). https://doi.org/10.1007/978-3-642-13089-2_24
9. Gruber, H., Holzer, M.: Finite automata, digraph connectivity, and regular expression size. In: Aceto, L., Damgård, I., Goldberg, L.A., Halldórsson, M.M., Ingólfsdóttir, A., Walukiewicz, I. (eds.) ICALP 2008. LNCS, vol. 5126, pp. 39–50. Springer, Heidelberg (2008). https://doi.org/10.1007/978-3-540-70583-3_4
10. Gruber, H., Holzer, M.: Provably shorter regular expressions from finite automata. Int. J. Found. Comput. Sci. **24**(8), 1255–1279 (2013). https://doi.org/10.1142/S0129054113500330
11. Gruber, H., Holzer, M.: From finite automata to regular expressions and back–a summary on descriptional complexity. Int. J. Found. Comput. Sci. **26**(8), 1009–1040 (2015). https://doi.org/10.1142/S0129054115400110

12. Hagenah, C., Muscholl, A.: Computing epsilon-free NFA from regular expressions in $O(n \log^2(n))$ time. RAIRO Inform. Théorique App. **34**(4), 257–278 (2000)
13. Hromkovič, J., Seibert, S., Wilke, T.: Translating regular expressions into small ε-free nondeterministic finite automata. J. Comput. Syst. Sci. **62**(4), 565–588 (2001). https://doi.org/10.1006/jcss.2001.1748
14. Ilie, L., Yu, S.: Follow automata. Inf. Comput. **186**(1), 140–162 (2003)
15. Leiss, E.L.: Constructing a finite automaton for a given regular expression. SIGACT News **12**(3), 81–87 (1980)
16. Leiss, E.L.: The complexity of restricted regular expressions and the synthesis problem for finite automata. J. Comput. Syst. Sci. **23**(3), 348–354 (1981)
17. Leung, H.: Separating exponentially ambiguous finite automata from polynomially ambiguous finite automata. SIAM J. Comput. **27**(4), 1073–1082 (1998)
18. Lifshits, Y.: A lower bound on the size of ε-free NFA corresponding to a regular expression. Inf. Process. Lett. **85**, 293–299 (2003)
19. Martinez, A.: Efficient computation of regular expressions from unary NFAs. In: Fourth International Workshop on Descriptional Complexity of Formal Systems (DCFS 2002, London, Canada, August 21–24, 2002) Department of Computer Science, University of Western Ontario, pp. 174–187 (2002)
20. McNaughton, R., Yamada, H.: Regular expressions and state graphs for automata. IRE Trans. Elect. Comput. EC. **9**(1), 39–47 (1960). https://doi.org/10.1109/TEC.1960.5221603
21. Schnitger, G.: Regular expressions and NFAs without ε-transitions. In: Durand, B., Thomas, W. (eds.) STACS 2006. LNCS, vol. 3884, pp. 432–443. Springer, Heidelberg (2006). https://doi.org/10.1007/11672142_35
22. Thompson, K.: Regular expression search algorithm. Commun. ACM **11**(6), 419–422 (1968). https://doi.org/10.1145/363347.363387

A First Taste of MeSCaL, a Tool for Solving Membership Problems for Regular Languages

Thomas Place and Marc Zeitoun[(✉)]

Univ. Bordeaux, CNRS, Bordeaux INP, LaBRI, UMR 5800, 33400 Talence, France
{tplace,mz}@labri.fr

Abstract. This article is the first in a two-part series presenting a software called "MeSCaL". This program is designed to test properties of regular languages. MeSCaL allows the user to define regular languages— either via regular expressions or finite automata—and to perform queries on them. These queries determine whether the input language belongs to a specific subclass of regular languages. While MeSCaL supports at present over one hundred such classes, this paper focuses on just twenty.

These membership tests present an inherent difficulty: they often require analyzing an algebraic structure derived from the input language, known as its *syntactic monoid*. Unfortunately, the size of this monoid can grow exponentially relative to the size of the minimal automaton of the language. Therefore, to handle nontrivial cases efficiently, the challenge is to highly optimize the tests performed on this monoid. In particular, naïvely applying algorithms from the literature is doomed to failure: careful optimizations are essential. The aim of this article is to provide an overview of some of the key optimizations implemented in MeSCaL.

Keywords: Regular languages · Finite monoids · Implementation · Membership problem · Separation problem · Star-free closure · First-order logic · Unary temporal logic

1 Introduction

Regular word languages play a fundamental role in theoretical computer science. They form a robust class, characterized by multiple natural definitions: they are those recognized by finite automata, described by regular expressions, defined via morphisms into finite monoids, or specified using monadic second-order logic. Among these formalisms, regular expressions and logical descriptions are syntactic in nature, which raises a key decision problem: determining how simply a given regular language can be expressed within such a syntax. For example, in logic, a first-order formula is typically regarded as simpler than a monadic

---

T. Place and M. Zeitoun—Work supported by ANR 24-CE48-1142, Project UnREAL.

G. Castiglione and S. Mantaci (Eds.): CIAA 2025, LNCS 15981, pp. 281–298, 2026.
https://doi.org/10.1007/978-3-032-02602-6_20

second-order formula. Within first-order logic itself, formulas with fewer quantifier alternations are considered simpler than those with many. By imposing syntactic restrictions—such as limiting the number of quantifier alternations—one defines subclasses within the broader class of regular languages.

The informal question of "how simple a regular language is" can now be made precise. Let $\mathcal{C}$ be a class of regular languages—typically defined by a syntactic fragment of a descriptive formalism. The $\mathcal{C}$ -*membership problem* asks whether a given regular language belongs to $\mathcal{C}$. Over the years, several such problems have been resolved, beginning with the class SF of star-free languages. Over an alphabet A, this class consists of all regular languages that can be built from singletons $\{a\}$ for $a \in A$, using only concatenation $K, L \mapsto KL$ and Boolean operations—including complement, which preserves regularity—while disallowing the Kleene star[1]. Schützenberger [37] provided an algorithm to decide SF-membership, laying the foundation for understanding the expressiveness of logical formalisms. Indeed, McNaughton and Papert later proved [19] that SF precisely corresponds to the class of languages definable in first-order logic FO($<$).

Two other notable language classes are worth mentioning. The class PT of piecewise-testable languages consists of all Boolean combinations of languages of the form $A^* a_1 A^* \cdots a_n A^*$. This class is exactly the set of languages definable in $\mathcal{B}\Sigma_1(<)$, that is, by first-order sentences without quantifier alternation [46]. The PT-membership problem was solved by Simon [38]. Finally, Thérien and Wilke [45] addressed the membership problem for the class of languages definable in $\text{FO}^2(<)$, *i.e.*, by first-order sentences that use only two variable names.

Although many membership problems have been solved, few corresponding algorithms have been implemented with performance in mind. PySemigroup [20] runs several but does not emphasize efficiency. Some tools compute an algebraic abstraction of the language, its *syntactic monoid*, but only derive a few membership tests as a byproduct. This includes Semigroupe [22,24], and, to some extent AUTOMATE [4], Automata [6] and AMoRE [18]. Others, such as LANGUAGE [3] and TESTAS [48], target a limited set of membership problems and implement them efficiently on automata, avoiding to compute the syntactic monoid.

The main factor that contributes to the lack of software support is that most known membership algorithms rely on computing the aforementioned *syntactic monoid* of the input language—a structure that can be exponentially larger than the size of the input automaton: algorithms working directly on automata, such as the one for PT-languages [40], are rare (see also [12] for the$<$ class of so-called locally testable languages). In fact, even over a binary alphabet, there exist deterministic automata with n-states whose syntactic monoids contain more than $n^n\left(1 - \frac{2}{\sqrt{n}}\right)$ elements [10]. For example, over a binary alphabet and with just 7 states, some deterministic automata have syntactic monoids of size 610871—in fact the maximum possible in this case. This exponential blowup makes any membership algorithm with quadratic or higher complexity impractical.

---

[1] We use a slightly different but equivalent definition of this original formulation of SF.

This issue can be tackled in two ways. One approach is to restrict attention to algorithms with linear complexity in the size of the syntactic monoid, which requires careful algorithm selection. Indeed, membership tests are often framed as equations this monoid must satisfy. Verifying an equation with k variables over a monoid of size m takes $\Theta(m^k)$ time in the worst case. For example, a language belongs to the class PT if and only if, for all s, t in the syntactic monoid, the identity $(st)^{\ell}s = (st)^{\ell} = t(st)^{\ell}$ holds (for some ℓ), yielding quadratic complexity. Fortunately, in some key cases (*e.g.*, the class PT), such equations are equivalent to properties that can be tested in *linear time* on the Cayley graph of the syntactic morphism. This is a common approach in the software mentioned above.

Regrettably, this strategy only applies in rare cases (such as the three mentioned above: languages definable in $\mathrm{FO}(<)$, $\mathcal{B}\Sigma_1(<)$ and $\mathrm{FO}^2(<)$). For most classes, algorithms must verify equations with variables under additional constraints, often requiring costly precomputations. As a result, existing software is either limited to a small set of membership tests or performs these precomputations in a straightforward but inefficient way, making it impractical even for small automata. In this paper, we focus on a specific bottleneck precomputation that suffices for most classes we consider: the *kernel* of the syntactic morphism—a standard, naturally defined subset of the syntactic monoid.

The second approach to mitigating the combinatorial explosion of the syntactic monoid plays a central role in this article. It involves significantly *restricting* the subset of the monoid where the equation must be tested, without losing correctness. To the best of our knowledge, this strategy has not been explored in the literature.

Contributions. This paper is the first in a two-part series, continued in [33]. It introduces the free software MeSCaL[2], which implements optimized membership algorithms for several classes of regular languages. We present the most important data structures and foundational algorithms used in MeSCaL.

MeSCaL provides a shell interface where users can define languages via regular expressions or automata, visualize their minimal automata and perform membership queries. It currently supports *over a hundred* classes. MeSCaL can also, for example, enumerate all deterministic automata up to a user-defined (reasonable) size to find a language belonging to one class but not to another.

This paper presents optimizations of the membership algorithms used in MeSCaL, targeting around twenty natural classes that extend the three discussed above. These classes can be described either by regular expressions or by logical formalisms. Nine of them arise by enriching the signatures of the classes defined in $\mathrm{FO}(<)$, $\mathcal{B}\Sigma_1(<)$ and $\mathrm{FO}^2(<)$ with additional predicates—for example, testing the value of a position in a word modulo some integer, or counting occurrences of a specific letter before a position modulo some integer. Additional classes, introduced in the body of the paper, are defined using unary temporal logic.

Organization. Section 2 introduces the core concepts and presents data structures for their efficient implementation in MeSCaL. Section 3 defines the classes

---

[2] https://github.com/thomas-place/mescal (see [34]).

of languages considered in this paper, reviews their corresponding membership algorithms, and outlines key optimizations. Finally, Sect. 4 focuses on the computation of kernels. Due to space constraints, proofs are deferred to [26].

2   Preliminaries

Words, Regular Languages, Classes, Membership. We fix a *finite alphabet* A. As usual, A^* is the set of all finite words over A, including the empty one ε. For $w \in A^*$, we write $|w|$ for the length of w. For $u, v \in A^*$, we let uv be the word obtained by concatenating u and v. A *language* is a subset of A^*. For $K, L \subseteq A^*$, we let $KL = \{uv \mid u \in K \text{ and } v \in L\}$.

A *class of languages* is a set of languages. A class $\mathcal{C}$ is a *Boolean algebra* if $\emptyset \in \mathcal{C}$, $A^* \in \mathcal{C}$ and $\mathcal{C}$ is closed under union, intersection and complement: for all $K, L \in \mathcal{C}$, we have $K \cup L \in \mathcal{C}$, $K \cap L \in \mathcal{C}$ and $A^* \setminus L \in \mathcal{C}$. Moreover, $\mathcal{C}$ is *closed under quotients* when for all $L \in \mathcal{C}$ and $u \in A^*$, the languages $\{v \in A^* \mid uv \in L\}$ and $\{v \in A^* \mid vu \in L\}$ both belong to $\mathcal{C}$. A *prevariety* is a Boolean algebra closed under quotients and containing only *regular languages*. Regular languages are those that can be equivalently defined by finite automata or finite monoids.

We are interested in implementing *membership algorithms* for several classes. Given a class $\mathcal{C}$, the $\mathcal{C}$ *-membership* problem asks whether an input regular language belongs to $\mathcal{C}$. We focus on specific classes $\mathcal{C}$, built by applying *operators* to *input classes*. An operator Op takes as input a class $\mathcal{D}$ and produces a larger class $\mathrm{Op}(\mathcal{D})$. Membership for these classes is handled via known transfer theorems, which imply that $\mathrm{Op}(\mathcal{D})$-membership reduces to a more involved problem called $\mathcal{D}$ *-separation*. It takes as input two regular languages $L_1, L_2 \subseteq A^*$, and asks whether there exists a language $K \in \mathcal{D}$ such that $L_1 \subseteq K$ and $L_2 \cap K = \emptyset$.

Monoids and morphisms. We define regular languages via monoids. A *monoid* is a set M endowed with an associative multiplication $(s, t) \mapsto st$ that has an identity element 1_M. Clearly, A^* forms a monoid under concatenation, with ε as the identity. In the paper, all monoids other than A^* are assumed to be finite. A *morphism* is a map $\alpha : A^* \to M$, where M is a monoid, such that $\alpha(\varepsilon) = 1_M$ and $\alpha(uv) = \alpha(u)\alpha(v)$ for all $u, v \in A^*$. The *right (resp. left) Cayley graph* of a morphism $\alpha : A^* \to M$ is the directed graph whose vertex set is M, and there is an edge $s \xrightarrow{a} t$ when $t = s\alpha(a)$ (resp. $t = \alpha(a)s$).

A language $L \subseteq A^*$ is *recognized* by a morphism $\alpha : A^* \to M$ when there exists a set $F \subseteq M$ such that $L = \alpha^{-1}(F)$. In this case, we also say that M *itself* recognizes L. It is well-known that a language is regular if and only if it is recognized by a *finite* monoid. Moreover, every regular language is recognized by a canonical morphism: its *syntactic morphism*. It can be computed from any recognizer, such as an automaton or an arbitrary recognizing morphism.

👂 MeSCaL Data Structures 1: representation of finite monoids

We represent a morphism $\alpha : A^* \to M$ (with M finite) by its Cayley graphs. This representation has the advantage that it requires $\Theta(|M||A|)$ space only, instead of $\Theta(|M|^2)$ for storing the full multiplication table. This is important, as syntactic monoids can be huge, even for simple languages. However, the drawback is that computing a product "st" is *not* $O(1)$ time: for instance, if we use the left Cayley graph and if $s = \alpha(a_1 \cdots a_k)$ with $a_i \in A$ for all $i \leq k$, the element st is obtained from vertex t by following the path labeled $a_k \cdots a_1$.

👂 MeSCaL Data Structures 2: representation of elements in a monoid

Each element is represented by a product of at most $|M|$ generators $\alpha(a)$ for $a \in A$. A naive approach would take $O(|M|^2)$ space. Instead, we use a spanning tree of the right Cayley graph of α rooted at 1_M, which takes $O(|M|)$ space.

Green's Relations. Over any monoid M, we define *preorders* $\leqslant_{\mathcal{R}}, \leqslant_{\mathcal{L}}, \leqslant_{\mathcal{J}}$ as follows. For $q, s \in M$, we write $q \leqslant_{\mathcal{R}} s$ (resp. $q \leqslant_{\mathcal{L}} s$, resp. $q \leqslant_{\mathcal{J}} s$) when there exist $p, r \in M$ such that $s = qr$ (resp. $s = pq$, resp. $s = pqr$). For $\mathcal{K} \in \{\mathcal{R}, \mathcal{L}, \mathcal{J}\}$, we let $\mathcal{K}$ be the equivalence relation associated with $\leqslant_{\mathcal{K}}$, i.e., defined by $s \mathrel{\mathcal{K}} t$ if and only if $s \leqslant_{\mathcal{K}} t$ and $t \leqslant_{\mathcal{K}} s$. Finally, $s \mathrel{\mathcal{H}} t$ when $s \mathrel{\mathcal{R}} t$ and $s \mathrel{\mathcal{L}} t$. The equivalence relations $\mathcal{H}, \mathcal{R}, \mathcal{L}, \mathcal{J}$ are standard and called the *Green's relations*. Sometimes, we consider a submonoid N of a monoid M. In this case, for $s, t \in N$ and $\mathcal{K} \in \{\mathcal{H}, \mathcal{R}, \mathcal{L}, \mathcal{J}\}$, the notation $s \mathrel{\mathcal{K}} t$ is ambiguous. In general, the relation $\mathcal{K}$ over N is *not* the restriction to $N \times N$ of the relation $\mathcal{K}$ over M. When such an ambiguity arises, we write $\mathcal{K}_N$ and $\mathcal{K}_M$ to distinguish both relations.

👂 MeSCaL Data Structures 3: computing Green's relations

From the Cayley graphs of a surjective morphism $\alpha : A^* \to M$, we can compute the Green's relations $\mathcal{R}$, $\mathcal{L}$ and $\mathcal{J}$ on M in $O(|M||A|)$ time.

Indeed, by definition, two elements of M are $\mathcal{R}$- (resp. $\mathcal{L}$-) equivalent if and only if they lie in the same strongly connected component in the right (resp. left) Cayley graph. Tarjan's algorithm [42] identifies these components in $O(|M||A|)$ time. The relation $\mathcal{J}$ is computed similarly in $O(|M||A|)$ time, using a merging of both Cayley graphs. Finally, to obtain $\mathcal{H}$, we sort each $\mathcal{J}$-class according to the $\mathcal{R}$- and $\mathcal{L}$-class of each element, so that $\mathcal{H}$-equivalent elements occur consecutively. This takes $O(|M||A| \log(|M|))$ time to compute $\mathcal{H}$.

Special Elements. Let M be a *finite* monoid. An *idempotent* in M is an element $e \in M$ such that $ee = e$. We write $E(M)$ for the set of idempotents of M. It is folklore that for every *finite* monoid M, there is a natural number $\omega(M) \geq 1$ (written ω when M is understood) such that s^ω is idempotent for all $s \in M$. An element is *regular* when it is $\mathcal{J}$-equivalent to some idempotent. It is standard that in this case, it is also $\mathcal{R}$- (resp. $\mathcal{L}$-) equivalent to some idempotent. Finally, we say that a $\mathcal{K}$-class is *regular* (for $\mathcal{K} \in \{\mathcal{H}, \mathcal{R}, \mathcal{L}, \mathcal{J}\}$) if it contains an idempotent.

3   Operators and Classes

We now introduce the classes studied in this paper, present the challenges of implementing their membership algorithms, and outline the strategy used in MeSCaL. These classes are built via *closure operators*, which we first describe.

3.1   Closure Operators

Star-Free Closure. Let $\mathcal{C}$ be a prevariety. Its *star-free closure*, denoted SF($\mathcal{C}$), is the least class containing $\mathcal{C}$ and closed under Boolean operations (if $K, L \in \mathcal{C}$, then $K \cup L \in \mathcal{C}$ and $A^* \setminus K \in$ SF($\mathcal{C}$)) and marked concatenation (if $K, L \in \mathcal{C}$ and $a \in A$, then $KaL \overset{\text{def}}{=} K\{a\}L \in$ SF($\mathcal{C}$)).

Just as SF consists in languages definable in first-order logic FO($<$) [19], the languages of SF($\mathcal{C}$) are captured by a variant of this logic, denoted FO($\mathbb{I}_\mathcal{C}$), whose signature depends on the class $\mathcal{C}$ [27]. The definition of $\mathbb{I}_\mathcal{C}$ is natural: each language L in $\mathcal{C}$ gives rise to a binary predicate I_L, where $x \; I_L \; y$ holds when $x < y$ and the infix strictly between x and y belongs to L. Star-free languages also have a well-known characterization in temporal logic [11]: they are exactly those definable in pure-future linear-time temporal logic LTL, and equivalently in full linear-time temporal logic. These characterizations extend to SF($\mathcal{C}$) as well [32].

Boolean Polynomial Closure. Let $\mathcal{C}$ be a prevariety. Its *polynomial closure* Pol($\mathcal{C}$) is the least class containing $\mathcal{C}$ closed under union and marked concatenation: if $K, L \in$ Pol($\mathcal{C}$), then $K \cup L \in$ Pol($\mathcal{C}$) and $KaL \in$ Pol($\mathcal{C}$) for $a \in A$. In general, Pol($\mathcal{C}$) is *not* closed under complement. This motivates the definition of BPol($\mathcal{C}$) as the least Boolean algebra containing Pol($\mathcal{C}$).

The class PT = BPol($\{\emptyset, A^*\}$) of piecewise testable languages corresponds to the class of languages definable in $\mathcal{B}\Sigma_1(<)$, *i.e.*, in the quantifier-alternation free fragment of FO($<$) [46]. As for the full class of star-free languages, this correspondence extends to classes of the form BPol($\mathcal{C}$). Indeed, when $\mathcal{C}$ is a prevariety, the classes BPol($\mathcal{C}$) and $\mathcal{B}\Sigma_1(\mathbb{I}_\mathcal{C})$ coincide [27].

Temporal Closures. Consider the class FO$^2(<)$ mentioned in the introduction. It consists of languages definable in FO($<$) using two variable names. Just as FO($<$) corresponds to LTL, the class FO$^2(<)$ corresponds to unary temporal logic TL [7]. We introduce three operators from [29] inspired by temporal logic: TL[$\mathcal{C}$], FL[$\mathcal{C}$] and PL[$\mathcal{C}$]. Let $\mathcal{C}$ be a prevariety. Formulas in TL[$\mathcal{C}$] are defined by the grammar $\varphi := a \mid \varphi \vee \varphi \mid \varphi \wedge \varphi \mid \neg\varphi \mid \mathrm{F}_L \, \varphi \mid \mathrm{P}_L \, \varphi$ where $a \in A$ and $L \in \mathcal{C}$. Formulas in FL[$\mathcal{C}$] (resp. in PL[$\mathcal{C}$]) are those that do not use P_L (resp. F_L).

A TL[$\mathcal{C}$]-formula is evaluated on a pair (w, i) where $w = a_1 \cdots a_n$ is a word ($a_i \in A$) and $i \in \{0, 1, \ldots, |w| + 1\}$ is a position in w. We view 0 and $|w| + 1$ as dummy positions, which carry no letter. In contrast, if $1 \leq i \leq |w|$, position i carries letter a_i. For two positions $i < j$, we write $w(i, j) = a_{i+1} \cdots a_{j-1}$. We define by induction what it means for the pair (w, i) *to satisfy* φ, denoted $w, i \models \varphi$. First, for $a \in A$, we have $w, i \models a$ if position i carries an "a". The semantics of the Boolean connectives $\neg$, $\wedge$, $\vee$ is standard. Finally, let $L \in \mathcal{C}$. We have $w, i \models F_L \psi$ if there exists a position $j > i$ such that $w(i, j) \in L$ and $w, j \models \psi$. Similarly, $w, i \models P_L \psi$ holds if there exists $j < i$ such that $w(j, i) \in L$ and $w, j \models \psi$.

For a formula $\varphi \in \mathrm{TL}[\mathcal{C}]$, we define $L_{min}(\varphi) = \{w \in A^* \mid w, 0 \models \varphi\}$ and $L_{max}(\varphi) = \{w \in A^* \mid w, |w| + 1 \models \varphi\}$. We are ready to define our three classes:

- $\mathrm{TL}(\mathcal{C})$ is the class consisting of all languages $L_{min}(\varphi)$ with $\varphi \in \mathrm{TL}[\mathcal{C}]$.
- $\mathrm{FL}(\mathcal{C})$ is the class consisting of all languages $L_{min}(\varphi)$ with $\varphi \in \mathrm{FL}[\mathcal{C}]$.
- $\mathrm{PL}(\mathcal{C})$ is the class consisting of all languages $L_{max}(\varphi)$ with $\varphi \in \mathrm{PL}[\mathcal{C}]$.

In particular, for $\mathcal{C} = \{\emptyset, A^*\}$ the standard unary temporal logic TL defines exactly $\mathrm{TL}(\{\emptyset, A^*\})$. Moreover, as for the two preceding operators, we have a correspondence result: $\mathrm{TL}(\mathcal{C})$ defines the same languages as $\mathrm{FO}^2(\mathbb{I}_\mathcal{C})$. Classes of the form $\mathrm{TL}(\mathcal{C})$ were first investigated for specific input classes $\mathcal{C}$, sometimes under the form $\mathrm{FO}^2(\mathbb{I}_\mathcal{C})$, see [5] and [13–15]. Finally, a generic membership algorithm for $\mathrm{TL}(\mathcal{C})$, based on a reduction to $\mathcal{C}$-separation, was given in [31].

3.2   Main Input Classes: Group Prevarieties

A monoid G is a *group* if every element $g \in G$ has an *inverse* g^{-1}, *i.e.*, such that $gg^{-1} = g^{-1}g = 1_G$. A *group language* is a language recognized by a finite group.

We consider input classes that are *group prevarieties*, *i.e.*, prevarieties consisting only of group languages. We are interested in four main group prevarieties:

- The prevariety GR of all *group languages*.
- The prevariety AMT of *alphabet modulo testable languages* consists of all finite Boolean combinations of languages $\{w \in A^* \mid \#_a w \equiv r \pmod q\}$, where $\#_a w$ is the number of occurrences of "a" in w. It is standard that AMT consists of all regular languages recognized by a finite *commutative group*.
- The prevariety MOD of *modulo languages* consists of all finite unions of languages of the form $(A^q)^* A^r = \{w \in A^* \mid |w| \equiv r \pmod q\}$ for $0 \le r < q$.
- Finally, the prevariety $\mathrm{ST} = \{\emptyset, A^*\}$ is the trivial class.

Remark 1. Considering group prevarieties as input of our operators is natural. Indeed, languages definable in first-order logic FO($<$) are characterized by the fact that their syntactic monoid is group-free (*i.e.*, it contains no nontrivial group).

Remark 2. Some classes we tackle in MeSCaL have nice interpretations in terms of first-order logic. Indeed, as discussed in Sect. 3.1, the classes $\mathrm{SF}(\mathcal{C})$, $\mathrm{BPol}(\mathcal{C})$ and $\mathrm{TL}(\mathcal{C})$ correspond to fragments of first order logic for a suitable signature $\mathbb{I}_\mathcal{C}$ (namely, to $\mathrm{FO}(\mathbb{I}_\mathcal{C})$, $\mathcal{B}\Sigma_1(\mathbb{I}_\mathcal{C})$ and $\mathrm{FO}^2(\mathbb{I}_\mathcal{C})$, respectively). In addition to the letter predicates $a(x), b(x), \ldots$ which are always present, and the linear order $<$:

- The signature $\mathbb{I}_{\mathrm{MOD}}$ contains all modular predicates. A *modular predicate* is of the form "$x \equiv r \pmod{q}$", where q, r are integers such that $0 \leq r < q$.
- The signature $\mathbb{I}_{\mathrm{AMT}}$ consists of all "*letter modular predicates*", of the form "$\#_a(x) \equiv r \pmod{q}$", where q, r are integers such that $0 \leq r < q$ and $\#_a(x)$ denotes the number of "a"'s that occur before position x, for $a \in A$.

> ➤ **MeSCaL features**
>
> We are ready to list the 20 classes discussed in this paper. We present optimized algorithms implemented in MeSCaL for deciding membership in classes of the form $\mathrm{Op}(\mathcal{C})$, where "$\mathrm{Op}()$" is one of the five operators $\mathrm{SF}()$, $\mathrm{BPol}()$, $\mathrm{TL}()$, $\mathrm{FL}()$ and $\mathrm{PL}()$, and $\mathcal{C}$ is one of the four classes GR, AMT, MOD and ST.

3.3 Characterizations and Membership

The membership tests implemented in MeSCaL for the aforementioned classes are based on generic characterizations of the operators SF, BPol, TL, FL and PL, which are specific to input classes $\mathcal{G}$ that are group prevarieties. For each operator Op, they imply that $\mathrm{Op}(\mathcal{G})$-membership reduces to $\mathcal{G}$-separation. We present these characterizations and discuss the challenges involved in implementing them.

Kernels. Most characterizations are based on a specific object: $\mathcal{G}$-kernels. Let $\mathcal{G}$ be a group prevariety and $\alpha : A^* \to M$ be a morphism. The $\mathcal{G}$ *-kernel of* α is the subset of M that consists of all elements $s \in M$ such that $\{\varepsilon\}$ is *not* $\mathcal{G}$-separable from $\alpha^{-1}(s)$. In other words, every group language of $\mathcal{G}$ that contains $\{\varepsilon\}$ must intersect $\alpha^{-1}(s)$. The computation of the GR-kernel was, for many years, a central question in semigroup theory, linked to a conjecture by Rhodes [9], eventually solved by Ash [1]. The following result is standard (see [29, Fact 5.5]).

Lemma 3. *Let $\mathcal{G}$ be a group prevariety and $\alpha : A^* \to M$ be a surjective morphism. Then, the $\mathcal{G}$-kernel of α is a submonoid of M containing $E(M)$.*

When their input is a group prevariety $\mathcal{G}$, the four operators SF, TL, FL and PL are characterized in terms of $\mathcal{G}$-kernels. Let us present these characterizations. For each of the Green's relations $\mathcal{K} \in \{\mathcal{J}, \mathcal{R}, \mathcal{L}, \mathcal{H}\}$, a monoid is said to be $\mathcal{K}$ *-trivial* if all of its $\mathcal{K}$-classes are singletons. In particular, $\mathcal{H}$-trivial monoids are known as *aperiodic* monoids. The following is proved in [32, Corollary 5.13].

Theorem 4. *Let $\mathcal{G}$ be a group prevariety. A regular language belongs to $\mathrm{SF}(\mathcal{G})$ if and only if the $\mathcal{G}$-kernel of its syntactic morphism is aperiodic.*

We now turn to the temporal closures. The characterization of $\mathcal{C} \mapsto \mathrm{TL}(\mathcal{C})$ relies on the class of monoids DA, typically defined by an equation: a finite monoid M belongs to DA if, for all $s, t \in M$, we have $(st)^\omega s(st)^\omega = (st)^\omega$. The following is proved in [29, Theorem 10.1] and [31, Theorems 37 and 38].

Theorem 5. *Let $\mathcal{G}$ be a group prevariety and L be a regular language.*

- *$L \in \mathrm{PL}(\mathcal{G})$ if and only if the $\mathcal{G}$-kernel of its syntactic morphism is $\mathcal{R}$-trivial.*
- *$L \in \mathrm{FL}(\mathcal{G})$ if and only if the $\mathcal{G}$-kernel of its syntactic morphism is $\mathcal{L}$-trivial.*
- *$L \in \mathrm{TL}(\mathcal{G})$ if and only if the $\mathcal{G}$-kernel of its syntactic morphism is in* DA.

✳ MeSCaL Optimization 1: regular elements suffice

In view of Theorems 4 and 5, we need to compute $\mathcal{G}$-kernels and check whether they are $\mathcal{K}$-trivial for $\mathcal{K} \in \{\mathcal{H}, \mathcal{R}, \mathcal{L}\}$. A simple but useful observation [23, Proposition 2.26], used for instance in [22], states that a monoid is $\mathcal{K}$-trivial if and only if all its *regular $\mathcal{K}$-classes* are trivial. This result plays a crucial role in our implementation: focusing only on the regular elements in the $\mathcal{G}$-kernel allows for a much more efficient computation than handling the full structure.

Similarly, in order to test whether the $\mathcal{G}$-kernel belongs to DA, it suffices to look at its regular elements. Indeed, it is known that a monoid belongs to DA if and only if all its *regular* elements are idempotent [23, Proposition 4.29].

✌ MeSCaL Data Structures 4: kernels and their Green's relations

According to Optimization 1, deciding membership tests for the above four operators when the input class is a group prevariety $\mathcal{G}$ presents two challenges.

- First, we must be able to represent $\mathcal{G}$-kernels efficiently. So far, we can only represent by a Cayley graph an *A-generated monoid*, but not a submonoid.
- Second, we have to compute $\mathcal{G}$-kernels (or at least their restriction to regular elements) and their Green's relations. Again, the difficulty is that we do not have in hand the generators of this $\mathcal{G}$-kernel.

The BPol Operator. There exists a generic characterization of the classes $\mathrm{BPol}(\mathcal{G})$ when $\mathcal{G}$ is a group prevariety. However, it depends on more information than just the $\mathcal{G}$-kernel of its syntactic morphism. To present it, we need an additional notion: $\mathcal{C}$ *-pairs*. Let $\mathcal{C}$ be a class of languages and $\alpha : A^* \to M$ be a morphism. We say that $(s, t) \in M^2$ is a $\mathcal{C}$ *-pair* (for α) when $\alpha^{-1}(s)$ is *not* $\mathcal{C}$-separable from $\alpha^{-1}(t)$. We rely on the following characterization from [28].

Theorem 6. *Let $\mathcal{G}$ be a group prevariety. Then, a regular language belongs to $\mathrm{BPol}(\mathcal{G})$ if and only if its syntactic morphism $\alpha : A^* \to M$ satisfies:*

$$(qr)^\omega st(st)^\omega = (qr)^\omega qt(st)^\omega \text{ for all } q, r, s, t \in M \text{ where } (q,s) \text{ is a } \mathcal{G}\text{-pair}. \quad (1)$$

Remark 7. In two cases, when $\mathcal{G} = \mathrm{ST}$ or $\mathcal{G} = \mathrm{GR}$, Theorem 6 is not needed. Firstly, $\mathrm{BPol}(\mathrm{ST})$ is the class of piecewise testable languages and Simon's original characterization [38] can be used: the syntactic monoid must be $\mathcal{J}$-trivial. Secondly, the class BPol(GR) admits an alternate characterization based on Green's relations [9,21]: a language belongs to BPol(GR) if every regular $\mathcal{R}$-class and every regular $\mathcal{L}$-class in its syntactic monoid contains exactly one idempotent.

✵ MeSCaL Optimization 2: bypassing the brute force computation

The naive BPol($\mathcal{G}$)-membership algorithm derived from Theorem 6 is inefficient: testing (1) requires a loop ranging over four variables. Instead, we rely on the following Corollary of Theorem 5. We say that a pair $(s,t) \in M^2$ is a $\mathcal{C}$-cancel for α if $\alpha^{-1}(s)\alpha^{-1}(t)$ is *not* $\mathcal{C}$-separable from $\{\varepsilon\}$. Although this notion is well-defined for any class $\mathcal{C}$, we use it when $\mathcal{C}$ is a group prevariety.

Corollary 8. *Let $\mathcal{G}$ be a group prevariety. Then, a regular language belongs to* BPol($\mathcal{G}$) *if and only if its syntactic morphism $\alpha : A^* \to M$ satisfies:*

$$ef = qt \text{ for all } e, f \in E(M) \text{ and } \mathcal{G} - cancel(q,t) \text{ where } e \mathrel{\mathcal{R}} q \text{ and } t \mathrel{\mathcal{L}} f. \tag{2}$$

Corollary 8 offers two main advantages. First, it narrows the range of two variables by requiring e and f to be idempotent, which are often much fewer than the set of all elements. Second, once e and f are fixed, q (resp. t) is constrained to the $\mathcal{R}$-class of q (resp. $\mathcal{L}$-class of f), significantly reducing the search space in practice, and improving the running time.

On the other hand, we have to determine whether (q,t) is a $\mathcal{G}$-cancel. This is linked to the computation of $\mathcal{G}$-kernels, addressed in the next section. More precisely, we use Proposition 11 to carry out this test.

---

4   Computing Kernels

Let us recapitulate our objectives in view of Theorem 5 and Corollary 8. Fix a morphism $\alpha : A^* \to M$ in this section. We need to design algorithms for computing the following information for the three group prevarieties $\mathcal{G} \in \{\text{MOD}, \text{AMT}, \text{GR}\}$:

- The $\mathcal{G}$-kernel of α restricted to *regular elements*, as well as its Green's relations.
- The $\mathcal{G}$-cancels $(q,t) \in M^2$ of α where q,t are *regular elements*.

Remark 9. For $\mathcal{G} = \text{ST}$, these computations are trivial. Indeed, since ST $= \{\emptyset, A^*\}$, two nonempty languages are not ST-separable. Since the syntactic morphism is surjective, the ST-kernel of a syntactic morphism $\alpha : A^* \to M$ is the whole syntactic monoid M, and the set of ST-cancels for α is $M \times M$.

We must *adapt* known separation algorithms to compute $\mathcal{G}$-kernels and $\mathcal{G}$-cancels. Indeed, the original algorithms can be slow for arbitrary input languages. For instance, AMT-separation is co-NP-complete [30]. Fortunately, we are only interested in *regular elements*. The first part of this section (Subsect. 4.1) is devoted to explaining how this assumption can be exploited to obtain significantly more efficient procedures. Then, in dedicated Subsects. 4.2, 4.3 and 4.4, we describe how to compute kernels and cancels for MOD, AMT and GR. Finally in Sect. 4.5, we discuss the computation of their Green's relations.

4.1 Separation and Regular Elements for MOD, AMT and GR

Our approach relies on nondeterministic finite automata (NFAs) in a crucial way. An NFA is a pair $\mathcal{A} = (Q, \delta)$ with Q a finite set of *states* and $\delta \subseteq Q \times A \times Q$ a set of *transitions*. Let $w \in A^*$ and $q, r \in Q$. There is a *run* labeled by w from q to r if either $w = \varepsilon$ and $q = r$, or $w = a_1 \cdots a_n$ for $n \geq 1$ and $a_1, \ldots, a_n \in A$, and there are $q_0, \ldots, q_n \in Q$ such that $q_0 = q$, $q_n = r$ and $(q_{i-1}, a_i, q_i) \in \delta$ for $1 \leq i \leq n$. For $q, r \in Q$, let $L(\mathcal{A}, q, r)$ be the language of words labeling a run from q to r.

When dealing with $\mathcal{G}$-cancels, we will consider *mirrors* of NFAs. For an NFA $\mathcal{A} = (Q, \delta)$, we define a transition relation $\gamma = \{(q, a, r) \in Q \times A \times Q \mid (r, a, q) \in \delta\}$. The *mirror* of $\mathcal{A}$ is then the NFA $mir(\mathcal{A}) = (Q, \gamma)$.

We come back to kernels and cancels. With each morphism $\alpha : A^* \to M$, we associate two NFAs $\mathcal{A}_R[\alpha] = (M, \delta_R)$ and $\mathcal{A}_L[\alpha] = (M, \delta_L)$ with the transitions,

$$\delta_R = \{(s, a, t) \mid s \,\mathcal{R}\, t \text{ and } s\alpha(a) = t\} \text{ and } \delta_L = \{(s, a, t) \mid s \,\mathcal{L}\, t \text{ and } \alpha(a)s = t\}.$$

Observe that $\mathcal{A}_R[\alpha]$ (resp. $\mathcal{A}_L[\alpha]$) is the restriction of the right (resp. left) Cayley graph of α to its strongly connected components: the $\mathcal{R}$-classes (resp. $\mathcal{L}$-classes). In particular, $\mathcal{A}_R[\alpha]$ and $\mathcal{A}_L[\alpha]$ are *not* connected in general: every run stays within a single strongly connected component of the corresponding Cayley graph.

We turn to the computation of the regular elements of $\mathcal{G}$-kernels. The main ingredient for speeding up this process is the following proposition.

Proposition 10. *Let $\mathcal{G}$ be a group prevariety and $\alpha : A^* \to M$ be a surjective morphism. Let $e \in E(M)$ and $s \in M$ such that $e \,\mathcal{R}\, s$. Then, s belongs to the $\mathcal{G}$-kernel of α if and only if $\{\varepsilon\}$ is not $\mathcal{G}$-separable from $L(\mathcal{A}_R[\alpha], e, s)$.*

✱ MeSCaL Optimization 3: computing kernels

Proposition 10 yields two optimizations, speeding up the algorithms in practice.

- First, it implies that to decide if a regular element $s \in M$ is in the $\mathcal{G}$-kernel of α, it suffices to consider a *small* NFA compared to the whole right Cayley graph of α: the strongly connected component of $\mathcal{A}_R[\alpha]$ of state s.
- Second, this NFA forms a *single strongly connected component*, a property that will be crucial for enhancing our $\mathcal{G}$-separation algorithms.

We turn to the computation of $\mathcal{G}$-cancels required by Corollary 8.

Proposition 11. *Let $\mathcal{G}$ be a group prevariety and $\alpha : A^* \to M$ be a morphism. Let $e, f \in E(M)$ and $q, t \in M$ such that $e \mathrel{\mathcal{R}} q$ and $t \mathrel{\mathcal{L}} f$. Then, (q,t) is a $\mathcal{G}$-cancel if and only if $L(\mathcal{A}_R[\alpha], e, q)$ is not $\mathcal{G}$-separable from $L(mir(\mathcal{A}_L[\alpha]), f, t)$.*

✡ MeSCaL Optimization 4: computing cancels

In this case as well, Proposition 11 implies that in order to decide whether a pair of regular elements $(q,t) \in M^2$ is a $\mathcal{G}$-cancel, it suffices to separate two *small* NFAs corresponding to the $\mathcal{R}$-class of q and the $\mathcal{L}$-class of t. Again, these NFAs both consist of a *single strongly connected component*.

4.2   Computing GR-Kernels and GR-Cancels

Both definitions of the $\mathcal{G}$-kernel and of $\mathcal{G}$-cancels are based on $\mathcal{G}$-separation. For $\mathcal{G} = \mathrm{GR}$, the decidability of GR-separation is a difficult result due to Ash [1]. Alternative proofs were found [2,25,35]. We follow the recent approach from [30]. It relies on a construction on automata that we introduce first.

Let $\mathcal{A} = (Q, \delta)$ be an NFA. We associate with it an NFA $\langle \mathcal{A} \rangle$ with transitions labeled by an extended alphabet $\tilde{A}$. For each $a \in A$, we create a fresh letter a^{-1} and define $A^{-1} = \{a^{-1} \mid a \in A\}$. We let $\tilde{A}$ be the disjoint union $\tilde{A} = A \cup A^{-1}$. We define $\langle \mathcal{A} \rangle = (Q, \langle \delta \rangle)$, where $\langle \delta \rangle$ is the following extended set of transitions:

$$\langle \delta \rangle = \delta \cup \left\{ (r, a^{-1}, q) \mid (q, a, r) \in \delta \text{ and } L(\mathcal{A}, r, q) \neq \emptyset \right\}.$$

In other words, for each transition $(q, a, r) \in \delta$ inside a strongly connected component of $\mathcal{A}$, we add to $\langle \delta \rangle$ the transition (r, a^{-1}, q) in the opposite direction.

By definition, all languages $L(\langle \mathcal{A} \rangle, q, r)$ are over the extended alphabet $\tilde{A}$. We use a standard rewriting rule on words in $\tilde{A}^*$. Given $w, w' \in \tilde{A}^*$, we write $w \to w'$ if there exist $x, y \in \tilde{A}^*$ and $a \in A$ such that either $w = xaa^{-1}y$ or $w = xa^{-1}ay$, and $w' = xy$. The reflexive transitive closure of "$\to$" is denoted by "$\overset{*}{\to}$". Finally, for every language $K \subseteq \tilde{A}^*$, we let,

$$\lfloor K \rfloor = \{ w \in \tilde{A}^* \mid \text{there exists } u \in K \text{ such that } u \overset{*}{\to} w \}.$$

The following result is proved in [30].

Theorem 12. *Let $\mathcal{A} = (Q, \delta)$ be an NFA and $q_1, r_1, q_2, r_2 \in Q$. Then, $L(\mathcal{A}, q_1, r_1)$ is $\underline{not}$ GR-separable from $L(\mathcal{A}, q_2, r_2)$ iff $\lfloor L(\langle \mathcal{A} \rangle, q_1, r_1) \rfloor \cap \lfloor L(\langle \mathcal{A} \rangle, q_2, r_2) \rfloor \neq \emptyset$.*

The criterion provided by Theorem 12 is decidable. In [30], the intersection is checked by computing an NFA (with ε-transitions) recognizing $\lfloor L(\langle \mathcal{A} \rangle, q_1, r_1) \rfloor$ and $\lfloor L(\langle \mathcal{A} \rangle, q_2, r_2) \rfloor$. This is costly: ε-transitions are *recursively* added between states connected by a word aa^{-1} or $a^{-1}a$. A rough analysis yields a $|Q|^3$ time complexity. Fortunately, we can do better for the *specific* NFAs that we consider.

Indeed, we only use the NFAs $\mathcal{A}_L[\alpha]$ and $\mathcal{A}_R[\alpha]$ via Propositions 10 and 11. Thanks to Optimization 1, it suffices to compute the *regular* elements of the GR-kernel of α. By Proposition 10, s belongs to the GR-kernel of α if and only if $\{\varepsilon\}$ is not GR-separable from $L(\mathcal{A}_R[\alpha], e, s)$, where e is any idempotent $\mathcal{R}$-equivalent to s. Since e and s lie in the same strongly connected component $\mathcal{A}_e$ of $\mathcal{A}_R[\alpha]$, we have $L(\mathcal{A}_R[\alpha], e, s) = L(\mathcal{A}_e, e, s)$. Hence, in view of Theorem 12, we have to test whether ε belongs to $\lfloor L(\langle \mathcal{A}_e \rangle, e, s) \rfloor$. Since $\mathcal{A}_e$ is strongly connected, one can verify that $\lfloor L(\langle \mathcal{A}_e \rangle, e, s) \rfloor$ is recognized by the *Stallings' folding* of $\mathcal{A}_e$ [39], obtained by repeatedly merging states with identical forward or backward transitions: we identify any two states r, s whenever there exist transitions $r \xrightarrow{a} t$ and $s \xrightarrow{a} t$, or transitions $t \xrightarrow{a} r$ and $t \xrightarrow{a} s$. The folding can be computed using the union-find algorithm [43], yielding an $O(|M||A| \cdot \text{Ackinv}(|M|))$ upper bound (see also [47]), where Ackinv is the inverse Ackermann function. The same kind of optimization is used in MeSCaL for computing GR-cancels.

4.3   Computing AMT-Kernels and AMT-Cancels

We first recall an AMT-separation procedure from [30]. It relies on the extended alphabet $\tilde{A}$ and the NFA construction $\mathcal{A} \mapsto \langle \mathcal{A} \rangle$ introduced above.

We start with a preliminary definition. Let $n = |A|$ and let $A = \{a_1, \ldots, a_n\}$. We define a map $\zeta : \tilde{A}^* \to \mathbb{Z}^n$ as follows:

$$\zeta(w) = (|w|_{a_1} - |w|_{a_1^{-1}}, \ldots, |w|_{a_n} - |w|_{a_n^{-1}}).$$

We extend ζ to languages by letting $\zeta(K) = \{\zeta(w) \mid w \in K\}$ for $K \subseteq A^*$. Our starting point is the following result [30], analogous to Theorem 12.

Theorem 13. *Let* $\mathcal{A} = (Q, \delta)$ *be an NFA and* $q_1, r_1, q_2, r_2 \in Q$. *Then,* $L(\mathcal{A}, q_1, r_1)$ *is* <u>*not*</u> *AMT-separable from* $L(\mathcal{A}, q_2, r_2)$ *iff* $\zeta(L(\langle \mathcal{A} \rangle, q_1, r_1)) \cap \zeta(L(\langle \mathcal{A} \rangle, q_2, r_2)) \neq \emptyset$.

In [30], it is argued that verifying the criterion provided by Theorem 13 amounts to checking whether an existential Presburger formula is true. Although such a formula can be constructed in polynomial time [8], the existential fragment of Presburger logic is NP-complete [36]. Unfortunately, this lower bound also applies to deciding AMT-separation, which is known to be co-NP-complete [30]. As a result, we avoid relying on the generic AMT-separation algorithm. Instead, we adopt an efficient, purpose-specific approach tailored to our needs. It applies to the computation of both AMT-kernels and AMT-cancels. We explain it for kernels.

According to Optimization 1, we only need to compute *regular* elements of the AMT-kernel of $\alpha : A^* \to M$. Let s be such an element. Proposition 10 states that s belongs to the AMT-kernel of α if and only if $\{\varepsilon\}$ is not AMT-separable from $L(\mathcal{A}_R[\alpha], e, s)$, where e is any idempotent $\mathcal{R}$-equivalent to s. By Theorem 13, this amounts to testing that $\bar{0} \notin \zeta(L(\langle \mathcal{A}_R[\alpha] \rangle, e, s))$, where $\bar{0} = (0, \ldots, 0) \in \mathbb{Z}^n$.

Let $SCC_R(e)$ be the strongly connected component of e in the graph induced by $\langle \mathcal{A}_R \rangle$: its vertices are the elements of the $\mathcal{R}$-class of e, and its edges are of the form $r \xrightarrow{a} t$ and $t \xrightarrow{a^{-1}} r$, with $r\alpha(a) = t$. Indeed, recall that if r, t lie in the same strongly connected component and if $r\alpha(a) = t$, we have transitions (r, a, t) and (r, a^{-1}, r) in $\langle \delta_R \rangle$. For a path P labeled by a word $w \in \tilde{A}^*$ in $SCC_R(e)$, we let $\zeta(P) = \zeta(w)$. Then, for any s $\mathcal{R}$-equivalent to e, deciding whether $\bar{0} \notin \zeta(L(\langle \mathcal{A}_R[\alpha] \rangle, e, s))$ amounts to finding a path P from e to s with $\zeta(P) = \bar{0}$.

We decide this question by computing a cycle basis. Let $T_R(e)$ be a spanning tree of $SCC_R(e)$. Let $r \xrightarrow{a} t$ be an edge of $SCC_R(e)$ with $a \in A$. If neither $r \xrightarrow{a} t$ nor $t \xrightarrow{a^{-1}} r$ is an edge of $T_R(e)$, we define a cycle by gluing the path from t to r in $T_R(e)$ with the edge $r \xrightarrow{a} t$. It is known [16,17] that the cycles $C_1, \ldots, C_k$ we obtain form an *integral cycle basis*. This means that for any cycle C in $SCC_R(e)$, the value $\zeta(C) \in \mathbb{Z}^n$ is a $\mathbb{Z}$-linear combination of the $\zeta(C_i)$ for $1 \leq i \leq k$.

Fix two paths P, P' from e to s in $SCC_R(e)$. The path obtained by following P forwards from e to s and then P' backwards from s to e is a cycle C, and we have $\zeta(C) = \zeta(P) - \zeta(P')$. Since $(C_1, \ldots, C_k)$ is an integral cycle basis, we have,

$$\zeta(P) = \zeta(P') + \sum_{i=1}^{k} \lambda_i \zeta(C_i), \quad \text{for some integers } \lambda_1, \ldots, \lambda_k.$$

Recall that deciding whether a (regular) element s of the $\mathcal{R}$-class of e is in the AMT-kernel is equivalent to determining whether there is a path P from e to s with $\zeta(P) = \bar{0}$. This leads to the following algorithm for answering this question:

- Pick $e \in E(M)$ such that $e \mathcal{R} s$.
- Compute a spanning tree of $SCC_R(e)$ and extract from this spanning tree:
  - a cycle basis $(C_i)_{1 \leq i \leq k}$ of $SCC_R(e)$,
  - the path P' connecting s to e in this spanning tree.
- Solve the system $\zeta(P') + \sum_{i=1}^{k} \lambda_i \zeta(C_i) = \bar{0}$ of unknowns $\lambda_1, \ldots, \lambda_k \in \mathbb{Z}$.

Altogether, this boils down to solving an integer system. We do this in MeSCaL using Hermite normal forms along with the Flint library [44]. A rough analysis yields an overall $O(|M||A| \log(|M|))$ time complexity, for treating all $\mathcal{R}$-classes.

Remark 14. In MeSCaL, one single idempotent e is chosen for each $\mathcal{R}$-class. The spanning tree is computed once, and used for all elements s of this $\mathcal{R}$-class.

4.4   Computing MOD-Kernels and MOD-Cancels

Finally, separation for MOD is handled by reduction to either GR or AMT. Let $U = \{\$\}$ be a *single-letter alphabet* and $\mu : A^* \to U^*$ be the morphism defined by $\mu(a) = \$$ for every $a \in A$. The following is proved in [30, Theorem 26].

Theorem 15. *Let $K, L \subseteq A^*$. The following conditions are equivalent:*

1. K is not MOD-*separable from* L.
2. $\mu(K)$ is not AMT-*separable from* $\mu(L)$.
3. $\mu(K)$ is not GR-*separable from* $\mu(L)$.

Clearly, if $L \subseteq A^*$ is recognized by an input NFA $\mathcal{A}$, one can compute an NFA recognizing $\mu(L)$ in linear time: one has to relabel all transitions with "\$". Therefore, we can compute MOD-kernels and MOD-cancels by using either the procedure for GR or that for AMT, yielding an efficient algorithm.

Remark 16. The MOD-kernel of a morphism $\alpha : A^* \to M$ was originally introduced by Straubing [41] under the name *stable monoid* using a specialized definition unrelated to MOD-separation. The image $\alpha(A)$ can be viewed as an element of the monoid 2^M, equipped with the natural multiplication. Observe that this monoid has exponential size in M, hence doubly exponential in the size of the corresponding minimal automaton. The stable monoid can then be defined as $(\alpha(A))^\omega \cup \{1_M\}$. While interesting, this original definition leads to a significantly less efficient algorithm than the one we presented.

4.5 Green's Relations

By Theorems 4 and 5, solving SF($\mathcal{G}$)-, PL($\mathcal{G}$)-, FL($\mathcal{G}$)- and TL($\mathcal{G}$)-membership requires computing the Green's relations of the $\mathcal{G}$-kernel. For the *full* syntactic monoid, the computation of its Green's relations amounts to computing the strongly connected components of its Cayley graphs. This approach is efficient because we have a small set of generators available for the syntactic monoid: the images of letters. Unfortunately, this is *not* the case for a submonoid.

In MeSCaL, we use an approach *specific to $\mathcal{G}$ -kernels*. As seen in Optimization 1, we only need the Green's relations over *regular elements. For $\mathcal{G}$ -kernels*, they are computed by restricting the Green's relations of the underlying monoid.

Lemma 17. *Let $\mathcal{G}$ be a group prevariety, $\alpha : A^* \to M$ be a surjective morphism and N be the $\mathcal{G}$-kernel of α. Let $s, t \in N$. Assume that s, t are regular __in M__. Then, for $\mathcal{K} \in \{\mathcal{H}, \mathcal{L}, \mathcal{R}\}$, we have $s \mathcal{K}_N t$ if and only if $s \mathcal{K}_M t$.*

5 Conclusion

We presented an overview of the MeSCaL software, which is designed to test properties of regular languages. We explained that directly implementing algorithms from the literature often proves inefficient, but that careful optimization can significantly improve performance.

A key insight is that certain tasks—such as kernel computations—can be restricted to *regular $\mathcal{R}$-* or *$\mathcal{L}$-classes*. This not only reduces the computational space but also limits processing to strongly connected components of the Cayley graphs, resulting in substantial algorithmic simplifications.

We applied these techniques to classes Op($\mathcal{G}$), for Op $\in \{$SF, BPol, TL, FL, PL$\}$ and $\mathcal{G} \in \{$ST, MOD, AMT, GR$\}$. MeSCaL currently supports around a hundred classes. A second paper will address additional classes, drawn from two distinct hierarchies, for which kernel and cancel precomputations are no longer sufficient.

Acknowledgments. We wish to thank the reviewers for some useful comments.

References

1. Ash, C.J.: Inevitable graphs: a proof of the type II conjecture and some related decision procedures. Internat. J. Algebra Comput. **1**(1), 127–146 (1991)
2. Auinger, K.: A new proof of the Rhodes type II conjecture. Internat. J. Algebra Comput. **14**(5–6), 551–568 (2004)
3. Caron, P.: LANGAGE: a Maple package for automaton characterization of regular languages. Theor. Comput. Sci. **231**(1), 5–15 (2000). https://doi.org/10.1016/S0304-3975(99)00013-4
4. Champarnaud, J.M., Hansel, G.: AUTOMATE, a computing package for automata and finite semigroups. J. Symb. Comput. **12**(2), 197–220 (1991). https://doi.org/10.1016/S0747-7171(08)80125-3
5. Dartois, L., Paperman, C.: Two-variable first order logic with modular predicates over words. In: Proceedings of the 30th International Symposium on Theoretical Aspects of Computer Science, STACS 2013, pp. 329–340. Leibniz International Proceedings in Informatics (LIPIcs), Schloss Dagstuhl–Leibniz-Zentrum fuer Informatik (2013). https://doi.org/10.4230/LIPIcs.STACS.2013.329
6. Delgado, M., Hoffmann, R., Linton, S., Morais, J.J.: Automata, a package on automata, Version 1.16 (2024). https://gap-packages.github.io/automata/. https://cmup.fc.up.pt/cmup/mdelgado/varia/package-automata.pdf
7. Etessami, K., Vardi, M.Y., Wilke, T.: First-order logic with two variables and unary temporal logic. Inf. Comput. **179**(2), 279–295 (2002). https://doi.org/10.1006/inco.2001.2953
8. Seidl, H., Schwentick, T., Muscholl, A., Habermehl, P.: Counting in trees for free. In: Díaz, J., Karhumäki, J., Lepistö, A., Sannella, D. (eds.) ICALP 2004. LNCS, vol. 3142, pp. 1136–1149. Springer, Heidelberg (2004). https://doi.org/10.1007/978-3-540-27836-8_94
9. Henckell, K., Margolis, S., Pin, J.É., Rhodes, J.: Ash's type II theorem, profinite topology and Malcev products. Internat. J. Algebra Comput. **1**, 411–436 (1991). https://doi.org/10.1142/S0218196791000298
10. Holzer, M., König, B.: On deterministic finite automata and syntactic monoid size. Theor. Comput. Sci. **327**(3), 319–347 (2004). https://doi.org/10.1016/J.TCS.2004.04.010
11. Kamp, H.W.: Tense logic and the theory of linear order. Ph.D. thesis, Computer Science Department, University of California at Los Angeles, USA (1968)
12. Kim, S.M., McNaughton, R., McCloskey, R.: A polynomial time algorithm for the local testability problem of deterministic finite automata. IEEE Trans. Comput. **40**(10), 1087–1093 (1991). https://doi.org/10.1109/12.93741
13. Krebs, A., Lodaya, K., Pandya, P.K., Straubing, H.: Two-variable logic with a between relation. In: Proceedings of the 31st Annual ACM/IEEE Symposium on Logic in Computer Science, LICS 2016, pp. 106–115 (2016). https://doi.org/10.1145/2933575.2935308
14. Krebs, A., Lodaya, K., Pandya, P.K., Straubing, H.: An algebraic decision procedure for two-variable logic with a between relation. In: 27th EACSL Annual Conference on Computer Science Logic, CSL 2018. Leibniz International Proceedings in Informatics (LIPIcs), Schloss Dagstuhl–Leibniz-Zentrum fuer Informatik (2018). https://doi.org/10.4230/LIPIcs.CSL.2018.28

15. Krebs, A., Lodaya, K., Pandya, P.K., Straubing, H.: Two-variable logics with some betweenness relations: expressiveness, satisfiability and membership. Log. Methods Comput. Sci. **16**(3) (2020). https://doi.org/10.23638/LMCS-16(3:16)2020
16. Liebchen, C., Peeters, L.: Integral cycle bases for cyclic timetabling. Discret. Optim. **6**(1), 98–109 (2009). https://doi.org/10.1016/j.disopt.2008.09.003
17. Liebchen, C., Peeters, L.: On cyclic timetabling and cycles in graphs. Technical report, Technische Universität Berlin (2009). https://api-depositonce.tu-berlin.de/server/api/core/bitstreams/64451d02-8d8e-40b2-8399-0b63f8f7dfbf/content
18. Matz, O., Miller, A., Potthoff, A., Thomas, W., Valkema, E.: Report on the program AMoRE. Technical report, Institut für informatik und Praktische Mathematik, Christian-Albrechts Universität, Kiel (1995). https://sourceforge.net/projects/amore/
19. McNaughton, R., Papert, S.A.: Counter-Free Automata. MIT Press (1971)
20. Paperman, C.: Python module Pysemigroup (2024). https://gitlab.inria.fr/cpaperma/pysemigroup, https://paperman.name/semigroup/
21. Pin, J.É.: PG = BG, a success story. NATO Advanced Study Institute, Semigroups, Formal Languages and Groups, pp. 33–47 (1995). https://www.irif.fr/~jep/PDF/BGPG.pdf
22. Pin, J.É.: Semigroup (2009). https://www.irif.fr/~jep/Logiciels/Semigroupe2.0/semigroupe2.html
23. Pin, J.É.: Mathematical foundations of automata theory (2025). http://www.irif.fr/~jep/PDF/MPRI/MPRI.pdf
24. Pin, J.É., Delcroix, V.: Algorithms for computing finite semigroups. In: Cucker, F., Shub, M. (eds.) Foundations of Computational Mathematics, pp. 112–126. Springer, Rio de Janeiro (1997). https://hal.science/hal-00143949v1/file/Rio97AMS.pdf
25. Pin, J.É., Reutenauer, C.: A conjecture on the Hall topology for the free group. Bull. Lond. Math. Soc. **23**(4), 356–362 (1991). https://doi.org/10.1112/blms/23.4.356
26. Place, T., Zeitoun, M.: A first taste of MeSCaL, a tool for solving membership problems for regular languages. https://www.labri.fr/perso/zeitoun/research/pdf/mescal1.pdf
27. Place, T., Zeitoun, M.: Generic results for concatenation hierarchies. Theory Comput. Syst. **63**(4), 849–901 (2018). https://doi.org/10.1007/s00224-018-9867-0
28. Place, T., Zeitoun, M.: Characterizing level one in group-based concatenation hierarchies. In: Kulikov, A.S., Raskhodnikova, S. (eds) CSR 2022. LNCS, vol. 13296, pp. 320–337. Springer, Cham (2022). https://doi.org/10.1007/978-3-031-09574-0_20
29. Place, T., Zeitoun, M.: All about unambiguous polynomial closure. TheoretiCS **2** (2023). https://doi.org/10.46298/THEORETICS.23.11
30. Place, T., Zeitoun, M.: Group separation strikes back. In: 38th Annual ACM/IEEE Symposium on Logic in Computer Science, LICS 2023, pp. 1–13. IEEE Computer Society (2023). https://doi.org/10.1109/LICS56636.2023.10175683, https://arxiv.org/pdf/2205.01632
31. Place, T., Zeitoun, M.: A generic characterization of generalized unary temporal logic and two-variable first-order logic. In: Murano, A., Silva, A. (eds.) Proceedings of the 32nd EACSL Annual Conference on Computer Science Logic, CSL 2024. LIPIcs, vol. 288, pp. 45:1–45:23. Schloss Dagstuhl - Leibniz-Zentrum für Informatik (2024). https://doi.org/10.4230/LIPICS.CSL.2024.45
32. Place, T., Zeitoun, M.: Closing star-free closure. ACM Trans. Comput. Log. **26**(3) (2025). https://doi.org/10.1145/3733831

33. Place, T., Zeitoun, M.: In orbit with MeSCaL: higher in concatenation and navigational hierarchies of regular languages. In: Castiglione, G., Mantaci, S. (eds.) CIAA 2025. LNCS, vol. 15981, pp. xx–yy. Springer, Cham (2025)

34. Place, T., Zeitoun, M.: MeSCaL, MEmbership and separation for ClAsses of languages (2025). https://github.com/thomas-place/mescal

35. Ribes, L., Zalesskii, P.A.: On the profinite topology on a free group. Bull. Lond. Math. Soc. **25**(1), 37–43 (1993). https://doi.org/10.1112/blms/25.1.37

36. Scarpellini, B.: Complexity of subcases of Presburger arithmetic. Trans. Am. Math. Soc. **284**, 203–218 (1984). https://doi.org/10.2307/1999283

37. Schützenberger, M.P.: On finite monoids having only trivial subgroups. Inf. Control **8**(2), 190–194 (1965). https://doi.org/10.1016/S0019-9958(65)90108-7

38. Simon, I.: Piecewise testable events. In: Brakhage, H. (ed.) GI-Fachtagung 1975. LNCS, vol. 33, pp. 214–222. Springer, Heidelberg (1975). https://doi.org/10.1007/3-540-07407-4_23

39. Stallings, J.R.: Topology of finite graphs. Invent. Math. **71**(3), 551–565 (1983). https://doi.org/10.1007/BF02095993

40. Stern, J.: Complexity of some problems from the theory of automata. Inf. Control **66**(3), 163–176 (1985). https://doi.org/10.1016/S0019-9958(85)80058-9

41. Straubing, H.: On logical descriptions of regular languages. In: Rajsbaum, S. (ed.) LATIN 2002. LNCS, vol. 2286, pp. 528–538. Springer, Heidelberg (2002). https://doi.org/10.1007/3-540-45995-2_46

42. Tarjan, R.: Depth-first search and linear graph algorithms. SIAM J. Comput. **1**(2), 146–160 (1972). https://doi.org/10.1137/0201010

43. Tarjan, R.E.: Efficiency of a good but not linear set union algorithm. J. ACM **22**(2), 215–225 (1975). https://doi.org/10.1145/321879.321884

44. The FLINT team: FLINT: Fast Library for Number Theory. Version 3.2.1 (2025). https://flintlib.org

45. Thérien, D., Wilke, T.: Over words, two variables are as powerful as one quantifier alternation. In: Proceedings of the 30th Annual ACM Symposium on Theory of Computing, STOC 1998, pp. 234–240. ACM (1998). https://doi.org/10.1145/276698.276749

46. Thomas, W.: Classifying regular events in symbolic logic. J. Comput. Syst. Sci. **25**(3), 360–376 (1982). https://doi.org/10.1016/0022-0000(82)90016-2

47. Touikan, N.W.M.: A fast algorithm for Stallings' folding process. Internat. J. Algebra Comput. **16**(06), 1031–1045 (2006). https://doi.org/10.1142/S0218196706003396

48. Trahtman, A.N.: A package TESTAS for checking some kinds of testability. CoRR abs/2105.12583 (2021). https://arxiv.org/abs/2105.12583

In Orbit with MeSCaL: Higher in Concatenation and Navigational Hierarchies of Regular Languages

Thomas Place and Marc Zeitoun[✉]

Univ. Bordeaux, CNRS, Bordeaux INP, LaBRI, UMR 5800, 33400 Talence, France
{tplace,mz}@labri.fr

Abstract. This paper is the second in a two-part series presenting "MeSCaL", a program dedicated to testing properties of regular languages. It provides users with an interface that allows them to define regular languages through regular expressions or finite automata. They can then perform property tests on the defined languages. These properties consist of checking membership in predefined classes of regular languages. MeSCaL supports about a hundred classes. The first paper in the series details some of its data structures, explains why existing algorithms from the literature are unlikely to run in reasonable time even for simple input languages, and describes optimizations that make the software more efficient than those algorithms. It applies to 20 different classes.

This second article continues that work, but remains largely independent of the first: it considers 44 new classes for which membership tests require the precomputation of new objects, called *orbits*. The computation of these orbits uses some results from the first part as a black box. The classes to which the presented optimizations apply are levels of language hierarchies, which come in two types: concatenation hierarchies and navigational hierarchies. While the first part focused on level one of these hierarchies, this second article addresses level two, as well as intermediate levels between one and two. It also explains how to handle hierarchies whose base level (level zero) was not covered in the first article.

Keywords: Regular languages · Finite monoids · Implementation · Membership problem · Separation problem · Concatenation hierarchies · Navigational hierarchies

1   Introduction

Context. This article is the second in a series of two, completing [25]. These papers introduce a recent software tool called MeSCaL[1] (MEmbership and Separation for ClAsses of Languages). MeSCaL implements "membership algorithms"

T. Place and M. Zeitoun—Work supported by ANR 24-CE48-1142, Project UnREAL.
[1] https://github.com/thomas-place/mescal (see [26]).

G. Castiglione and S. Mantaci (Eds.): CIAA 2025, LNCS 15981, pp. 299–315, 2026.
https://doi.org/10.1007/978-3-032-02602-6_21

for subclasses of regular languages of finite words. Each membership algorithm is tailored to a specific class $\mathcal{C}$, and determines whether a given regular language belongs to the class $\mathcal{C}$. The associated "$\mathcal{C}$-membership problem" amounts, in terms of computability theory, to asking whether the class $\mathcal{C}$ is recursive. Such questions are fundamental in computer science. In the context of regular languages, they provide a way to measure the "complexity" of a language, by determining whether it lies within simple subclasses. This second paper focuses on the implementation of membership algorithms for classes that belong to natural hierarchies within the class of regular languages. More precisely, we consider two types of hierarchies:

1. *Concatenation hierarchies*, introduced by Brzozowski and Cohen [1].
2. *Navigational hierarchies*, a closely related framework introduced in [27].

Concatenation hierarchies are a foundational and historically significant concept in automata theory. Each hierarchy is defined uniformly from a single parameter: a class $\mathcal{C}$, called its *basis*. The hierarchy consists of classes of regular languages—its *levels*—indexed by natural numbers. Level 0 is the base class $\mathcal{C}$, and level $n + 1$ is obtained by forming all Boolean combinations of languages of the form $L_0 a_1 L_1 \cdots a_n L_n$, where each a_i is a letter of the alphabet and each L_i is a language of level n. Intuitively, each level in the hierarchy reflects the number of required alternations between concatenation and Boolean operations.

The interest in concatenation hierarchies stems from the fact that they correspond, level by level, to the quantifier alternation hierarchies in first-order logic. It is well known that regular languages are those defined by monadic second-order logic, and that first-order logic defines a strict subclass [11,28]. Thomas [30] established that level n of the concatenation hierarchy of basis $\{\emptyset, \{\varepsilon\}, A^+, A^*\}$ (where A denotes the alphabet) corresponds to the logical fragment $\mathcal{B}\Sigma_n(<, +1)$ of first-order logic, which allows at most n blocks of quantifiers $\exists^*/\forall^*$. More generally, for any (well-behaved) class $\mathcal{C}$, there exists a set of predicates $\mathbb{I}_\mathcal{C}$ such that level n in the concatenation hierarchy of basis $\mathcal{C}$ contains exactly the languages defined in $\mathcal{B}\Sigma_n(\mathbb{I}_\mathcal{C})$ [18]. Deciding membership in levels of concatenation hierarchies remains one of the last "old" problems still open for regular languages.

Navigational hierarchies are a more recent development [27]. Like concatenation hierarchies, they are built from a base class $\mathcal{C}$ by repeatedly applying an operator that produces increasingly large levels. They come in three variants. In the first one, the operator is inspired by full unary temporal logic [6,29], using the *finally* (F) and *past* (P) modalities. The operator $\mathcal{C} \mapsto \mathrm{TL}(\mathcal{C})$ extends $\mathcal{C}$ by allowing the use of similar modalities, each guarded by a language of $\mathcal{C}$. The semantics is natural: a language used as a guard restrict the jumps of the modality. For instance, a formula $\mathrm{F}_L\, \varphi$ holds in w at position i if there exists a position $j > i$ such that φ holds in w at position j, *and* the infix between i and j belongs to L. The other variants are based on restricted operators $\mathcal{C} \mapsto \mathrm{FL}(\mathcal{C})$ and $\mathcal{C} \mapsto \mathrm{PL}(\mathcal{C})$: the FL- (resp. PL-)hierarchy disallows P_L (resp. F_L) modalities.

Related Work. Several programs in the literature do implement membership tests. For example, PySemigroup [12] offers a range of such tests, though efficiency is not its main concern. Others tools, such as Semigroupe [14,16] and, to some extent, AUTOMATE [3], AMoRE [10] and Automata [5], are built around the computation and manipulation of the syntactic monoid of a language, with a few membership results derived as a byproduct. In contrast, tools like LANGUAGE [2] and TESTAS [31] concentrate on a limited set of membership problems, which they handle efficiently directly on automata, bypassing the computation of the syntactic monoid. However, none of these tools cover the classes we target here—first, because the corresponding algorithms are relatively recent, and second, because they are difficult to implement efficiently: due to the size of the monoids in which computations are performed, naive implementations would be impractical.

Contributions. We address computational challenges for several known membership algorithms, optimizing their performance in order to implement them in MeSCaL. We consider 44 specific classes $\mathcal{C}$. All of them are the form $\mathrm{Op}(\mathcal{G})$ or $\mathrm{Op}(\mathcal{G}^+)$, where $\mathcal{G}$ spans four standard classes of languages, $\mathcal{G}^+$ denotes an extension of such a class, and Op denotes specific levels of concatenation hierarchies, or levels one or two in the TL, FL or PL hierarchies. More precisely, the four base classes $\mathcal{G}$ in our hierarchies are: the trivial class $\mathrm{ST} = \{\emptyset, A^*\}$, the class MOD of all unions of languages of the form $(A^q)^* A^r$, the class AMT of Boolean combinations of languages counting letter occurrences modulo some integer, and the class GR of group languages, recognized by DFAs where each letter induces a permutation on states. The inclusion of MOD and AMT is justified from the logical perspective, as they yield natural predicates in the aforementioned signature $\mathbb{I}_{\mathcal{C}}$ constructed from $\mathcal{C}$.

Some levels we study are based on additional operators. The class $\mathrm{Pol}(\mathcal{C})$ consists of unions of languages of the form $L_0 a_1 L_1 \cdots a_n L_n$, where the a_i's are letters and the L_i's belong to $\mathcal{C}$. The class $\mathrm{Bool}(\mathcal{C})$ is the Boolean closure of $\mathcal{C}$, and we let $\mathrm{BPol}(\mathcal{C}) = \mathrm{Bool}(\mathrm{Pol}(\mathcal{C}))$. Moreover, we inductively set $\mathrm{BPol}_0(\mathcal{C}) = \mathcal{C}$ and, for $n \geq 1$, define $\mathrm{BPol}_n(\mathcal{C}) = \mathrm{BPol}(\mathrm{BPol}_{n-1}(\mathcal{C}))$. Furthermore, $\mathrm{IPol}(\mathcal{C})$ is the greatest Boolean algebra included in $\mathrm{Pol}(\mathcal{C})$. At last, we define $\mathrm{TL}_2(\mathcal{C}) = \mathrm{TL}(\mathrm{TL}(\mathcal{C}))$. We are ready to present the 44 classes studied in this paper. For $\mathcal{G} \in \{\mathrm{ST}, \mathrm{MOD}, \mathrm{AMT}, \mathrm{GR}\}$,

- The 8 classes of the form $\mathrm{IPol}(\mathrm{BPol}_2(\mathcal{G}))$ and $\mathrm{IPol}(\mathrm{BPol}_2(\mathcal{G}^+))$,
- The 12 classes $\mathrm{TL}(\mathcal{G}^+)$, $\mathrm{FL}(\mathcal{G}^+)$ and $\mathrm{PL}(\mathcal{G}^+)$. The non-+ variants were already treated in [25].
- The 24 classes $\mathrm{TL}_2(\mathcal{G})$, $\mathrm{FL}_2(\mathcal{G})$, $\mathrm{PL}_2(\mathcal{G})$, $\mathrm{TL}_2(\mathcal{G}^+)$, $\mathrm{FL}_2(\mathcal{G}^+)$ and $\mathrm{PL}_2(\mathcal{G}^+)$.

A crucial point is that the $\mathrm{Op}(\mathcal{C})$-membership algorithms implemented in MeSCaL (for the involved operators Op) rely on the precomputation of subsets of the syntactic monoid of the input language, called its $\mathcal{C}$ -*orbits*. Such subsets are expensive to compute, and much of the work presented in this article consists of optimizing this calculation.

Remark 1. As stated above, this paper is the second one introducing the MeSCaL software in a two-part series. It is written to be self-contained and can be read independently of the first [25]. That said, some of the algorithms presented here build on simpler procedures described in [25]. However, these are used as black boxes, so that the two papers remain largely independent.

Organization. Section 2 introduces the fundamental objects manipulated in MeSCaL. In Sect. 3, we define the classes discussed in the paper, describe the corresponding membership algorithms, and highlight key optimizations. Finally, Sect. 4 and Sect. 5 focus on the computation of orbits. Due to space constraints, proofs are deferred to [17].

2   Preliminaries

Words, Regular Languages, Classes, Membership. We fix a *finite alphabet* A. As usual, A^* is the set of all finite words over A, including the empty one ε. For $w \in A^*$, we write $|w|$ for the length of w. For $u, v \in A^*$, we let uv be the word obtained by concatenating u and v. A *language* is a subset of A^*. For $K, L \subseteq A^*$, we let $KL = \{uv \mid u \in K \text{ and } v \in L\}$.

A *class of languages* is a set of languages. A class $\mathcal{C}$ is a *Boolean algebra* if $\emptyset \in \mathcal{C}$, $A^* \in \mathcal{C}$ and $\mathcal{C}$ is closed under union, intersection and complement: for all $K, L \in \mathcal{C}$, we have $K \cup L \in \mathcal{C}$, $K \cap L \in \mathcal{C}$ and $A^* \setminus L \in \mathcal{C}$. Moreover, $\mathcal{C}$ is *closed under quotients* when for all $L \in \mathcal{C}$ and $u \in A^*$, the languages $\{v \in A^* \mid uv \in L\}$ and $\{v \in A^* \mid vu \in L\}$ both belong to $\mathcal{C}$. A *prevariety* is a Boolean algebra closed under quotients and containing only *regular languages*. Regular languages are those that can be equivalently defined by finite automata or finite monoids.

MeSCaL implements *membership algorithms* for several classes. For a class $\mathcal{C}$, $\mathcal{C}$ *-membership* asks whether an input regular language belongs to $\mathcal{C}$.

We focus on specific classes $\mathcal{C}$, built by applying *operators* to *input classes*. An operator Op takes as input a class $\mathcal{D}$ and produces a larger class $\mathrm{Op}(\mathcal{D})$. Membership for these classes is handled via known transfer theorems, which imply that $\mathrm{Op}(\mathcal{D})$-membership reduces to a more involved problem called $\mathcal{D}$ *-separation*. The $\mathcal{D}$-separation problem takes as input two regular languages $L_1, L_2 \subseteq A^*$, and asks whether there exists a language $K \in \mathcal{D}$ such that $L_1 \subseteq K$ and $L_2 \cap K = \emptyset$.

Monoids and Morphisms. We define regular languages via monoids. A *monoid* is a set M endowed with an associative multiplication $(s, t) \mapsto st$ that has an identity element 1_M. Clearly, A^* forms a monoid under concatenation, with ε as the identity. In the paper, all monoids other than A^* are assumed to be finite. A morphism is a map $\alpha : A^* \to M$, where M is a monoid, such that $\alpha(\varepsilon) = 1_M$ and $\alpha(uv) = \alpha(u)\alpha(v)$ for all $u, v \in A^*$. The *right (resp. left) Cayley graph* of a morphism α is the directed graph whose vertex set is M, and there is an edge $s \xrightarrow{a} t$ when $t = s\alpha(a)$ (resp. $t = \alpha(a)s$).

A language $L \subseteq A^*$ is *recognized* by a morphism $\alpha : A^* \to M$ when there exists a set $F \subseteq M$ such that $L = \alpha^{-1}(F)$. In this case, we also say that M *itself* recognizes L. It is well known that a language is regular if and only if it is recognized by a *finite* monoid. Moreover, every regular language is recognized by a canonical morphism: its *syntactic morphism*. It can be computed from every recognizer, such as an automaton or an arbitrary recognizing morphism.

⚐ MeSCaL Data Structures 1

In MeSCaL, syntactic morphisms are represented using Cayley graphs instead of explicitly storing (potentially large) multiplication tables. As a result, element multiplication is not constant-time, so we strive to limit such multiplications. For further details on how MeSCaL represents morphisms, see [25].

Green's Relations. Over any monoid M, we define *preorders* $\leqslant_{\mathcal{R}}, \leqslant_{\mathcal{L}}, \leqslant_{\mathcal{J}}$ as follows. For $q, s \in M$, we write $q \leqslant_{\mathcal{R}} s$ (resp. $q \leqslant_{\mathcal{L}} s$, resp. $q \leqslant_{\mathcal{J}} s$) when there exist $p, r \in M$ such that $s = qr$ (resp. $s = pq$, resp. $s = pqr$). For $\mathcal{K} \in \{\mathcal{R}, \mathcal{L}, \mathcal{J}\}$, we let $\mathcal{K}$ be the equivalence relation associated with $\leqslant_{\mathcal{K}}$, *i.e.*, defined by $s \mathcal{K} t$ if and only if $s \leqslant_{\mathcal{K}} t$ and $t \leqslant_{\mathcal{K}} s$. Finally, $s \mathcal{H} t$ when $s \mathcal{R} t$ and $s \mathcal{L} t$. The equivalence relations $\mathcal{H}, \mathcal{R}, \mathcal{L}, \mathcal{J}$ are standard and called the *Green's relations*. Sometimes, we consider a submonoid N of a monoid M. In this case, for $s, t \in N$ and $\mathcal{K} \in \{\mathcal{H}, \mathcal{R}, \mathcal{L}, \mathcal{J}\}$, the notation $s \mathcal{K} t$ is ambiguous. In general, the relation $\mathcal{K}$ over N is *not* the restriction to $N \times N$ of the relation $\mathcal{K}$ over M. When such an ambiguity arises, we write $\mathcal{K}_N$ and $\mathcal{K}_M$ to distinguish both relations.

Special Elements. Let M be a *finite* monoid. An *idempotent* in M is an element $e \in M$ such that $ee = e$. We write $E(M)$ for the set of idempotents of M. It is folklore that for every *finite* monoid M, there is a natural number $\omega(M) \geq 1$ (written ω when M is understood) such that s^{ω} is idempotent for all $s \in M$. An element is *regular* when it is $\mathcal{J}$-equivalent to some idempotent. It is standard that in this case, it is also $\mathcal{R}$- (resp. $\mathcal{L}$-) equivalent to some idempotent. Finally, we say that a $\mathcal{K}$-class is *regular* (for $\mathcal{K} \in \{\mathcal{H}, \mathcal{R}, \mathcal{L}, \mathcal{J}\}$) if it contains an idempotent.

3   Concatenation Hierarchies and Navigational Hierarchies

We look at two types of hierarchies: the *concatenation hierarchies* introduced by Brzozowski and Cohen [1], and the more recently defined navigational hierarchies [27]. Each hierarchy is built using a specific operator. Starting from an initial class $\mathcal{C}$, referred to as the basis (this is level zero), subsequent levels are built iteratively: level $n + 1$ is obtained by applying the operator to level n.

For most of the well-known bases studied in the literature, MeSCaL provides membership tests for the lower levels of both hierarchy types—typically up to level two. The algorithms rely on algebraic characterizations of the operators. In this section, we review the definitions of these hierarchies along with their characterizations, and we discuss the challenges posed by their implementation.

3.1   Operators and Hierarchies

Concatenation Hierarchies. For every class $\mathcal{C}$, the *polynomial closure of $\mathcal{C}$*, written $\mathrm{Pol}(\mathcal{C})$, is the least class containing $\mathcal{C}$ closed under *union* and *marked* *concatenation*: if $K, L \in \mathrm{Pol}(\mathcal{C})$, then $K \cup L \in \mathrm{Pol}(\mathcal{C})$ and $KaL \stackrel{\text{def}}{=} K\{a\}L \in \mathrm{Pol}(\mathcal{C})$ for all $a \in A$. In general, $\mathrm{Pol}(\mathcal{C})$ is *not* closed under Boolean operations. We write $\mathrm{BPol}(\mathcal{C})$ for the least *Boolean algebra* containing $\mathrm{Pol}(\mathcal{C})$.

We now define concatenation hierarchies. Consider an arbitrary class $\mathcal{C}$. With each $n \in \mathbb{N}$, we associate a class $\mathrm{BPol}_n(\mathcal{C})$. First, we let $\mathrm{BPol}_0(\mathcal{C}) = \mathcal{C}$. Then, we let $\mathrm{BPol}_{n+1}(\mathcal{C}) = \mathrm{BPol}(\mathrm{BPol}_n(\mathcal{C}))$. Clearly, $\mathrm{BPol}_n(\mathcal{C}) \subseteq \mathrm{BPol}_{n+1}(\mathcal{C})$ for all n. Hence, this defines a hierarchy of classes: the *concatenation hierarchy* of basis $\mathcal{C}$.

Membership is difficult to handle for $\mathrm{BPol}_n(\mathcal{C})$. For t<his reason, we also look at intermediary levels. For a class $\mathcal{D}$, we let *co-$\mathcal{D}$* be the class of all *complements* of languages in $\mathcal{D}$. That is, $\text{co-}\mathcal{D} = \{A^* \setminus L \mid L \in \mathcal{D}\}$. For every class $\mathcal{C}$, $\mathrm{IPol}(\mathcal{C}) = \mathrm{Pol}(\mathcal{C}) \cap \text{co-}\mathrm{Pol}(\mathcal{C})$ is the greatest Boolean algebra contained in $\mathrm{Pol}(\mathcal{C})$. The classes $\mathrm{IPol}(\mathrm{BPol}_n(\mathcal{C}))$ are called *intermediary levels* in the concatenation hierarchy of basis $\mathcal{C}$, and are simpler to handle with respect to membership.

Remark 2. For a prevariety $\mathcal{D}$, $\mathrm{IPol}(\mathcal{D})$ is exactly the *unambiguous polynomial closure* $\mathrm{UPol}(\mathcal{D})$ of $\mathcal{D}$ [13,22]. UPol is defined from Pol by restricting the marked concatenations to those that are *unambiguous*: KaL is unambiguous if and only if every word $w \in KaL$ admits a *unique* decomposition witnessing this membership.

Historically, the first concatenation hierarchy is the *dot-depth* [1] of basis $\{\emptyset, \{\varepsilon\}, A^+, A^*\}$. In particular, Thomas [30] proved that the dot-depth corresponds exactly to the *quantifier alternation hierarchy* of first-order logic equipped with predicates for the linear order and the successor $\mathrm{FO}(<, +1)$. Specifically, $\mathrm{BPol}_n(\{\emptyset, \{\varepsilon\}, A^+, A^*\}) = \mathcal{B}\Sigma_n(<, +1)$ for all $n \in \mathbb{N}$. This extends to arbitrary concatenation hierarchies [18]: for every prevariety $\mathcal{C}$, there exists a suitable set $\mathbb{I}_\mathcal{C}$ of first-order predicates such that $\mathrm{BPol}_n(\mathcal{C}) = \mathcal{B}\Sigma_n(\mathbb{I}_\mathcal{C})$ for all $n \in \mathbb{N}$ (see [25]).

Navigational Hierarchies. This notion was introduced in [27]. Navigational hierarchies are built with three operators from [22] inspired by temporal logic: $\mathrm{TL}(\mathcal{C})$, $\mathrm{FL}(\mathcal{C})$ and $\mathrm{PL}(\mathcal{C})$. Let $\mathcal{C}$ be a prevariety. Formulas in $\mathrm{TL}[\mathcal{C}]$ are defined by the grammar $\varphi := a \mid \varphi \wedge \varphi \mid \varphi \vee \varphi \mid \neg \varphi \mid \mathrm{F}_L\, \varphi \mid \mathrm{P}_L\, \varphi$ where $a \in A$ and $L \in \mathcal{C}$. Formulas in $\mathrm{FL}[\mathcal{C}]$ (resp. in $\mathrm{PL}[\mathcal{C}]$) are those that do not use P_L (resp. F_L).

A $\mathrm{TL}[\mathcal{C}]$-formula is evaluated on a pair (w, i) where $w = a_1 \cdots a_n$ is a word ($a_i \in A$) and $i \in \{0, 1, \ldots, |w| + 1\}$ is a position in w. We view 0 and $|w| + 1$ as dummy positions, which carry no letter. In contrast, if $1 \leq i \leq |w|$, position i carries letter a_i. For two positions $i < j$, we write $w(i, j) = a_{i+1} \cdots a_{j-1}$. We define by induction what it means for the pair (w, i) to satisfy φ, denoted $w, i \models \varphi$. First, for $a \in A$, we have $w, i \models a$ if position i carries an a. The semantics of the Boolean connectives $\neg, \wedge, \vee$ is standard. Finally, let $L \in \mathcal{C}$. We have $w, i \models \mathrm{F}_L\, \psi$ if there exists a position $j > i$ such that $w(i, j) \in L$ and $w, j \models \psi$. Similarly, $w, i \models \mathrm{P}_L\, \psi$ holds if there exists $j < i$ such that $w(j, i) \in L$ and $w, j \models \psi$.

For a formula $\varphi \in \mathrm{TL}[\mathcal{C}]$, we define $L_{min}(\varphi) = \{w \in A^* \mid w, 0 \models \varphi\}$ and $L_{max}(\varphi) = \{w \in A^* \mid w, |w| + 1 \models \varphi\}$. We are ready to define our three classes:

- $\mathrm{TL}(\mathcal{C})$ is the class consisting of all languages $L_{min}(\varphi)$ with $\varphi \in \mathrm{TL}[\mathcal{C}]$.

- FL($\mathcal{C}$) is the class consisting of all languages $L_{min}(\varphi)$ with $\varphi \in$ FL[$\mathcal{C}$].
- PL($\mathcal{C}$) is the class consisting of all languages $L_{max}(\varphi)$ with $\varphi \in$ PL[$\mathcal{C}$].

A correspondence result between the operator TL and the two-variable fragment of first-order logic is known [22]. We have TL($\mathcal{C}$) = FO$^2(\mathbb{I}_\mathcal{C})$ for each prevariety $\mathcal{C}$.

We are ready to define three *navigational hierarchies*. Their construction parallels that of concatenation hierarchies. Given a base class $\mathcal{C}$, we define three levels for each $n \in \mathbb{N}$: TL$_n(\mathcal{C})$, FL$_n(\mathcal{C})$, and PL$_n(\mathcal{C})$.

- We let TL$_0(\mathcal{C})$ = FL$_0(\mathcal{C})$ = PL$_0(\mathcal{C})$ = $\mathcal{C}$
- For every $n \in \mathbb{N}$, we let TL$_{n+1}(\mathcal{C})$ = TL(TL$_n(\mathcal{C})$), FL$_{n+1}(\mathcal{C})$ = FL(FL$_n(\mathcal{C})$) and PL$_{n+1}(\mathcal{C})$ = PL(PL$_n(\mathcal{C})$).

These hierarchies are closely related to concatenation hierarchies. It is shown in [27] that $\bigcup_{n \in \mathbb{N}}$ BPol$_n(\mathcal{C})$ = $\bigcup_{n \in \mathbb{N}}$ TL$_n(\mathcal{C})$ = $\bigcup_{n \in \mathbb{N}}$ FL$_n(\mathcal{C})$ = $\bigcup_{n \in \mathbb{N}}$ PL$_n(\mathcal{C})$ for all prevarieties $\mathcal{C}$. However, the levels themselves differ: in general, BPol$_n(\mathcal{C})$ is *strictly* contained in TL$_n(\mathcal{C})$, FL$_n(\mathcal{C})$, and PL$_n(\mathcal{C})$. Although navigational hierarchies were formally introduced in [27], specific levels had already been studied in earlier works [4,7–9,29].

3.2 Prominent Bases: Group prevarieties and well-suited Extensions

Prominent hierarchies in the literature use specific bases. A *group* G is a monoid in which every element $g \in G$ has an *inverse* g^{-1}, *i.e.*, such that $gg^{-1} = g^{-1}g = 1_G$. A *group language* is a language recognized by a finite group. Our bases are either *group prevarieties* (*i.e.*, prevarieties consisting only of group languages) or extensions of such classes. We are interested in four main group prevarieties:

- The prevariety GR of all *group languages*.
- The prevariety AMT of *alphabet modulo testable languages* consists of all finite Boolean combinations of languages $\{w \in A^* \mid \#_a w \equiv r \pmod{q}\}$, where $\#_a w$ is the number of occurrences of "a" in w. It is standard that AMT consists of all languages recognized by a finite *commutative group*.
- The prevariety MOD of *modulo languages* consists of all finite unions of languages of the form $(A^q)^* A^r = \{w \in A^* \mid |w| \equiv r \pmod{q}\}$ for $0 \leq r < q$.
- Finally, the prevariety ST = $\{\emptyset, A^*\}$ is the trivial class.

Remark 3. The significance of these base classes is underscored by their connection to quantifier alternation hierarchies in first-order logic. The bases ST, MOD, and AMT correspond to natural logical predicates:

- ST yields the hierarchy of first-order logic with linear order only, FO($<$).
- MOD extends this with *modular predicates* of the form "$x \equiv r \pmod{q}$."
- AMT further adds *letter modular predicates* of the form "$\#_a(x) \equiv r \pmod{q}$," where $\#_a(x)$ counts the occurrences of letter $a \in A$ before position x.

We also look at well-suited extensions. For a class $\mathcal{G}$, the *well-suited extension of* $\mathcal{G}$ is the class $\mathcal{G}^+$ consisting of all languages $L \cap A^+$ and $L \cup \{\varepsilon\}$ with $L \in \mathcal{G}$.

Remark 4. From a logical perspective, using a well-suited-extension as the basis of a concatenation hierarchy corresponds to enriching the logic with the successor predicate "+1." For example, the basis $\mathrm{ST}^+ = \{\emptyset, \{\varepsilon\}, A^+, A^*\}$ of the dot-depth hierarchy yields the quantifier alternation hierarchy of $\mathrm{FO}(<, +1)$.

In the case of temporal closures, a well-suited-extension expands the set of temporal modalities by adding the classical ones *tomorrow* (X) and *yesterday* (Y).

⚒ MeSCaL features

MeSCaL supports membership for level one in all hierarchies discussed, using bases $\mathcal{G}$ or $\mathcal{G}^+$, with $\mathcal{G} \in \{\mathrm{GR}, \mathrm{AMT}, \mathrm{MOD}, \mathrm{ST}\}$. The case of bases $\mathcal{G}$ is detailed in [25]. Here, we focus the classes $\mathrm{TL}(\mathcal{G}^+)$, $\mathrm{FL}(\mathcal{G}^+)$, and $\mathrm{PL}(\mathcal{G}^+)$. MeSCaL also includes procedures for the classes $\mathrm{BPol}(\mathcal{G}^+)$, but we omit them here.

MeSCaL also handles level two membership for all three navigational hierarchies over the bases $\mathcal{G}$ and $\mathcal{G}^+$ for $\mathcal{G} \in \{\mathrm{GR}, \mathrm{AMT}, \mathrm{MOD}, \mathrm{ST}\}$. We detail the implementation in this paper.

For concatenation hierarchies, level two is implemented only for $\mathrm{BPol}_2(\mathrm{ST})$ (we omit the details in this paper). However, MeSCaL supports all classes of the form $\mathrm{IPol}(\mathrm{BPol}_2(\mathcal{G}))$ and $\mathrm{IPol}(\mathrm{BPol}_2(\mathcal{G}^+))$, for $\mathcal{G} \in \{\mathrm{GR}, \mathrm{AMT}, \mathrm{MOD}, \mathrm{ST}\}$; we present the algorithms here.

Remark 5. In the above discussion, we omit the intermediate levels $\mathrm{IPol}(\mathrm{BPol}(\mathcal{G}))$ and $\mathrm{IPol}(\mathrm{BPol}(\mathcal{G}^+))$. This is because when $\mathcal{G}$ is a group prevariety, they correspond exactly [22,29] to $\mathrm{TL}(\mathcal{G})$ and $\mathrm{TL}(\mathcal{G}^+)$, respectively.

3.3   Algebraic Characterizations and Membership

Membership tests implemented in MeSCaL are based on characterization theorems. For each operator Op, they imply that $\mathrm{Op}(\mathcal{C})$-membership reduces to a more general problem for $\mathcal{C}$, typically, $\mathcal{C}$-separation.

$\mathcal{C}$**-orbits.** Most characterizations rely on a key object: $\mathcal{C}$-orbits. In order to define them, we need a preliminary notion: $\mathcal{C}$-pairs. Let $\mathcal{C}$ be a prevariety and $\alpha : A^* \to M$ a morphism. A $\mathcal{C}$ *-pair* (for α) is a pair $(s, t) \in M^2$ such that $\alpha^{-1}(s)$ is *not* $\mathcal{C}$-separable from $\alpha^{-1}(t)$. The next lemma is proved in [22, Lemma 5.12].

Lemma 6. *Let $\mathcal{C}$ be a prevariety and $\alpha : A^* \to M$ be a morphism. Moreover, let $(s, t), (s', t') \in M^2$ be two $\mathcal{C}$-pairs. Then (ss', tt') is a $\mathcal{C}$-pair.*

We are ready to define $\mathcal{C}$-orbits. Let $\alpha : A^* \to M$ be a morphism. Let $e \in E(M)$ be an idempotent. Clearly, eMe is a monoid (its identity is e). For each prevariety $\mathcal{C}$, we define a submonoid M_e of eMe: the $\mathcal{C}$ *-orbit of e for α.*

We let $M_e = \{s \in eMe \mid (e, s) \text{ is a } \mathcal{C} - pair \text{ for } \alpha\}$. Lemma 6 implies that M_e is indeed a submonoid of eMe. Computing $\mathcal{C}$-orbits clearly reduces to solving $\mathcal{C}$-separation.

Most of the operators defined in the section can be characterized using $\mathcal{C}$-orbits (the only exception is BPol). We start with the temporal closures. The characterization of $\mathcal{C} \mapsto \mathrm{TL}(\mathcal{C})$ uses the class of monoids DA. Typically, it is defined by an equation: a finite monoid M belongs to DA if for all $s, t \in M$, we have $(st)^{\omega} s(st)^{\omega} = (st)^{\omega}$. The theorem is proved in [24, Theorems 37 and 38].

Theorem 7. *Let $\mathcal{C}$ be a prevariety and L be a regular language:*

- *$L \in \mathrm{PL}(\mathcal{C})$ if and only if all $\mathcal{C}$-orbits for its syntactic morphism are $\mathscr{L}$-trivial.*
- *$L \in \mathrm{FL}(\mathcal{C})$ if and only if all $\mathcal{C}$-orbits for its syntactic morphism are $\mathscr{R}$-trivial.*
- *$L \in \mathrm{TL}(\mathcal{C})$ if and only if all $\mathcal{C}$-orbits for its syntactic morphism are in DA.*

The operator $\mathcal{C} \mapsto \mathrm{IPol}(\mathrm{BPol}(\mathcal{C}))$ is characterized by an equation involving the $\mathcal{C}$-orbits of the morphism. The following result is proved in [22, Theorem 6.7].

Theorem 8. *Let $\mathcal{C}$ be a prevariety, L be a regular language and $\alpha : A^* \to M$ be its syntactic morphism. For every $e \in E(M)$, let $M_e \subseteq eMe$ be the $\mathcal{C}$-orbit of e for α. Then, $L \in \mathrm{IPol}(\mathrm{BPol}(\mathcal{C}))$ if and only if α satisfies the following property:*

$$(st)^{\omega+1} = (st)^{\omega} t(st)^{\omega} \text{ for all } e \in E(M), \text{ all } s \in M_e \text{ and all } t \in eMe. \quad (1)$$

Remark 9. We omit the algorithms used in MeSCaL to decide the classes built with BPol, as this involves objects that go beyond $\mathcal{C}$-orbits and would require introducing too much material. We come back to BPol in the conclusion.

Computing Orbits. MeSCaL includes efficient algorithms that compute $\mathcal{C}$-orbits when $\mathcal{C}$ is level *zero* or *one* in the concatenation and navigational hierarchies of bases $\mathcal{G}$ or $\mathcal{G}^+$ where $\mathcal{G}$ is ST, MOD, AMT or GR. The membership procedures tests from Theorem 7 and Theorem 8. Let us first make an important remark concerning the information that we have to compute.

✵ MeSCaL Optimization 1: Green's relations for regular elements

Implementing Theorem 7 requires checking whether the $\mathcal{C}$-orbits are $\mathscr{R}$-trivial, $\mathscr{L}$-trivial, or belong to DA. A standard result (see, e.g., [15, Proposition 2.26]) states that for $\mathscr{K} \in \{\mathscr{R}, \mathscr{L}\}$, a monoid is $\mathscr{K}$-trivial if and only if all its *regular* $\mathscr{K}$-classes are trivial. Similarly, a monoid lies in DA if and only if all its *regular* elements are idempotent. This is crucial: to implement Theorem 7, it suffices to compute the *regular elements* of the $\mathcal{C}$-orbits and their Green's relations.

In contrast, implementing Theorem 8 requires *all* elements in the $\mathcal{C}$-orbits, though the Green's relations are not needed.

We now concentrate on computing of $\mathcal{C}$-orbits. We start with a key optimization, common to all classes $\mathcal{C}$.

Lemma 10. *Let $\alpha : A^* \to M$ be a morphism and $e, f \in E(M)$ such that $e \; \mathcal{J} \; f$. There is an isomorphism $\gamma : eMe \to fMf$ such that for all prevarieties $\mathcal{C}$, if $M_e \subseteq eMe$ and $M_f \subseteq fMf$ are the $\mathcal{C}$-orbits of e and f for α, then $\gamma(M_e) = M_f$.*

🏃 MeSCaL Optimization 2: Factorizing the $\mathcal{C}$-orbits

When a new syntactic morphism $\alpha : A^* \to M$ is computed in MeSCaL, a single witness idempotent is chosen in each regular $\mathcal{J}$-class of M. For each input class $\mathcal{C}$, MeSCaL computes only the $\mathcal{C}$-orbits of these witnesses. By Lemma 10, this is sufficient to determine which properties hold for *all* $\mathcal{C}$-orbits: every omitted $\mathcal{C}$-orbit is isomorphic to one that has been computed.

4   Computing $\mathcal{G}$-Orbits and $\mathcal{G}^+$-Orbits

We now turn to the computations of $\mathcal{G}$- and $\mathcal{G}^+$-orbits, which rely on an important object already discussed in [25]: the $\mathcal{G}$ *-kernel*.

$\mathcal{G}$**-kernel.** Let $\mathcal{G}$ be a group prevariety and $\alpha : A^* \to M$ be a morphism. The $\mathcal{G}$-kernel of α is the set $N \subseteq M$ that consists of all elements $s \in M$ such that $\{\varepsilon\}$ is *not* $\mathcal{G}$-separable from $\alpha^{-1}(s)$. We have the following lemma (see [22, Fact 5.5]).

Lemma 11. *Let $\mathcal{G}$ be a group prevariety and $\alpha : A^* \to M$ be a morphism. Then, the $\mathcal{G}$-kernel of α is a submonoid of M containing $E(M)$.*

🏃 MeSCaL Optimization 3: Computing regular elements in $\mathcal{G}$-kernels

MeSCaL implements efficient procedures for computing the $\mathcal{G}$-kernel when $\mathcal{G}$ is either ST, MOD, AMT or GR, as detailed in [25]. Here, we treat these procedures as *black boxes*. Crucially, computing only the *regular elements* of the $\mathcal{G}$-kernel is *significantly more efficient* than computing the entire kernel—and this is always sufficient for our purposes.

$\mathcal{G}$ **-orbits.** When $\mathcal{G}$ is a group prevariety, we do not need the $\mathcal{G}$-orbits. Instead, the $\mathcal{G}$-kernel alone suffices. Indeed, we have the following lemma.

Lemma 12. *Let $\mathcal{G}$ be a group prevariety, $\alpha : A^* \to M$ be an onto morphism and $N \subseteq M$ be its $\mathcal{G}$-kernel. For all $e \in E(M)$, the $\mathcal{G}$-orbit of e for α is $eNe \subseteq N$. In particular, the $\mathcal{G}$-orbit of 1_M for α is N.*

By Lemma 12, if $\mathcal{G}$ is a group prevariety, we only need *a single $\mathcal{G}$ -orbit*: that of 1_M, which is also the $\mathcal{G}$-kernel. This forms the foundation of how MeSCaL implements membership for classes $\mathrm{Op}(\mathcal{G})$. Kernel computations are detailed in [25].

$\mathcal{G}^+$**-orbits.** We now consider the computation of the $\mathcal{G}^+$-orbits for the well-suited extension $\mathcal{G}^+$ of a group prevariety $\mathcal{G}$. This computation is based on the $\mathcal{G}$-kernel.

Proposition 13. *Let $\mathcal{G}$ be a group prevariety, $\alpha : A^* \to M$ be an onto morphism, $S = \alpha(A^+)$ and $N \subseteq M$ be the $\mathcal{G}$-kernel of α. For all $e \in E(S)$, the $\mathcal{G}^+$-orbit of e for α is $eNe \subseteq N$. Moreover, if $1_M \notin S$, then the $\mathcal{G}$-orbit of 1_M for α is $\{1_M\}$.*

Remark 14. When $\mathcal{G} = \mathrm{ST} = \{\emptyset, A^*\}$, the ST-kernel is the whole monoid M. By Proposition 13, it follows that for $e \in S$, the ST^+-orbit of e is the monoid eSe. This is a well-known object in semigroup theory, called the *local monoid of e*.

✵ MeSCaL Optimization 4: Computation of $\mathcal{G}^+$-orbits

Once we have the $\mathcal{G}$-kernel N in hand, Proposition 13 shows that for $e \in E(S)$, the $\mathcal{G}^+$-orbit of e can be computed by intersecting N with eM and Me. Computing eM and Me is straightforward and done in MeSCaL via traversals of the right and left Cayley graphs of α, which takes $O(|M||A|)$ time.

Finally, since the $\mathcal{G}^+$-orbit of e is included in N, the restriction of N to its regular elements is sufficient to compute all regular elements of this $\mathcal{G}^+$-orbit.

Moreover, once a $\mathcal{G}^+$-orbit is obtained, computing its Green's relations restricted to regular elements becomes straightforward (recall from Optimization 1 that this information is sufficient to decide level 1 in navigational hierarchies). The next lemma shows that, for regular elements, this amounts to restricting the Green's relations of the underlying monoid M.

Lemma 15. *Let $\mathcal{G}$ be a group prevariety, $\alpha : A^* \to M$ be an onto morphism, $e \in E(M)$ and M_e be the $\mathcal{G}^+$-orbit of e for α. Let $s, t \in M_e$ be regular in M. Then, for $\mathcal{K} \in \{\mathcal{H}, \mathcal{L}, \mathcal{R}\}$, we have $s \, \mathcal{K}_{M_e} \, t$ if and only if $s \, \mathcal{K}_M \, t$.*

Finally, yet another important optimization is available for $\mathcal{G}^+$-orbits. We already mentioned after Lemma 10 that for an arbitrary prevariety $\mathcal{C}$, computing one $\mathcal{C}$-orbit per regular $\mathcal{J}$-class suffices. This can be pushed further when $\mathcal{C} = \mathcal{G}^+$.

Let $\alpha : A^* \to M$ be a morphism. Let $S = \alpha(A^+)$. A *root set* (of idempotents) for α is a set $T \subseteq E(S)$ that satisfies the two following properties:

1. $e_1 \not\leqslant_{\mathcal{J}} e_2$ for all $e_1, e_2 \in T$, and,
2. for all $f \in E(S)$, there exists $e \in T$ such that $e \leqslant_{\mathcal{J}} f$.

By definition, root sets are typically small in practice. Importantly, we only need the $\mathcal{G}^+$-orbits of the idempotents they contain.

Proposition 16. < *Let $\mathcal{G}$ be a group prevariety, $\alpha : A^* \to M$ be an onto morphism and $S = \alpha(A^+)$. Let $e, f \in E(S)$, M_e be the $\mathcal{G}^+$-orbit of e, M_f be the $\mathcal{G}^+$-orbit of f. Assume that there exists $q, r \in M$ such that $f = qer$ (i.e., $e \leqslant_{\mathcal{J}} f$). Then, M_f is isomorphic to a subsemigroup of M_e and $M_f \subseteq qM_er$.*

✼ **MeSCaL Optimization 5: Factorizing the $\mathcal{G}^+$-orbits**

When a new syntactic morphism $\alpha : A^* \to M$ is computed in MeSCaL, a root set is computed for α. Given a group prevariety $\mathcal{G}$, MeSCaL computes only the $\mathcal{G}^+$-orbits of the idempotents in this root set. By Proposition 16, this is sufficient to determine which properties hold for all $\mathcal{G}^+$-orbits: every omitted $\mathcal{G}^+$-orbit is isomorphic to a subsemigroup of one that has been computed.

5   Orbits for More Complex Classes

As announced in Sect. 3, MeSCaL includes membership tests for the classes $\mathrm{TL}_2(\mathcal{C})$, $\mathrm{FL}_2(\mathcal{C})$, $\mathrm{PL}_2(\mathcal{C})$ and $\mathrm{IPol}(\mathrm{BPol}_2(\mathcal{C}))$ when $\mathcal{C} \in \{\mathcal{G}, \mathcal{G}^+\}$ for one of the group prevarieties ST, MOD, Ǎ̆AMT and GR. By Theorem 7 and Theorem 8, this means that MeSCaL computes orbits for the classes $\mathrm{TL}(\mathcal{C})$, $\mathrm{FL}(\mathcal{C})$, $\mathrm{PL}(\mathcal{C})$ and $\mathrm{BPol}(\mathcal{C})$ when $\mathcal{C}$ is among the aforementioned bases. Fortunately, there are actually fewer classes to consider in practice, thanks to the following proposition.

Proposition 17. *Let $\mathcal{G}$ be a group prevariety and $\alpha : A^* \to M$ be a surjective morphism. Let $\mathcal{C} \in \{\mathcal{G}, \mathcal{G}^+\}$ and $\mathcal{D} \in \{\mathrm{BPol}(\mathcal{C}), \mathrm{FL}(\mathcal{C}), \mathrm{PL}(\mathcal{C}), \mathrm{TL}(\mathcal{C})\}$. For every $e \in E(M)$ the $\mathcal{D}$-orbit of e for α is exactly the co-$\mathrm{Pol}(\mathcal{C})$-orbit of e for α.*

MeSCaL implements algorithms for computing co-$\mathrm{Pol}(\mathcal{G})$- and co-$\mathrm{Pol}(\mathcal{G}^+)$-orbits when $\mathcal{G} \in \{\mathrm{ST}, \mathrm{MOD}, \mathrm{AMT}, \mathrm{GR}\}$. For co-$\mathrm{Pol}(\mathrm{ST})$-orbits, a simple specialized algorithm is used. For the other classes, MeSCaL relies on generic algorithms based on $\mathcal{G}$-kernels and $\mathcal{G}^+$-orbits.

5.1   Computing co-Pol(ST)-orbits

Let $\alpha : A^* \to M$ be a morphism. For each idempotent $e \in E(M)$, we associate a subalphabet of A. The *alphabet of e* (for α) consists of all letters $b \in A$ which label an edge inside the strongly connected component of e within the right Cayley graph of α. The following proposition describes the co-$\mathrm{Pol}(\mathrm{ST})$-orbits.

Proposition 18. *Let $\alpha : A^* \to M$ be a surjective morphism, $e \in E(M)$ and B be the alphabet of e. The co-$\mathrm{Pol}(\mathrm{ST})$-orbit of e for α is $M_e = \{e\alpha(w)e \mid w \in B^*\}$. Moreover, for every $s, t \in M_e$,*

- *we have $s \leqslant_\mathcal{R} t$ in M_e if and only if there exists $u \in B^*$ such that $s\alpha(u) = t$.*
- *we have $s \leqslant_\mathcal{L} t$ in M_e if and only if there exists $v \in B^*$ such that $\alpha(v)s = t$.*

✱ MeSCaL Optimization 6: Computation of *co*-Pol(ST)-orbits

Proposition 18 provides an efficient procedure for computing the *co*-Pol(ST)-orbits. Let $e \in E(M)$. We want to compute its *co*-Pol(ST)-orbit $M_e \subseteq eMe$.

Computing the alphabet $B \subseteq A$ of e is straightforward provided that we have the strongly connected components of the right Cayley graph in hand. This is the case in MeSCaL, as these are the $\mathcal{R}$-classes of M.

We restrict the left and right Cayley graphs of α by keeping the edges labeled by a letter in B and discarding the others. We are able to compute the sets $\{e\alpha(w) \mid w \in B^*\}$ and $\{\alpha(w)e \mid w \in B^*\}$ by exploring these restrictions. By Proposition 18, M_e is their intersection. The Green's relations of M_e are given by the strongly connected components (obtained via Tarjan's algorithm).

5.2   General Case

It was shown in [19, 21] that if *separation* is decidable for a group prevariety $\mathcal{G}$, then it is also decidable for Pol($\mathcal{G}$), Pol($\mathcal{G}^+$). This extends to *co*-Pol($\mathcal{G}$) and *co*-Pol($\mathcal{G}^+$) since separation for a class *co*-$\mathcal{D}$ reduces to separation for $\mathcal{D}$ (H is *co*-$\mathcal{D}$-separable from L if and only if L is $\mathcal{D}$-separable from H). Hence, in this case, we have algorithms that compute the *co*-Pol($\mathcal{G}$)- and *co*-Pol($\mathcal{G}^+$)-pairs and, in turn, the *co*-Pol($\mathcal{G}$)- and *co*-Pol($\mathcal{G}^+$)-orbits.

In fact, the results of [19, 21] imply that one can directly compute *co*-Pol($\mathcal{G}$)- and *co*-Pol($\mathcal{G}^+$)-pairs from the $\mathcal{G}$-kernel and the $\mathcal{G}^+$-orbits respectively. Here, we present these procedures and explain how they are implemented in MeSCaL.

The classes *co*-Pol($\mathcal{G}$). Consider a monoid M. A set of pairs $P \subseteq M^2$ is said to be *stable* if it is closed under multiplication: for all $(s, t), (s', t') \in P$, we have $(ss', tt') \in P$. The next theorem follows from results of [19, 21].

Theorem 10. *Let $\mathcal{G}$ be a group prevariety and $\alpha : A^* \to M$ be an onto morphism. The set of co-Pol($\mathcal{G}$)-pairs for α is the least stable set $P \subseteq M^2$ such that,*

- *$(\alpha(a), \alpha(a)) \in P$ for every $a \in A$, and,*
- *$(t, 1_M) \in P$ for all t in the $\mathcal{G}$-kernel of α.*

By Theorem 19, computing *co*-Pol($\mathcal{G}$)-pairs reduces to graph traversal. The graph has vertices in M^2, with edges from each $(q, r) \in M^2$ to all $(q\alpha(a), r\alpha(a))$ for $a \in A$, and to (qt, r) for each t in the $\mathcal{G}$-kernel. According to Theorem 19, a pair (q, r) is a *co*-Pol($\mathcal{G}$)-pair if and only if there is a path from $(1_M, 1_M)$ to (q, r) in this graph. Yet, there are *two drawbacks to this naive approach in practice*:

1. There are $|M|^2$ vertices and $|M|^2 \times (|N| + |A|)$ edges (where N is the $\mathcal{G}$-kernel).
2. Seemingly, we need the whole $\mathcal{G}$-kernel whereas the algorithms of MeSCaL are only efficient when the computation is restricted to regular elements.

Fortunately, we are not interested in *all* co-Pol($\mathcal{G}$)-pairs, only in those required to compute orbits. For this reason, we may improve the procedure significantly.

Consider a group prevariety $\mathcal{G}$, a morphism $\alpha : A^* \to M$ and let $N \subseteq M$ be the $\mathcal{G}$-kernel of α. For each idempotent $e \in E(M)$, we associate a graph $V_{\mathcal{G},\alpha}(e)$. Let $R_e \subseteq eM$ be the $\mathcal{R}$-class of e. The vertices of $V_{\mathcal{G},\alpha}(e)$ are the pairs in $R_e \times eM$. There are two kinds of edges. Let $(q,r) \in R_e \times eM$ be a vertex.

– For all $a \in A$ such that $q\alpha(a) \; \mathcal{R} \; e$, we add an edge $(q,r) \to (q\alpha(a), r\alpha(a))$.
– For all $t \in N$ such that $t \; \mathcal{J} \; e$ and $qt \; \mathcal{R} \; e$, we add an edge $(q,r) \to (qt,r)$.

Corollary 20. *Let $\mathcal{G}$ be a group prevariety, $\alpha : A^* \to M$ be an onto morphism and $e \in E(M)$. The co-Pol($\mathcal{G}$)-orbit of e for α consists of all elements $s \in eMe$ such that there is a path from (e,e) to (e,s) in $V_{\mathcal{G},\alpha}(e)$.*

✸ MeSCaL Optimization 7: Computation of *co*-Pol($\mathcal{G}$)-orbits

Corollary 20 improves significantly on the above drawbacks. Computing the co-Pol($\mathcal{G}$)-orbit of $e \in E(M)$ is achieved by traversing a graph with $|R_e| \times |eM|$ vertices (R_e is the $\mathcal{R}$-class of e). This is much smaller than $|M|^2$. In particular, recall that MeSCaL only computes one orbit per $\mathcal{J}$-class (see Lemma 10).

In order to compute the edges of $V_{\mathcal{G},\alpha}(e)$, MeSCaL precomputes all products qt with $q \; \mathcal{R} \; e$ and $t \; \mathcal{J} \; e$. This is achieved in time $O(|R_e| \times |J_e|)$ (where J_e is the $\mathcal{J}$-class of e). This involves some Green's theory, since MeSCaL does not store multiplication tables: we only have the right Cayley graph.

Finally, observe that computing the *whole* co-Pol($\mathcal{G}$)-orbit of e only requires the elements t of the $\mathcal{G}$-kernel such that $t \; \mathcal{J} \; e$: these are *regular* elements.

The classes *co*-Pol($\mathcal{G}^+$). The computation of *co*-Pol($\mathcal{G}^+$)-orbits is based on the next theorem which also follows from results of [19,21].

Theorem 21. *Let $\mathcal{G}$ be a group prevariety and $\alpha : A^* \to M$ be an onto morphism. The set of co-Pol($\mathcal{G}^+$)-pairs for α is the least stable set $P \subseteq M^2$ such that,*

1. *$(\alpha(a), \alpha(a)) \in P$ for every $a \in A$, and,*
2. *$(t, f) \in P$ for all $f \in E(M)$ and t in the $\mathcal{G}^+$-orbit of f for α.*

A naive implementation of the algorithm from Theorem 21 suffers from the same drawbacks noted after Theorem 19 in the context of computing *co*-Pol($\mathcal{G}$)-orbits. We address these issues using the same strategy.

Let $\mathcal{G}$ be a group prevariety, $\alpha : A^* \to M$ be a morphism, and for each $f \in E(M)$, let $M_f \subseteq fMf$ denote the $\mathcal{G}^+$-orbit of f for α. For every idempotent $e \in E(M)$ and each root set T for α, we define a graph $V_{\mathcal{G},\alpha}^+(e,T)$. Let $R_e \subseteq eM$ be the $\mathcal{R}$-class of e. The vertices of $V_{\mathcal{G},\alpha}^+(e,T)$ are pairs in $R_e \times eM$, and edges are defined as follows for each vertex $(q,r) \in R_e \times eM$:

– For all $a \in A$, if $q\alpha(a) \; \mathcal{R} \; e$, add an edge $(q,r) \to (q\alpha(a), r\alpha(a))$.

– For all $f \in T$ and $t \in M_f$, if $t \mathcal{G} e$ and $qt \mathcal{R} e$, add an edge $(q, r) \rightarrow (qt, rf)$.

Corollary 22. *Let $\mathcal{G}$ be a group prevariety, $\alpha : A^* \rightarrow M$ be an onto morphism, T be a root set for α and $e \in E(M)$. The co-$\mathrm{Pol}(\mathcal{G}^+)$-orbit of e for α consists of all elements $s \in eMe$ such that there is a path from (e, e) to (e, s) in $V_{\mathcal{G},\alpha}^+(e, T)$.*

✻ MeSCaL Optimization 8: Computation of co-$\mathrm{Pol}(\mathcal{G}^+)$-orbits

The implementation of Corollary 22 follows the same approach as discussed above for Corollary 20. Notably, it only requires the *regular elements* of the $\mathcal{G}^+$-orbits. Moreover, we only need the $\mathcal{G}^+$-orbits of the idempotents in a root set—which is exactly what MeSCaL computes (see Proposition 16).

✻ MeSCaL Optimization 9: Computation of Green's relations

Computing Green's relations for co-$\mathrm{Pol}(\mathcal{G})$- and co-$\mathrm{Pol}(\mathcal{G}^+)$-orbits is non-trivial. They do not, in general, align with those of the original monoid—two elements may be $\mathcal{R}$-equivalent and regular in the monoid but not within a given orbit.

A naive approach is expensive: unlike syntactic monoids, we lack generators for the orbits. However, we want to implement two specific characterizations:

– For Theorem 7, the Green's relations restricted to *regular* elements suffice.
– For Theorem 8, we do not need the Green's relations.

Using Green's theory, one can efficiently identify which elements remain regular within a subsemigroup and compute the Green's relations restricted to them. In MeSCaL, this is implemented with time complexity $O(|M| \times \max)$, where M is the original monoid and max is the maximum size of an $\mathcal{R}$- or $\mathcal{L}$-class.

---

6 Conclusion

In this paper, we presented optimizations for algorithms that compute information associated with a morphism, named its $\mathcal{C}$-orbits, where $\mathcal{C}$ is a prevariety. This information is instrumental in solving various membership problems. In MeSCaL, such precomputations are applied to test membership of classes of the form $\mathrm{Op}(\mathcal{G})$ or $\mathrm{Op}(\mathcal{G}^+)$, where $\mathcal{G}$ belongs to $\{\mathrm{ST}, \mathrm{MOD}, \mathrm{AMT}, \mathrm{GR}\}$ and where $\mathrm{Op} \in \{\mathrm{IPol}(\mathrm{BPol}), \mathrm{TL}, \mathrm{FL}, \mathrm{PL}, \mathrm{IPol}(\mathrm{BPol}_2), \mathrm{TL}_2, \mathrm{FL}_2, \mathrm{PL}_2\}$.

This paper omits the classes built using the BPol operator. However, MeSCaL does implement membership tests for $\mathrm{BPol}(\mathcal{G})$ with $\mathcal{G} \in \{\mathrm{ST}, \mathrm{MOD}, \mathrm{AMT}, \mathrm{GR}\}$, as discussed in [25]. It also supports membership tests for $\mathrm{BPol}(\mathcal{G}^+)$, based on a generic characterization from [20]. The implementation combines techniques from [25] (for $\mathrm{BPol}(\mathcal{G})$) and from Sect. 4 (for computing $\mathcal{G}^+$-orbits). Also, MeSCaL implements membership for $\mathrm{BPol}_2(\mathrm{ST})$, using a characterization of [23]. This generalizes the approach used to compute co-$\mathrm{Pol}(\mathrm{ST})$-orbits in Sect. 5.

Acknowledgments. We wish to thank the reviewers for some useful comments.

References

1. Brzozowski, J.A., Cohen, R.S.: Dot-depth of star-free events. J. Comput. Syst. Sci. **5**(1), 1–16 (1971). https://doi.org/10.1016/S0022-0000(71)80003-X
2. Caron, P.: LANGAGE: a Maple package for automaton characterization of regular languages. Theoret. Comput. Sci. **231**(1), 5–15 (2000). https://doi.org/10.1016/S0304-3975(99)00013-4
3. Champarnaud, J.M., Hansel, G.: AUTOMATE, a computing package for automata and finite semigroups. J. Symb. Comput. **12**(2), 197–220 (1991). https://doi.org/10.1016/S0747-7171(08)80125-3
4. Dartois, L., Paperman, C.: Two-variable first order logic with modular predicates over words. In: Proceedings of the 30th International Symposium on Theoretical Aspects of Computer Science, STACS'13 pp. 329–340. Leibniz International Proceedings in Informatics (LIPIcs), Schloss Dagstuhl–Leibniz-Zentrum fuer Informatik (2013). https://doi.org/10.4230/LIPIcs.STACS.2013.329
5. Delgado, M., Hoffmann, R., Linton, S., Morais, J.J.: Automata, a package on automata, Version 1.16. https://gap-packages.github.io/automata/ (2024), GAP package. See also https://cmup.fc.up.pt/cmup/mdelgado/varia/package-automata.pdf
6. Etessami, K., Vardi, M.Y., Wilke, T.: First-order logic with two variables and unary temporal logic. Inf. Comput. **179**(2), 279–295 (2002). https://doi.org/10.1006/inco.2001.2953
7. Krebs, A., Lodaya, K., Pandya, P.K., Straubing, H.: Two-variable logic with a between relation. In: Proceedings of the 31st Annual ACM/IEEE Symposium on Logic in Computer Science, LICS'16, pp. 106–115 (2016). https://doi.org/10.1145/2933575.2935308
8. Krebs, A., Lodaya, K., Pandya, P.K., Straubing, H.: An algebraic decision procedure for two-variable logic with a between relation. In: 27th EACSL Annual Conference on Computer Science Logic, CSL'18. Leibniz International Proceedings in Informatics (LIPIcs), Schloss Dagstuhl–Leibniz-Zentrum fuer Informatik (2018). https://doi.org/10.4230/LIPIcs.CSL.2018.28
9. Krebs, A., Lodaya, K., Pandya, P.K., Straubing, H.: Two-variable logics with some betweenness relations: Expressiveness, satisfiability and membership. Log. Methods Comput. Sci. **16**(3) (2020). https://doi.org/10.23638/LMCS-16(3:16)2020
10. Matz, O., Miller, A., Potthoff, A., Thomas, W., Valkema, E.: Report on the program AMoRE. Tech. rep., Institut für informatik und Praktische Mathematik, Christian-Albrechts Universität, Kiel (1995). https://sourceforge.net/projects/amore/
11. McNaughton, R., Papert, S.A.: Counter-Free Automata. MIT Press (1971)
12. Paperman, C.: Python module Pysemigroup (2024). https://gitlab.inria.fr/cpaperma/pysemigroup, online version at https://paperman.name/semigroup/
13. Éric Pin, J., Weil, P.: Polynomial closure and unambiguous product. Theory Comput. Syst. **30**(4), 383–422 (1997)
14. Pin, J.É.: Semigroup (2009). https://www.irif.fr/~jep/Logiciels/Semigroupe2.0/semigroupe2.html
15. Pin, J.É.: Mathematical foundations of automata theory (2025). http://www.irif.fr/~jep/PDF/MPRI/MPRI.pdf, in preparation

16. Pin, J.É., Delcroix, V.: Algorithms for computing finite semigroups. In: Cucker, F., Shub, M. (eds.) Foundations of Computational Mathematics, pp. 112–126. Springer, Rio de Janeiro, Brazil (1997). https://hal.science/hal-00143949v1/file/Rio97AMS.pdf

17. Place, T., Zeitoun, M.: In orbit with MeSCaL: higher in concatenation and navigational hierarchies of regular languages. https://www.labri.fr/perso/zeitoun/research/pdf/mescal2.pdf, long version

18. Place, T., Zeitoun, M.: Generic Results for Concatenation Hierarchies. Theory Comput. Syst. **63**(4), 849–901 (2018). https://doi.org/10.1007/s00224-018-9867-0

19. Place, T., Zeitoun, M.: Separation and covering for group based concatenation hierarchies. In: Proceedings of the 34th Annual ACM/IEEE Symposium on Logic in Computer Science, pp. 1–13. LICS'19 (2019). https://doi.org/10.1109/LICS.2019.8785655

20. Place, T., Zeitoun, M.: Characterizing level one in group-based concatenation hierarchies. In: Proceeding of the 17th International Computer Science Symposium in Russia, CSR'22. LNCS (2022). https://doi.org/10.1007/978-3-031-09574-0_20

21. Place, T., Zeitoun, M.: A generic polynomial time approach to separation by first-order logic without quantifier alternation. In: Dawar, A., Guruswami, V. (eds.) 42nd IARCS Annual Conference on Foundations of Software Technology and Theoretical Computer Science, FSTTCS'22. Leibniz International Proceedings in Informatics (LIPIcs), IARCS (2022). https://doi.org/10.4230/LIPIcs.FSTTCS.2022.43

22. Place, T., Zeitoun, M.: All about unambiguous polynomial closure. TheoretiCS **2** (2023). https://doi.org/10.46298/THEORETICS.23.11

23. Place, T., Zeitoun, M.: Dot-depth three, return of the J-class. In: Sobocinski, P., Lago, U.D., Esparza, J. (eds.) Proceedings of the 39th Annual ACM/IEEE Symposium on Logic in Computer Science, LICS 2024, Tallinn, Estonia, July 8-11, 2024, pp. 64:1–64:15. ACM (2024). https://doi.org/10.1145/3661814.3662082

24. Place, T., Zeitoun, M.: A generic characterization of generalized unary temporal logic and two-variable first-order logic. In: Murano, A., Silva, A. (eds.) Proceedings of the 32nd EACSL Annual Conference on Computer Science Logic, CSL'24. LIPIcs, vol. 288, pp. 45:1–45:23. Schloss Dagstuhl - Leibniz-Zentrum für Informatik (2024). https://doi.org/10.4230/LIPICS.CSL.2024.45

25. Place, T., Zeitoun, M.: A first taste of MeSCaL, a tool for solving membership problems for regular languages. In: Castiglione, G., Mantaci, S. (eds.) Proceedings of the 29th International Conference on Implementation and Application of Automata, CIAA 2025, pp. 281–298. LNCS. Springer, Palermo, Italy (2025)

26. Place, T., Zeitoun, M.: MeSCaL, MEmbership and Separation for ClAsses of Languages (2025). https://github.com/thomas-place/mescal

27. Place, T., Zeitoun, M.: Navigational hierarchies of regular languages. In: Proceedings of the 40th Annual ACM/IEEE Symposium on Logic in Computer Science, LICS 2025 (2025), full version at https://arxiv.org/abs/2402.10080

28. Schützenberger, M.P.: On finite monoids having only trivial subgroups. Inf. Control **8**(2), 190–194 (1965). https://doi.org/10.1016/S0019-9958(65)90108-7

29. Thérien, D., Wilke, T.: Over words, two variables are as powerful as one quantifier alternation. In: Proceedings of the 30th Annual ACM Symposium on Theory of Computing, pp. 234–240. STOC'98, ACM (1998). https://doi.org/10.1145/276698.276749

30. Thomas, W.: Classifying regular events in symbolic logic. J. Comput. Syst. Sci. **25**(3), 360–376 (1982). https://doi.org/10.1016/0022-0000(82)90016-2

31. Trahtman, A.N.: A package TESTAS for checking some kinds of testability. CoRR abs/2105.12583 (2021). https://arxiv.org/abs/2105.12583

A Hierarchy of Reversible Finite Automata

Maria Radionova$^{(\boxtimes)}$ and Alexander Okhotin

Department of Mathematics and Computer Science, St. Petersburg State University,
7/9 Universitetskaya nab., Saint Petersburg 199034, Russia
st084881@student.spbu.ru, alexander.okhotin@spbu.ru

Abstract. In this paper, different variants of reversible finite automata are compared, and their hierarchy by the expressive power is established. It is shown that one-way reversible automata with multiple initial states (MRFA) recognize strictly more languages than sweeping reversible automata (sRFA), which are in turn stronger than one-way reversible automata with a single initial state (1RFA). It is also shown that the hierarchy of sRFA by the number of turns over the input string collapses: two head reversals are always enough. On the other hand, MRFA form a hierarchy by the number of initial states: their subclass with at most k initial states (MRFA^k) recognizes strictly fewer languages than MRFA^{k+1}, and also MRFA^k are incomparable with sRFA. The last result is that MRFA can be simulated by reversible multi-head automata.

1 Introduction

Classical one-way deterministic automata (1DFA) have an interesting variant: *reversible automata* with one or multiple initial states (1RFA, MRFA), which were first studied by Angluin [1] and by Pin [17]. Such automata model reversible computation, which is important to study, because reversibility is the only potential way to create computer hardware which does not heat up during work. Indeed, according to Landauer's principle [13], irreversible erasure of one bit of information causes generation of a certain amount of heat, and this physical limit can be overcome only by logically reversible devices. Among recent studies of 1RFA, there are papers by Héam [5], Lavado et al. [14] and Lavado and Prigioniero [15] on the relative succinctness of 1DFA and 1RFA. Holzer and Jakobi [6] obtained results about computational complexity of minimization and hyper-minimization problems for 1RFA. The minimal reversible automata were also studied by Holzer et al. [7]. Other decision problems for reversible automata were investigated by Birget et al. [3], and recently by the authors [20].

Another model of reversible automata studied in the literature are the *one-way permutation automata* (1PerFA), in which the transition function by each symbol is a bijection. One-way permutation automata were first studied by

This work was supported by the Russian Science Foundation, project 23-11-00133.

G. Castiglione and S. Mantaci (Eds.): CIAA 2025, LNCS 15981, pp. 316–329, 2026.
https://doi.org/10.1007/978-3-032-02602-6_22

Thierrin [24]. Recently, Jecker, Mazzocchi and Wolf [10] investigated the computational complexity of the decomposition problem for permutation automata. Also, Hospodár and Mlynárčik [9] studied the state complexity of operations on permutation automata, Rauch and Holzer [21] obtained the complexity of operations on permutation automata in the number of accepting states, while Okhotin, Radionova and Sazhneva [16] proved that this family is closed under the new GF(2)-concatenation operation.

One-way reversible automata recognize strictly fewer languages than classical deterministic automata: for example, they cannot recognize a regular language a^*b^*. At the same time, permutation automata turn out to be an even more constrained model: in particular, they cannot recognize any finite language, whereas reversible automata recognize all finite languages. In this paper, this hierarchy is completed by other variants of reversible and permutation automata.

Among the models considered, there are two-way reversible automata. As soon as reversible automata are allowed to read a string in both directions, they can recognize all regular languages, as was shown by Kondacs and Watrous [11], which exceeds the expressive power of one-way reversible automata (1RFA). Between one-way and two-way automata, an intermediate model of *sweeping automata* [23] was studied. These automata alternate between full passes over the input string from left to right and from right to left. Recently, a sweeping variant of permutation automata (2PerFA) was studied by the authors [19], and it was shown that the new model recognizes exactly the same languages as 1PerFA.

Another variant of automata is that with multiple initial states: these are one-way automata that can choose non-deterministically one of their initial states before reading an input string. Deterministic automata with multiple initial states (MDFA), studied by Holzer et al. [8], recognize exactly the family of regular languages, that is, their expressive power is the same as that of 1DFA. A similar result holds for permutation automata: permutation automata with multiple initial states (MPerFA) cannot recognize more languages than 1PerFA, and this directly follows from the closure of the group languages under the union operation. In the case of reversible automata, as a matter of fact, Pin [17] studied the model with multiple initial states (MRFA).

In the paper, we compare the expressive power of these variants of reversible and permutation automata and build a hierarchy. The resulting hierarchy is shown in Fig. 1, in which $1RFA_1$ denotes reversible automata with a unique accepting state, as studied by Angluin [1] (Pin [18] called them *bideterministic*), $1PerFA_1$ are permutation automata with the same restriction, while $MRFA^k$ denotes MRFA with k initial states. The last result, not reflected in the figure, is a simulation of MRFA by one-way reversible multi-head automata, which were recently studied by Kutrib and Malcher [12].

2   Definitions

In the literature, different variants of one-way reversible automata were investigated: Angluin [1] studied reversible automata with one initial and one accepting

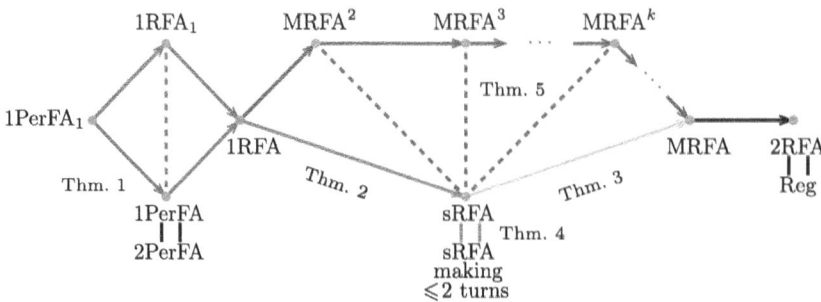

Fig. 1. The hierarchy of reversible and permutation automata (solid arrows indicate proper inclusions, dashed lines indicate incomparability).

state (1RFA$_1$), Pin [17] allowed reversible automata to have an arbitrary number of initial and accepting states (MRFA), and Holzer et al. [7] studied reversible automata with one initial and any number of accepting states (1RFA). As long as the initial state is unique, reversible automata are a special case of 1DFA.

Definition 1. *A one-way deterministic automaton is a quintuple* $(\Sigma, Q, q_0, \langle \delta_a \rangle a, F)$, *which includes*

- *an input alphabet* Σ;
- *a set of states* Q;
- *an initial state* $q_0 \in Q$;
- *a transition function* $\delta_a \colon Q \to Q$ *by each symbol a from the alphabet* Σ;
- *a set of accepting states* $F \subseteq Q$.

    A computation of a 1DFA on a string $s = a_1 a_2 \ldots a_k$ *is a sequence of states* $q_0, q_1, \ldots, q_k$, *in which every next state is obtained by applying the transition function to the previous one:* $q_{i+1} = \delta_{a_{i+1}}(q_i)$. *A one-way deterministic automaton (1DFA) accepts a string if its computation ends in an accepting state* $q_k \in F$.

    In the literature, partial deterministic finite automata, with partial transition functions, are also common. Their computation on a string can end in advance by an undefined transition, and in this case the string is rejected.

Definition 2. *A one-way reversible automaton (1RFA) is a partial one-way deterministic automaton with an injective transition function* δ_a *for each symbol a.*

Definition 3. *A one-way permutation automaton (1PerFA) is a one-way reversible automaton (1RFA) with a bijective (and hence fully defined) transition function* δ_a *by each symbol a.*

    Sweeping automata, unlike their one-way counterparts, can read a string several times, from left to right and from right to left.

Definition 4. *A sweeping deterministic automaton (sDFA) is a 9-tuple* $(\Sigma, Q_+, Q_-, q_0, \langle \delta_a^+ \rangle_a, \langle \delta_a^- \rangle_a, \delta_\vdash, \delta_\dashv, F)$*, in which there are*

- *an input alphabet* Σ*;*
- *a set of states* $Q_+ \cup Q_-$ *split into two disjoint sets, in one of which the automaton reads the input from left to right, and in the other from right to left;*
- *an initial state* $q_0 \in Q_+$*, in which the automaton starts to read the input;*
- *partial transition functions by each symbol* $a \in \Sigma$ *inside a string:* $\delta_a^+ : Q_+ \to Q_+$ *for reading from left to right, and* $\delta_a^- : Q_- \to Q_-$ *for reading from right to left;*
- *partial transition functions at the end-markers,* $\delta_\vdash : \{q_0\} \cup Q_- \to Q_+$ *and* $\delta_\dashv : Q_+ \to Q_-$*;*
- *a set of accepting states* $F \subseteq Q_+$ *effective at the right end-marker.*

A computation of a sweeping deterministic automaton on the string $s = \vdash a_1 a_2 \ldots a_k \dashv$ *is a sequence of pairs* (q_j, i_j)*, where* q_j *is a state, and* i_j*, with* $i_j \in \{0, 1, \ldots, k, k+1\}$*, is a position of the automaton's head in the string. The elements of the sequence are defined in the following way: the first pair is* $(q_0, 0)$*, and every* $(j+1)$*-th one is obtained from the* j*-th pair by changing the state and the position of the head according to the transition function by the current symbol or end-marker.*

$$
(q_{j+1}, i_{j+1}) = \begin{cases} (\delta_{a_{i_j}}^+(q_j), i_j + 1), & \text{if } 1 \leqslant i_j \leqslant k \text{ and } q_j \in Q_+; \\ (\delta_{a_{i_j}}^-(q_j), i_j - 1), & \text{if } 1 \leqslant i_j \leqslant k \text{ and } q_j \in Q_-; \\ (\delta_\vdash(q_j), 1), & \text{if } i_j = 0 \text{ and } q_j \in Q_- \cup \{q_0\}; \\ (\delta_\dashv(q_j), k), & \text{if } i_j = k+1 \text{ and } q_j \in Q_+. \end{cases}
$$

A sweeping deterministic automaton accepts the string, if the computation ends at the right end-marker in an accepting state, that is, in a pair $(q, k+1)$ *with* $q \in F$*.*

As in the case of deterministic automata, it is possible to define a reversible sweeping automaton. Such an automaton is an sDFA with the restriction that its transition functions are injective.

Definition 5. *A sweeping reversible automaton (sRFA) is a sweeping deterministic automaton, with injective transition functions on symbols of an input string,* δ_a^+ *and* δ_a^-*, for each* $a \in \Sigma$*, and with injective transition functions by the end-markers,* $\delta_\vdash$ *and* $\delta_\dashv$*.*

The last model of sweeping automata considered in the paper are *sweeping permutation automata*.

Definition 6. *A sweeping permutation automaton (2PerFA) is a sweeping reversible automaton (sRFA) with bijective (and hence fully defined) transition functions* δ_a^+ *and* δ_a^-*, for each symbol* $a \in \Sigma$*. Transition functions at both end-markers can still be any injective functions, as in sRFA.*

A few results in this paper refer to the classical transformation of sweeping automata to one-way automata [22], which works as follows.

Proposition 1. *For a sweeping deterministic automaton (sDFA) with a set of states $Q_- \cup Q_+$, a one-way deterministic automaton (1DFA) is constructed, which has pairs (p, f) for its states, with $p \in Q_+$, and a partial function f mapping Q_- to Q_+. The transitions of the 1DFA are defined so that the reachability of a state (p, f) by some string w is equivalent to the next two conditions.*

1. *After reading the string w from left to right starting in the initial state, the given sweeping automaton comes to the state p.*
2. *If the given sDFA starts to read the string w at its last symbol in a state $q \in Q_-$, reads it from right to left, makes a transition by the left end-marker, and next reads w from left to right, then it finishes in the state $f(q) \in Q_+$ (and the value $f(q)$ must be defined in this case). If the computation of the sDFA on the string w ends prematurely, then $f(q)$ is undefined.*

Finally, the last type of automata considered in the paper are reversible automata with multiple initial states.

Definition 7 (Pin [17]). *A one-way reversible automaton with multiple initial states (MRFA) is a quintuple $(\Sigma, Q, Q_0, \langle \delta_a \rangle_a, F)$, in which*

- *Σ is an input alphabet;*
- *Q is a finite set of states;*
- *$Q_0 \subseteq Q$ is a set of initial states;*
- *$\delta_a \colon Q \to Q$, for each symbol $a \in \Sigma$, is an injective partially defined transition function by a;*
- *$F \subseteq Q$ is a set of accepting states.*

The computation of an MRFA on a string $s = a_1 a_2 \ldots a_k$ starting in a state $q_0 \in Q_0$ is a sequence of states q_0, q_1, $\ldots$, q_k, in which every next state is obtained from the previous one by using the transition function: $q_i = \delta_{a_i}(q_{i-1})$ for $1 \leqslant i \leqslant k$. The computation is accepting, if it ends in an accepting state from the set F. A string is accepted by an MRFA, if there exists an accepting computation on that string starting in some initial state.

This is the most general variant of one-way reversible automata, and the following lemma for showing non-representability of languages by these automata is known.

Lemma 1 (Pin [17]). *Let a language L be recognized by some MRFA, and let $xy^+z \subseteq L$ for some strings $x, z \in \Sigma^*$ and $y \in \Sigma^+$. Then $xz \in L$.*

For example, for the language a^+, the lemma is applied to $x = \varepsilon$, $y = a$ and $z = \varepsilon$, and it provides a contradiction. Hence, this language is not recognized by any MRFA (as well as by 1RFA and by weaker models).

One of the results of this paper (Theorem 3) will imply that Lemma 1 also applies to all languages recognized by sRFA.

3   Main Hierarchy

Let us begin to build the hierarchy with results for the weaker models.

Theorem 1. *One-way permutation automata (1PerFA), one-way reversible automata (1RFA), and their variants with one accepting state (1PerFA$_1$, 1RFA$_1$) form the following proper inclusions.*

$$\mathcal{L}(\text{1PerFA}_1) \subsetneq \mathcal{L}(\text{1PerFA}) \subsetneq \mathcal{L}(\text{1RFA}),$$
$$\mathcal{L}(\text{1PerFA}_1) \subsetneq \mathcal{L}(\text{1RFA}_1) \subsetneq \mathcal{L}(\text{1RFA}),$$

where $\mathcal{L}(\mathcal{X})$ stands for the family of languages recognized by automata from a class $\mathcal{X}$. Also, 1PerFA and 1RFA$_1$ are incomparable in the expressive power.

Proof (a sketch). Reversible automata can recognize finite languages, whereas permutation automata cannot. Hence, it is possible to separate these classes by the language $\{a\}$. Automata with one accepting state and automata with multiple accepting states are separated by the language $(a^3)^* \cup a(a^3)^*$. The incomparability of 1PerFA and 1RFA$_1$ is by the same two languages.   □

Recently, it was shown that sweeping permutation automata (2PerFA) are equal in their expressive power to their one-way variant (1PerFA) [19]. Could it also be possible to transform every sRFA to a 1RFA? The answer turns out to be negative. What is the difficulty of transforming sRFA to 1RFA? That is, why does the usual transformation of sRFA to 1DFA, in which the constructed one-way automaton computes a certain behavior function (see Proposition 1), sometimes produce an irreversible automaton? This is because a reversible sweeping automaton can reject in the middle of a string, thus invalidating some earlier computed values of the behavior function. Therefore, the same behavior function can be obtained from two different behavior functions upon reading a symbol with undefined transitions. The following example shows that a transformation of sRFA to 1RFA is impossible in the general case.

Example 1. The language $(aa)^* \cup \{a\}$ cannot be recognized by any 1RFA, but it is recognized by a sRFA.

Proof (a sketch). Indeed, to recognize the language, a 1RFA must have a cycle by the symbol a, which contains the initial state. The initial state is accepting, because the automaton accepts the empty string. Hence, the length of the cycle must be even, because no long string with an odd number of symbols should be accepted. But, at the same time, the state $\delta_a(q_0)$ is accepting too, as the string a is in the language. Thus, the automaton also accepts a long string with an odd number of symbols, a contradiction.

An sRFA recognizing this language first computes the length of the string modulo 2, and accepts if it is even. If the length is odd, it makes another pass through the string to check whether it consists of one symbol.   □

From the example above, the next theorem is obtained.

Theorem 2. *One-way reversible automata (1RFA) recognize a proper subfamily of the languages recognized by sweeping reversible automata (sRFA).*

Turning back to the transformation of sRFA to one-way automata, why does the classical transformation create irreversible transitions for the sRFA from Example 1? On the string a, this sRFA will return to the left end-marker and make a turn there. When the resulting 1DFA starts to read a string, it cannot know in advance whether the sRFA is going to make such a turn. The possibility of return cannot be excluded, hence, the 1DFA should remember the computation involving this transition. However, after reading two symbols a it becomes clear that such a transition is not used in the computation. Then the 1DFA forgets it, and thereby it remembers, that the string contains more than one symbol. After two more as, the 1DFA will return to the same state, and this transition will be irreversible.

How to avoid such problems with irreversibility? To relieve a one-way automaton from the burden of forgetting undefined transitions of sRFA, let it guess non-deterministically the domain of a behavior function before starting a computation, and then watch that its size never decreases. If the domain is ever reduced upon reading a symbol, then the automaton just rejects at that point. This observation allows an sRFA to be transformed to a one-way reversible automaton with multiple initial states (MRFA), which guesses a behavior function's domain at the beginning of its computation, and then proceeds with reversibly calculating the function, maintaining the size of its domain.

Lemma 2. *A sweeping reversible automaton (sRFA) with sets of states P_+ and P_- can be transformed to a one-way reversible automaton with multiple initial states (MRFA) with the set of states*

$$Q = \{ (p, f) \mid p \in P_+, \ f \colon P_- \to P_+ \text{ is partial injective, } p \notin \operatorname{Im} f \}$$

Proof. Given a sweeping reversible automaton $\mathcal{A} = (\Sigma, P_+, P_-, p_0, \langle \gamma_a^+ \rangle_a, \langle \gamma_a^- \rangle_a, \gamma_\vdash, \gamma_\dashv, E)$, a one-way reversible automaton with multiple initial states $\mathcal{B} = (\Sigma, Q, Q_0, \langle \delta_a \rangle_a, F)$ is constructed, which guesses the domain of the behavior function of the sweeping automaton on a string while choosing the initial state. The MRFA calculates this function while reading the string, and in the end determines the result of the sweeping automaton's computation using that function, and accordingly makes a decision to accept the string or not. The automaton manages to be reversible through using multiple initial states. In each computation, it now has no need to reduce the domain of the constructed behavior function: if such a moment comes, then the automaton can just reject by an undefined transition.

In each of its states, the MRFA stores a state of the given sRFA and its behavior function (which must be injective, because the sRFA $\mathcal{A}$ is reversible). Such a pair has the same meaning as in the classical transformation to 1DFA, except that the behavior function need not reflect *all* possible computations on the prefix, only a subset of them. That is to say, if the MRFA comes to a state (p, f) by some string w, then the sRFA comes to the state $p \in P_+$ after reading

w for the first time from left to right, and in the states from the domain of f, the given sRFA operates as described in f (see the explanation of the behavior function in Proposition 1); however, for all other states not in the domain of f, the MRFA knows nothing.

Note that, for some string w, there may be multiple states (p, f) of the MRFA satisfying the above condition (with different subsets of possible computations undefined). An essential trick in this construction is that the MRFA can reach *all* such states, if it starts from different initial states.

The MRFA's initial states must satisfy the above property for the empty string $w = \varepsilon$. The first component of every initial state is always set to the state $\gamma_\vdash(p_0)$, in which sRFA comes after the transition at the left end-marker from its initial state; the second component, a function, is always defined to act as the transition function at the left end-marker $\gamma_\vdash$. Moreover, the MRFA chooses the domain of a behavior function of the sRFA in the initial states, hence, for their second component, they have all possible restrictions of the function $\gamma_\vdash$ to subsets of its domain.

$$Q_0 = \left\{\, (\gamma_\vdash(p_0), f) \mid f = \gamma_\vdash|_S \text{ for some } S \subseteq (\operatorname{Dom}\gamma_\vdash) \setminus \{p_0\} \,\right\}$$

While making a transition by a symbol a, the MRFA must change both components of its state. And it changes them so that the new state of the MRFA again describes the behavior of the given sRFA, but on the string extended with the symbol a. For this, the transition functions of the sRFA are applied to the pair (p, f), from which the MRFA makes its transition, as shown below. Moreover, if the domain of the function f is reduced after applying the symbol, then the transition is left undefined in order to avoid irreversible behavior.

$$\delta_a((p, f)) = \begin{cases} \left(\gamma_a^+(p), \gamma_a^+ \circ f \circ \gamma_a^-\right), & \text{if } \gamma_a^+(p) \text{ is defined,} \\ & \text{and } |\operatorname{Dom}\gamma_a^+ \circ f \circ \gamma_a^-| = |\operatorname{Dom} f|; \\ \text{is undefined,} & \text{otherwise.} \end{cases}$$

In the definition of the MRFA's set of accepting states, for each state (p, f), consider the sequence of the sRFA's states, in which it visits the right end-marker in the computation described by the function f. The first state from the sequence, p_1, is equal to p. To compute the second one, starting in the state p_1, apply the transition function at the right end-marker $\gamma_\dashv$ and then the function f, obtaining $p_2 = f(\gamma_\dashv(p_1))$, etc. If the resulting sequence $p_1, p_2, \ldots$ ends in an accepting state of the sRFA, then (p, f) will be accepting in the MRFA.

$$F = \left\{\, (p, f) \mid (f \circ \gamma_\dashv)^k(p) \in E \text{ for some } k \geqslant 0 \,\right\}.$$

First of all, it should be checked that the initial states and all states reached by transitions belong to the set Q. Each initial state $(\gamma_\vdash(p_0), f)$ has $\gamma_\vdash(p_0) \notin \operatorname{Im} f$ and has f injective, because f is a restriction of the injective transition function $\gamma_\vdash$ to a subset of P_-. The correctness of transition function δ_a for every $a \in \Sigma$ is shown in the following claim.

Claim 1. *Let $(p, f) \in Q$ and $a \in \Sigma$, and assume that $\gamma_a^+(p)$ is defined. Then the function $g = \gamma_a^+ \circ f \circ \gamma_a^-$ is injective and $\gamma_a^+(p) \notin \operatorname{Im} g$.*

The function g is injective, because such is the function f and the transition functions γ_a^+, γ_a^-. The function g does not contain $\gamma_a^+(p)$ in its image, because γ_a^+, which is applied to both the function f and the state p to obtain the current pair $(\gamma_a^+(p), g)$, is injective and maps all the values from the image of f and the state p to pairwise distinct states.

The constructed automaton can be proved to be reversible, because its transitions maintain fixed size of the domain of the function f, and hence a previous state can always be reconstructed.

Claim 2. *If a transition $\delta_a((p, f)) = (q, g)$ is defined, then $p = (\gamma_a^+)^{-1}(q)$, and $f = (\gamma_a^+)^{-1} \circ g \circ (\gamma_a^-)^{-1}$. In particular, the constructed automaton is reversible.*

To see that the constructed MRFA indeed recognizes the same language as the given sRFA, the following correctness statement is proved by induction on the length of the string.

Claim 3. *On a string w, the constructed MRFA may come to a state (p, f) from any initial state if and only if*

1. *the sRFA comes to the state $p \in P_+$ after the first pass on w from left to right;*
2. *the partial injective function $f : P_- \to P_+$ maps each state $q \in P_-$ from its domain to such a state $f(q) \in P_+$ that when the sRFA starts to read w from right to left in the state q, then turns at the left end-marker and reads w once again from left to right, it finishes in the state $f(q)$. Also, p is not in the image of f.*

The proof that the constructed MRFA recognizes the same language as the given sRFA follows from Claim 3 and the definition of accepting states of the MRFA.   □

To show that sRFA are strictly weaker than MRFA, consider the following example.

Example 2. The language $a^* \cup b^*$ is recognized by an MRFA, but no sRFA can recognize it.

Proof (a sketch). Assume, that there is an sRFA recognizing $a^* \cup b^*$. Let n be the total number of the sRFA's states. Consider the computation on a string $a^{n!}b^{n!}$, which is not in the language. First, the sRFA reads a prefix $a^{n!}$ from the initial state q_0. The sRFA cannot reject by an undefined transition while reading, and since it is reversible, it has to return to the state q_0 after some m symbols, where $1 \leqslant m \leqslant n$. Since m divides $n!$, the automaton finishes reading $a^{n!}$ in the state q_0. By the same argument, it finishes reading the suffix $b^{n!}$ also in q_0. It can neither accept nor reject, because $a^{n!}b^{n!}$ should be rejected, whereas $a^{n!}$ and $b^{n!}$ should be accepted. So it turns at the right end-marker in some state from Q_-. But, in the same way, at every subsequent traversal of the string the automaton cannot distinguish it from a string from the language, and any acceptance or rejection decision would lead to a contradiction.   □

Now, from Lemma 2 and Example 2 follows a theorem about comparing the expressive power of sRFA and MRFA.

Theorem 3. *Sweeping reversible automata (sRFA) recognize strictly fewer languages than one-way reversible automata with multiple initial states (MRFA).*

4   Is There a Hierarchy of sRFA by the Number of Turns over an Input String?

The number of turns, also called the number of head reversals, is a widely studied complexity measure of automata; in particular, Balcerzak and Niwiński [2] studied the descriptional complexity of 2DFA with different number of turns. For reversible automata, as shown in Example 1, 1RFA making one left-to-right pass without turns are weaker than sRFA making two turns. However, it turns out that any additional head reversals do not increase the power of this model any further.

Theorem 4. *Sweeping reversible automata (sRFA) recognize the same family of languages as sweeping reversible automata making at most two turns over a string.*

Proof (a sketch). The plan is that the new automaton $\mathcal{B}$, instead of making many passes one by one, will compute some kind of behavior function of the original automaton $\mathcal{A}$. However, as argued in front of Example 1, the behavior function of an sRFA cannot be reversibly computed as long as any transitions of the sRFA are undefined, because this would effectively erase some values of the behavior function (and this is irreversible). To circumvent this problem, while computing the behavior function at its first pass, the new sRFA $\mathcal{B}$ will consider the given sRFA $\mathcal{A}$ as a sweeping permutation automaton $\widetilde{\mathcal{A}}$, completing its transition function in each direction to a bijection in an arbitrary way. In other words, $\mathcal{B}$ will act as if the undefined transitions of $\mathcal{A}$ were defined. Thus, $\mathcal{B}$ makes its first left-to-right pass in states of the form (p, f), with $p \in Q_+$ and $f: Q_- \to Q_+$, and in the end of the pass $\mathcal{B}$ will have computed the behavior function of $\widetilde{\mathcal{A}}$ on the input string.

At this point, $\mathcal{B}$ will see the outcome of the computation of $\widetilde{\mathcal{A}}$ on this string. If $\widetilde{\mathcal{A}}$ rejects it, then $\mathcal{A}$ is known to reject as well, because either it can make all the steps of $\widetilde{\mathcal{A}}$ and reject, or it will reach an undefined transition on the way; and in this case $\mathcal{B}$ immediately rejects. However, if $\widetilde{\mathcal{A}}$ accepts, this does not guarantee that $\mathcal{A}$ accepts. What $\mathcal{B}$ knows, is the set of states that $\mathcal{A}$ would have to pass through at the right end-marker in order to accept, if it accepts at all (these are the same states that $\widetilde{\mathcal{A}}$ passes through).

In this case, $\mathcal{B}$ makes some further computations to test whether a computation passing through those states is possible. On its way from right to left, it uses states of the form (f, R), where f is the behavior function of $\widetilde{\mathcal{A}}$ on the prefix ending in the current position, whereas $R \subseteq Q_-$ is the set of states that $\widetilde{\mathcal{A}}$ passes through at the current position. At each step, the behavior function on a prefix shorter by one symbol is made by applying the inverse of the transition function of $\widetilde{\mathcal{A}}$, while R maintains the current states of all computations of $\mathcal{A}$ necessary to reach acceptance. At the same time, $\mathcal{B}$ verifies that all the required transitions of $\mathcal{A}$ are defined: it tests the transitions in all states from R and in all states from $f(R)$; and if any of them is undefined, $\mathcal{B}$ rejects.

By the end of its right-to-left pass, $\mathcal{B}$ will have *uncomputed* the behavior function to its initial value (which is the transition function by the left end-marker), and also all the required transitions will have been verified. All that remains is to move to the right end-marker for acceptance.   □

5   Hierarchy of MRFA with a Bounded Number of Initial States

In this section, a hierarchy of MRFA in the number of initial states is obtained. The bottom level of this hierarchy is 1RFA, that is, MRFA with one initial state. The k-th level contains MRFA with at most k initial states (MRFA^k). The entire hierarchy is contained in the class of MRFA with unbounded number of initial states. And, as shown in the following lemma, every additional initial state increases the expressive power of the model.

Lemma 3. *For each $k \geqslant 2$, the language $L_k = \bigcup_{i=1}^{k}(ab^i)^*$ is recognized by an MRFA with k initial states (and with $\frac{k(k+3)}{2}$ total states), but no MRFA with fewer than k initial states recognizes this language.*

Proof. An MRFA recognizing this language has a separate cycle of length $i+1$ to accept each subset $(ab^i)^*$.

For the second part, let the language be recognized by an MRFA with n states. The idea is that the strings $(ab^i)^{n!}$, for different i, must be accepted from pairwise distinct initial states. Indeed, if the automaton reads $(ab^i)^{n!}$ starting from some initial state q_0, then it finishes reading it in q_0 as well. Hence, if $(ab^j)^{n!}$ with $j \neq i$ is also accepted from q_0, then the automaton accepts $(ab^i)^{n!}(ab^j)^{n!}$, which is a contradiction.   □

Earlier, it has been proved that the class of sRFA is included in MRFA. Now, the intermediate classes of MRFA^k have been added to the hierarchy, and a question arises, how do they compare to sRFA. It is already known from Example 2 that MRFA^2 are not contained in sRFA. But could all sRFA be simulated by MRFA with at most k initial states? The answer is given in the next lemma.

Lemma 4. *For each $k \geqslant 2$, the language $L_k = \bigcup_{i=0}^{k-1}(ab^i)^* ba^i$ is recognized by sRFA, but it is not recognized by any MRFA with fewer than k initial states.*

Proof (a sketch). An sRFA first moves to the right end-marker, and then recognizes the language from right to left. The non-existence of an MRFA with $k-1$ initial states is proved by an argument similar to that it Lemma 3.

Theorem 5. *For each $k \geqslant 1$, one-way reversible automata with k initial states (MRFA^k) recognize a proper subfamily of the family recognized by one-way reversible automata with $k + 1$ initial states (MRFA^{k+1}). Also, MRFA^k, for each $k \geqslant 2$, are incomparable with sRFA in the expressive power.*

The proof follows from Lemmata 3 and 4 and from the earlier Example 2.

6   One-Head Vs Multi-Head Reversible Automata

One-way reversible multi-head automata were recently studied by Kutrib and Malcher [12], who established many properties of this model, but left one important question open: it remains unknown whether these automata can recognize every regular language. While no solution to this problem is attempted in this paper, it shall be proved that the entire hierarchy of one-head reversible automata studied here is contained in the family recognized by multi-head reversible automata.

Theorem 6. *For every MRFA with n states and k initial states, there exists a one-way $(k + 1)$-head reversible automaton with $(|\Sigma| + 2)n^k$ states recognizing the same language.*

Proof (a sketch). The idea is to simulate in parallel k computations of the MRFA starting from each of its initial states, moving head 1 across the string. If any of these computations fail in the middle the string, the automaton must remember this fact, and a one-head 1RFA cannot generally do this. A multi-head 1RFA will employ a special head for each of the k computations: head $i + 1$ moves by one symbol to the right each time the i-th computation is about to reject, while the computation itself is continued by simulating an artificially added transition. □

It could also be noted that the regular language $a^* b^*$ that is not recognized by any MRFA can be recognized by a one-way two-head reversible automaton, the problem whether *all* regular languages are recognized by multi-head reversible automata stays open. Furthermore, it is left open whether every MRFA can be simulated by a k-head reversible automaton for any fixed k.

7   Conclusion

A hierarchy of several natural variants of reversible automata has been established. However, some reversible models of interest were not included in this study. The most notable omission is that of k-*reversible automata*, in which the previous state can be determined from the current state and the last $k + 1$ symbols read. This model was introduced by Angluin [1], who additionally imposed a condition that the accepting state in which a string is accepted can be determined by its last k symbols; under this definition, she established a strict hierarchy of k-reversible automata, as well as proper containment of the entire hierarchy in the class of regular languages. Later, Guillon et al. [4] reinvestigated k-reversible automata under a classical definition with no restrictions on accepting states, and also established their strict hierarchy by k, as well as presented a regular language outside this hierarchy. In our notation, the hierarchy of Guillon et al. [4] stems from 1RFA, which are the 0-reversible automata. Comparing the k-reversible automata of Guillon et al. [4] to sRFA and MRFA, as well as possibly to MRFA^i, is left as a problem for future research.

Another problem worth investigating is the state complexity of the transformations between the classes of automata studied here. For instance, how many states in MRFA are needed to simulate an n-state sRFA, and how many initial states are required? Are the constructions presented in this paper any close to optimal in terms of automaton size? Could they be improved?

References

1. Angluin, D.: Inference of reversible languages. J. ACM **29**(3), 741–765 (1982). https://doi.org/10.1145/322326.322334
2. Balcerzak, M., Niwiński, D.: Two-way deterministic automata with two reversals are exponentially more succinct than with one reversal. Inf. Process. Lett. **110**(10), 396–398 (2010). https://doi.org/10.1016/j.ipl.2010.03.008
3. Birget, J.-C., Margolis, S.W., Meakin, J.C., Weil, P.: PSPACE-complete problems for subgroups of free groups and inverse finite automata. Theoret. Comput. Sci. **242**(1–2), 247–281 (2000). https://doi.org/10.1016/S0304-3975(98)00225-4
4. Guillon, B., Lavado, G.J., Pighizzini, G., Prigioniero, L.: Weakly and strongly irreversible regular languages. Int. J. Found. Comput. Sci. **33**(3–4), 263–284 (2022). https://doi.org/10.1142/S0129054122410052
5. Héam, P.-C.: A lower bound for reversible automata. RAIRO Theoret. Inform. App. **34**(5), 331–341 (2000). https://doi.org/10.1051/ita:2000120
6. Holzer, M., Jakobi, S.: Minimal and hyper-minimal biautomata. Int. J. Found. Comput. Sci. **27**(2), 161–186 (2016). https://doi.org/10.1142/S0129054116400050
7. Holzer, M., Jakobi, S., Kutrib, M.: Minimal reversible deterministic finite automata. Int. J. Found. Comput. Sci. **29**(2), 251–270 (2018). https://doi.org/10.1142/S0129054118400063
8. Holzer, M., Salomaa, K., Yu, S.: On the state complexity of k-entry deterministic finite automata. J. Autom. Lang. Comb. **6**(4), 453–466 (2001). https://doi.org/10.25596/jalc-2001-453
9. Hospodár, M., Mlynárčik, P.: Operations on permutation automata. In: Jonoska, N., Savchuk, D. (eds.) DLT 2020. LNCS, vol. 12086, pp. 122–136. Springer, Cham (2020). https://doi.org/10.1007/978-3-030-48516-0_10
10. Jecker, I., Mazzocchi, N., Wolf, P.: Decomposing permutation automata. In: CONCUR 2021, LIPIcs, vol. 203, pp. 18:1–18:19 (2021). https://doi.org/10.4230/LIPIcs.CONCUR.2021.18
11. Kondacs, A., Watrous, J.: On the power of quantum finite state automata. In: 38th Annual Symposium on Foundations of Computer Science (FOCS 1997, Miami Beach, Florida, USA, 19–22 October 1997), pp. 66–75. IEEE (1997). https://doi.org/10.1109/SFCS.1997.646094
12. Kutrib, M., Malcher, A.: One-way reversible multi-head finite automata. Theoret. Comput. Sci. **682**, 149–164 (2017). https://doi.org/10.1016/j.tcs.2016.11.006

13. Landauer, R.: Irreversibility and heat generation in the computing process. IBM J. Res. Dev. **5**(3), 183–191 (1961). https://doi.org/10.1147/rd.53.0183
14. Lavado, G.J., Pighizzini, G., Prigioniero, L.: Minimal and reduced reversible automata. J. Autom. Lang. Comb. **22**(1–3), 145–168 (2017). https://doi.org/10.25596/jalc-2017-145
15. Lavado, G.J., Prigioniero, L.: Concise representations of reversible automata. Int. J. Found. Comput. Sci. **30**(6–7), 1157–1175 (2019). https://doi.org/10.1142/S0129054119400331
16. Okhotin, A., Radionova, M., Sazhneva, E.: GF(2)-operations on basic families of formal languages. Theoret. Comput. Sci. **995**, 114489 (2024). https://doi.org/10.1016/j.tcs.2024.114489
17. Pin, J.-É.: On the languages accepted by finite reversible automata. In: Ottmann, T. (ed.) Automata, Languages and Programming. ICALP 1987. LNCS, vol. 267, pp. 237–249. Springer, Heidelberg (1987). https://doi.org/10.1007/3-540-18088-5_19
18. Pin, J.-E.: On reversible automata. In: Simon, I. (ed.) LATIN 1992. LNCS, vol. 583, pp. 401–416. Springer, Heidelberg (1992). https://doi.org/10.1007/BFb0023844
19. Radionova, M., Okhotin, A.: Sweeping permutation automata. In: Proceedings of the 13th International Workshop on Non-Classical Models of Automata and Applications (NCMA 2023, Famagusta, North Cyprus, 18–19 September 2023), EPTCS, vol. 388, pp. 110–124 (2023). https://doi.org/10.4204/EPTCS.388.11
20. Radionova, M., Okhotin, A.: Decision problems for reversible and permutation automata. In: Fazekas, S.Z. (ed.) Implementation and Application of Automata. CIAA 2024. LNCS, vol. 15015, pp. 302–315. Springer, Cham (2024). https://doi.org/10.1007/978-3-031-71112-1_22
21. Rauch, C., Holzer, M.: On the accepting state complexity of operations on permutation automata. RAIRO Theoret. Inform. App. **57**, 9 (2023). https://doi.org/10.1051/ita/2023010
22. Shepherdson, J.C.: The reduction of two-way automata to one-way automata. IBM J. Res. Dev. **3**, 198–200 (1959). https://doi.org/10.1147/rd.32.0198
23. Sipser, M.: Lower bounds on the size of sweeping automata. J. Comput. Syst. Sci. **21**(2), 195–202 (1980). https://doi.org/10.1016/0022-0000(80)90034-3
24. Thierrin, G.: Permutation automata. Math. Syst. Theory **2**(1), 83–90 (1968). https://doi.org/10.1007/BF01691347

Author Index

A
Alber, Franziska   1
Attou, Samira   15

B
Berglund, Martin   27
Boigelot, Bernard   41
Borsotti, Angelo   55
Braipson, Thomas   41
Breveglieri, Luca   55

C
Clara, Tom   41
Clerici Lorenzini, Alessandro   73
Crespi Reghizzi, Stefano   55

D
De Giacomo, Giuseppe   129
Dieck, Simon   99
Drewes, Frank   115
Duret-Lutz, Alexandre   129

F
Fazekas, Szilárd Zsolt   148

G
Gallot, Paul   165
Geffert, Viliam   180

H
Herrmann, Roland   193
Herwig, Maurice   208
Holzer, Markus   223
Hundeshagen, Norbert   208

J
Jonsson, Anna   27

K
Khodier, Mazen   237
Klein, Béla   148
Koß, Tore   148
Kuhlmann, Marco   115
Kutrib, Martin   252

L
Lange, Martin   208

M
Malcher, Andreas   252
Manea, Florin   148
Maneth, Sebastian   165
Martens, Willeke   27
Martynova, Olga   267
Mercaş, Robert   148
Merwe, Brink van der   27
Mignot, Ludovic   15
Miklarz, Clément   15
Morzenti, Angelo   55

N
Nicart, Florent   15

O
Okhotin, Alexander   180, 267, 316

P
Pighizzini, Giovanni   73
Piterman, Nir   129
Place, Thomas   281, 299
Prigioniero, Luca   73

R
Radionova, Maria   316
Raucci, Priscilla   252
Rauch, Christian   223
Rümmer, Philipp   1, 193

G. Castiglione and S. Mantaci (Eds.): CIAA 2025, LNCS 15981, pp. 331–332, 2026.
https://doi.org/10.1007/978-3-032-02602-6

S
Schaeffer, Luke   237
Shallit, Jeffrey   237
Specht, Timo   148

T
Tianxiang, Tang   86
Torstensson, Olle   115

V
Vardi, Moshe Y.   129
Verwer, Sicco   99

Z
Zakharov, Vladimir A.   86
Zeitoun, Marc   281, 299
Zhibo, Deng   86
Zhu, Shufang   129

The manufacturer's authorised representative in the EU is Springer
Nature Customer Service Centre GmbH, Europaplatz 3, 69115 Heidelberg,
Germany. If you have any concerns regarding our products, please
contact ProductSafety@springernature.com

Printed and bound by CPI Group (UK) Ltd, Croydon, CR0 4YY

28/04/2026

02098515-0006